Microwaves Made Simple: Principles and Applications

the Staff of the Microwave Training Institute

Edited by
W. Stephen Cheung and
Frederic H. Levien

International Standard Book Number: 1-58053-121-0
Library of Congress Catalog Card Number: 85-047748

Printed and manufactured in the United States by Bookcrafters, Inc., Chelsea, MI

CONTENTS

LIST OF CONTRIBUTORS

THE AUTHORS

Salvatore Algeri	Engineering Manager, Solid State Devices Division, Watkins-Johnson Company, 3333 Hillview Avenue, Palo Alto, CA 94304
W. Stephen Cheung, *editor*	Research Scientist, NASA-Gravity Probe Program, Stanford University, Stanford, CA 94305
George Gillies	Dept. of Physics, University of Virginia, Charlottesville, VA 22901
Ronald Lawler	Manufacturing Manager, Microwave Tube Division, Varian Associates, 611 Hansen Way, Palo Alto, CA 94303
Frederic Levien, *editor*	President, CTT, Inc., 3005 Democracy Way, Santa Clara, CA 95050
Robert Owens	Senior Staff Engineer, Peninsula Engineering Group, Inc., 390 Convention Way, Redwood City, CA 94063
Allan Scott	Director, Microwave Training Institute, 444 Castro Street, Mountain View, CA 94041
Don Sharp	Associate Director of News Engineering, KRON, 1001 Van Ness Avenue, San Francisco, CA 94119
Lawrence Stark	Product Marketing Engineer, Hewlett Packard Company, 1501 Page Mill Road, Palo Alto, CA 94304
Eiji Tanabe	Manager of Microwave Research, Radiation Division, Varian Associates, 611 Hansen Way, Palo Alto, CA 94303

FOREWORD

This book, *Microwaves Made Simple,* at last provides the easy to understand explanation of the technical details of microwaves that the industry has desperately needed. Microwave equipment for communication, radar, and electronic warfare systems amounts to a $21 billion annual market, and approximately one-sixth of all electronic equipment is microwave equpment. Yet to many managers, engineers, and technicians working in the industry, all of this microwave equipment remains a mystery because conventional wiring, transistors, integrated circuits, and tubes cannot be used at microwave frequencies. Instead, unique microwave devices—impatts, transferred electron devices, YIG filters, ferrite circulators, paramps, magnetrons, TWTs—are required. Existing textbooks, which try to explain this unique equipment, begin with complicated mathematical formulas which allow the physical understanding of the devices to be lost in the equations. In contrast, *Microwaves Made Simple* explains even the most complicated microwave devices in complete detail using physical explanations with only the simple mathematics, without calculus, necessary to explain the devices.

Much of the material in this book was developed in a practical way during the teaching of microwave classes for technicians, engineers, and managers at the Microwave Training Institute. The Microwave Training Institute was established to meet the same needs as this new book, that is, to provide in-depth but easily understood instruction in all aspects of microwave systems and devices in order to train managers, engineers, and technicians to meet the need of the microwave industry. Most of the authors of this new book have taught microwave subjects in this way at the Microwave Training Institute and *Microwaves Made Simple* clearly fulfills the training needs of the microwave industry.

Microwaves Made Simple Begins with explanations of the theory of microwaves, the terminology of microwaves, and microwave transmission lines. Gain attenuation, insertion loss, and return loss are described as well as an explanation of the Smith chart and matching. After these fundamentals are covered in the first eight chapters of the book, microwave devices, including low-noise receivers, microwave tubes, solid-state devices, and antennas are covered in Chapters 9–13. Chapters 14–19 cover microwave systems including radar, satellite communication, electronic warfare, and the uses of microwaves in television and in the health sciences. Chapter 17 covers the manufacturing of microwave devices and Chapter 20 looks at the future of microwaves. These last four chapters of the book provide unique and valuable information for anyone working in the microwave field.

Microwaves Made Simple is the microwave textbook that the industry has been waiting for. I commend Dr. Cheung and his co-authors for their contribution to the microwave industry.

Allan W. Scott
Director
Microwave Training Institute
February 1985

PREFACE

The structure of this book evolved from the Microwave Technician Training Program and the Microwave Seminars for Managers offered at the Microwave Training Institute (MTI), in Mountain View, California (*Silicon Valley*). Like the courses, this book is intended for technical managers, working technicians, prospective technicians, and anyone who wishes to achieve a basic understanding of the advancing microwave field. This book excludes the first quarter of the MTI technician program, which covers elementary mathematics, basic electronics, and analog and digital electronics, because a great number of books on these subjects are already available.

The curriculum at MTI is designed to be taught by different instructors who are themselves microwave engineers. Each of the instructors teaches different sections of the program so that the students can enjoy different approaches of technical presentation. The curriculum places more stress on the practical and conceptual aspects of the field, as opposed to mathematical formulations. This emphasis gives the instructor the challenge of explaining complicated material in non-mathematical terms. Mathematics is resorted to only when necessary to aid the student.

The contributors bring to the reader a wellspring of experience and know-how in the microwave field. With a combined total of over 215 years of involvement, spanning three decades, they share a prodigious amount of useful information with the reader. Their present employers cover the spectrum, from giants in the microwave field, such as Hewlett-Packard, Varian Associates, and Watkins-Johnson, to young start-up companies in their infancy, such as CTT, Inc.

The readers of this book are assumed to have some basic knowledge of mathematics and electronics, but not calculus. Numerous examples based on the instructor's field experience have been included to highlight real-life situations encountered in the microwave field.

The book covers a wide range of topics in this exciting field. Conventional topics such as transmission line theory, matching, noise, microwave tubes, FETs, satellites, and radars are covered in detail. In addition, special topics included are microwave applications to television, electronic warfare, and health sciences as well as manufacturing microwave devices. The broad coverage of this book and simplicity of understanding is what makes this book unique.

The success of putting this book together has been made possible by the full cooperation of the MTI consulting staff and their affiliated organizations, as named in the list of contributors. In particular, the editors wish to thank Mr. Allan Scott, Director of the Microwave Training Institute, for starting the training institute, which inspired the writing of this book.

The editor, Stephen Cheung owes special gratitude to his wife, Annette, who drafted all the drawings in his chapters, and Susan Cheramy and Michael Mathews for their prodigious efforts during the final preparation of most of the manuscript. Additional thanks are given to Ken Johnson of MTI for his support throughout the writing of this book.

CHAPTER 1

INTRODUCTION

W. Stephen Cheung

The science of microwaves owes its origin to the development of radar and received its biggest push during World War II. The invention of microwave generators such as the magnetron and the klystron opened the gigahertz (GHz = 10^9Hz) frequency to communication engineers. The microwave field became vitally important as men reached out to space. A very important spin-off of the space program was the commercialization of satellites. Almost overnight, we no longer had to wait a few days for news tapes to be flown in by an aircraft from a news site to the local television station. We began watching news as it happened via satellites. Also, microwave links have replaced miles of intercontinental telephone cables. Today, many institutions such as network television, hotel franchises, and newspapers have access to satellite transponders which provide effective customer services.

Microwaves refer to electromagnetic radiation of frequencies from several hundred MHz to several hundred GHz. For comparison, the signal from an AM radio station is about 1 MHz, i.e., 1 million cycles per second and the signal from an FM station is about 100 MHz, i.e., 100 million or 0.1 billion cycles per second. Figure 1.1 shows the locations of different means of broadcasting in the frequency spectrum. Figure 1.2 emphasizes the microwave spectrum. The readers may be familiar with items such as microwave cooking at about 2.2 GHz, microwave relay (telephone) at about 4.0 GHz, satellite television at 4 GHz (downlink) and 6 GHz (uplink), and police radar at about 22 GHz.

The electronic technology for microwaves seems to have lagged behind its low frequency counterpart. For example, the low frequency electronics industry took off at such a fast pace after the invention of transistors in the late 1940s that we have witnessed the incredible advances of digital electronics, integrated circuits, very high speed computers with huge memory capabilities, and inexpensive personal computers. Today, low frequency tubes are rarely used and even television tubes have begun to see their solid-state replacements. The microwave industry did not receive such a blessing. The main reason is that life at high frequency is very difficult.

The knowledge for low frequency electronics is rich and relatively easy; engineers are equipped with Ohm's law and other network theorems for analyzing low frequency circuits. What makes life easy at low frequencies and difficult at microwave frequencies is the size of a signal's wavelength. The wavelength of a low frequency signal is very large compared to the physical dimensions of the processing equipment. For example, the wavelength of the 60 Hz ac power line is 5,000 km (3,000 miles) and that of a 1 MHz signal is 300 meters. As will be discussed in Chapter 3, the large wavelength-to-equipment dimension ratio results in an extremely small phase difference between signals at different test points. More importantly, the small phase angle means that a standing wave, which is the result of the interference between a forward moving signal and its reflection,

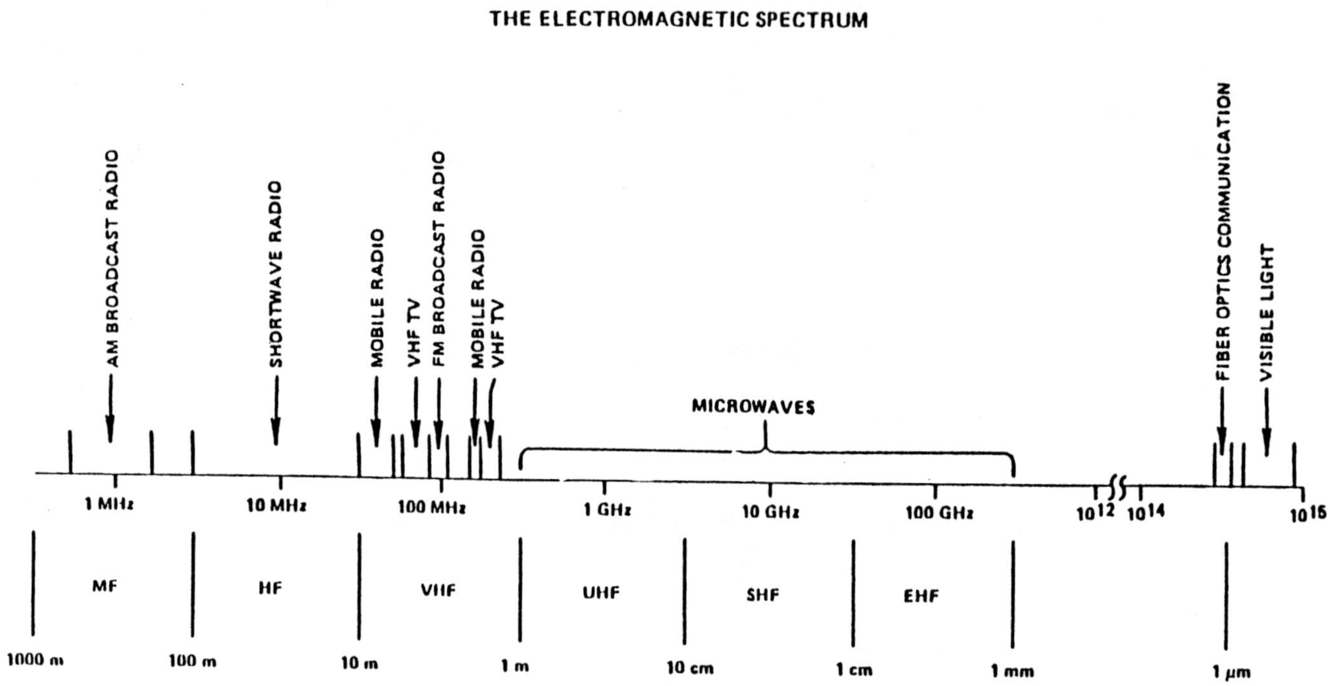

Figure 1.1 The electromagnetic spectrum from MHz to PHz (10^{15}Hz) and their applications.

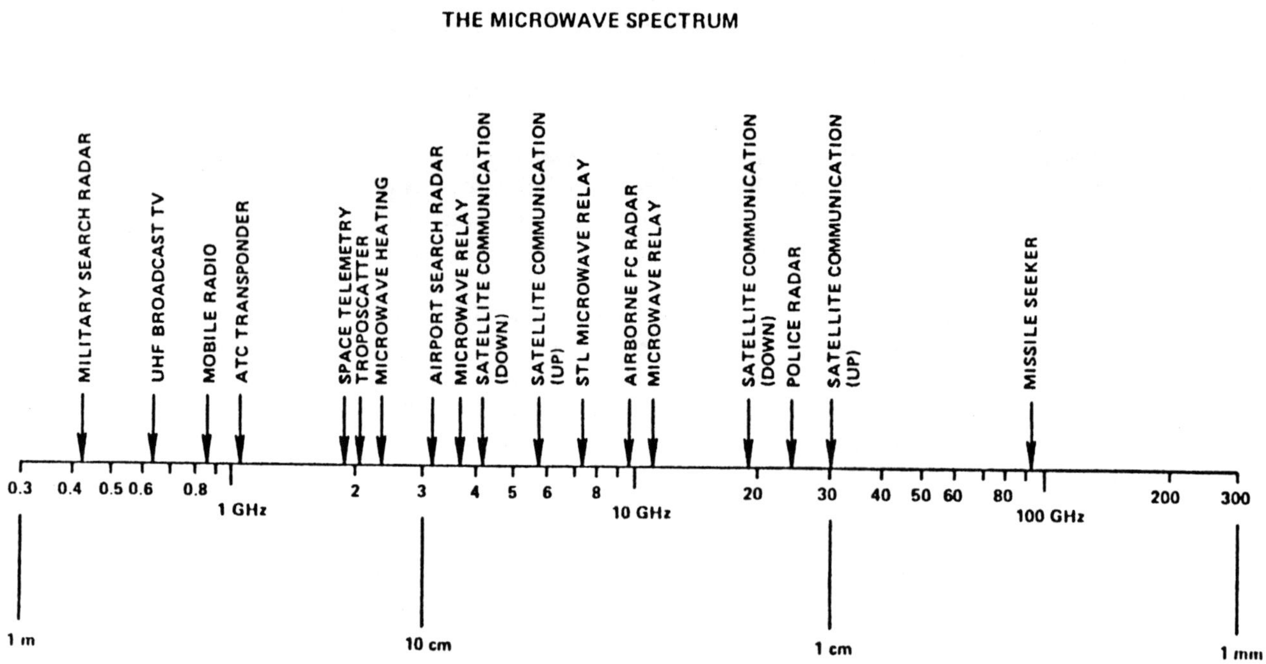

Figure 1.2 The microwave spectrum is from sub-GHz (centimeter wave) to hundreds of GHz (millimeter wave) and their applications.

cannot be formed. Consequently, the solutions of the so-called Maxwell's equations can be greatly simplified, and handy results such as Ohm's law and other network theorems make low frequency circuit design relatively easy.

A microwave engineer cannot take his low frequency formulas and plug in the high frequency values to analyze or design a microwave circuit. The only reliable resource he has is the set of Maxwell's equations which, in principle, can solve any problem in electromagnetics. In fact, many low frequency network theorems are derivable from the Maxwell's equations. At microwave frequencies, however, the wavelength of a signal is comparable to, or even smaller than, the physical dimension of the processing equipment. For example, the wavelength of a 10 GHz signal is 3 cm. Hence, two nearby test points can have a significant phase difference. Similarly, because a forward moving wave may be partially reflected due to circuit mismatch, a test point in the circuit is a superposition of the forward and the reflected wave. These two waves are different from each other by a sizeable phase angle as well as their directions of propagation.

The finite phase difference between the forward wave and the reflected wave results in the formation of a standing wave. Also, high frequency effects such as radiation loss, dielectric loss, and capacitive couplings make microwave circuit design rather tedious. We can no longer use ordinary wires and low frequency components such as resistors, capacitors, and inductors. These components have different behaviors at high frequencies which lead to their likely failure in the high frequency regime. High frequency network construction is very sensitive to the shape and surface flaws (discontinuities) of a conductor.

The maturity of the theoretical and the engineering development of high frequency electronics has taken a longer and more complicated path. Selected techniques developed for (low frequency) integrated circuits are playing important roles in hybrid and monolithic microwave integrated circuits. Microwave circuit designs need no longer be the "cut-and-try" approach, but rather CAD/CAM (computer-aided-design and computer-aided-manufacturing) can be successfully utilized to ensure minimum high frequency problems and maximize compactness. In fact, some of the techniques and solutions to certain high frequency problems have now been successfully applied back to high speed digital signal processing and computer engineering.

An important characteristic of a microwave signal is that it can propagate through the ionosphere with minimum loss. However, water vapor, rain, ozone, and oxygen absorb microwaves of certain frequencies. The earth's ionosphere is composed of ionized atoms and electrons (mostly due to cosmic rays and solar radiations) which are collectively known as plasma. Electromagnetic radiation of frequency in the tens of megahertz range and below cannot penetrate the ionosphere, but is reflected back. Microwaves can, however, go through the ionosphere and only suffer some refraction. Hence, space-bound communications must employ microwaves.

Another feature of using high frequency electromagnetic waves in communications is that a high frequency wave as an information carrier can pack more information within its bandwidth. For example, the typical carrier frequency of an AM signal is 1,000 kHz (or 1 MHz), and the musical and voice information (collectively known as the audio information) are contained over a bandwidth of 40 kHz, i.e., 4% of the carrier. If a 10 GHz microwave is used as a carrier, then a bandwidth of 1% will be 100 MHz, which means that as many as 100 MHz/40 kHz = 25000 separate groups of audio information can be theoretically accommodated (without guard bands). Therefore, exploring the microwave spectrum and making it available to communications are important. The megahertz range was quickly filled up as the demand for more means of communications increased. Opening up the microwave spectrum could ease the demand, although it would also eventually lead to even more demand.

The first portion of this book is devoted to the general understanding of microwave electronics and the background topics. Transmission line theory is essential in understanding electronics in this regime. The formation of standing waves due to circuit mismatches can be conveniently characterized by the Smith chart. The graphical information contained in the Smith chart can then be used to match the circuit by special matching techiques.

Almost every device and component in low frequency electronics has a microwave counterpart. The general rule of thumb is that vacuum tubes are high power devices while solid-state devices such as transistors are used whenever possible because they have many more working hours and less noise. The construction of microwave tube devices such as the magnetron and the klystron is discussed in Chapter 10.

Bipolar transistors cannot operate beyond several GHz. Microwave transistors are usually field-effect transistors (FETs). Also, low frequency semiconductors such as silicon and germanium cannot be used in most microwave circuits because of their low electronic mobility. Gallium arsenide (GaAs) is usually used to fabricate field-effect transistors, hence the name GaAs FET. The circuit designer must learn how to characterize his active devices in order to effecively design a microwave circuit. A set of parameters, known as the *S*-parameters, are extremely useful in circuit designs. The basic principle of microwave transistor circuit design and the general understanding of solid-state devices such as PIN diodes, IMPATT diodes, Gunn diodes, *et cetera* are covered in Chapter 11.

Printed circuit boards are common in every aspect of electronics. Microwave circuits also employ printed circuit technology. Special materials must be considered in making the circuit boards because losses that are insignificant at low frequencies become quite large at microwave frequencies. The technique of microwave printed circuitry falls under the topic of miniature circuits, which has become very important as technologists attempt to realize monolithic microwave integrated circuits (MMICs). The subject of miniature circuits is briefly covered in Chapter 12.

Today, a television reporter can use a mobile van or helicopter to "beam" back his news report to the television station via a designated microwave frequency. This technology, which coincided with the replacement of rolls of film by video recording tapes, has made on-site news transmission possible. Another microwave-related television event becoming a reality in the 1980s is the direct broadcast satellite (DBS). An average household will have a dish receiver and be able to directly pick up microwave television signals. The application of microwaves to the television industry is discussed in Chapter 19.

The general public is familiar with microwaves mostly because of microwave cooking. Indeed, a microwave oven has become an essential part of a modern kitchen. Some microwave ovens have labels which warn pace-maker wearers of the potential hazard. To an extent, a beam of concentrated microwave, as for that matter is a laser beam, is harmful to many parts of the human body. Hence, education for the general public concerning microwave safety is inevitable. Microwaves also have an important role in medicine, both in direct application and as a support component in medical instrumentation. Chapter 18 describes how microwaves are used in x-ray machines and hyperthermia cancer therapies.

Satellites, radars, and huge earth stations tend to increase the mysticism of microwaves. The topics of radars and satellites are covered in Chapters 14 and 15. Electronic warfare (EW) has become a specialized area in the field of microwaves and will be covered in Chapter 16.

Chapter 17 covers the manufacturing of microwave devices, a topic that is rarely discussed. This chapter will mainly discuss concerns involved in manufacturing tube devices such as klystrons. Choice of materials, machining specifications, tolerances, and the manufacturing environment are discussed.

More and more applications of microwaves have been discovered. For example, recent research by the US Department of Energy (DOE) showed that treating coal with microwaves can remove organic sulfur and other potential pollutants which are molecularly bound to the coal molecular structure (see the December 1984 issue of *High Technology*). This technique was capable of producing coals that can meet or exceed federal standards for sulfur dioxide emission. Another industrial application of microwaves is the non-destructive testing of material for determining reliability. The eddy current patterns of a piece of metal with fatigue or cracks can be detected by using microwaves. The applications of microwaves outside of communications are enormous and continue to grow.

The field of microwaves has come to a stage of rapid growth because of the demand to enter the high frequency regime, and other supporting technologies such as computers and integrated circuitry are ready to overcome many old design difficulties. The entire microwave industry has come a long way. In 1984, the sales of microwave-related equipment and services in the US was over $20 billion. It is beyond any doubt that the microwave industry will see more and faster growth as human civilization approaches the year 2000.

CHAPTER 2

MATHEMATICS OF DECIBEL SCALE

W. Stephen Cheung and George T. Gillies

2.1 INTRODUCTION

In Ch. 1, the spectrum of electromagnetic radiation was discussed, and shown to range from very low frequencies (slower than kHz) to cosmic ray frequencies and beyond (faster than 10^{16} Hz). All of these frequencies were "squeezed" onto one page. While this methoid of scaling helps us to visualize a large part of the spectrum, it is, however, a nonlinear scale. This will become apparent later in this chapter. In order to more easily deal with this nonlinearity, logarithmic scales and units of decibels will be introduced. These scaling techniques will then be applied to the characterization of microwave power levels.

2.2 LINEAR SCALES

A linear scale is an inherently uniform way of scaling data, and such scales are used frequently. The scale's origin generally starts at zero. Figures 2.1A and 2.1B show two examples of linear scales.

The greatest disadvantage of linear scaling is that it is impossible to incorporate data encompassing several orders of magnitude on one sheet of paper.

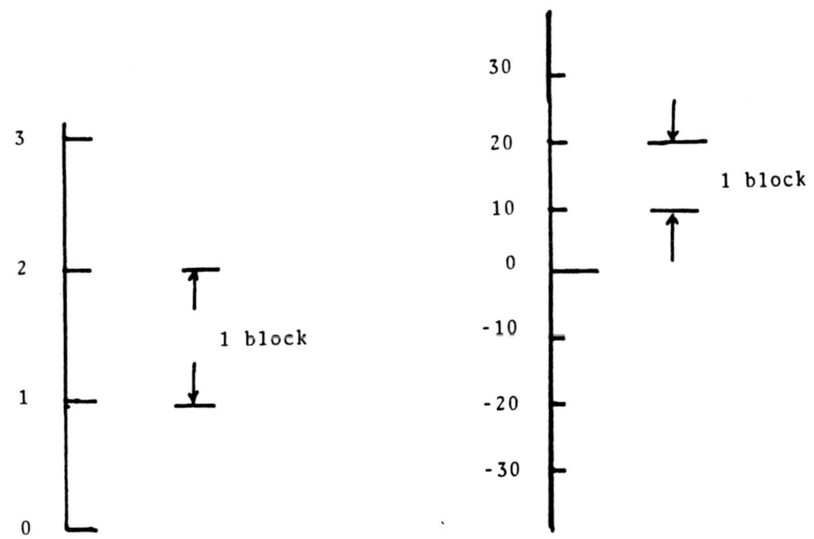

Figure 2.1 Examples of linear scales: (A) the scale is positive and the origin is zero; (B) the scale is bipolar.

2.3 LOGARITHMIC SCALES AND DECIBEL SCALES

Every block of a logarithmic scale increases or decreases by a factor of 10 as shown in Fig. 2.2. It is impossible to include zero or negative numbers in this type of scale.

Let us magnify, for example, the block between 1 and 10 as seen in Fig. 2.3. Where is the number 2? The number 2 is no longer located in the position that we would normally expect. The mathematics of logarithm must be used to determine the relative locations of the numbers from 1 to 10 in this kind of block.

As a preliminary step, the readers must become familiar with the concept of *scientific notation* and have a good working knowledge of their calculators.

What is 10^0? Try it on your calculator and it will give you the answer $10^0 = 1$. Similarly, verify that $10^1 = 10$. If $10^0 = 1$ and $10^1 = 10$, is there some power of 10, between 0 and 1, that is equal to 2? In other words, we are interested in finding an unknown x such that,

$$10^x = 2$$

To find x, we must take the logarithm of 2. We write this mathematically as

$$x = \log 2$$

Numerically, this value is 0.301, or approximately 0.3. The readers should use their calculators to verify that $10^{0.3}$ is indeed approximately equal to 2. Note that

$$10^{0.2} \neq 2$$

as some might intuitively expect. In fact,

$$10^{0.2} = 1.58 \text{ (the reader should check this)}$$

Similarly, $10^{0.5}$ is not 5 but 3, and $10^{0.7}$ is not 7 but 5, *et cetera*. We shall apply the mathematics of logarithms next.

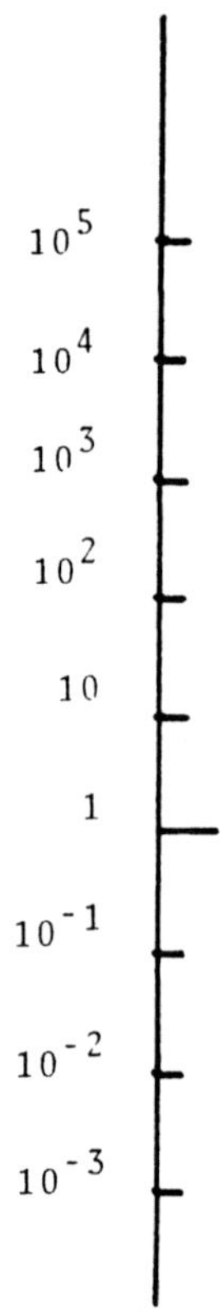

Figure 2.2 The logarithmic scale. Each block is a factor of 10.

2.3.1 Decibel Scales

This type of scale was originally used in the comparison of sound intensities. Modern usage of the decibel scale is primarily employed to comparison of power or voltage levels. We shall concern ourselves only with electrical power here.

The definition of the decibel, abbreviated dB, is

$$\text{dB} = 10 \times \log(P_1/P_2) \quad (2.3.1)$$

For example, if $P_1 = 20\text{W}$ and $P_2 = 5\text{W}$, then P_1 is four times larger than P_2. In a decibel scale, we have

$$\begin{aligned} \text{dB} &= 10 \times \log(20/5), \\ &= 10 \times \log(4), \\ &= 10 \times 0.602, \\ &= 6.02\,\text{dB, or} \\ &= 6\,\text{dB}. \end{aligned}$$

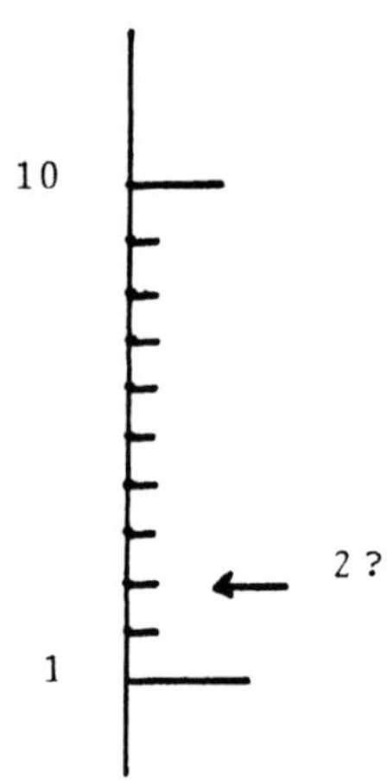

Figure 2.3 Close examination of the logarithmic scale. The location of a number such as 2 is not as straightforward as that in the linear scale.

The following power ratios and their decibel values are quite helpful, and the ones marked with asterisks should be memorized. The readers should verify these conversions with their calculators. Figure 2.4 shows how decibel scales are constructed.

Ratio	dB	Ratio	dB
* 1	0 dB	* 8	9 dB
* 2	3 dB	* 9	9.5 dB
* 3	5 dB	* 10	10 dB
4	6 dB	100	20 dB
* 5	7 dB	10^3	30 dB
6	8 dB	10^4	40 dB
* 7	8.5 dB	10^5	50 dB

It should be emphasized that unlike watts, volts, and amperes, the dB is *not* a physical quantity. Rather, a dB represents a ratio of two physical quantities, typically power; and it is itself a dimensionless number much like the units "radian" and "degree" which are used to measure plane angles.

We now present some rules useful in the manipulation of decibel quantities. These rules, or theorems, as the mathematicians call them, should be carefully studied by the reader.

THEOREM I: The *product* of two pure numbers (or ratios) A and B is equivalent to their *sum* when their values are expressed in dB.

Example (2.3.1)

$$\begin{aligned} A &= 1 & &(0\text{ dB})\\ B &= 2 & &(3\text{ dB})\\ A\times B &= 1\times 2 & &0\text{ dB}+3\text{ dB}\\ &= 2 & &= 3\text{ dB} \end{aligned}$$

Example (2.3.2)

$$\begin{aligned} A &= 2 & &(3\text{ dB})\\ B &= 3 & &(5\text{ dB})\\ A\times B &= 2\times 3 & &3\text{ dB}+5\text{ dB}\\ &= 6 & &= 8\text{ dB} \end{aligned}$$

Example (2.3.3)

$$\begin{aligned} A &= 10 & &(10\text{ dB})\\ B &= 100 & &(20\text{ dB})\\ A\times B &= 10\times 100 & &10\text{ dB}+20\text{ dB}\\ &= 1000 & &= 30\text{ dB} \end{aligned}$$

THEOREM II: The *division* of two pure numbers (or ratios) A and B is equivalent to their *difference* when their values are expressed in dB.

Example (2.3.4)

$$\begin{aligned} A &= 4 & &(6\text{dB})\\ B &= 2 & &(3\text{dB})\\ A/B &= 4/2 & &6\text{dB}-3\text{dB}\\ &= 2 & &= 3\text{ dB} \end{aligned}$$

Example (2.3.5)

$$\begin{aligned} A &= 10 & &(10\text{ dB})\\ B &= 5 & &(7\text{ dB})\\ A/B &= 10/5 & &10\text{ dB}-7\text{ dB}\\ &= 2 & &= 3\text{ dB} \end{aligned}$$

SEC. 2.3

LOG & dB SCALES

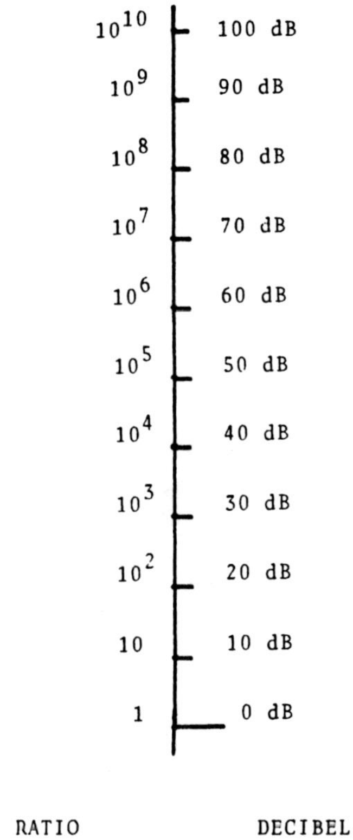

Figure 2.4 From logarithmic to decibel scale.

Example (2.3.6)

A = 1000	(30 dB)
B = 10	(10 dB)
A/B = 1000/10	30 dB − 10 dB
= 100	= 20 dB

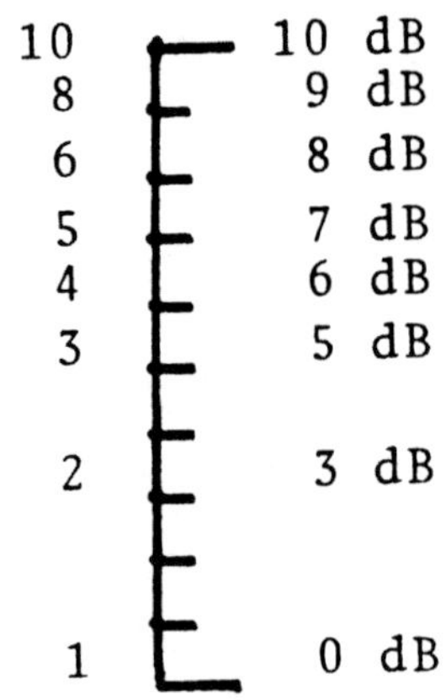

Figure 2.5 Layout of the decibel scale in the 10-block. The number 3, for example, is 5 dB, and the number 5 is 7 dB.

Theorems I and II are quite useful in determining the decibel values of pure numbers. We recommend that the reader memorize the dB values for all the integers from 1 to 10. Then, with the help of a calculator, practice guessing the dB value of non-integer real numbers. For example, you have memorized that 2 is 3 dB and 3 is 5 dB, but what about the number 2.5? If your guess is 4 dB, it is very close. Usually, the chance of your guess being way off is high when the number is between 1 and 2. This is due to the fact that 1 = 0 dB, but already 2 = 3 dB. Figure 2.5 shows the relative locations of the integers 1 to 10 and their dB values.

Let us practice finding the dB values of a few numbers using Theorem I. We assume that the readers are familiar with scientific notation.

Example (2.3.7)

Number = 20
= 2×10^1
= 3 dB + 10 dB (Theorem I)
= 13 dB

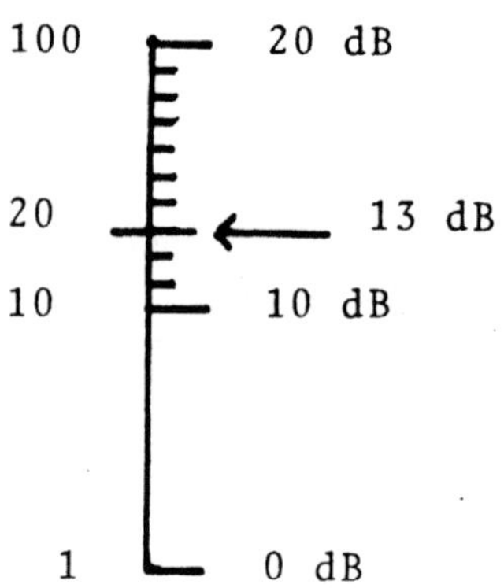

Figure 2.6 The number 20 is 13 dB.

The reader should mentally picture that the number 20 is somewhere between 10 and 100. Since 10 = 10 dB and 100 = 20 dB, we can see that 20 is somewhere between 10 dB and 20 dB, as illustrated in Fig. 2.6.

Example (2.3.8)

Number = 500
= 5×10^2
= 7 dB + 20 dB (Theorem I)
= 27 dB

This is illustrated in Fig. 2.7.

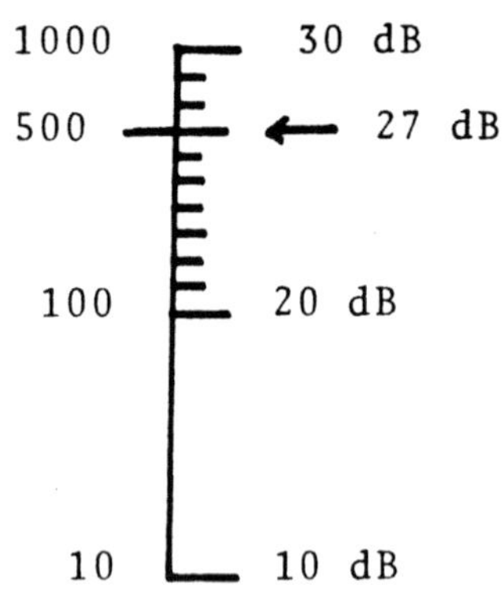

Figure 2.7 The number 500 is 27 dB.

Example (2.3.9)

Number = 23000
= 2.3×10^4
= 3.6 dB + 40 dB
= 43.6 dB

2.4 FRACTIONAL NUMBERS IN DECIBELS

Figure 2.8 shows the continuation of the scale shown in Fig. 2.2. As we go downward, each block decreases in size by a factor of 10.

To show that 0.1 is −10 dB, we invoke Theorem II in the following manner.

Since	0.1 = 1/10
recall that	1 = 0 dB
and	10 = 10 dB
Theorem II gives	

0.1 = 1/10
= 0 dB − 10 dB
= −10 dB

Similarly, we can show that 10^{-2}, 10^{-3}, *et cetera* are −20 dB, −30 dB, and so on.

We should now be able to find the dB equivalent of any *positive* fractional numbers. Remember that negative numbers cannot be expressed in dB. The reader should verify this important point for himself.

A few examples will assist the reader in familiarizing himself with finding the dB equivalents of positive fractional numbers.

Example (2.4.1)

$$\begin{aligned} \textit{Number} &= 0.2 \\ &= 2/10 \\ &= 3\text{ dB} - 10\text{ dB} \qquad \text{(Theorem II)} \\ &= -7\text{ dB} \end{aligned}$$

Alternately,

$$\begin{aligned} \textit{Number} &= 0.2 \\ &= 2 \times 10^{-1} \\ &= 3\text{ dB} + (-10\text{ dB}) \qquad \text{(Theorem I)} \\ &= -7\text{ dB} \end{aligned}$$

Example (2.4.2)

$$\begin{aligned} \textit{Number} &= 0.004 \\ &= 4 \times 10^{-3} \\ &= 6\text{ dB} + (-30\text{ dB}) \qquad \text{(Theorem I)} \\ &= -24\text{ dB} \end{aligned}$$

Example (2.4.3)

$$\begin{aligned} \textit{Number} &= 7 \times 10^{-9} \\ &= 8.5\text{ dB} + (-90\text{ dB}) \qquad \text{(Theorem I)} \\ &= -81.5\text{ dB} \end{aligned}$$

Example (2.4.4)

$$\begin{aligned} \textit{Number} &= 3.5 \times 10^{-6} \\ &= 5.4\text{ dB} + (-60\text{ dB}) \qquad \text{(Theorem I)} \\ &= -54.6\text{ dB} \end{aligned}$$

Figure 2.9 shows the locations of the numbers and their dB equivalents from the examples we have just investigated.

SEC. 2.5

CONVERTING dB VALUES TO PURE NUMBERS

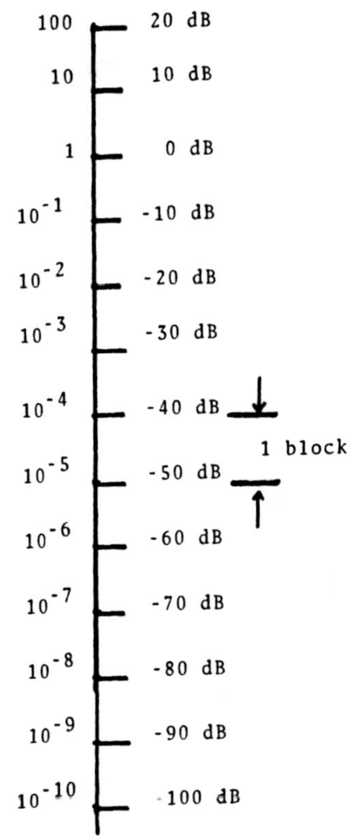

Figure 2.8 Layout of the fractional numbers and their corresponding decibel values.

2.5 CONVERTING dB VALUES TO PURE NUMBERS

The mathematical function "anti-logarithm" is used to find the numerical value of quantities expressed in dB. On most modern calculators, the "anti-log" function corresponds to the key sequence of either INV log or 10^x.

The procedure is the following:

Step a) Divide the given dB value by 10.

Step b) Use anti-log of (dB/10) to obtain the number.

Example (2.5.1)

Find the number corresponding to 5 dB.

Step a) 5/10 = 0.5

Step b) *Number* = anti-log (0.5) = 3.16 = 3 (approximately).

SEC. 2.6

ELECTRIC POWER IN dB

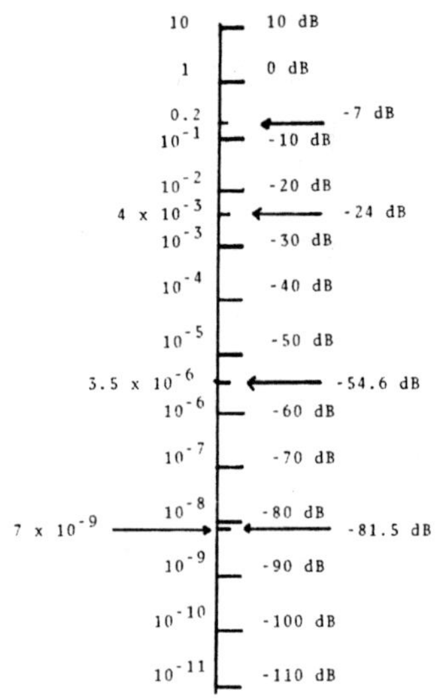

Figure 2.9 Locations of several fractional numbers and their corresponding decibel values.

Example (2.5.2)

Find the number corresponding to 27 dB.
Step a) $27/10 = 2.7$.
Step b) *Number* = anti-log $(2.7) = 501.2 = 500$ (approximately).

Example (2.5.3)

Find the number corresponding to -27 dB.
Step a) $-27/10 = -2.7$.
Step b) *Number* = anti-log $(-2.7) = 1.99 \times 10^{-3} = 2 \times 10^{-3}$ (approximately).

Example (2.5.4)

Find the number corresponding to -91 dB.
Step a) $-91/10 = -9.1$.
Step b) *Number* = anti-log $(-9.1) = 7.94 \times 10^{-10} = 8 \times 10^{-10}$ (approximately).

Example (2.5.5)

Find the number corresponding to -8 dB.
Step a) $-8/10 = -0.8$
Step b) *Number* = anti-log $(-0.8) = 0.158 = 0.16$ (approximately).

EXERCISES

1. Convert the following numbers to dB.
   a) 2000; b) 5 million; c) 3×10^4; d) 7×10^5; e) 3.4×10^2; f) 0.035; g) 2×10^4; h) 1.3×10^{-6}; i) 0.5; j) 9×10^{-8}.
2. Convert the following dB values to numbers.
   a) 8 dB; b) 9.8 dB; c) 24 dB; d) 16.5 dB; e) 47 dB; f) 66 dB; g) 128 dB; h) -37 dB; i) -76 dB; j) -114.5 dB; k) -80.6 dB; l) -2.6 dB; m) -0.0001 dB; n) -0 dB.
3. Perform the following multiplications and divisions by using Theorems I and II.
   a) 40×5; b) 300×6;
   c) 9000×2; d) $20/4$;
   e) $6 \times 10^6/30$; f) $2200/11$;
   g) $8 \times 6 \times 3$; h) $90 \times 600 \times 4000$;
   i) $35 \times 760 \times 2400$; j) $100 \times 20/4$;
   k) $(500 \times 4 \times 6)/(2 \times 3 \times 4)$; l) 0.03×0.4;
   m) $6 \times 10^{-4} \times 5 \times 10^{-2}$; n) $7 \times 10^{-3} \times 3.3 \times 10^{-5}$;
   o) $9 \times 0.03 \times 6 \times 10^{-4}$; p) $0.08 \times 0.0006/0.04$.

2.6 ELECTRIC POWER IN DECIBELS

The standard unit of electrical power is the watt (W). Operationally, it is the product of the voltage (V) across and the current (A) through some circuit. Figure 2.10 shows the relative sizes of different levels of power.

As will be made clear in succeeding chapters, it is often convenient to express power levels using the decibel scale. Furthermore, we distinguish between high and low power levels. For instance, watt becomes dBW and mW (milliwatt) becomes dBm. Here we have appended a "W" and an "m" to the dB symbol as appropriate to the power level being measured.

While the watt (W) is the standard unit for the measurement of power, the milliwatt (mW) is widely used in the field of microwave engineering. Therefore, we have chosen mW and dBm as the standard power units in this book.

The reader should be aware, however, that neither dBW nor dBm is widely accepted by the natural sciences or other engineering curricula. They were invented more for convenience than for scientific needs.

The conversions of typical power units to mW are given in the following.

$$1\text{MW} = 10^{9}\text{mW} \qquad 1\text{kW} = 10^{6}\text{mW}$$
$$1\text{W} = 10^{3}\text{mW} \qquad 1\mu\text{W} = 10^{-3}\text{mW}$$
$$1\text{nW} = 10^{-6}\text{mW} \qquad 1\text{pW} = 10^{-9}\text{mW}$$

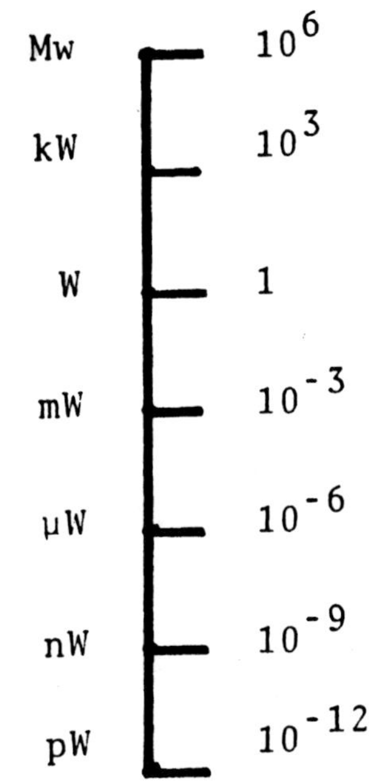

Figure 2.10 Prefixes of power levels and their multiples.

Finding a given power in dBm is straightforward. The procedure is explained below, accompanied by an example.

Procedure:	*Example:*
a) Express the given power in mW.	2W = 2000mW
b) Take the numerical part of the power in mW and convert it to dB as before.	2000 = 33 dB
c) Write the power using dBm.	2W = 33 dBm

Now the reader should practice this technique to study the following examples.

Example 2.6.1

a) 3mW = 5 dBm
b) 50mW = 17 dBm
c) $60\text{W} = 6 \times 10^{4}\text{mW} = 48$ dBm
d) $4\text{kW} = 4 \times 10^{6}\text{mW} = 66$ dBm
e) $2.5\text{MW} = 2.5 \times 10^{9}\text{mW} = 94$ dBm
f) $5\mu\text{W} = 5 \times 10^{-3}\text{mW} = -23$ dBm
g) $8\text{nW} = 8 \times 10^{-6}\text{mW} = -51$ dBm
h) $20\text{pW} = 2 \times 10^{-8}\text{mW} = -77$ dBm

Figure 2.11 shows the mW and dBm scale. The reader should always remember that the symbol dB is used to represent pure numbers or ratios, and that dBm always represents power levels only.

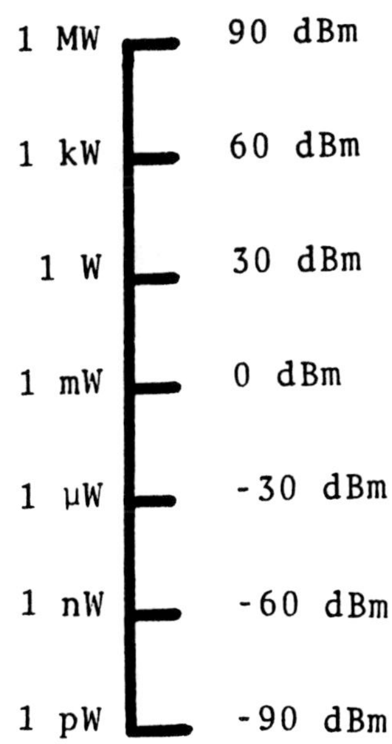

Figure 2.11 Prefixes of power levels and their dBm values.

2.7 MATHEMATICAL MANIPULATIONS USING dBm

Because dBm represents power levels, such quantities must be manipulated with care. While pure numbers can be multiplied repeatedly without limit (always with meaningful results), we cannot multiply together two power levels, because such a product has no physical meaning. Consequently, an operation such as dBm + dBm has no meaning.

It is, however, permissible to compare one power level with another. Therefore, dBm − dBm is allowed, and the resulting unit is dB insead of dBm. This stems from the fact that comparing two power levels results in a ratio (which can be expressed in dB) and no longer a power level.

Table 2.1 presents a summary of the above discussion.

Table 2.1

Operation	Resulting Units	Meaning	Allowed?
1. dB + dB	dB	product of two numbers	yes
2. dB − dB	dB	quotient of two numbers	yes
3. dBm + dBm	XX	product of two power levels	no
4. dBm − dBm	dB	comparing two power levels	yes

We shall use some examples to illustrate case 4.

Example (2.7.1)

Power A = 1W (30 dBm)
Power B = 1mW (0 dBm)

$$\frac{Power\ A}{Power\ B} = \frac{1\text{W}}{1\text{mW}} = \frac{10^3\text{mW}}{1\text{mW}} = 10^3$$

This means that power A is 10^3 times larger than power B. When computing in decibel scale,

30 dBm − 0 dBm = 30 dB

Example (2.7.2)

Power A = 10mW (10 dBm)
Power B = 500mW (27 dBm)

$$\frac{Power\ A}{Power\ B} = \frac{10\text{mW}}{500\text{mW}} = \frac{1}{50} = 0.02; \qquad 10\text{ dBm} - 27\text{ dBm} = -17\text{ dB}$$

This means that power A is 50 times *smaller* than power B. Another way of stating this is that power A is 0.02 times power B.

Example (2.7.3)

Power A = 10mW (10 dBm)
Power B = $1\mu\text{W} = 10^{-3}\text{mW}$ (−30 dBm)

$$\frac{Power\ A}{Power\ B} = \frac{10\text{mW}}{10^{-3}\text{mW}} = 10^4 \qquad 10\text{ dBm} - (-30\text{ dBm}) = 40\text{dB}$$

The readers should carefully note that the double negative in Example (2.7.3) implies only that two powers are being compared. Do not rewrite 40 dBm − (−30 dBm) as 40 dBm + 30 dBm. Remember that the latter expression means you are trying to multiply two power levels, which is physically impossible, and so disallowed.

2.8 COMBINING dBm AND dB

Just like voltage and current, power can be *amplified* by a power amplifier. Power amplification means that the input power to an amplifier is enlarged by the amplifier such that the output power is larger than the input power by some factor. The amplification factor is known as power gain, G. For instance, an input power of 1mW to an amplifier of gain 100 results in an output power of 1mW × 100 = 100mW.

Power amplification can be described in terms of dBm and dB, with the former referring to the input and output power, and the latter to the gain of the amplifier. Hence, dBm + dB has physical meaning, namely amplification of a power by some numerical factor over its initial level. The resulting quantity is, therefore, also a power level, namely dBm.

Important: dBm + dB = dBm Power amplification

Just like the division of a certain voltage by a resistor network, power can be divided (or attenuated) as well. When expressed in terms of dBm and dB, power attentuation is dBm − dB. Because an attenuated power is still power, we have the following important relation:

Important: dBm − dB = dBm Power attentuation

Table 2.2 summarizes all possible combinations of dB and dBm. Power gain and attenuation will be studied in greater detail in Ch. 5.

Table 2.2

Operation	Resulting Unit	Physical meaning	Allowed?
1. dB + dB	dB	product of two numbers	yes
2. dB − dB	dB	comparing two numbers	yes
3. dBm + dBm	XX	multiplying two powers	no
4. dBm − dBm	dB	comparing two powers	yes
5. dBm + dB	dBm	power amplification	yes
6. dBm − dB	dBm	power attenuation	yes

EXERCISES

Perform the following operations. If the expression describes a physically disallowed situation, write "NA". Also write down the case number from Table 2.2 under the "Remarks" column.

Remarks

1. 5 dB + 19 dB =
2. 26 dB + 57 dB =
3. 64 dB − 3 dB =
4. 29 dB − 44 dB =
5. −7 dB − 20 dB =
6. −24 dB − 30 dB =
7. −20 dB − (−60 dB) =
8. −12 dB − (−90 dB) =
9. 3 dBm + 4 dBm =
10. 2 dBm + (−6 dBm) =
11. −23 dBm + (−40 dBm) =
12. 9 dBm − 0 dBm =
13. 24 dBm − 3 dBm =
14. −6 dBm − 5 dBm =
15. −15 dBm − (−10 dBm) =
16. −24 dBm − (−24 dBm) =
17. −30 dBm − (−60 dBm) =
18. 10 dBm − (−80 dBm) =
19. 4 dBm + 10 dB =
20. −4 dBm + 20 dB =
21. −46 dBm + 10 dB =
22. −90 dBm + 100 dB =
23. 3 dBm − 10 dB =
24. −14 dBm − 20 dB =
25. −62 dBm − 3 dB =

SEC. 2.9
VOLTAGE RATIO

2.9 VOLTAGE RATIO

So far we have concerned ourselves only with power ratios and power amplification. However, in low frequency electronics, voltage amplification (i.e., voltage gain) is an important factor. Although voltage gain can also be expressed as a ratio, this particular ratio cannot be immediately converted to dB.

Remember that decibels originated in power comparisons. It is possible in many cases, however, to express powers in terms of the associated voltages. Since *power* = (*voltage*)^2/*resistance,* the following can be done:

since

$$Power\ A = V_A^2/R_A \tag{2.9.1}$$

and

$$Power\ B = V_B^2/R_B \tag{2.9.2}$$

then

$$\begin{aligned} \text{dB} &= 10 \times \log (P_A/P_B) \\ &= 10 \times \log \frac{V_A^2 \cdot R_B}{V_B^2 \cdot R_A} \end{aligned} \tag{2.9.3}$$

Normally, the two resistances R_A and R_B are chosen to be equal for purposes of comparison. This means that the power generated by two separate voltages, V_A and V_B, can be compared when applied across resistors of equal values (or across one standard resistor).

Hence, Eq.(2.9.3) can be simplified to

$$\text{dB} = 10 \times \log (V_A^2/V_B^2) \tag{2.9.4}$$

$$= 20 \times \log (V_A/V_B) \tag{2.9.5}$$

Notice that in Eq.(2.9.5), a factor of 20 multiplies the logarithm of the voltage ratio. Although this is mathematically correct, it is unfortunately somewhat confusing.

According to the original decibel concept, a 10 dB power gain means that power *A* is 10 times power *B*. However, if voltage *A* is 10 times voltage *B*, this results in a power gain of 20 dB. This is because a 10 times voltage increase across a resistor results in a ten times current increase and therefore a total of a hundred times power increase.

Because microwave technology is primarily concerned with power levels, voltage gain will seldom be discussed, and confusion on this point should then be minimized. Some microwave devices measure power in terms of voltages, however, and this makes comparison in dB somewhat inconvenient. The authors will be as consistent as possible, however, in maintaining that all quantities expressed in dB refer only to *power* levels.

2.10 NEPER

The decibel and its associated logarithmic scales are referenced to the number 10 (called base 10). For instance, remember that the base of the exponent x in the example $10^x = 2$ is the number 10. The choice of 10 is more for convenience than anything else. Just like measuring distance in feet or in meters, the reference is not absolute.

SEC. 2.10

NEPER

The number $e = 2.718\ldots$ plays an important role in calculus and is considered as a "natural" number in mathematics; another example of a natural number is $\pi = 3.1416$.

Let us say powers A and B are 10mW and 1mw, respectively. To express the ratio $P_A:P_B$ in dB, we use Eq.(2.3.1),

$$\text{dB} = 10 \times \log (P_A/P_B) \tag{2.10.1}$$

To express the ratio in base e, the definition is

$$\text{Neper*} = \ln (P_A/P_B) \tag{2.10.2}$$

where ln stands for natural logarithm. Most calculators have the [ln x] button.

Example (2.10.1)

$$\begin{aligned} P_A &= 4\text{mW}, P_B = 8\text{mW}. \\ \text{Neper} &= \ln (8/4) = \ln 2 = 0.69 \text{ Neper} \end{aligned}$$

Example (2.10.2)

$$\begin{aligned} P_A &= 100\text{mW}, P_B = 10\text{mW} \\ \text{Neper} &= \ln (100/10) = \ln 10 = 2.30 \text{ Neper} \end{aligned}$$

Example (2.10.3)

$$\begin{aligned} P_A &= 1\text{mW}, P_B = 10\text{mW} \\ \text{Neper} &= \ln (1/10) = \ln 0.1 = -2.3 \text{ Neper} \end{aligned}$$

Example (2.10.4)

$$\begin{aligned} P_A &= 6\text{mW}, P_B = 1000\text{mW} \\ \text{Neper} &= \ln (6/1000) = \ln 0.006 = -5.11 \text{ Neper} \end{aligned}$$

Despite its importance in calculus and mathematics, in general, the natural logarithm and the Neper are much less important than the dB in engineering practice. The Neper was discussed here only in order to provide conveniently available references for those who may come across this term.

* The term Neper was chosen (and spelled incorrectly) to honor the Scottish mathematician, John Napier, for his first proposed use of logarithms.

CHAPTER 3

ELECTROMAGNETICS

W. Stephen Cheung

3.1 INTRODUCTION

In basic low frequency electronics, electrical signals travel from one component to another via connecting wires. The power loss in the wires is usually ohmic, i.e., heat dissipation due to the ohmic resistance of the wires.

As the frequency increases, other power losses become important. These losses include radiation loss and dielectric loss. Radiation loss is energy radiated away as a result of the wire acting like an antenna (see Ch. 13). Radiation loss is especially significant for openly hanging wires, and is functionally related to the frequency and the power of the signals carried on the wires.

In the dielectric loss, the rearrangement of induced surface charges on the dielectric due to the voltage change between, for example, the center conductor and the outer conductor of a coaxial cable, would dissipate energy. The dielectric loss is functionally related to the dielectric constant of the dielectric, the voltage between the inner and the outer conductors, and the frequency of the signal.

To minimize the radiation and dielectric losses, ordinary open wires are not suitable for conducting microwave signals. The dielectric loss in coaxial cables becomes significant for frequencies above 10 GHz.

To correctly describe the behavior of electrical and associated magnetic phenomena, Maxwell's equations must be used. This chapter qualitatively describes the set of four Maxwell's equations and their consequences.

3.2 THE CONCEPT OF FIELD

The concept of a force field was originally invented as a visual aid to describe the spatial distribution of the gravitational, electrical, and magnetic forces. One familiar example of such visualization is given in Fig. 3.1, where some iron filings are shown on a piece of paper held above a permanent magnet. The filings follow a certain pattern which, if traced out, is representative of the distribution of the magnetic field lines produced by the permanent magnet.

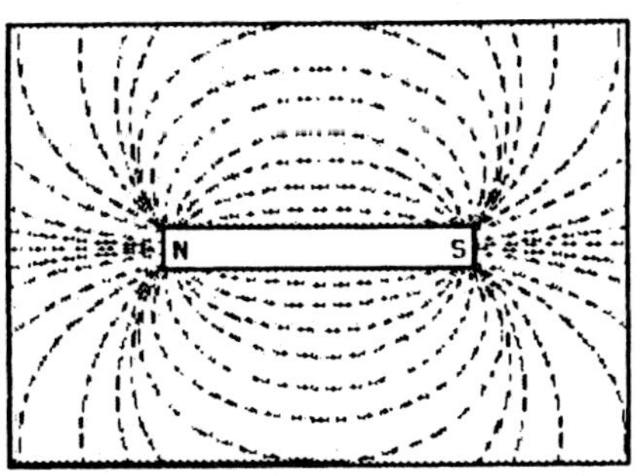

Figure 3.1 The magnetic field pattern of a bar magnet.

Modern physics holds that fields are more than just a visual aid. Their actual presence can be detected as a form of energy. In other words, part of the energy needed to "create," let us say, a magnet goes into the establishment of the resulting magnetic field pattern. The electromagnetic wave, as we shall see, is a form of time-varying field, electric and magnetic combined, which carries energy away from its source.

3.2.1 Electric Field (E)

Naturally occurring individual electrical charges are monopolar, i.e., they are always either positive or negative. The readers should note that the choice of polarity, i.e., positive or negative, is just a matter of convention.

An important feature of electrical charges is that like charges repel and unlike charges attract. The electric field patterns of a single positive charge, a single negative charge, a positive-positive charge arrangement, and a positive-negative charge arrangement are shown in Fig. 3.2.

Maxwell Equation #1 (Coulomb's law): The overall electric field pattern, strength, and direction are determined by the geometrical distribution of the electrical charges producing the field.

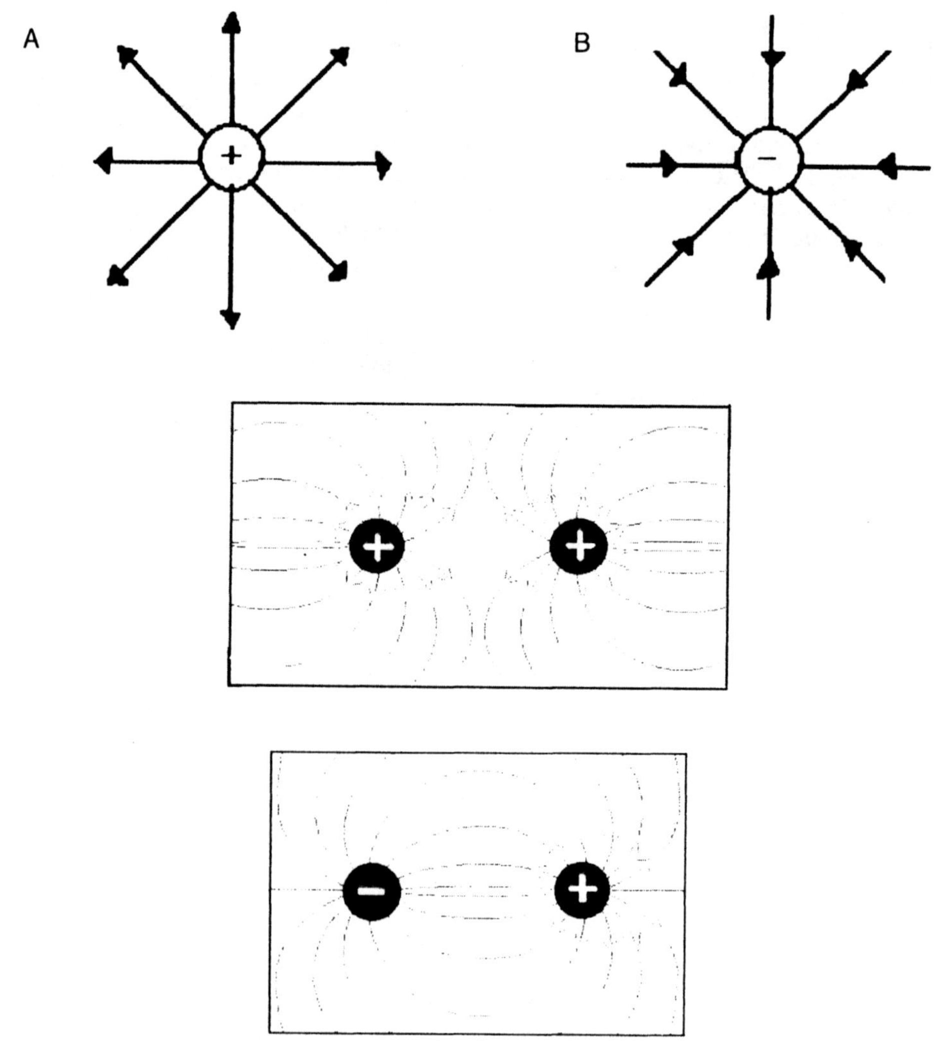

Figure 3.2 The electric field distributions of (A) a single positive charge, (B) a single negative charge, a positive-positive charge combination, and a positive-negative charge combination.

3.2.2 Magnetic Field (H)

Permanent magnets or current-carrying wires produce magnetic fields. Unlike electric fields, however, the fact that magnets take their basic form as a dipole, rather than a monopole, demands that the magnetic field must describe a closed loop. Examples of three situations are shown in Fig. 3.3.

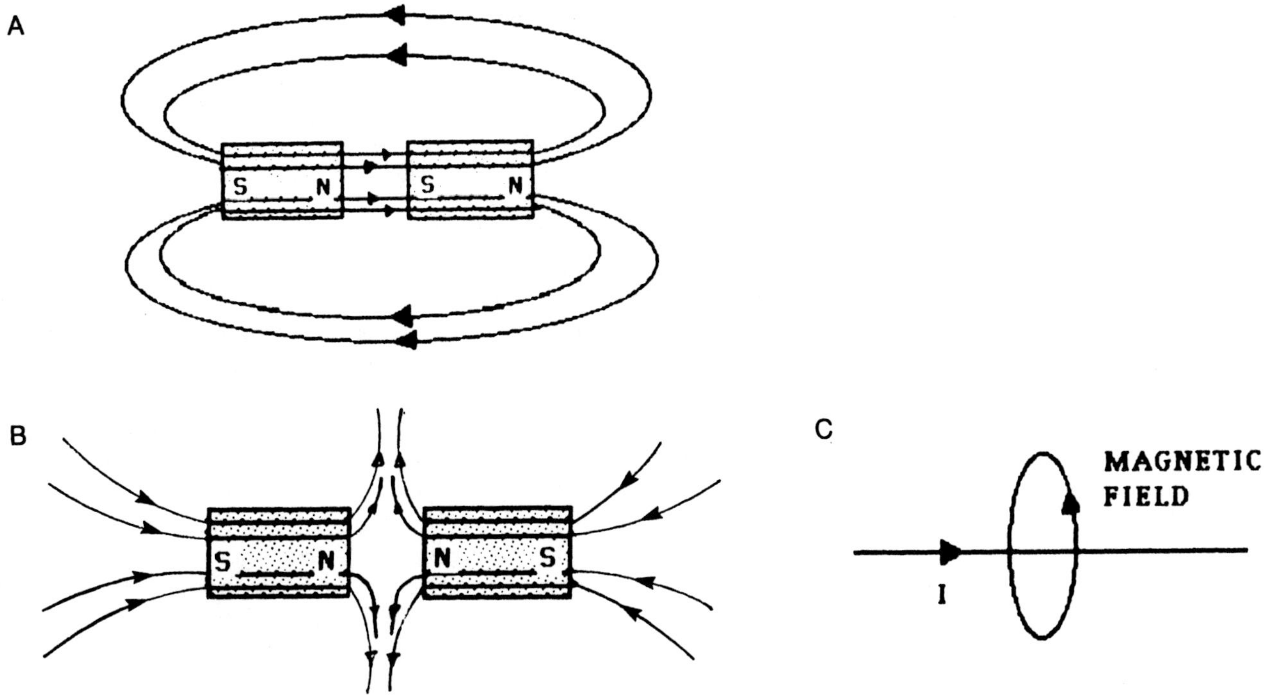

Figure 3.3 The magnetic field distributions of (A) two magnets with opposite poles facing each other, (B) two magnets with the same poles facing each other, and (C) a current-carrying wire.

The condition that the magnetic field must follow a closed loop is described mathematically as *Maxwell's Equation #2*, which will not be written here.

An electrical current is actually electrical charges in motion. Charges produce electric fields, so moving charges produce a changing electrical field. Therefore, a changing electric field gives rise to magnetic field. This is mathematically given by *Maxwell's Equation #3*. If the electric field is changing at a steady rate, i.e., a direct current, the magnetic field at an arbitrary location is constant both in magnitude and direction. If the rate is not constant, e.g., it follows a sinuosoidal pattern, then the resulting magnetic field is also changing sinuosoidally with time.

What about a changing magnetic field? The most common application of a changing magnetic field is that of a transformer. If an alternating current enters the primary coil, the resulting magnetic field is also alternating. Some of the magnetic field from the primary coil penetrates the secondary coil. Hence, the secondary coil "sees" a changing magnetic field. In general, a system is "happy" with a steady (i.e., unchanged) state and opposes any changes imposed on the system. Therefore, the secondary coil must react in a certain way to oppose the changing magnetic field. The easiest way is for the secondary coil to generate its own magnetic field to compensate for the change. To do that, a current must flow in the secondary coil. Hence, the overall effect is that an alternating current applied to the primary coil results in an alternating current flow in the secondary coil.

SEC. 3.3
PERMITTIVITY & PERMEABILITY

A few points about the transformer just discussed must be noted. First, the secondary coil gets its energy from the magnetic field in the primary coil. Second, if the two ends of the secondary coil are open, no current will flow because it will take infinite energy for a current to flow across an open circuit and such an opposition is therefore "uneconomical." Third, current is driven by an electric field (electric field is proportional to voltage). Therefore, the alternating magnetic field from the primary coil results in an alternting electric field in the secondary coil, which then drives the secondary current to produce an opposing magnetic field.

Maxwell's Equation #4 (Faraday's law): A changing magnetic field applied to a system results in a changing electric field within the system to counteract the change. This equation also completes the symmetry between electric and magnetic field, as can be seen by comparing Maxwell's equations #3 and #4.

3.3 PERMITIVITY AND PERMEABILITY

The effectiveness of producing an electric or magnetic field inside a certain medium depends on the dielectric or magnetic properties of the individual materials involved. Dielectric properties, given by the permitivity ϵ, quantitatively describes how easily the material can be polarized, i.e., how easily the charges inside the material can be displaced upon application of a voltage (or electric field). In a vacuum or in air, the value of the permitivity is $\epsilon_0 = 8.85 \times 10^{-12}$ F/m*. The permitivities of all other substances are usually expressed as a product of ϵ_0 and a numerical factor which is known as the dielectric constant k ($k = 1$ for a vacuum). Therefore, the dielectric constant of a certain material is $k = \epsilon/\epsilon_0$.

In the same way, the magnetic properties of a substance are quantitatively described by its permeability, μ. For a vacuum or in air, $\mu = \mu_0 = 4 \times 10^{-7}$ H/m. The permeabilities of other substances are, like permitivities, given as $\mu = \mu_r \mu_0$ where μ_r is a pure number known as the relative permeability ($\mu_r = 1$ for a vacuum).

Dielectric constants and relative permeabilities of some common substances are listed in Appendix A.

3.4 SPEED OF PROPAGATION

Suppose an electrical charge is suddenly created. How fast will its presence be known by other electrical charges?

As soon as the charge is created, an electric field starts propagating outward. Contrary to the old belief that gravitational, electrical, and magnetic forces are "felt" instantaneously, the propagation travels at a finite speed which is the speed of light. In vacuum or air, the speed of light, denoted as c, is approximately 3×10^8 m/s (or 186,000 miles/second). The speed of light is the upper limit on the speed of all physically moving objects. In a medium, the speed of light changes somewhat, and is generally denoted as v ($v = c$ when the medium is a vacuum).

As will be seen in the chapter on antennas, when a charge is made to vibrate at some frequency f, both the electric field and the magnetic field (a moving charge is a current and a current creates magnetic field) are changing their respective directions. As a result, a wave, with electric and magnetic fields changing direction at frequency f, is radiated outward.

The speed of an electromagnetic wave in a medium, whose permitivity and permeability are ϵ and μ, respectively, is obtained by simultaneously solving the Maxwell's equations. We find that

$$v = 1/\sqrt{\epsilon \mu} \tag{3.4.1}$$

*The units of ϵ and μ, as illustrated in Appendix D.2, belong to the Standard International or *Système Internationale* (SI) units to consistently describe the quantities in an electromagnetic system. Examples of SI units are kilogram (kg), meter (m), second (s), volt (V), and ampere (A).

In a vacuum, the speed of light, according to Eq. (3.4.1), is

$$v = c = 1/\sqrt{\epsilon_0 \mu_0} \tag{3.4.2}$$

It is left as a quick exercise for the reader to verify that $c = 3 \times 10^8$ m/s by using $\mu_0 = 4\pi \times 10^{-7}$ H/m and $\epsilon_0 = 8.85 \times 10^{-12}$ F/m.

It is often convenient to express the speed of light v in a certain medium as a fraction of c. Therefore,

$$v = 1/\sqrt{\epsilon\, \mu}$$

$$= 1/\sqrt{k\epsilon_0\, \mu_r\, \mu_0}$$

$$= c/\sqrt{k\, \mu_r} \tag{3.4.3}$$

As an example, Teflon® has a dielectric constant equal to 2.0 and is non-magnetic, i.e., it behaves like a vacuum magnetically ($\mu_0 = 1$). Therefore, the propagation speed of an electromagnetic wave in Teflon is

$$v = \frac{c}{\sqrt{2 \times 1}} = 0.7 \times c$$

$$= 2.1 \times 10^8 \text{ meter/second}$$

3.5 CHARACTERISTIC IMPEDANCE Z_0

Suppose an electromagnetic wave in a medium could be "frozen" at some instant, and the electric and magnetic field could be "measured." This is a thought experiment that cannot be performed in reality. The ratio of the electric field strength E to the magnetic field strength H is known as the characteristic impedance (or wave impedance), Z_0, of the medium.

$$Z_0 = E/H \tag{3.5.1}$$

In Standard International (SI) units, the electric field E is in volt/meter, the magnetic field H is in ampere/meter, and the characteristic impedance Z_0 is in ohms. Note that Z_0 is the ratio of E to H and is independent of their absolute amplitudes, as shown in Fig. 3.4. Mathematically, the characteristic impedance is expressed as

$$Z_0 = \frac{E \text{ (volt/meter)}}{H \text{ (ampere/meter)}}$$

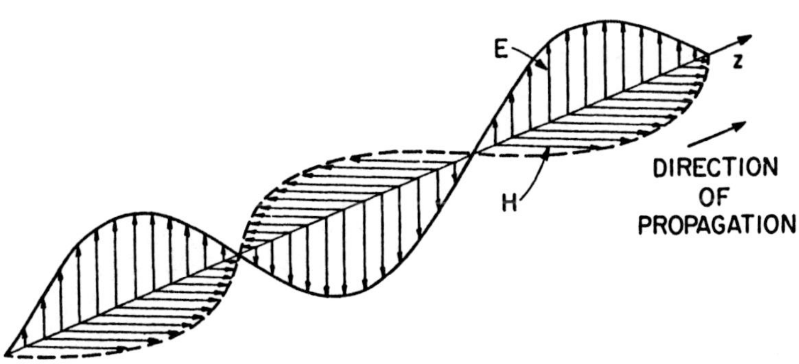

Figure 3.4 The characteristic impedance Z_0 of an electromagnetic wave is the ratio of the electric field E to the magnetic field H. Both E and H change their magnitudes so that their ratio is always the same.

Because the strengths of electric fields and magnetic fields are related to the permitivity and permeability of the medium, Z_0 can be expressed alternately as

$$Z_0 = \sqrt{\mu/\epsilon} \tag{3.5.2}$$

For free space, $\mu = \mu_0 = 4\pi \times 10^{-7}$ H/m and $\epsilon = \epsilon_0 = 8.85 \times 10^{-12}$ F/m,

$$Z_0 \text{ (free space)} = \sqrt{\mu_0/\epsilon_0} = 377 \text{ ohms}$$

The characteristic impedance of other media can be computed if their ϵ and μ are known. Note that characteristic impedance can be expressed in different but consistent forms. We shall see still another form in the chapter on transmission lines (Ch. 4).

Free space, of course, has no mechanical resistance to motion, but it does have impedance to electromagnetic wave propagation. If space behaved like an electrical short circuit (i.e., no resistance), no radiation could leave an antenna and 100% reflection will result. Nor does free space have infinite resistance (i.e., like an open circuit) to electromagnetic waves. It is, however, an open circuit to the flow of electrons (electron flow is mechanical motion).

3.6 SKIN DEPTH

One consequence of Maxwell's equation #4 is the existence of eddy currents. We shall explore this phenomenon briefly.

Imagine that we have a magnetic field detector behind a block of metal, in front of which a magnet is placed. Will the detector sense the magnetic field? Note that the metal may or may not be magnetic, like iron.

For most materials, the answer to the above question is yes. Magnetic field can penetrate most substances. Now, imagine that the magnet is replaced by a solenoid in which an alternating current flows. Will the detector sense the alternating magnetic field? The magnetic field now attempts to go in and out of the metal block. According to Maxwell's equation #4, the changing magnetic field creates an opposition from the neighboring system, in this case, the metal. The opposition manifests itself as a circulating (and alternating) current, known as eddy current. The eddy current would then generate an alternating magnetic field to oppose any change. The effectiveness of such opposition depends on the magnitude of the eddy currents, which in turn depends on the resistivity of the metal block. In other words, if the block has poor conductivity, e.g., wood or plastic, the eddy current distribution would be very small and, therefore, the opposition would also be very small. If the block is absolutely without resistance, e.g., a perfectly conducting metal, the eddy current flow will be maximum and the opposition will exactly cancel the penetrating field. The latter case is, of course, unrealistic. Another factor is the question of whether the medium is ferromagnetic. The effectiveness of the opposition increases as the permeability of the metal increases. This point can be easily understood from the fact that iron can be magnetized much more easily than aluminum, as an example, because the permeability of iron is hundreds of times larger than that of aluminum.

The last factor to be considered is the frequency of the alternating field. The higher the frequency, the larger the rate of change of the magnetic field. Therefore, the opposition increases as the frequency increases.

In essence, the opposition results in a reduction of the penetrating magnetic field inside the material. Skin depth is defined as the distance from the metal surface beyond which the overall magnitude of the penetrating magnetic field falls below 30% of its original magnitude. Mathematically, the skin depth, δ, for a conductor is expressed as

$$\delta = \frac{1}{\sqrt{\pi f \mu \sigma}} = \sqrt{\frac{\rho}{\pi f \mu}} \qquad (3.6.1)$$

where

ρ = resistivity of the metal,
σ = $1/\rho$ = conductivity of the metal,
f = frequency of the alternating magnetic field,
μ = permeability of the metal.

For microwave applications, copper is usually used as a conducting medium. The electrical conductivity of copper is about the same order of magnitude as silver or any other good conductor (gold, silver, aluminum, *et cetera*). Copper is also non-magnetic. A more useful expression for skin depth in copper is

$$\delta(\mu\text{m}) = \frac{2}{\sqrt{f\,(\text{GHz})}} \qquad (3.6.2)$$

where f is given in GHz and δ is given in microns (10^{-6} m), μm. For example, the skin depth in a copper medium is approximately 2 microns for a 1 GHz magnetic field.

The reader should bear in mind as a rule of thumb: GHz frequency implies a skin depth of micron size.

From Eq. (3.6.1) for skin depth, we conclude the following:

1. δ increases with resistivity;
2. δ decreases with frequency;
3. δ decreases with permeability.

Skin depth is basically the effective distance of penetration of an electromagnetic wave into a metal. This concept can be applied to a conductor carrying high frequency signals, the self-inductance of the conductor effectively limits the conduction of the signal to its outer shell. The shell's thickness is given by the skin depth. The effective area of conduction is considerably smaller than the cross section area of the solid conductor. Consequently, the effective resistance of this conductor is higher, and is no longer the resistance as seen by a direct current.

In waveguides, where microwave signals propagate by reflecting off the walls of the guide, its inner surfaces are sometimes coated with silver or copper layers about several mils (1 mil = 1/1000 of an inch = 25 μm) thick. This practice enhances the signal conduction and reduces power absorption of the signal by the waveguide. The coating thickness is typically a few times larger than the skin depth.

Lastly, if the metal is a perfect conductor, i.e., $\rho = 0$ meaning $\delta = 0$, the skin depth is zero. This means that the incoming magnetic field cannot penetrate into the metal. The overall magnetic field perpendicular to the metal surface is nil. If the alternating magnetic field is that of an electromagnetic wave, the alternating electric field must be taken into account. For a perfect conductor, the overall effect is that 100% of the incoming wave is reflected and the metal does not absorb any power. If the metal has some resistance, reflection is no longer 100%. The incoming wave can penetrate into the metal by an effective distance of δ and some power is absorbed by the metal to sustain the eddy current which, together with the metal resistance, then generate heat. This power absorption accounts for the attenuation of a waveguide. Such attenuation is similar to the ohmic resistance of an ordinary wire in low frequency electronics.

SEC. 3.7
PHASE DIFFERENCE

3.7 PHASE DIFFERENCE

The reader may recall from low frequency electronics that two waves of the same frequency, but not in a perfectly synchronous alignment with respect to each other, are said to have a difference in their phases. Figure 3.5 illustrates that two waves *A* and *B* of the same frequency but having a phase of 90° ($\pi/2$ radian); wave *A* leads wave *B* by 90°.

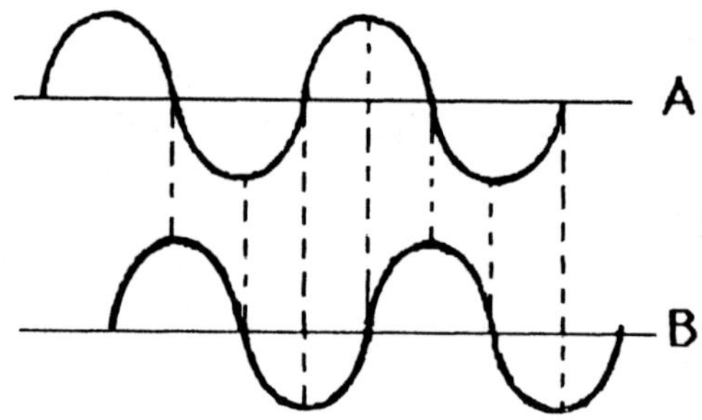

Figure 3.5 Two sine waves A and B with a phase difference of 90°. Wave A leads wave B.

Consider an electromagnetic wave traveling toward two locations *A* and *B* as shown in Fig. 3.6. A particular point of the wave, call it *P*, passes location *A* first, then a moment later passes location *B*. A finite travel time is needed for *P* to go from *A* to *B*. Therefore, by the time location *B* receives the information described by *P*, point *Q* on the waveform is now passing *A*.

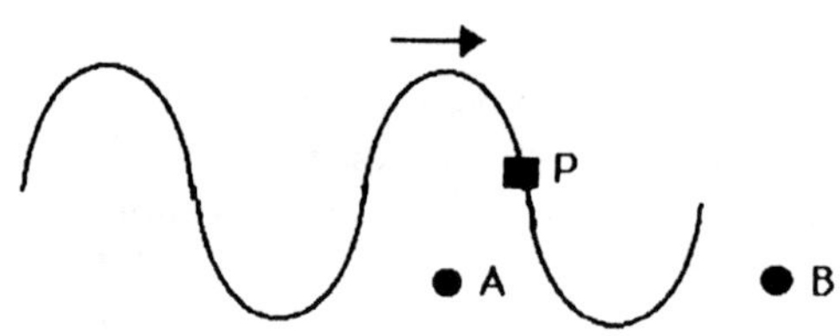

Figure 3.6 A particular point P of a wave propagating from left to right passes location A first, then location B.

Referring to Fig. 3.6, if we could freeze time at some instant t_0, it becomes evident that there is a phase difference between the wave passing *A* and the wave passing *B*. Phase difference is characterized by a measured quantity we call the phase angle. The phase angle θ can be calculated in either of the following expressions:

$$\theta = \frac{D}{\lambda} \times 360° \tag{3.7.1}$$

or

$$\theta = \frac{\Delta t}{T} \times 360° \tag{3.7.2}$$

where

D = distance between point A and point B,
λ = wavelength of the electromagnetic wave in the same unit as D,
Δt = time of travel by signal from point A to point B,
T = period of the signal in the same unit as D.

Note that the travel time is related to the distance between A and B by the following relationship:

$$\Delta t = D/v \qquad (3.7.3)$$

where v = velocity of the electromagnetic wave = c, if the medium is a vacuum.

Example (3.7.1)

Two points, A and B, are 3cm apart and the frequency of an electromagnetic wave is 1 GHz, find the phase difference A and B. Assume the medium is free space.

In free space,

$$v = c = 3 \times 10^8 \text{ m/s}$$
$$= 3 \times 10^{10} \text{ cm/s}$$

Frequency $f = 1$ GHz implies that $\lambda = 30$cm and $T = 10^{-9}$ second.
According to Eq. (3.7.1),

$$\theta = \frac{3\text{cm}}{30\text{cm}} \times 360° = 36°$$

In order to use Eq. (3.7.2), we must first calculate the time of travel.

$$\Delta t = \frac{D}{C} = \frac{3\text{cm}}{3 \times 10^{10} \text{ cm/s}} = 10^{-10} \text{ seconds}$$

According to Eq. (3.7.2).

$$\theta = \frac{10^{-10} \text{ s}}{10^{-9} \text{ s}} \times 360° = 36°$$

Example (3.7.2)

Two points, A and B, are 3m apart and the signal is an AM radio wave of 1 MHz, find the phase angle. Assume the medium to be free space.
Frequency $f = 1$ MHz implies that $\lambda = 300$m, and $T = 10^{-6}$ s.
Also,

$$\Delta t = \frac{D}{\text{C}} = \frac{3\text{m}}{3 \times 10^8 \text{ m/s}} = 10^{-8} \text{ seconds}$$

According to Eq. (3.7.1),

$$\theta = \frac{3\text{m}}{300\text{m}} \times 360° = 3.6°$$

According to Eq. (3.7.2),

$$\theta = \frac{10^{-8}\ \text{s}}{10^{-6}\ \text{s}} \times 360° = 3.6°$$

Example (3.7.3)

The frequency of the ac line in the US is 60 Hz. The typical distance between an average household and the power plant is 10km. Find the phase angle between the plant and the household for the ac.

Frequency $f = 60$ Hz implies that

$$\lambda = \frac{3 \times 10^8\ \text{m/s}}{60\ \text{Hz}} = 5 \times 10^6\ \text{meters}$$

$$T = \frac{1\ \text{s}}{60\ \text{Hz}} = 0.0166\ \text{second}$$

and

$$\Delta t = \frac{10 \times 10^3\ \text{m}}{3 \times 10^8\ \text{m}} = 3.33 \times 10^{-5}\ \text{seconds}$$

According to Eq. (3.7.1),

$$\theta = \frac{10^4\ \text{m}}{5 \times 10^6\ \text{m}} \times 360° = 0.72°$$

According to Eq. (3.7.2),

$$\theta = \frac{3.33 \times 10^{-5}\ \text{s}}{0.0166\ \text{s}} \times 360° = 0.72°$$

From the above examples, the following conclusions can be drawn. The wavelengths of low and medium frequency signals (dc–MHz) are much larger than the physical dimensions of typical laboratory equipment. Hence, the phase difference between the signal source and a load for a low frequency signal is negligibly small.

In the case of the 60 Hz ac line, the maximum distance of a power line is about 100km long. The wavelength of a 60 Hz signal, however, is about 3000 miles (the width of the US from coast to coast). Therefore, the phase difference of an ac signal between the generator and the local household is again negligible.

The extremely small phase angle for low frequency signals make circuit analysis rather simple. Each piece of equipment is basically processing the "same" piece of information due to the tremendously small phase differences.

At microwave frequencies, the wavelengths are comparable to, if not smaller than, the laboratory cables and the connected equipment. The phase angle between two locations can be quite large. We shall see that it is this sizable phase difference that compounds the difficulties encountered in the behavior of microwave electronics. Unlike low frequency electronics where the lengths of connecting wires are almost irrelevant (phase angle $\cong 0$), the lengths of the microwave conductors and waveguides play important roles in determining the overall behavior of a microwave circuit.

CHAPTER 4

TRANSMISSION LINES

W. Stephen Cheung

4.1 INTRODUCTION

In its simplest form, a transmission line is a pair of conductors linking together two electrical systems, components, or devices. Otherwise stated, transmission lines conduct electronic signals from a source component to a load component. Low frequency circuitry uses simple wires and cables with or without insulated coatings as transmission lines. For noise reduction purposes, twisted pairs and coaxial cables are sometimes used. Figure 4.1 shows a few common transmission lines.

The basic components of a low frequency transmission line consist of a forward path and a return path. The return path is sometimes grounded and acts as a shield. For cases where the return path must float, a third conductor, usually braided and completely embracing the forward and return path conductors, is introduced as the grounding shield.

A waveguide is a transmission line for microwaves, as shown in Fig. 4.1C and Fig. 4.1D. Waveguides are usually metallic tubes with rectangular, circular, elliptical, and ridged cross sections. Rectangular waveguides are the most popular type used today. The physical size of any particular waveguide determines the frequency, bandwidth, power handling capability, and impedance of the line. These properties will be discussed. Waveguides are different from coaxial cables or other transmission lines in that the tube, while resembling the outer shield of a coaxial cable, has no center conductor.

We shall study waveguides in more detail later. Other transmission lines can usually be analyzed from a knowledge of their voltage, current, and impedance characteristics. Consider the connection in Fig. 4.2A, the voltage and current wavefronts travel along the forward and return conductors. Because voltages are directly proportional to electric fields, and currents are directly related to magnetic fields, we can envision that the alternative to Fig. 4.2A is that of an electromagnetic field as illustrated in Fig. 4.2B. This field must follow the physical layout of the transmission lines. In essence, the two conductors "guide" the electromagnetic field's propagation from the generator to the load. Some of the electromagnetic fields are also radiated away.

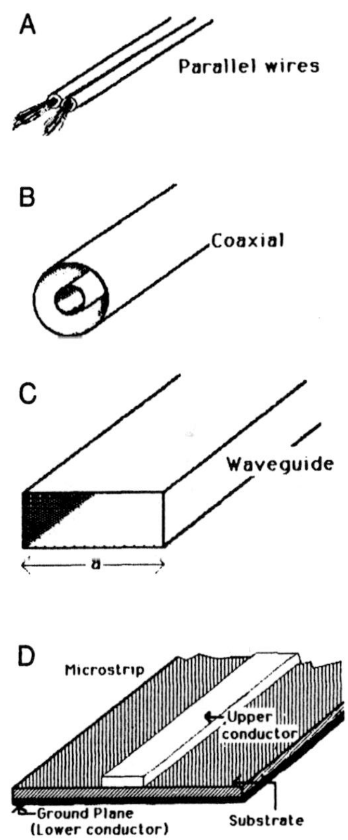

Figure 4.1 Four common transmission lines: (A) parallel wires; (B) coaxial cable; (C) rectangular waveguide; and (D) microstrip.

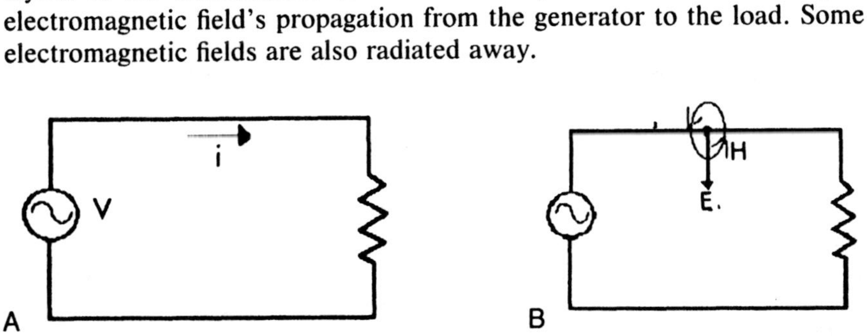

Figure 4.2 A simple electronic circuit can be expressed in terms of (A) voltage V and current i, or (B) electric field E and magnetic field H.

SEC. 4.2

QUARTER-WAVE SHORT CIRCUIT

4.2 QUARTER-WAVE SHORT CIRCUIT

Consider a signal generator connected to a C-shaped *conductor* of width s and height b as shown in Fig. 4.3. For simplicity, the C-shaped conductor has no ohmic resistance. Our understanding of basic electronics would suggest that this is a short circuit and the generator is severely loaded.

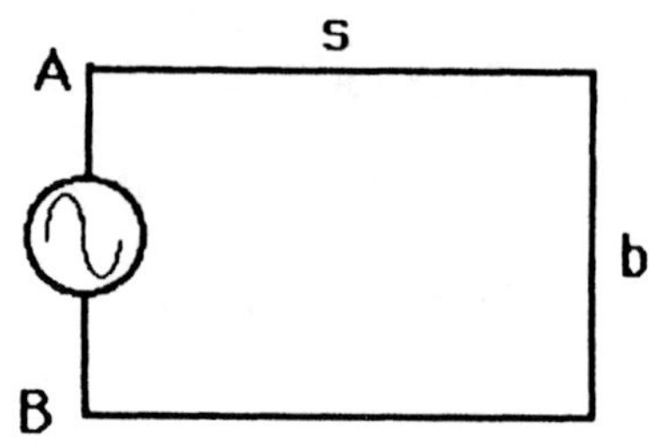

Figure 4.3 *A C-shaped conductor is connected across a generator. The conductor is a short circuit for low frequency, but not so for high frequency. If the width of the short circuit, s, is one-quarter of the signal wavelength, the impedance is actually infinite, i.e., a perfect insulation.*

The connection in Fig. 4.3 is *not* a short circuit, if the width s is a quarter of a wavelength of the signal. This can be seen by the following qualitative reasoning. Upon connection, there is a current wavefront moving from the generator toward the load (the short circuit). The short circuit shunts *all* the current back to the generator. Because it is a quarter of a wavelength each way, the returned current is a half-wavelength, or 180° phase difference, in comparison with the current that is just coming out of the generator. There is total current cancellation which results in zero net current. Hence, a quarter-wave ($\lambda/4$) short circuit is essentially an "insulator" to a signal of wavelength.

If the width s is approximately $\lambda/4$, the C-shaped conductor will be considered as high impedance, and only a small current will flow. Quantitatively, the impedance of a short as a function of s is given without proof to be

$$Z(s) \propto \tan(\beta s)$$

where

$$\beta = 360°/\lambda$$

If s is much smaller than λ, as is the case for low frequency electronics, tan (βs) and therefore $Z(s)$ is practically zero, which is consistent with our understanding in basic electronics that this conductor is a short circuit. On the other hand, if s is equal to $\lambda/4$, the trigonometrical function tan 90° is equal to infinity, i.e., a perfect insulator.

It is worth pointing out that the quarter-wave short circuit is not a practical insulator for low frequency. For example, the wavelength of the 60 Hz power line is 5000km, approximately 3000 miles. Using a quarter-wave short circuit would need an impractical length of 1250km. Therefore, low frequency lines use dielectrics like porcelain, Teflon®, *et cetera* as insulators. As frequency increases, however, the dielectric losses of these insulators become significant. Moreover, the required lengths of quarter-wave short circuits at high frequencies are much shorter. For example, the wavelength of a 1 GHz signal is 30cm, so the length of a quarter-wave short circuit is only 7.5cm.

4.3 RECTANGULAR WAVEGUIDES

We shall now show that a rectangular waveguide can be regarded as a series of quarter-wave short circuits. Figure 4.4A shows two bus-bars of width d at a distance b apart. A quarter-wave short circuit of length s, i.e., $s=\lambda/4$, is connected across the two bus-bars. Because the impedance of a quarter-wave short circuit is infinite, its presence should not change anything, i.e., the quarter-wave short circuit is electrically "invisible." Now imagine that more of such quarter-wave short circuits are added, as shown in Fig. 4.4B. No net change in impedance will occur. As the number of quarter-wave short circuits becomes infinite, a rectangular waveguide is formed.

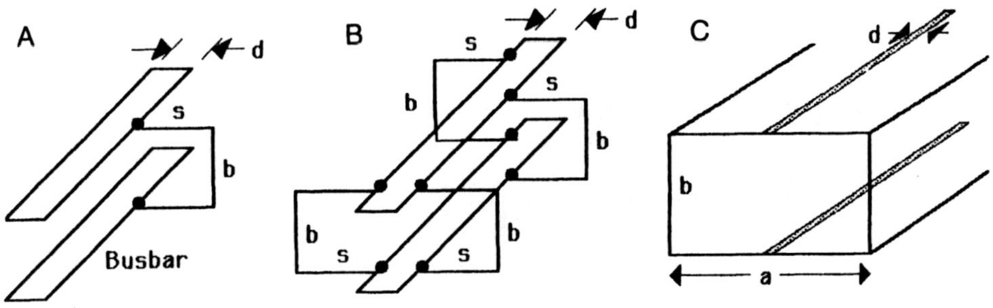

Figure 4.4 Construction of a rectangular waveguide by (A) starting with two parallel busbars and a C-shaped quarterwave-short connected, (B) adding more quarterwave short circuits on both sides of the bus bars, (C) increasing the number of quarterwave short circuits to infinity.

The interesting features of the rectangular waveguide are now described. At first glance, the total width of a rectangular waveguide, a, is simply equal to 2 s. This assumes that the bus-bars have no width, i.e., $d=0$ in Fig. 4.4C. Since $s=\lambda/4$, it is clear that the total width of the rectangular waveguide a is half the wavelength of the signal. Hence, the given rectangular waveguide sets an upper limit on the wavelength, or a lower limit on the frequency, of a signal to be conducted. This upper limit of wavelength (the cut-off wavelength) is given by $\lambda_c=2\ a$. Any signal whose wavelength is larger than this cut-off wavelength is not allowed. A numerical example helps illustrate this point.

Consider the WR90 rectangular waveguide whose inner cross section dimensions are 0.9″ × 0.4″. We shall set $a=0.9''=2.286\text{cm}$ and $b=0.4''=1.016\text{cm}$.

The cut-off wavelength λ_c can be calculated from the equation:

$$1/2\ \lambda_c = a = 2.286\text{cm}$$

i.e.,

$$\lambda_c = 4.572\text{cm}$$

This cut-off wavelength corresponds to a cut-off frequency

$$f_c = \frac{3\times 10^{10}\text{cm/s}}{4.572\text{cm}} = 6.56\times 10^9\text{Hz}$$

i.e.,

$$f_c = 6.56\ \text{GHz}$$

SEC. 4.4
MODES

Any signal whose frequency is less than 6.56 GHz will not propagate in the WR90 waveguide. If a 60 Hz signal is connected to the WR90 waveguide, as shown in Fig. 4.5, conduction will be lateral (transverse) rather than longitudinal. This is because the lateral impedance is very small. This can be shown numerically by putting $s = 1/4\ \lambda_c = 1.143\text{cm}$, and the wavelength of the 60 Hz signal is 5×10^8 cm.

$$Z \propto \tan\left(\frac{360° \times 1.143\text{cm}}{5 \times 10^8\text{cm}}\right)$$

$$= \tan(8.23 \times 10^{-7})$$

$$= 1.43 \times 10^{-8}$$

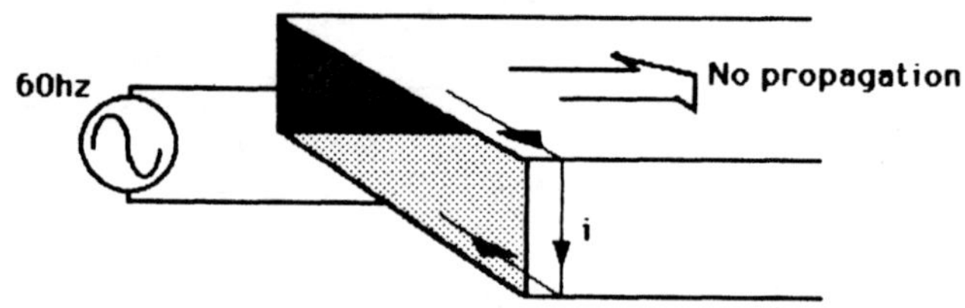

Figure 4.5 A 60 Hz generator connected to an ordinary sized rectangular waveguide would result in lateral but not longitudinal conduction. This is because the wavelength of the 60 Hz signal is much larger than the width of the waveguide.

It appears that a given rectangular waveguide works only for one particular frequency because the corresponding impedances of other frequencies would not yield infinite values. This is not so. The subtlety is that we have overlooked the width of our "imaginary" bus-bars as shown in Fig. 4.4B. We can argue that for frequencies above the cut-off frequency, the width of the bus-bars will increase from very narrow ($d = 0$ at f_c) to a sizable portion of the total width, such that the remaining width is exactly what is required for the quarter-wave short circuits on each side. Hence, a rectangular waveguide is not a band-pass filter, but a high-pass filter.

We shall return to the rectangular waveguide in the later sections.

4.4 MODES

4.4.1 TEM Mode

With reference to Fig. 4.2B, the electromagnetic wavefront has certain characteristics. The direction of propagation, the electric field, and the magnetic field are mutually perpendicular to one another. This propagation mode is known as the *transverse electromagnetic* (TEM) mode.

TEM modes are the principal modes of propagation for all low frequency transmission lines. TEM modes are not allowed in waveguides. In fact, we have just shown that waveguides do not conduct electromagnetic signals for frequencies below a certain cut-off value.

It is a general and valid statement that the TEM mode is a low frequency mode, and is related to the fact that the wavelengths of low frequency signals are usually much larger than the physical dimensions (length and width) of the transmission line. As frequency increases, the shrinking wavelength of a signal is now comparable to the length and width of the line. Hence, TEM dominates in the low frequency regime, and other modes become viable at high frequencies.

4.4.2 Other Modes

Unlike the TEM mode, the electric fields E or the magnetic fields H, or both, of other modes have some components along the effective direction of propagation. We shall use a rectangular waveguide to help define other modes. Imagine three forms of electromagnetic waves entering a rectangular waveguide. The first form is the direct entry as shown in Fig. 4.6A. This is a TEM mode and will be shown below to be disallowed.

The second and third forms are such that the electromagnetic wave enters the waveguide at an angle and propagates along the waveguide in a zig-zag pattern. In the second form, the electric field is perpendicular to the effective direction of propagation, which is along the length of the waveguide (Fig. 4.6B). This is known as the *transverse electric* (TE) mode. In the third form, the magnetic field is perpendicular to the effective direction of propagation (Fig. 4.6C). This is known as the *transverse magnetic* (TM) mode.

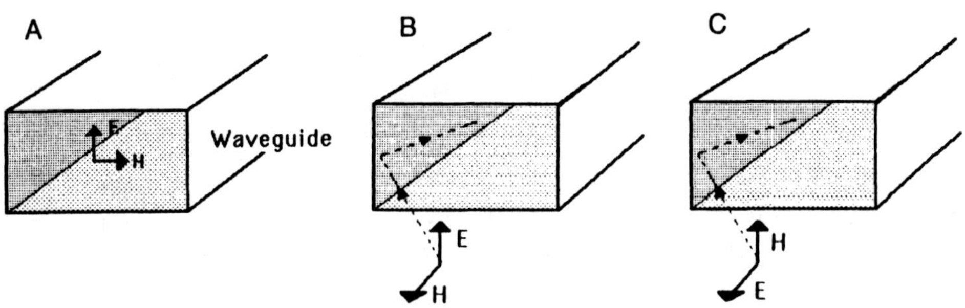

Figure 4.6 Three configurations are investigated when an electromagnetic wave enters a rectangular waveguide: (A) TEM mode (direct entry); this mode is not allowed; (B) TE mode, the electric field E is perpendicular to the effective direction of propagation; and (C) TM mode, the magnetic field is perpendicular to the effective direction of propagation.

We now show that the TEM mode is disallowed in single-enclosure waveguides by means of self-contradiction. Let us suppose that the TEM mode is indeed possible. At some instant, the electric field E of the signal induces charges on the inner surfaces of the upper and bottom plates as shown in Fig. 4.7A. The momentary gathering of charges results in a voltage to form a current. The shortest path for the current to flow is along the sidewalls as shown in Fig. 4.7B. This current flow will generate a magnetic field pattern, part of which runs parallel to the direction of propagation. The overall magnetic field (magnetic field of the signal and that produced by the current) now has some component along the direction of propagation, and hence contradicts the definition of a transverse electromagnetic mode. A similar conclusion is reached if the electric field is oriented along the width of the waveguide.

Consider an electromagnetic wave entering a waveguide at an angle as shown in Fig. 4.8. The wave makes contact with a perfect conductor. The alternating fields (magnetic or electric) result in oppositions generated by the eddy currents on the conductor's wall. The overall result is that there is a reflected wave, as shown in Fig. 4.8, and the angle of incidence is equal to the angle of reflection, i.e., $\theta_1 = \theta_2$. That the conductor is perfect, i.e., resistance $= 0$, means that the opposing eddy currents require no power to generate and the skin depth is zero.

As pointed out in the previous chapter, if the conductor has a non-zero resistivity, the skin depth is finite, and a portion of the power of the incident wave is absorbed by the conductor to generate the eddy currents, which in turn produces heat. This contributes to the attenuation property of the waveguide. Another important feature is that the zig-zag pattern effectively slows down the signal propagation along the line. Under the limitation that the angle of incidence θ in Fig. 4.8 is 0°, the reflected wave coincides with the incident wave and the signal will never proceed forward. The effective speed of propagation along the waveguide is known as the *group velocity*.

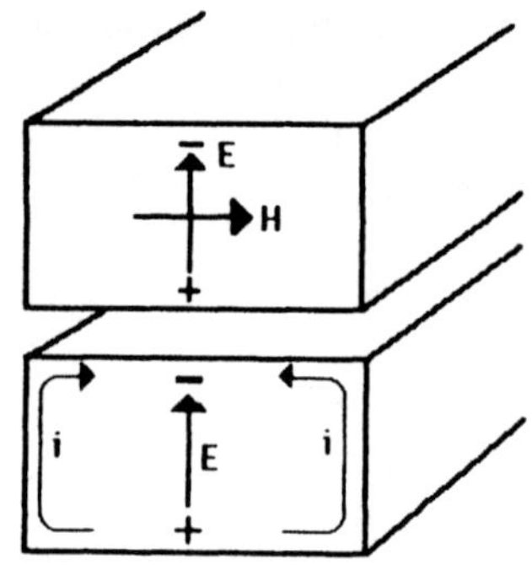

Figure 4.7 A possible scenario showing why TEM mode is not allowed in a rectangular waveguide. Lateral currents produce magnetic fields that point along the effective direction of propagation, which therefore contradicts the definition of TEM mode.

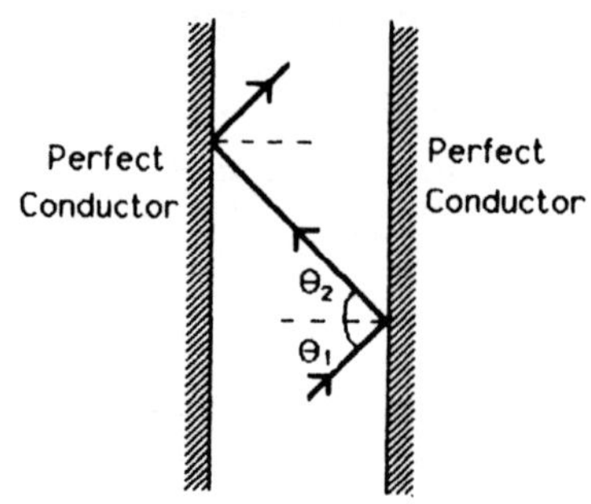

Figure 4.8 The zig-zag propagation pattern of an electromagnetic wave inside a rectangular waveguide. The angle of incidence is equal to the angle of reflection.

4.5 THEORY OF TRANSMISSION LINES

We are now in a position to develop a general theoretical foundation for transmission lines. The theory is applicable to transmission lines of all forms: parallel lines, twisted pair, coaxial cable, and even waveguides.

Figure 4.9 Two parallel conducting wires separated by one or two dielectric layers. A dielectric is electronically equivalent to a capacitor.

Consider two parallel lines as shown in Fig. 4.9. Between the two conductors are the insulation coatings or air, both of which are dielectrics. All conductors have some ohmic resistance, no matter how high their conductivity is, and we shall call this ohmic resistance r'. When the wire is used to conduct ac, the magnetic field generated by the alternating current is changing all the time. The system responds to the change in magnetic field as a current opposing the original alternating current. It takes energy to form the magnetic field. This energy is proportional to the rate of change of the incoming current. The proportionality constant is called L', the inductance of the conductor.

SEC. 4.5

THEORY OF TRANSMISSION LINES

The dielectric medium serves as an insulation of currents between the two wires. When subjected to an electric field, a dielectric is polarized with positive charges on one side and negative charges on the opposite side. This phenomenon is responsible for the capability of capacitors to "pass" ac currents. We shall label the capacitance in our circuit with C.

Lastly, all dielectric media permit leakage current to flow, especially at high voltage. It is, therefore, necessary to assign a resistance R (which may be 10^{12} ohms or more) to the leakage path between the two conductors. Sometimes R is replaced by its reciprocal quantity, conductance G (the smaller G is, the better the insulation).

The overall circuit and its equivalent are shown in Figs. 4.10A and 4.10B. As a current (ac or dc) flows down the pair of transmission lines, energy is distributed partly to the ohmic resistance r (dissipated), partly to overcome the local inductive opposition due to L, partly to charge up the capacitor C, and partly to compensate the leakage from one line to the other across R.

If the two lines extend to infinity, the equivalent circuit will extend to infinity as well. The quantities r, R, C, and L shown in Fig. 4.10B are specified as quantities per unit length. In other words, these electrical quantities are distributed rather than lumped (total) values. The equivalent impedance of the infinite lines, however, can be calculated.* The result is

$$Z_0 = \sqrt{\frac{r+jL\omega}{G+jC\omega}} \qquad (4.5.1)$$

where $G = 1/R$ = conductance of the dielectric medium, j is the imaginary number $\sqrt{-1}$ (see Appendix B), and ω is 2π multiplied by the frequency of the signal concerned.

For a *lossless* and *infinitely long* line, $r = 0$ and $G = 0$; the impedance Z_0 in Eq. (4.5.1) is reduced to

$$Z_0 = \sqrt{L/C} \qquad (4.5.2)$$

where L and C are the inductance per unit length and capacitance per unit length, respectively. The quantity Z_0 is known as the characteristic impedance of the transmission line. Quantities L and C are individually measurable (in most cases calculable), and are generally specified by the manufacturer for a particular transmission line.

As an example, the inductance per unit length of a RG59 cable is 370 nH/m and its capacitance per unit length is 67 pF/m. The cable's characteristic impedance is

$$Z_0 = \sqrt{\frac{370 \times 10^{-9}}{67 \times 10^{-12}}} \text{ ohms}$$

$$= 74.3 \text{ ohms}$$

The characteristic impedance of a given line of infinite length is Z_0 as calculated from inductance and capacitance per unit length. The infinite length is immaterial in the calculation because both the total inductance and the total capacitance increase proportionally, so the ratio is always the same.

In deriving Eq. (4.5.2), inductance and capacitance are used. This definition of characteristic impedance is consistent with the definition of characteristic impedance for a medium in which an electromagnetic wave travels [Eq. (3.5.1.)]. When a signal travels inside a cable, the capacitance C of the cable helps to determine the electric field strength E of the signal. The inductance L of the

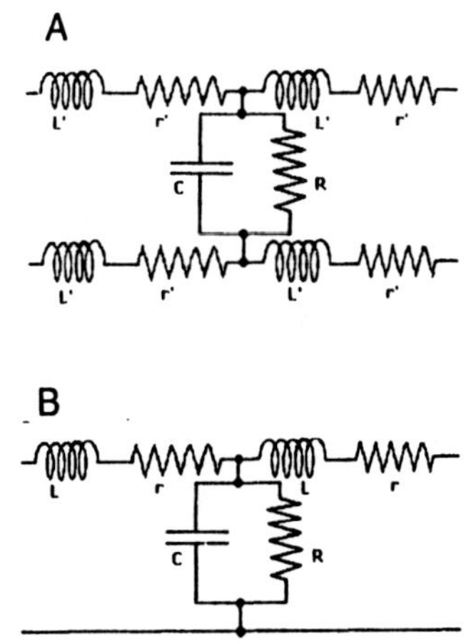

Figure 4.10 (A) The equivalent circuit of a transmission line and (B) its simplified version.

* The reader should not be surprised by such calculations. In dc basic electronics, the problem of calculating an infinite series of series or parallel resistors is frequently encountered.

cable, on the other hand, is related to the magnetic field H of the signal. Hence, characteristic impedance, defined as

$$Z_0 = E/H \tag{3.5.1}$$

or as

$$Z_0 = \sqrt{L/C} \tag{4.5.2}$$

describes the strength of electric field in relation to that of the magnetic field for an electromagnetic signal propagating along a transmission line.

The characteristic impedance given in the format of Eq. (3.5.1) is more appropriate for a beam of electromagnetic wave traveling inside a waveguide. Clearly, both the electric and magnetic fields of the wave interact with the walls of the waveguide. We, therefore, expect either the electric or magnetic field, or both, will change somewhat, as compared to when the signal travels in free space. Consequently, the characteristic impedance of a waveguide is different from that of free space. The difference will be even more significant if the inside of the waveguide is stuffed with a dielectric instead of air or vacuum.

4.6 MATCHED TERMINATION

The concept of an infinitely long transmission line is extremely useful. Suppose we start out with an infinitely long line whose characteristic impedance is Z_0. Now we keep one segment of the cable as per our model, and replace the rest of the infinitely long line by a load resistor whose impedance is also Z_0 as shown in Fig. 4.11. Hence, *a finite transmission line terminated by an impedance Z_0 is electrically equivalent to an infinitely long transmission line.*

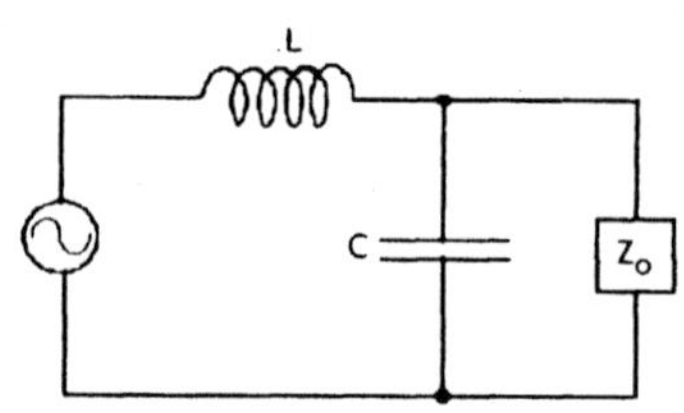

Figure 4.11 *A segment of a transmission line of characteristic impedance Z_0 terminated by a resistor of resistance Z_0 is equivalent to an infinitely long transmission line. This is a perfect match.*

What is the advantage of using an infinitely long transmission line? Any signal leaving the generator will keep going down the infinite line and will *never* return. In other words, *no reflection* occurs. Reflections inside a transmission line mean that we have inefficient transfer of power (and information) from the source to the receiving end. Reflection also causes distortion of information.

There is no self-absorption in a lossless line of finite length. Because the signal is no longer required to travel down an infinite length, it must, therefore, be totally absorbed by the terminating load. This can only happen if the terminating load impedance is equal to the characteristic impedance of the line. The generator cannot tell the difference between an infinitely long line and a finite line terminated by a matched load. No reflection results in either case.

The reader may recall from low frequency electronics that maximum power is transferred to the load when the output resistance of the generator is exactly equal to the load resistance. If complex impedances are involved, then the resistive parts must be identical, and the reactive parts must be of opposite sign with respect to each other. Similarly, if a single generator is connected to a device with input resistance R via a lossless cable of characteristic impedance Z_0, we must have $R = Z_0$ in order to achieve maximum power transfer. If the input impedance of the device is complex, however, its reactive part is not compensated by the lossless cable whose characteristic impedance has only resistive but no reactive part, then reflection is bound to occur. External measures must be taken to match the connection.

4.7 SPEED OF PROPAGATION

In free space, an electromagnetic wave travels at a speed equal to 3×10^8 m/s. Inside a transmission line, the voltage (electric field) and current (magnetic field) propagations are delayed somewhat due to interaction with the line. The inductance along the line and the capacitance across the line tend to slow things down. The speed of propagation in a transmission line of known inductance and capacitance per unit length is given to be

$$v = \frac{1}{\sqrt{LC}} \qquad (4.7.1)$$

If the inductance is in henrys/meter (H/m) and capacitance in farads/meter (F/m), then the unit for Eq. (4.7.1) is meters/second (m/s). We shall use the data given in the example of Sec. 4.5 to calculate the speed of propagation of a RG59 cable.

Because $L = 370$ nH/m and $C = 67$ pF/m, Eq. (4.7.1) yields

$$v = \frac{1}{\sqrt{370 \times 10^{-9} \times 67 \times 10^{-12}}} \text{ m/s}$$

$$= 2.0 \times 10^8 \text{ m/s}$$

(approximately 67% of c)

4.8 STANDING WAVE

Consider a general case in which the line's impedance is not the same as that of the load. Wavefront A hits the load Z_L and when not all of its energy is absorbed by Z_L, the remaining energy is reflected. This reflected wave, traveling from right to left in Fig. 4.12 meets another wavefront B (which was generated some time after wavefront A). The forward wave and the reflected wave can coexist. A directional coupler positioned one way can detect the forward wave and while positioned another way can detect the reflected wave. If a detector, or probe, is inserted into the line, the forward and reflected waves will register effects on the probe. Due to the phase difference between the two waves, the overall registration varies along the line from one location to another. A standing wave, or interference pattern, is said to have formed.

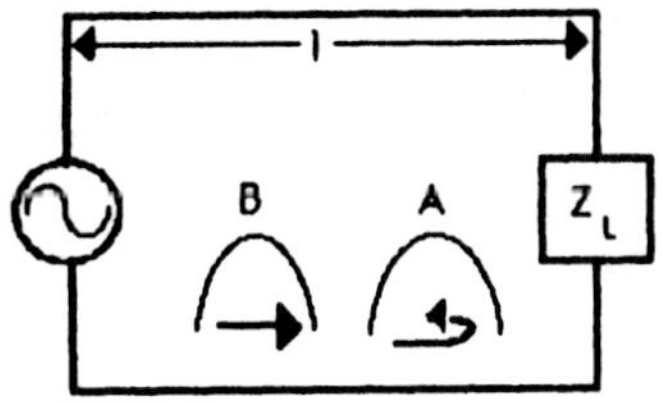

Figure 4.12 A standing wave is formed when the load impedance does not match that of the line. Part of the incident wave is reflected, which then combines with the upcoming wave to form the standing wave.

SEC. 4.8 STANDING WAVE

At a particular point P, there is, generally speaking, some definite phase angle (difference) between the forward wave and the reflected wave. This phase angle is determined by the wavelength of the wave, the exact length of the line, and the location of point P relative to the source (or the load), as shown in Sec. 3.7. Standing waves occur for both the voltage and current wavefronts. Note that if there is no reflection, the voltage and the current wavefronts are in phase. This is no longer true when reflection occurs.

The nature of the interference pattern depends upon the location of the wave forms' crests and troughs with respect to each other. We must also note that the direction of current flow depends on the polarity of the wavefront at the time of observation. The result is that if two positive directed wavefronts (one forward and one reflected) meet, the current wavefronts subtract but the voltage wavefronts add, as shown in Fig. 4.13A. Likewise, if a positive directed wavefront meets a negative directed wavefront, the current will add and the voltage will subtract.

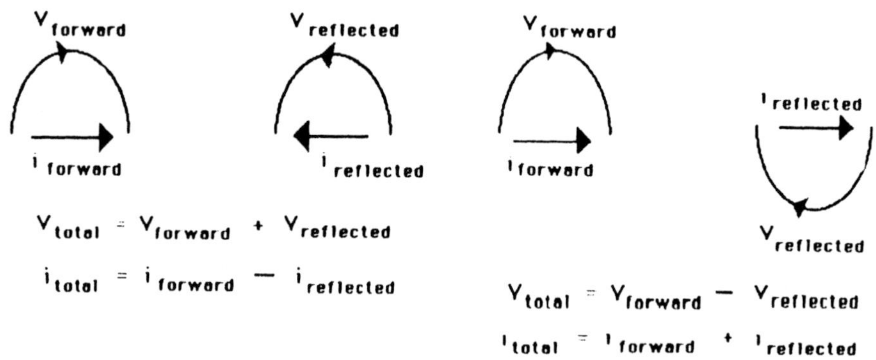

Figure 4.13 Two possible interactions between the voltage (current) of the forward wave and the voltage (current) of the reflected wave.

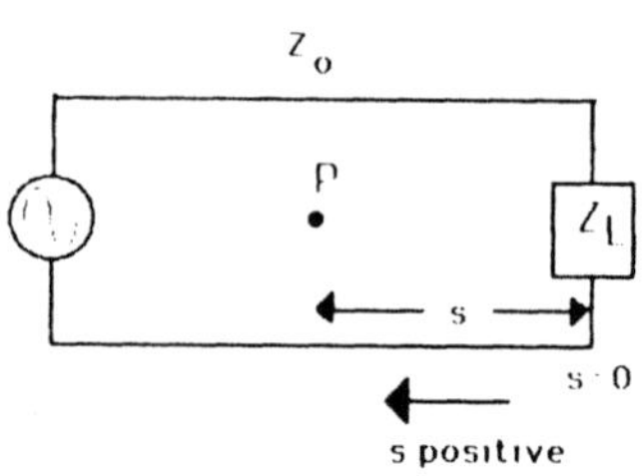

Figure 4.14 Parameters necessary to formulate the standing wave equation.

For our purposes here, the ratio of V_{total} to i_{total} at any point is called the (apparent) impedance at that point. The exact value depends on the relative position of the waveform from some reference point. As a convention, the load is chosen to be the reference position. The expression for the apparent impedance is given below, and the respective distance s is measured from the load as shown in Fig. 4.14. Note the order of the signs in the equations.

$$Z(s) = \frac{V_{total}(s)}{i_{total}(s)}$$

$$= \frac{V_f(s) \pm V_r(s)}{i_f(s) \mp i_r(s)} \qquad (4.8.1)$$

As s varies, the phase relationship between the forward and reflected wavefronts changes. This means that the total voltage and the total current vary with s, as will the impedance. Without going into the mathematical detail, the exact expression for Eq. (4.8.1) is

$$Z(s) = \frac{Z_0(Z_L + Z_0 \tan \beta s)}{Z_0 + Z_L \tan \beta s} \qquad (4.8.2)$$

where

$$\beta = 360°/\lambda_g \qquad (4.8.3)$$

We define λ_g as the wavelength of the signal inside the waveguide.* It is worth pointing out that the quantity

$$\beta s = \frac{360° \times s}{\lambda_g} \tag{4.8.4}$$

is for calculating the phase difference as shown in Sec. 3.7.

The imaginary number j in Eq. (4.8.2) occurs because there is a phase difference between the total voltage and the total current. This phase difference is different from that between the forward and reflected waves. This varying phase angle results in a lead-lag relationship analogous to the lag-lead relationship acting upon the voltages and currents in inductors and capacitors in basic electronics.

Eq. (4.8.2) applies also to an electromagnetic wave traveling in a waveguide. When traveling freely, the electric and magnetic fields are perpendicular to each other, but their magnitudes are in phase. Specifically, E is maximum when H is maximum, and E is zero when H is zero. The electric field of the forward wave together with that of the reflected wave register a signal on a probe in a manner dependent upon their relative magnitudes and phase difference. Similar abrupt interaction happens with the magnetic field. Hence, the total electric field and the total magnetic field are no longer necessarily in phase. Consequently, the overall wave impedance, which is E_{total} divided by H_{total}, varies in the manner described by Eq. (4.8.2).

Equation (4.8.2) is complicated because it is in its most general form. This is the "price" we pay for having a mismatched configuration. It describes the impedance (voltage to current, or electric to magnetic field) behavior caused by the standing wave set up between the forward and reflected waves.

4.9 NORMALIZED IMPEDANCE AND NORMALIZED DISTANCE

4.9.1 Impedance Normalization

Let us take a second look at Eq. (4.8.2).

$$Z(s) = Z_0 \times \left[\frac{Z_L + jZ_0 \tan \beta s}{Z_0 + jZ_L \tan \beta s}\right] \tag{4.9.1.1}$$

If we divide both sides by Z_0, we obtain

$$Z_n(s) = \frac{Z(s)}{Z_0} = \frac{Z_L + jZ_0 \tan \beta s}{Z_0 + Z_L \tan \beta s} \tag{4.9.1.2}$$

The impedance $Z(s)$ is now "normalized" against the line impedance, i.e., it is expressed as a numerical factor of Z_0. Note that $Z(s)$ and Z_0 have the ordinary unit of ohms, but $Z_n(s)$ is dimensionless, i.e., no unit.

Example (4.9.1)

Given that

$$Z(s) = 100 + j0 \quad \text{ohms}$$
$$Z_0 = 50 \quad \text{ohms}$$

Find $Z_n(s)$.

The solution is

$$Z_n(s) = Z(s)/Z_0 = 2 + j0.$$

* The guide wavelength, λ_g, is, in general, different from the free space wavelength. This difference will be elaborated later.

Example (4.9.2)

Given that

$$\begin{aligned} Z(s) &= 300 + j150 \quad \text{ohms} \\ Z_0 &= 75 \quad \text{ohms} \end{aligned}$$

Find $Z_n(s)$.

The solution is

$$Z_n(s) = 4 + j2$$

4.9.2 Length Normalization

Quite often, it is desirable to express a given length or distance in terms of a convenient unit of length and a multiplier. In microwave electronics, the guide wavelength of a signal is chosen as the standard length against which all other lengths are compared. Like normalized impedance, normalized length is dimensionless.

Example (4.9.3)

Given the guide wavelength $\lambda_g = 2\text{cm}$, find the normalized lengths of the following lengths.

Solution:

Length	Normalized Length
5cm	2.5
10cm	5.0
7.4cm	3.7
0.1cm	0.05

4.10 STANDING WAVE AND STANDING WAVE RATIO

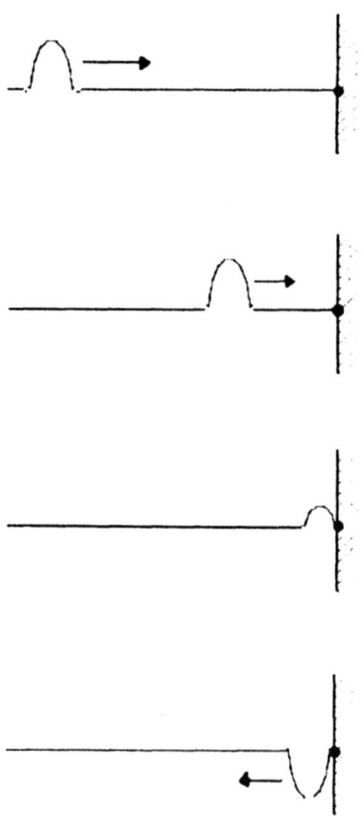

Figure 4.15 The propagation of a pulse along a rope fixed on one end is illustrated from A to D.

A standing wave is formed when two waves traveling in opposite directions interfere with each other. To make an analogy, we can imagine a rope with one end nailed to a wall and the other end held by a person. The person shakes the rope once and a wave package is sent down the rope toward the wall. Figure 4.15 (A—D) illustrates the progressive sequence of the wave. The important feature here is that one end of the rope is fixed to the wall and the rope cannot set that joint into motion. Consequently, all of the wave is reflected, and this example is analogous to a waveguide terminated by a short circuit.

Now, imagine the person shaking his end of the rope continuously in a sinusoidal manner. The reflected wave, whose ideal magnitude is equal to that of the incident wave, will form a standing wave pattern as shown in Fig. 4.16.

An important property of a standing wave is that there exist nodes and antinodes. Nodes are locations along the standing wave pattern where there is no net motion (all the *N*'s in Fig. 4.16). Antinodes are those locations whose amplitudes of motion are maximum.

Let us carry our imagination further and connect our rope's end to a gear at the wall's end which is movable in the vertical direction. Also, another rope of different mass is connected to this gear behind the wall, as shown in Fig. 4.17. There is a mismatch between the left rope and right rope, similar to our mismatch in line impedance. The incident wave will now set the gear and the right rope in motion. There is also reflection, although not 100%. The degree of reflection to the left depends on the degree of mismatch. If the mismatch is slight, the reflected wave is small in magnitude, and thus the standing wave pattern is small.

SEC. 4.10

STANDING WAVE & SWR

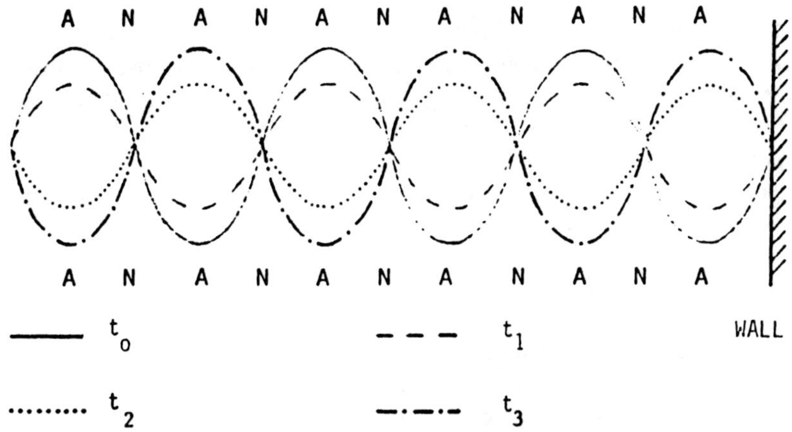

Figure 4.16 The standing wave pattern of a rope shaken continuously at one end and fixed at the other. Four time instants t_0, t_1, t_2, and t_3 are shown.

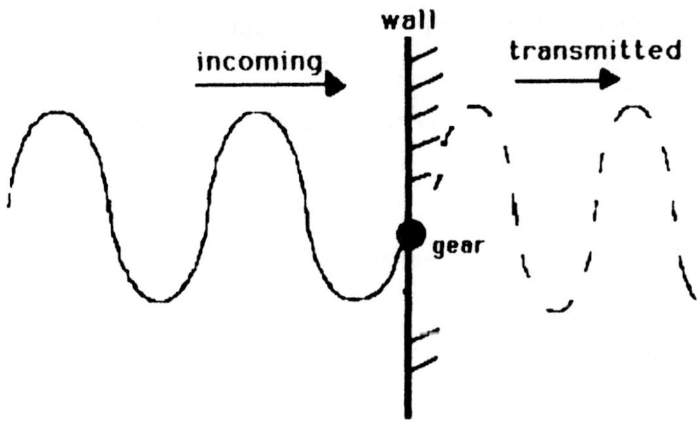

Figure 4.17 A more appropriate analogy of a moderate mismatch is that the rope is not fixed at the wall but attached to a gear. Some of the wave's energy is transferred to another rope behind the wall through the gear.

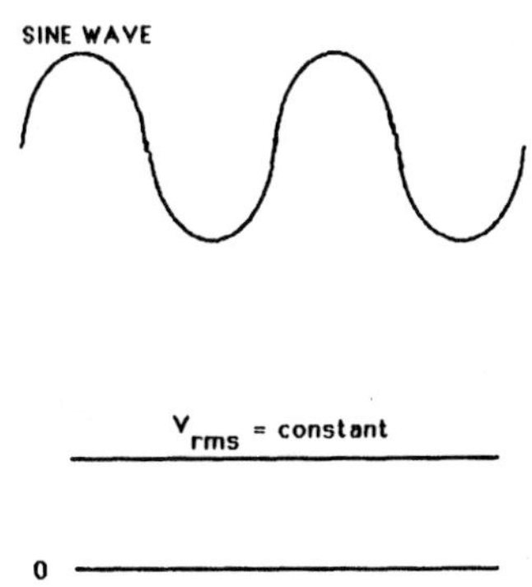

Figure 4.18 The rms value of a normal sine wave is constant.

Similar arguments can be made for standing waves in electronics. In a transmission line, voltage and current standing waves are formed. For waveguides, the electric fields of the forward and reflected electromagnetic waves form a standing wave; the same holds true for the magnetic fields.

A useful quantity for characterizing a standing wave is called the *standing wave ratio* (SWR). Because voltage is generally the most important parameter to be measured, we shall approach the *voltage standing wave ratio* (VSWR) as follows.

Consider a forward moving sine wave as shown in Fig. 4.18A. The root-mean-square (rms) value of the forward wave is V_f. If this wave travels forward with no reflection, the rms value is constant. The reader should note that the rms value of a wave is its power capability.* A uniform sine wave means that each cycle carries the same energy. A device that measures the rms of a wave will register a constant value along the transmission line (Fig. 4.18B).

Now, suppose this forward wave results in a reflected wave upon encountering a load. This reflected wave alone has its own rms value, let us say, V_r. In a

* Power is proportional to V_{rms}^2.

voltage standing wave, the voltage of the forward wave combines, constructively or destructively, with the voltage of the reflected wave. Consequently, a device that can detect rms voltage, will register maximum rms value at some locations, minimum rms at some other locations, and values between the maximum and minimum at the remaining locations.

The ratio of the maximum rms voltage to the minimum rms value is the voltage standing wave ratio (VSWR); i.e.,

$$VSWR = V_{max}/V_{min} \tag{4.10.1}$$

where rms values for the maximum and the minimum are understood.

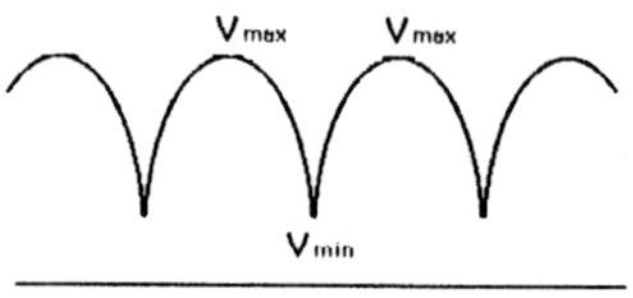

Figure 4.19 The rms value variation of a standing wave. V_{max} is where there is total reinforcement while V_{min} is where there is total cancellation.

Fig. 4.19 shows the variation in the rms voltage, i.e., power content, of the standing wave. An important property of the standing wave, as shown in Fig. 4.19. is that the minimum (or maximum) repeats every half-wavelength (guide wavelength, if the standing wave is formed inside a transmission line). This can be explained by the fact that the power content of a wave is independent of the sign of the wave, i.e., the positive half of a sine wave carries the same power as the negative half.

The reader should note that some authors define VSWR as the ratio of maximum voltage amplitude (not rms) to the minimum voltage amplitude. Such definition is consistent with that given by this author. The amplitude of a wave is a measure of size, with no sign attached, and the amplitude is related to the rms value by a factor of 0.707.

4.11 GUIDE WAVELENGTHS

When an electromagnetic wave travels in free space, there is neither electric nor magnetic interaction between the wave and the environment. Inside a medium, there is interaction between either the electric field or the magnetic field, or both, and the atoms constituting the medium. As a result, the speed of electromagnetic wave propagation is no longer $c = 3 \times 10^8$ m/s, but given by Eq. (3.4.3).

The change in the propagating speed leads to a change in the wavelength of the signal. The guide wavelengths for three types of transmission lines are given here. The symbols for free-space wavelength and guide wavelength are labelled as λ_0 and λ_g, respectively. The geometries of the transmission lines are given in Fig. 4.20.

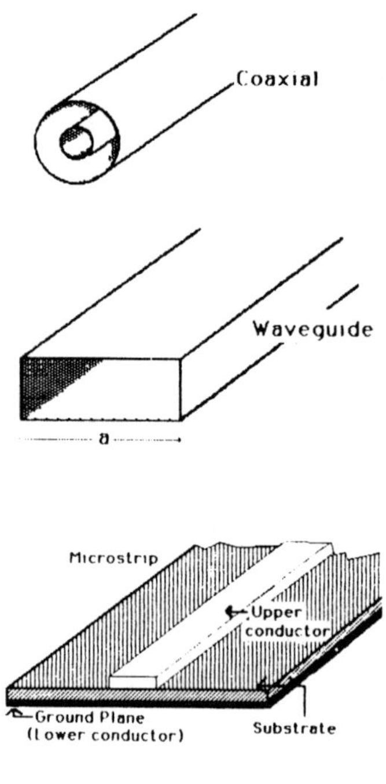

Figure 4.20 Geometries of three transmission lines for calculating guide wavelengths in these lines.

For coaxial cable

$$\lambda_g = \frac{\lambda_0}{\sqrt{k}} \tag{4.11.1}$$

rectangular waveguide

$$\lambda_g = \frac{\lambda_0}{\sqrt{k}\sqrt{1-\left(\frac{\lambda_0}{2a}\right)^2}} \tag{4.11.2}$$

microstrip

$$\lambda_g = \frac{\lambda_0}{k_{eff}} \tag{4.11.3}$$

$$k_{eff} = \frac{k_1 + k_2}{2} \tag{4.11.4}$$

where

k = the dielectric constant of the medium filling the waveguide or between conductors;

a = width of the rectangular waveguide;

k_{eff} = effective dielectric constant of the microstrip;

k_1 = dielectric constant of the microstrip substrate between the two conducting planes or lines;

k_2 = dielectric constant of the medium enclosing the microstrip.

Example (4.11.1)

A 10 GHz signal enters a coaxial cable filled with polyethylene whose dielectric constant is 2.2. find the guide wavelength.

Solution: The free-space wavelength $\lambda_0 = 3.0\text{cm}$.
The guide wavelength, according to Eq. (4.11.1) is

$$\lambda_g = \frac{3.0\text{cm}}{\sqrt{2.2}} = 2.02\text{cm}$$

Example (4.11.2)

A 10 GHz signal enters an air-filled rectangular waveguide whose width a is 2.1cm. Find the guide wavelength.

Solution: The waveguide is air-filled so the dielectric constant k is 1.0; the free space wavelength is 3.0cm.
According to Eq. (4.11.2),

$$\lambda_g = \frac{3.0\text{cm}}{\sqrt{1.0}\sqrt{1-\left(\frac{3}{2\times 2.1}\right)^2}} = 4.29\text{cm}$$

Example (4.11.3)

A microstrip circuit is produced on an alumina substrate whose dielectric constant is 10.0. The surrounding medium is air whose dielectric constant is 1.0. A 10 GHz signal is being processed. Find the guide wavelength.

Solution: The free-space wavelength is 3.0cm.
The effective dielectric constant is

$$k_{eff} = \frac{10.0 + 1.0}{2} = 5.5$$

The guide wavelength is

$$\lambda_g = \frac{3.0\text{cm}}{\sqrt{5.5}} = 1.28\text{cm}$$

4.12 COMPARISON OF DIFFERENT TRANSMISSION LINES

In choosing the type of transmission line used in a circuit or system design, we must keep several factors in mind. They are power handling, temperature (which can be directly or indirectly a result of power), flexibility of the line, power attenuation, and dimensional confinement.

SEC. 4.12

COMPARISON OF DIFFERENT TRANSMISSION LINES

In general, waveguides can handle large amounts of power and have low attenuation. When the signal power is large, the electric field is strong. The strong electric field can cause electrical arcing or breakdown of the dielectric or the air in the waveguide. Therefore, taller waveguides can handle higher powers. However, waveguides are inflexible, bulky, and have narrow bandwidths. Coaxial lines, except for the thick ones, can handle medium amounts of power, operate from dc to a few GHz, and are flexible.

The attenuation of a coaxial line, however, increases with frequency due to dielectric loss of the dielectric medium. Striplines are common for both hybrid and monolithic integrated circuits because of their space compactness. The attenuation of striplines is significant, but the signals only travel short distances.

Appendix A lists the temperature and flexibility properties of several common dielectrics. Table 4.1 lists the physical dimensions and other features of several transmission lines. The attenuation (dB/100ft) properties of these lines are plotted in Fig. 4.21 *versus* frequency. The reader should note that the attenuation of coaxial cables increases with frequency. Also, waveguides show low attenuation but narrow frequency bandwidths.

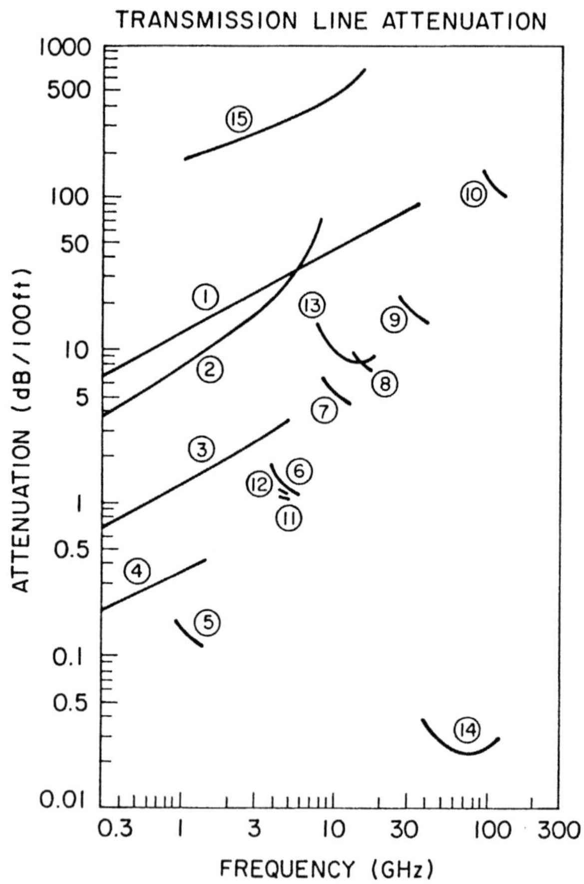

Figure 4.21 The attenuation properties of various types and sizes of transmission lines versus frequency.

SEC. 4.12

COMPARISON OF DIFFERENT TRANSMISSION LINES

Table 4.1
Comparison of Transmission Lines

	Description	Material	Type #	Outer Dimensions (inches)	CW Power Handling	Flexibility
1.	Coaxial	Teflon	0.141″ dia	0.14 dia	50W	semi-flexible
2.	Coaxial	Polyethylene	RG 8	0.42 dia	30W	flexible
3.	Coaxial	Helical polystyrene	7/8″ HELIAX	1.0 dia	700W	semi-flexible
4.	Coaxial	Air	3-1/8″ RIGID	3.5 dia	12kW	rigid
5.	Rectangular WG	Aluminum	WR 770	8×4	57MW	rigid
6.	Rectangular WG	Aluminum	WR 187	2×1	3MW	rigid
7.	Rectangular WG	Brass	WR 90	1×0.5	730kW	rigid
8.	Rectangular WG	Brass	WR 62	0.7×0.4	440kW	rigid
9.	Rectangular WG	Silver	WR 28	0.36×0.22	95kW	rigid
10.	Rectangular WG	Silver	WR 8	0.16 dia	1.8kW	rigid
11.	Elliptical WG	Aluminum	RG 379	2.5×1.5	20kW	semi-rigid
12.	Flexible Elliptical WG	Copper	WE 44	2.5×1.5	4kW	flexible
13.	Ridged WG	Aluminum	WRD 750 D24	0.7×0.4	100kW	rigid
14.	Overmoded Circular WG	Copper with Teflon Liner	60mm dia	4 dia	—	rigid
15.	Microstrip	Gold on Alumina	0.025″	—	50W	rigid

CHAPTER 5

INSERTION LOSS, POWER GAIN, AND RETURN LOSS

W. Stephen Cheung and George T. Gillies

5.1 INTRODUCTION

In microwave systems, we use the concepts of insertion loss and return loss to quantify absorption and reflection of the microwave power (or signal). Both reflection and attenuation have been covered in greater detail in Ch. 3. We shall accept the concepts of reflection and attenuation, and proceed to study it in a more quantitative way.

5.2 INSERTION LOSS AND ATTENUATION

Wires and cables used in low frequency circuits are usually assumed to be without resistance. In reality, the small but finite cable resistance consumes some of the original power and dissipates it as heat.

As we have seen in Ch. 4, microwave signals in a particular configuration are connected by cables, waveguides, or miniature lines. These components ideally have no dc resistance or impedance at low or high frequencies. In other words, they do not absorb any of the microwave power as it passes through. In practice, some fraction of the microwave power is absorbed due to the skin effect. As a result, the power coming out of a cable is less than that entering the cable.

Absorption of power is usually called *attenuation* and the absorbed power is converted into heat. Another reason that the output is less than the input power is *reflection*. Hence, a component or device may reflect some of the input power and allow the rest to pass without absorbing any. Insertion loss is a more general quantity which takes reflection into account. In the following mathematical analysis, attenuation and insertion loss have the same form. The reader must keep in mind that attenuation is strictly due to absorption and dissipation, while insertion loss accounts for reflection as well as attenuation. In particular, insertion loss is the same as attenuation if there is no power reflection, only absorption.

The insertion loss (*IL*) for a given component is defined as the ratio of the input power to the output power.

$$IL \text{ (Insertion Loss)} = P_{in}/P_{out} \tag{5.2.1}$$

Note that insertion loss is a ratio (i.e., a pure number) of two power levels. For example, if for some circuit element the input power is 10mW and the output power is 8mW, then 2mW of the input power is absorbed or reflected, and the corresponding insertion loss is

$$IL = \frac{10\text{mW}}{8\text{mW}} = 1.25$$

SEC. 5.2

INSERTION LOSS & ATTENUATION

It is important to note that insertion loss only applies to passive components. Active devices, such as an amplifier, add power to the input so that the output power is greater than the input power. A truly lossless component would have an insertion loss equal to 1.0 because the output power exactly equals the input power. On the other hand, a component of infinite insertion loss means that it absorbs or reflects 100% of the input power, i.e., P_{out} is zero. Hence, *the larger the insertion loss, the more absorptive the component.*

The insertion loss of a particular component can be expressed in dB. The method of conversion is covered in Ch. 2. We can, therefore, convert the insertion loss given by Eq. (5.2.1) to dB; or, alternatively, it can be calculated in the following way:

$$\text{Insertion Loss} \quad (\text{dB}) = P_{in}\,(\text{dBm}) - P_{out}\,(\text{dBm}) \tag{5.2.2}$$

Example (5.2.1)

$$P_{in} = 17\text{ dBm (50mW)},\ P_{out} = 10\text{ dBm (10mW)}$$

$$IL\,(\text{dB}) = 17\text{ dBm} - 10\text{ dBm} = 7\text{ dB}$$

Using Eq.(5.2.1), we see that this result is equivalent to the dimensionless value of 5.

When expressed in dB, an ideal component ($IL = 1$) has an insertion loss of 0 dB.

It is standard practice for manufacturers to mark the attenuation of their products in dB. In the case of a cable or waveguide, the attenuation per unit length is usually given. As an example, a cable marked 0.1 dB/ft means that one foot of such cable has a 0.1 dB attenuation and five feet of such cable will have a total of 0.5 dB attenuation.

If any two of the three quantities (input power, output power, and insertion loss) are known, the third can be easily computed.
Numerically,

$$IL = P_{in}/P_{out} \tag{5.2.3A}$$

$$P_{out} = P_{in}/IL \tag{5.2.3B}$$

$$P_{in} = P_{out} \times IL \qquad \text{5.2.3C)}$$

In terms of dBm and dB,

$$IL\,(\text{dB}) = P_{in}\,(\text{dBm}) - P_{out}\,(\text{dBm}) \tag{5.2.4A}$$

$$P_{out}\,(\text{dBm}) = P_{in}\,(\text{dBm}) - IL\,(\text{dB}) \tag{5.2.4B}$$

$$P_{in}\,(\text{dBm}) = P_{out}\,(\text{dBm}) + IL\,(\text{dB}) \tag{5.2.4C}$$

Example (5.2.2)

a) Given $P_{in} = 100\text{mW}$ (20 dBm) and $P_{out} = 50\text{mW}$ (17 dBm), find IL in number and in dB.
Using Eq.(5.2.3A), $IL = 100\text{mW}/50\text{mW} = 2$.
Using Eq.(5.2.4A), $IL\,(\text{dB}) = 20\text{ dBm} - 17\text{ dBm} = 3\text{ dB}$.

b) Given $P_{in} = 5\text{mW}$ (7 dBm) and $IL = 5$ (7 dB), find P_{out}.
Using Eq.(5.2.3B), $P_{out} = 5\text{mW}/5 = 1\text{mW}$.
Using Eq.(5.2.4B), $P_{out}\,(\text{dBm}) = 7\text{ dBm} - 7\text{ dB} = 0\text{ dBm}$.

c) Given $P_{out} = 3\text{mW}$ (5 dBm) and $IL = 2$ (3 dB), find P_{in}.
Using Eq,(5.2.3C), $P_{in} = 3\text{mW} \times 2 = 6\text{mW}$.
Using Eq.(5.2.4C), P_{in} (dBm) = 5 dBm + 3 dB = 8 dBm.

5.3 CASCADED INSERTION LOSS

Figure 5.1 shows an example of two circuit elements connected in a cascaded (series) manner. Rectangles *A* and *B* (with the resistance symbols) indicate that *A* and *B* are absorptive. The power leaving *A* becomes the input power for *B*. What is the overall insertion loss of the connection if the individual insertion loss of *A* and *B* are known?

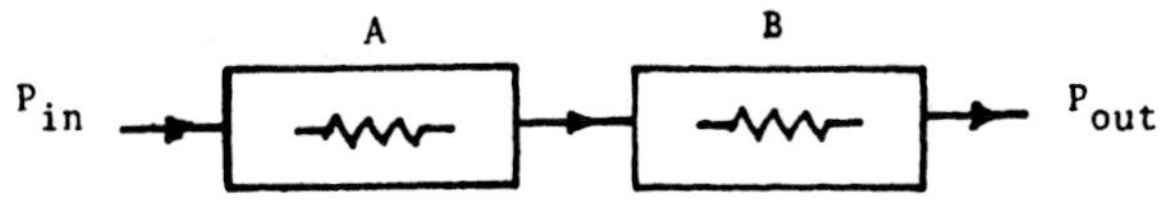

Figure 5.1 Two absorptive elements in series.

The connection may remind the reader of resistors in series, such as we normally find in low frequency electronics. While there is a definite resemblance, nonetheless, we must be careful in deriving the correct formula.

In low frequency electronics, the *current* through two (or more) resistors in series is the same for each one, as shown in Fig. 5.2A. The two resistors act as a voltage divider, and the total power dissipated is given by the sum of the individual power dissipation in resistors *A* and *B*. In a series of power absorbers, current, voltage, and power do not remain constant.

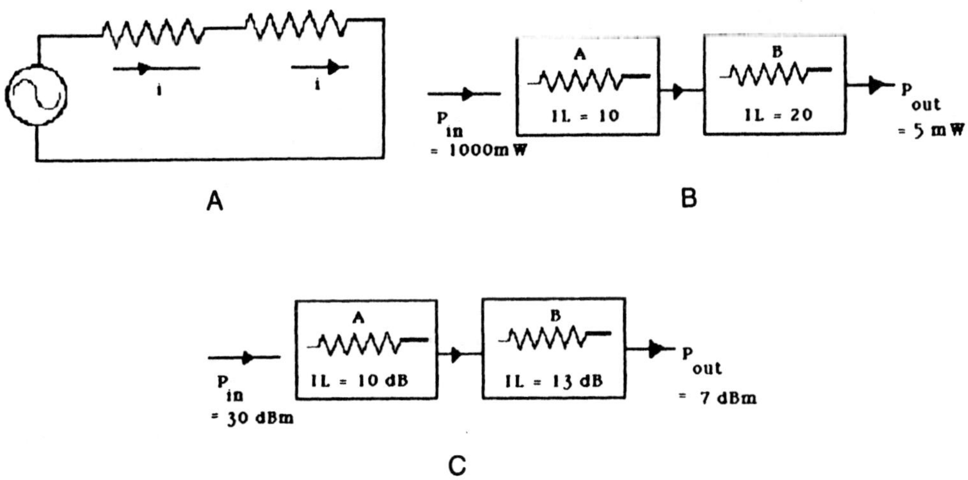

Figure 5.2 (A) Two resistive elements in series in a low frequency circuit. The current in each element is the same. (B) Two absorptive elements in series in a microwave circuit. Power is the only relevant parameter. The output power is 1/200 of the input power. (C) The example in B done in dB.

Now, consider two absorbers (with known insertion losses) in series, as in Fig. 5.2B. Suppose the input power is 1000mW. The power coming out of absorber *A*, whose insertion loss is 10, is P_M and is given by

$$
\begin{aligned}
P_M &= P_{in}/IL \text{ of } A \\
&= 1000\text{mW}/10 \\
&= 100\text{mW}
\end{aligned}
$$

This 100mW of power is now the input power to absorber *B*, whose insertion loss is 20. The cascade's output power, P_{out}, is

$$
\begin{aligned}
P_{out} &= P_M/IL \text{ of } B \\
&= 100\text{mW}/20 \\
&= 5\text{mW}
\end{aligned}
$$

The overall insertion loss of *A* and *B* together can be calculated by knowing P_{in} and P_{out}.

$$
\begin{aligned}
IL\ (overall) &= \frac{1000\text{mW}}{5\text{mW}} = 200 \\
&= (IL \text{ of } A) \times (IL \text{ of } B)
\end{aligned}
$$

The above example can be generalized to state that the overall insertion loss of *n* comonents in series is given by the *product* of the individual insertion losses when these values are expressed in dimensionless form. Mathematically, this means

$$IL\ (overall) = IL\ (\#1) \times IL\ (\#2) \times \ldots \times IL\ (\#n) \tag{5.3.1}$$

The insertion loss for a cascaded connection can also be computed in dB,

$$IL\ (overall)\ (\text{dB}) = IL\ (\#1, \text{dB}) + IL\ (\#2, \text{dB}) + \ldots + IL\ (\#n, \text{dB}) \tag{5.3.2}$$

It is important to note that when using Eq.(5.3.2) the individual insertion losses must always be in dB.

Let us reconsider the above example in dB. Figure 5.2C shows the insertion losses for the components *A* and *B* now expressed in dB. According to Eq.(5.3.2), the overall insertion loss is

$$IL\ (overall)\ (\text{dB}) = 10\text{ dB} + 13\text{ dB} = 23\text{ dB}$$

Note that 23 dB is equivalent to a dimensionless insertion loss of 200.

The convenience of expressing the insertion loss in dB can be seen easily in the above example. Let us carry the example one step further and calculate the output power if the input power is known. Let us say that the input power is 1000mW, or 30 dBm, then the output power in dB is

$$
\begin{aligned}
P_{out}\ (\text{dBm}) &= P_{in}\ (\text{dBm}) - IL\ (A, \text{dB}) - IL\ (B, \text{dB}) \\
&= 30\text{ dBm} - 10\text{ dB} - 13\text{ dB} \\
&= 7\text{ dBm (5mW)}.
\end{aligned}
$$

The answer is in agreement with our earlier calculation.

5.4 POWER GAIN

An amplifier enlarges signals. The signal might be a voltage, current, or power level. A bipolar transistor is an example of a current amplifier. Here, the collector current is many times larger than the base current. With suitable resistors connected at appropriate places, a bipolar transistor can be a voltage amplifier. In low frequency electronics, a power amplifier is usually controlled by some input voltage at its base, and its output has both higher voltage and current levels, i.e.,

SEC. 5.4

POWER GAIN

a greater power. Note that the extra power is not created by the transistor, but rather is derived from the transistor's power supply.

In microwave electronics, an amplifier *amplifies* power. The output power is larger than the input power by a factor called the (power) gain, G. Once again, the extra power comes from the supply operating the amplifier.

Formally, the amplifier's gain G is defined as

$$G\ (\text{gain}) = P_{out}/P_{in} \qquad (5.4.1)$$

Example (5.4.1)

An amplifier has 10mW input power and the output power is measured to be 100mW. what is the gain?

The gain of this amplifier, according to Eq.(5.4.1), is

$$G = 100\text{mW}/10\text{mW} = 10$$

If two of the three quantities (P_{in}, P_{out}, and G) are known, the third can be calculated easily.

$$G = P_{out}/P_{in} \qquad (5.4.2A)$$

$$P_{out} = P_{in} \times G \qquad (5.4.2B)$$

$$P_{in} = P_{out}/G \qquad (5.4.3C)$$

Example (5.4.2)

It is known that an amplifier has a gain of 50. What is the output power corresponding to an input power of 0.3mW?

According to Eq.(5.4.2 B), the output power is

$$P_{out} = 0.3\text{mW} \times 50 = 15\text{mW}$$

Example (5.4.3) What input power to an amplifier with a gain of 1000 would give an output power of 500 mW?

According to Eq.(5.4.2 C), the input power is

$$P_{in} = \frac{500\text{mW}}{1000} = 0.5\text{mW}$$

EXERCISES

Complete the following table.

	P_{in}	P_{out}	G
1.	1mW	4W	
2.	30nW	3mW	
3.	4.5pW	450W	
4.	3mW		200
5.	5.4W		500
6.	7.6mW		10,000
7.	0.4W		10^6
8.		4W	10^5
9.		6mW	6×10^6
10.		1mW	10^9

SEC. 5.4

POWER GAIN

Power gain can also be expressed in dB. This procedure is standard, but we must be sure that the input and output powers are in dBm. the decibel version of Eq. (5.4.1) is

$$G\ (\text{dB}) = P_{out}\ (\text{dBm}) - P_{in}\ (\text{dBm}) \tag{5.4.3}$$

Example (5.4.4)

a) Given $P_{in} = 10$ dBm (10mW) and $P_{out} = 20$ dBm (100mW), find gain in dB.

$$G\ (\text{dB}) = 20\ \text{dBm} - 10\ \text{dBm} = 10\ \text{dB}$$

b) Given $P_{in} = -45$ dBm (3×10^{-5}mW) and $P_{out} = 5$ dBm (3mW), find gain in dB.

$$\begin{aligned} G\ (\text{dB}) &= 5\ \text{dBm} - (-45\ \text{dBm}) \\ &= 50\ \text{dB} \end{aligned}$$

The decibel versions of Eq. (5.4.2 A, B, C) are

$$G\ (\text{dB}) = P_{out}\ (\text{dBm}) - P_{in}\ (\text{dBm}) \tag{5.4.4A}$$

$$P_{out}\ (\text{dBm}) = P_{in}\ (\text{dBm}) + G\ (\text{dB}) \tag{5.4.4B}$$

$$P_{in}\ (\text{dBm}) = P_{out}\ (\text{dBm}) - G\ (\text{dB}) \tag{5.4.4C}$$

Example (5.4.5)

The gain of an amplifier is 25 dB and the input power is 0 dBm, i.e., 1mW. What is the output power?

According to Eq. (5.4.4B), the output power is

$$P_{out}\ (\text{dBm}) = 0\ \text{dBm} + 25\ \text{dB} = 25\ \text{dBm}$$

Example (5.4.6)

The gain of an amplifier is 40 dB and its output power is measured to be 26 dBm. What is the input power?

According to Eq. (5.4.4 C), the input power is

$$P_{in}\ (\text{dBm}) = 26\ \text{dBm} - 40\ \text{dB} = -14\ \text{dBm}$$

EXERCISES

Complete the following exercises in dB and dBm.

	P_{in} (dBm)	P_{out} (dBm)	G (dB)
1.	4	64	
2.	−27	33	
3.	−66	34	
4.	−2		30
5.	−35		50
6.	−66.5		40
7.	−51.5		70
8.		25	50
9.		33	60
10.		42.5	45

5.5 CASCADED AMPLIFIERS

Amplifiers can be cascaded together to provide stage-by-stage power amplification. Let us say that two power amplifiers with dimensionless or numerical gains G_1 and G_2 are cascaded. The total power gain is then $G_1 \times G_2$.

In general, if n amplifiers whose numerical power gains are $G_1, G_2, \ldots, G_n$ are cascaded, the total gain is

$$G_T \text{ (total gain)} = G_1 \times G_2 \times \ldots \times G_n \qquad (5.5.1)$$

The decibel version of Eq. (5.5.1) is

$$G_T \text{ (total gain, dB)} = G_1 \text{ (dB)} + G_2 \text{ (dB)} + \ldots + G_n \text{ (dB)} \qquad (5.5.2)$$

Note again that in using Eq. (5.5.2), the individual gains must be expressed in dB.

Example (5.5.1)

Three amplifiers, whose numerical power gains are 2, 5, and 10, are connected in series. What is the total gain?

According to Eq.(3.5.1), the total gain is

$$G_T = 2 \times 5 \times 10 = 100$$

Example (3.5.2)

Repeat the previous example in dB.

The gains in dB are 3 dB, 7 dB, and 10 dB. Therefore, the total gain in dB, according to Eq. (3.5.2), is

$$G_T \text{ (dB)} = 3 \text{ dB} + 7 \text{ dB} + 10 \text{ dB} = 20 \text{ dB}$$

5.6 GAIN AND INSERTION LOSS COMBINED

We are now in a position to use the concepts of gain and insertion loss in practical applications. We shall do all the calculations in dB and dBm. This mode of thinking is used by virtually all microwave professionals.

Figure 5.3 shows an input signal P_{in} applied to one end of a transmission line. The line has a certain insertion loss. Therefore, the signal power, after passing through the cable, is attenuated (or weakened). It is then amplified by an amplifier with a given fixed gain. What is the output power?

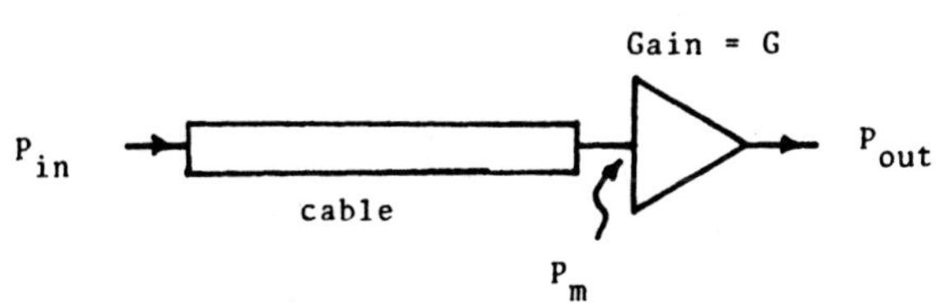

Figure 5.3 A situation with attenuation and gain combined.

We will put in some numbers in order to help us understand the calculation. Let us say that the input power is 10 dBm (10mW) and the line has a total insertion loss of 1 dB. The signal power coming out of the cable (before amplification) is denoted by P_m and can be computed using Eq. (5.2.4B).

SEC. 5.6

GAIN & INSERTION LOSS COMBINED

$$P_m \text{ (dBm)} = 10 \text{ dBm} - 1 \text{ dB} = 9 \text{ dBm}$$

Now, this power P_m is amplified by the amplifier whose gain is, for example, 20 dB. Then, the output power, according to Eq. (5.4.4B), is

$$P_{out} \text{ (dBm)} = 9 \text{ dBm} + 20 \text{ dB} = 29 \text{ dBm}$$

We can summarize the procedure for finding P_{out} starting from P_{in} by the following expressions.

$$P_{out} \text{ (dBm)} = P_{in} \text{ (dBm)} - IL \text{ (dB)} + G \text{ (dB)} \tag{5.6.1}$$

The numerical version of Eq. (5.6.1) is also presented, although the readers are encouraged to use the decibel version,

$$P_{out} = P_{in} \times G/IL \tag{5.6.2}$$

where the powers are in any consistent units, and the gain and insertion loss are dimensionless. We shall continue to use the decibel version in the following practical examples.

Example (5.6.1)

The signal power leaving a generator is 3 dBm. It is connected to an amplifier by a cable 4ft long. The attenuation per foot of cable is 0.2 dB/ft. The gain of the amplifier is 30 dB. What is the output power?

The total attenuation of the cable, 4ft long, is 4×0.2 dB $= 0.8$ dB.

The output power of the amplifier is

$$P_{out} \text{ (dBm)} = 3 \text{ dBm} - 0.8 \text{ dB} + 30 \text{ dB} = 32.2 \text{ dBm}$$

Example (5.6.2)

Figure 5.4 shows three amplifiers connected in series by cables. The gains and lengths of the individual components are shown. The cables have 1 dB/ft attenuation. The signal power from the generator is −5 dBm. What is the final output power?

The attentuations of the three cables are 0.3 dB, 1 dB, and 5 dB, respectively. The final output power is

$$P_{out} \text{ (dBm)} = -5 \text{ dBm} - 0.3 \text{ dB} + 20 \text{ dB} - 1 \text{ dB} + 15 \text{ dB} - 5 \text{ dB} + 10 \text{ dB} = 33.7 \text{ dBm}$$

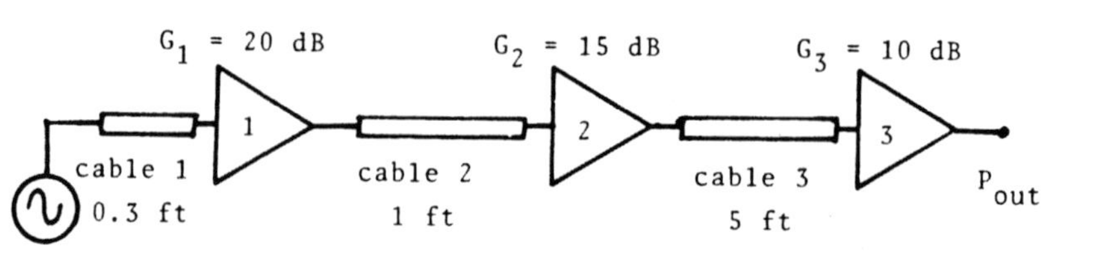

Figure 5.4 A signal generator is connected to a series of amplifiers via cables. The net gain is the sum of individual gains in dB minus the cable attenuation values.

SEC. 5.6

GAIN & INSERTION LOSS COMBINED

Example (5.6.3)

A microwave signal is to be broadcast from an earth station as shown in Fig. 5.5. The output of the amplifier is 90 dBm (1MW) and is delivered to the transmitting antenna by 300ft of cable whose attenuation per foot is 0.01 dB/ft. What is the actual power available for transmission at the antenna?

The total insertion loss of a 300ft length of this cable is $300 \times 0.01\ \text{dB} = 3\ \text{dB}$. The final power, therefore, is $90\ \text{dBm} - 3\ \text{dB} = 87\ \text{dBm}$.

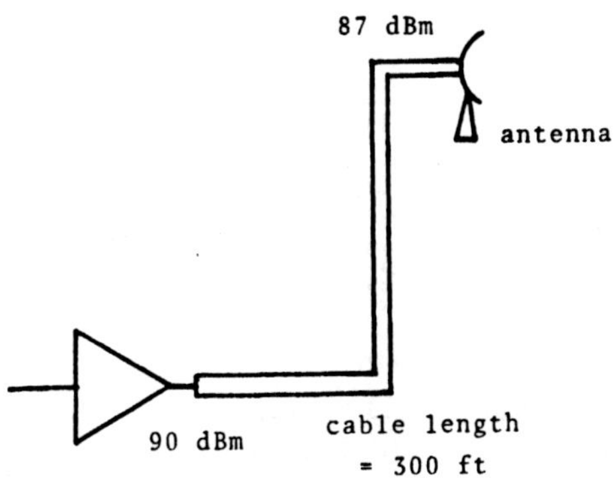

Figure 5.5 The 90 dBm output power of a power amplifier is delivered to a transmitting antenna via a 300ft cable, which absorbs half of the power.

Comment: The power loss to the cable is actually quite significant. The starting power is 90 dBm (1MW) and the final power is 87 dBm (0.5MW). Half of the starting power is dissipated in the process of delivery. It should be noted that the cable used in this example is actually quite good, with attenuation 0.01 dB/ft. The great length which the signal must travel, however, accounts for the tremendous loss.

Example (5.6.4)

The power amplifier in the previous example is now designed to be located right at the transmitter, as shown in Fig. 5.6. This amplifier can provide an 80 dB gain. The distance beween the final amplifier and the transmitter is only 10ft. We want the transmitted power to be 90 dBm. What is the signal strength at the ground floor? The same type of cable is used.

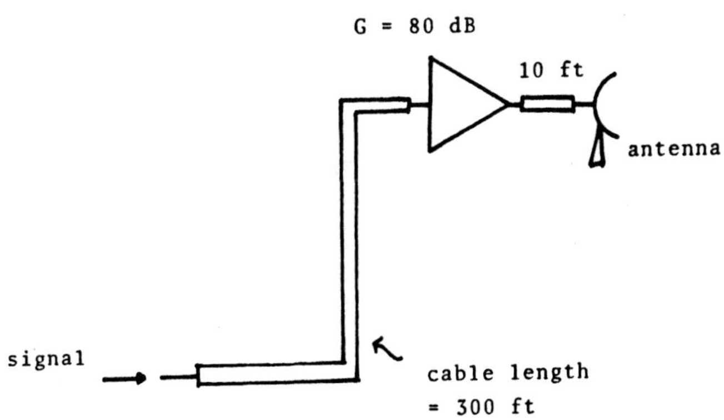

Figure 5.6 The power amplifier in the previous example is now put next to the antenna. Much less power is absorbed.

Let the signal power at the ground floor be P_{in} (dBm). The following equation must be solved:

$$90 \text{ dBm} = P_{in} \text{ (dBm)} - 300 \times 0.01 \text{ dB} + 80 \text{ dB} - 10 \times 0.01 \text{ dB}$$
$$P_{in} \text{ (dBm)} = 12.9 \text{ dBm}$$

Comment: The input power at the ground floor is about 20mW. Half of this amount is dissipated after going through 300ft of cable, but the absolute amount of power lost here (10mW) is much less than that in the previous example (0.5MW).

Example (5.6.6)

A mobile television van transmits 37 dBm (5W) of signal back to its station over a microwave link. By the time the station receives the signal, its power is only −60 dBm (1nW). An amplifier is used to bring the received signal up to an operational level of 0 dBm (1mW). Assume that all the cables used in the station have negligible losses, and consider the atmosphere between the van and the station to be a big attenuator (absorber). Draw a block diagram depicting the entire situation. Next, calculate the equivalent insertion loss of the atmosphere and the amplifier gain necessary to restore the received signal to an operational level.

The desired block diagram is shown in Fig. 5.7. The insertion loss of the atmosphere is

$$(-60 \text{ dBm}) - 37 \text{ dBm} = -97 \text{ dB}$$

The gain of the amplifier is

$$G \text{ (dB)} = 0 \text{ dBm} - (-60 \text{ dBm}) = 60 \text{ dB}$$

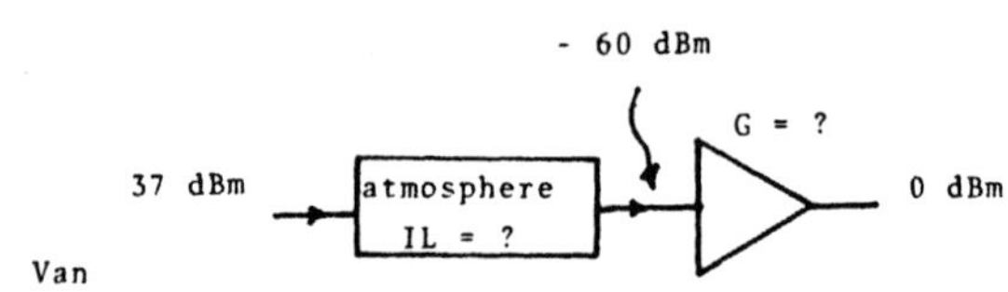

Figure 5.7 A block diagram describing the signal transmission from a van to the station. The atmosphere can be regarded as a large attenuator.

5.7 RETURN LOSS

When a beam of microwaves enters a system that is not perfectly matched, some reflection generally occurs. The phenomenon is similar to the reflection we see by shining a flashlight at a piece of glass. Part of the input power is absorbed or transmitted to the system and the rest is reflected back.

When the reflected wave meets the incident wave, interference occurs, which results in a standing wave as discussed in Ch. 4. Methods for minimizing reflection exist and will be covered in a separate chapter.

Let us consider a general case where an arrangement reflects part of an incident signal. Return loss is defined as the ratio of the power of the incident wave to the power of the reflected wave.

$$RL \text{ (return loss)} = P_{in}/P_{ref} \tag{5.7.1}$$

Example (5.7.1)

The power of an incident microwave is 30mW and the power of the reflected wave is measured to be 10^{-3}mW. What is the return loss?

According to Eq.(5.7.1),

$$RL = \frac{30\text{mW}}{10^{-3}\text{mW}} = 3 \times 10^4$$

An ideal arrangement, i.e., one that is perfectly matched, should have no reflection, i.e., $P_{ref}=0$ regardless of the value of P_{in}. This yields an infinite return loss. The worst case of return loss occurs when 100% of the incident power is reflected, i.e., $P_{ref}=P_{in}$. This makes the worst return loss equal to 1.0. In general, *the larger the return loss, the less reflective is the arrangement.*

Return loss, like insertion loss, can also be expressed in dB. The corresponding formula is

$$RL\ (\text{dB}) = P_{in}\ (\text{dBm}) - P_{ref}\ (\text{dBm}) \tag{5.7.2}$$

Example (5.7.2)

a) P_{in} = 33 dBm (2W), P_{ref} = 5 dBm (3mW)
RL (dB) = 33 dBm − 5 dBm = 28 dB

b) P_{in} = 5 dBm (3mW), P_{ref} = −27 dBm (0.002mW)
RL (dB) = 5 dBm − (−27 dBm) = 32 dB

An ideal arrangement has infinite dB return loss when expressed in dB and the poorest arrangement has a 0 dB return loss.

If any two of the input power, reflected power, or return loss are known, the third can be calculated.

Numerically:

$$RL = P_{in}/P_{ref} \tag{5.7.3A}$$

$$P_{ref} = P_{in}/RL \tag{5.7.3B}$$

$$P_{in} = P_{ref} \times RL \tag{5.7.3C}$$

In dB:

$$RL\ (\text{dB}) = P_{in}\ (\text{dBm}) - P_{ref}\ (\text{dBm}) \tag{5.7.4A}$$

$$P_{ref}\ (\text{dBm}) = P_{in}\ (\text{dBm}) - RL\ (\text{dB}) \tag{5.7.4B}$$

$$P_{in}\ (\text{dBm}) = P_{ref}\ (\text{dBm}) + RL\ (\text{dB}) \tag{5.7.4C}$$

If the incident power is known and the reflected power is calculated or measured, the power transmitted to the arrangement can then be found.

$$\text{Transmitted Power} = P_{in} - P_{ref} \tag{5.7.5}$$

The reader must be cautioned that Eq.(5.7.5) is *not* applicable for powers expressed in dBm.

SEC. 5.8
REFLECTION COEFFICIENT

EXERCISES

Complete the following table.

P_{in}		P_{ref}		RL	
Number	dBm	Number	dBm	Number	dB
1W		20mW			
40W		10W			
	−25				63
	−33				30
		30W		5000	
		4nW			70
			−13	10^5	
		5W			45
500mW				10^4	
3mW					55

5.8 REFLECTION COEFFICIENT

An alternate, reciprocal quantity which can be used instead of return loss is called the voltage reflection coefficient ρ. If the microwave power is measured by a device, e.g., a diode detector, which converts power to a voltage reading, the voltage reflection coefficient is defined as

$$\rho = V_r/V_i \tag{5.8.1}$$

where V_r = voltage reading corresponding to P_{ref}, and V_i = voltage reading corresponding to P_{in}.

Because power is related to the square of voltage, a relationship between the voltage reflection coefficient and return loss can be derived.

From the definition of return loss

$$\begin{aligned} RL &= P_{in}/P_{ref} \qquad (5.8.2) \\ &= V_{in}^2/V_{ref}^2 \end{aligned}$$

in effect,

$$RL = 1/\rho^2 \tag{5.8.3}$$

and

$$\rho = 1/\sqrt{RL} \tag{5.8.4}$$

Note that both the RL and ρ in Eqs. (5.8.3) and (5.8.4) are in *numerical* form, *not dB.*

An ideal arrangement has a return loss of infinity (or ∞ dB), this corresponds to a reflection coefficient equal to zero. Similarly, an arrangement with 100% refelection, i.e., $RL = 1$ (0 dB), gives $\rho = 1$.

Example (5.8.1)

If $\rho = 0.2$, what is RL?

$$RL = 1/0.2^2 = 1/0.04 = 25 \text{ (14 dB)}$$

Example (5.8.2)

If $RL = 100$ (20 dB), what is ρ?

$$\rho = 1/\sqrt{100} = 0.1$$

5.9 RETURN LOSS, REFLECTION COEFFICIENT, AND VSWR

The concept of voltage standing wave ratio (VSWR) was introduced in Ch. 4 and defined as

$$VSWR = V_{max}/V_{min} \tag{5.9.1}$$

where V_{max} = maximum (rms) voltage of a standing wave, and V_{min} = minimum (rms) voltage of a standing wave.

Because return loss, reflection coefficient, and VSWR are all quantities describing the phenomenon of reflection, they are, therefore, related to one another. In Sec. 5.8, we derived the relationship between RL and ρ, namely,

$$RL = 1/\rho^2 \tag{5.9.2a}$$

and

$$RL\ (\text{dB}) = 20 \log (1/\rho) \tag{5.9.2b}$$

We shall derive the relationship between reflection coefficient and VSWR. Recall that V_{max} is the sum of the incident wave's voltage and the reflected wave's voltage, and V_{min} is the difference between the incident wave's voltage and the reflected wave's voltage. Therefore,

$$VSWR = \frac{V_i + V_r}{V_i - V_r} \tag{5.9.3}$$

A little algebra shows that

$$VSWR + 1 = \frac{2V_i}{V_i - V_r} \tag{5.9.4}$$

and

$$VSWR - 1 = \frac{2V_r}{V_i - V_r} \tag{5.9.5}$$

Hence,

$$\frac{VSWR + 1}{VSWR - 1} = \frac{V_i}{V_r}$$

However, the ratio V_i to V_r is the reciprocal of the definition of reflection coefficient ρ ($\rho = V_r/V_i$). Therefore, we have arrived at the relationship between $VSWR$ and ρ.

$$\rho = \frac{VSWR - 1}{VSWR + 1} \tag{5.9.6}$$

SEC. 5.9

RETURN LOSS, REFLECTION COEFFICIENT, & VSWR

It will be left as an exercise for the reader to show that

$$VSWR = \frac{1 + \rho}{1 - \rho} \tag{5.9.7}$$

The relationship between the return loss and the reflection coefficient has already been given in Eq. (5.9.2). We can, therefore, show that

$$RL = \left(\frac{VSWR + 1}{VSWR - 1}\right)^2 \tag{5.9.8}$$

and

$$VSWR = \frac{\sqrt{RL} + 1}{\sqrt{RL} - 1} \tag{5.9.9}$$

It must be emphasized again that in using all of the equations given above, all quantities must be in numerical form, not dB.

We shall summarize the interrelations between *RL*, ρ, and VSWR as follows.

$$RL = 1 / \rho^2 = \left(\frac{VSWR + 1}{VSWR - 1}\right)^2 \tag{5.9.10}$$

$$\rho = 1 / \sqrt{RL} = \frac{VSWR - 1}{VSWR + 1} \tag{5.9.11}$$

and

$$VSWR = \frac{\sqrt{RL} + 1}{\sqrt{RL} - 1} = \frac{1 + \rho}{1 - \rho} \tag{5.9.12}$$

Appendix B tabulates the conversions of return loss, voltage reflection coefficient, and voltage standing wave ratio.

Example (5.9.1)

Show that the Eqs. (5.9.10) to (5.9.12) are valid for a perfect system with no reflection, and for a poor arrangement with 100% reflection.

a) No reflection means that $RL = \infty$. Eq. (5.9.11) gives

$$\rho = 1/\infty = 0$$

and Eq. (5.9.12) gives

$$VSWR = \frac{\sqrt{RL} + 1}{\sqrt{RL} - 1} = 1$$

b) An arrangement with 100% reflection means $RL = 1$. Eq. (5.9.11) gives

$$\rho = 1/\sqrt{1} = 1$$

and Eq. (5.9.12) gives

$$VSWR = \frac{1+1}{1-1} = \frac{2}{0} = \infty$$

Example (5.9.2)

Given that 2% of the input power to an arrangement is reflected, find RL, ρ, and $VSWR$.

From the definition of return loss,

$$RL = \frac{P_{in}}{P_{ref}} = \frac{100\%}{2\%} = 50$$

According to Eq. (5.9.11),

$$\rho = 1/\sqrt{RL} = 1/\sqrt{50} = 0.141$$

According to Eq. (5.9.12),

$$VSWR = \frac{\sqrt{50}+1}{\sqrt{50}-1} = 1.33$$

EXERCISES

Complete the following table.

RL	ρ	$VSWR$
2.4		
	0.006	
		3.60
	3×10^{-4}	
		1.44
80		
400		
	15%	

5.10 ISOLATION

An isolator is a passive component which allows microwaves to go in one direction, but not in the opposite direction. An ideal isolator, therefore, has zero insertion loss in the forward path, but has infinite loss in the reverse path. The low frequency analog of an isolator is a diode.

If located between a signal generator and the rest of the circuit, an isolator can reduce any reflected wave toward the generator to an even lower level, so that the amount of reflected wave entering the generator is minimum. How well the isolator can "block" the reflected wave is quantitatively given by its isolation. We shall use two examples to illustrate this point.

Figure 5.8 shows an isolator with 40 dB isolation and 0 dB forward insertion loss. The signal power leaving the generator is 10 dBm. Because the isolator has 0 dB forward insertion loss, the power level of the forward wave is still 10 dBm after passing through the isolator. Now, this forward wave encounters a short circuit and all the forward wave is reflected back, i.e., the power of the reflected wave is 10 dBm. As the reflected wave goes through the isolator on its way back to the generator, its power level is reduced by 40 dB, i.e., −30 dBm. As far as the generator is concerned, it is connected to an arrangement of 40 dB return loss, despite the fact that 100% reflection is actually taking place to the right of the isolator. Hence, the isolator acts as a buffer and protects the generator from being damaged by the reflected wave.

As a second example, an arrangement without an isolator is measured to have a return loss of 10 dB, which produces an unacceptable effect to the generator. An isolator can be inserted right after the generator. Let us say that the isolation of the isolator is 30 dB. The reflected wave will suffer a 30 dB loss when passing through the isolator. Together with the 10 dB return loss already taking place in the circuit, the generator is now "seeing" a 40 dB (30 dB + 10 dB) return loss.

Basically, an isolator absorbs any electromagnetic wave travelling in the reverse direction. The working principle of an isolator will be discussed in Ch. 6.

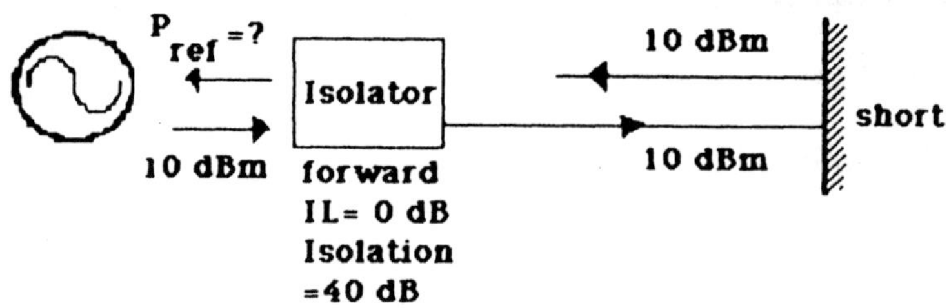

Figure 5.8 An isolator can greatly reduce the reflected power from a short to the signal generator.

5.11 ONE-dB COMPRESSION POINT

A power amplifier, whether it is solid-state or vacuum tube type, is evaluated in many ways, including linearity, power level, noise, *et cetera*. Linearity for an amplifier is discussed here, and the concept of one-dB compression point will emerge from this discussion.

An amplifier's linearity can be illustrated from the following data obtained from an amplifier:

Input power (dBm)	*Output power (dBm)*
−30	0
−20	10
−10	20
0	30
10	39
20	40

These data are plotted in Fig. 5.9. It can be seen that the output is linearly proportional to the input with a gain of 30 dB when the input power values are small. As the input power increases and the electronic components are driven near saturation, further increase in power will not produce a proportionately increased output. This is illustrated by the leveling of the output curve. The dotted line is the extrapolated output, i.e., if there were no saturation. The one-dB compression point is where the real output deviates from the extrapolated output by 1.0 dB. In Fig. 5.9, the one-dB compression point is when the input power is 10.0 dBm. The significance of the one-dB compression point can be qualitatively appreciated by the fact that any further increase in input after the compression point will not produce any proportional increase of output power.

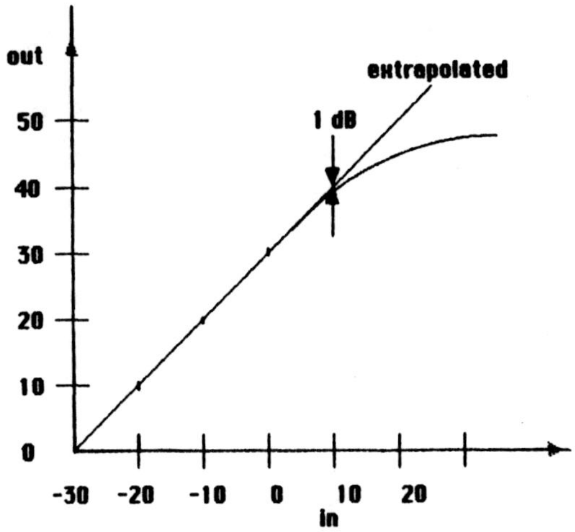

Figure 5.9 The one-dB compression point is where the amplifier's output power deviates from the extrapolated value by one dB.

PROBLEMS

1. A scientific satellite beams data down to an earth station on a microwave telemetry downlink. The satellite's transmitter has a power of 10W, but the atmosphere's ionsphere reflects 1W of the beam back into space. What is the reflection coefficient of the ionsphere in dB?
2. A cascade of amplifiers has a gain of 10^3. If the output power is 10W and the input power is 0.1W, what is the overall insertion loss in dB of the cables interconnecting the amplifiers?
3. A transmitting antenna is located on a mountain top overlooking the sea. The power os the signal leaving the antenna is 500W. A ship far away receives the signal whose power is now 25nW. The signal is immediately amplified by an amplifier with power gain 50 dB. The amplified signal is then passed to the control room via a 100ft cable. The cable has 0.05 dB/ft. insertion loss. A second amplifier is needed in order to bring the signal up to 0 dBm level. Draw a block diagram dcpicting the entire situation, and calculate the equivalent insertion loss of the atmosphere between the transmitter and the ship's receiver. Also, calculate the gain of the second amplifier.

CHAPTER 6

BASIC MICROWAVE COMPONENTS

W. Stephen Cheung

6.1 INTRODUCTION

Microwave components are passive and active devices commonly used in the laboratory or on test benches. Passive components include antennas, attenuators, couplers, frequency meters, and non-reciprocal devices. Thermistors and diode detectors are passive components, but they must be operated with active electronics such as amplifiers. The principles of operation for these two components will be discussed. Active devices covered in this chapter include power meters, SWR meters, and two microwave generators commonly used in a laboratory setup. The internal electronic networks of these meters will not be covered here. Also, advanced microwave equipment such as sweep generators and network analyzers will not be covered in this book.

Three important actions are introduced before any component is covered. They are the coupling action, the gyromagnetic action, and the cavity action. Many of the components discussed in this chapter base their operations on one of these actions.

6.2 COUPLING ACTION

Imagine two pendulums of identical length but independently hung. Pendulum 1 is made to swing while pendulum 2 is stationary (Fig. 6.1). It is observed that the swinging action of pendulum 1 will cause pendulum 2 to swing, although no contact is made. Physically, pendulum 1 excites the surrounding air, which then drives pendulum 2 into oscillation. In other words, some of the energy of pendulum 1 is now shared by pendulum 2.

Generally speaking, mechanical and electrical oscillations in one system can induce oscillations in another system of the same nature. This property will be very useful in various microwave couplers.

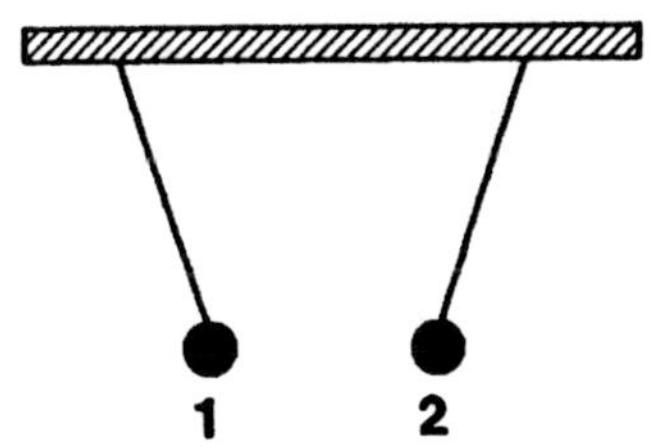

Figure 6.1 The mechanical energy of a swinging pendulum is coupled to another pendulum.

6.3 GYROMAGNETIC ACTION

Ferrite material originated in the communications field as powdered iron cores for RF transformers. It is magnetic and has low electrical conductivity, so the magnetic action can remain with minimum induced eddy currents.

The application of ferrite in microwaves is unique in its *non-reciprocal* property. When a piece of ferrite is sandwiched between two permanent magnetic poles, i.e., north and south, microwave signals entering the ferrite from one direction will go through with little attenuation, while those entering from the opposite direction will be absorbed.

When "biased" by an external static magnetic field, i.e., a permanent magnet, the spin vectors of the valence electrons of the iron atoms in the ferrite material will attempt to align with the external field H_1, as shown in Fig. 6.2. However, perturbations usually cause the electron spin to *precess* about H_1. If an electromagnetic wave is present, its alternating magnetic field will drive the precession

SEC. 6.3
GYROMAGNETIC ACTION

more and more severely. The precession of the electrons requires energy which is supplied by the alternating field. The precession, however, will not go unlimited because other atomic couplings will dampen the precession.

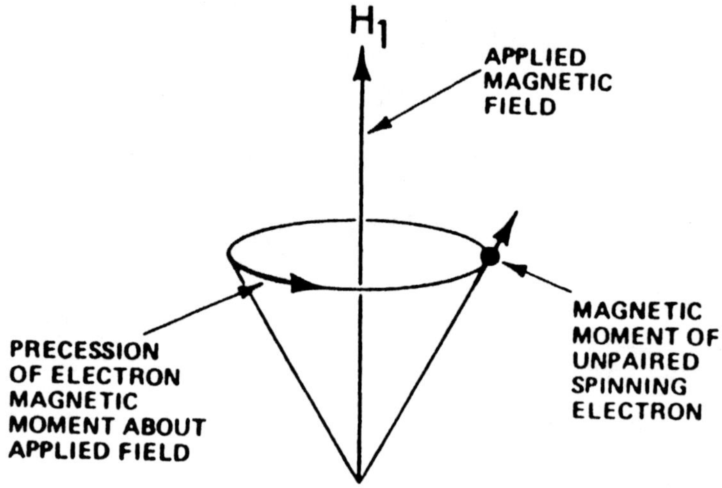

Figure 6.2 Gyromagnetic interaction in ferrite. The applied magnetic field H_1 causes the electron's spin axis to precess.

The electron precession only takes place in one direction, but not the other. The analogy is that of the conical pendulum shown in Fig. 6.3. If the pivot point is jerked in one direction at the right frequency, the conical motion will increase while jerking in the opposite direction. This will result in great damping of the conical motion. This analogy shows the non-reciprocity.

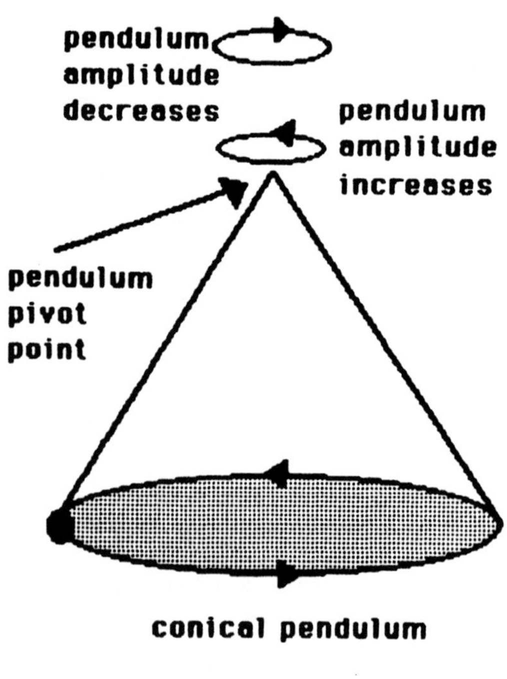

Figure 6.3 The amplitude of the electron precession either increases or decreases as a result of jerking the pivot point.

6.4 CAVITY ACTION

A cavity is a metallic enclosure that can store electromagnetic energy injected into the cavity through a small hole. It is the high frequency version of a LC tank circuit. The dimensions of a microwave cavity must be equal to the half-wavelength of the signal concerned.

How well a cavity can store microwave energy is determined by its quality factor Q, which is the ratio of stored energy to energy dissipated due to resistance or leakage. The quality factor is also related to the mechanical stability, e.g., expansion and contraction as a function of temperature. Any change in dimension due to temperature variation will result in a shift in resonant frequency. Consequently, the higher the quality factor, the more mechanically stable is the cavity. The quality factor is also different when the cavity is loaded, as opposed to unloaded, because a loaded cavity means energy is taken away from the cavity and can be viewed as a dissipating mechanism. An unloaded cavity may attain a Q as high as 10^5, and much effort must be expended to ensure the mechanical and dimensional stability of the cavity.

It is clear that an adjustable cavity cannot possess high quality factors. Cavities are used as energy storage parts in microwave tubes and as filters. They are also used in frequency meters and as crucial parts in microwave generators.

6.5 ATTENUATORS

Attenuators are components which can reduce the microwave power. Attenuators are especially important in laboratory experiments and bench tests. There are two types of attenuators: fixed and variable. Terminators belong to a special kind of attenuators.

Attenuation can be accomplished in two ways. One is using graphitized sand. When a microwave signal encounters the sand, the current generated turns the signal energy into heat. This method is most useful for terminators. An alternative to graphitized sand is using a resistive rod or resistive vane placed at the center of the electric field. The electric field induces a current flow resulting in an ohmic power loss. The vane method is useful in variable attenuators.

A fixed attenuator reduces the input signal power by a fixed portion, e.g., 3 dB, 10 dB, et cetera. Hence, an input signal power of 10 dBm (10mW) passing through a 3 dB fixed attenuator will exit with a power of 10 dBm − 3 dB = 7 dBm (5mW).

A terminator totally absorbs the input signal power (resulting in heat). Therefore, one end of a terminator is closed because no output power is expected. A terminator is shown in Fig. 6.4. For a terminator using a resistive rod or vane, its internal geometry is usually tapered to minimize reflection. Terminators are extremely useful in a bench test as the last stage to terminate the transmission line, thereby producing little or no reflection.

Figure 6.4 A waveguide terminator. Courtesy of Matcom.

A waveguide variable attenuator is shown in Fig. 6.5. The vane is manipulated by a knob so that a portion of the vane is lowered into the waveguide, thereby the amount of attenuation is varied. Variable attenuators are employed as power level adjustors, either incorporated in microwave signal generators or as discrete components in set-ups where data of some sort are taken with different signal powers.

Attenuation involves placing resistive material to absorb the signal's electric field. Some reflection will occur, so attenuators must be designed to minimize reflection. The important quantity that the user should look for in the data sheet of an attenuator is the VSWR. The VSWR for a variable attenuator increases slightly as the vane penetrates more and more into the waveguide, i.e., as the attenuation value is increasing. Also, if the data sheet of an attenuator is 1.10, we can calculate the return loss (in number) to be 441 (26.4 dB). The input power is 441 times larger than the reflected power, i.e., 0.23% of the input power is

SEC. 6.6
DIRECTIONAL COUPLERS

reflected. Finally, in choosing an attenuator, the user must select attenuators in the proper frequency range. The attenuation range, e.g., 0 to 60 dB, and the accuracy of the dial reading of a variable attenuator are also factors to be considered.

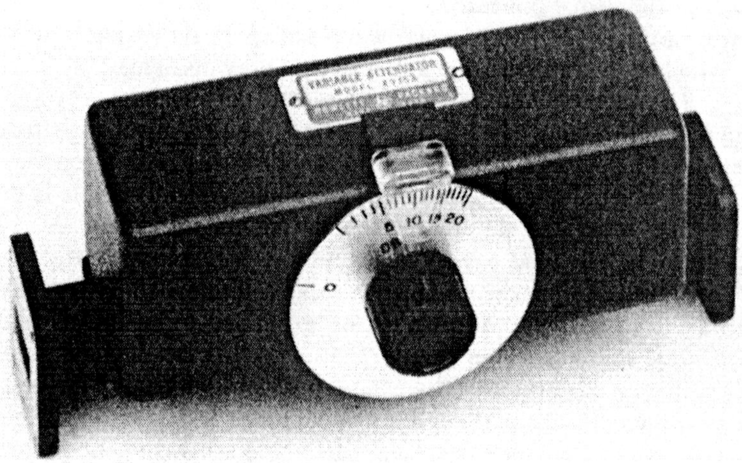

Figure 6.5 Waveguide variable attenuators. Courtesy of Hewlett Packard Company.

6.6 DIRECTIONAL COUPLERS

Functionally, a directional coupler allows a designated portion of the microwave power traveling in the main line to be coupled to the secondary arm in a preferred direction of flow. Hence, a 20 dB directional coupler is one that 0.01, or 1%, of the main line power is coupled to the secondary. Common coupling factors are 10 dB, 20 dB, and 30 dB.

Directional couplers can take form in waveguides, coaxial lines, and integrated circuits. Figure 6.6 shows several common waveguide directional couplers. In this figure, coupling occurs only when power enters the waveguide from right to left. Ideally, no coupling will occur if power enters from left to right. Two holes located approximately a quarter of a (guide) wavelength apart, are located along the center line of the waveguide. Some of the signal entering the main line from the right is coupled to the secondary arm through either hole A or hole B because the secondary arm is of similar structure as the main line, therefore, satisfying the criterion of coupling as discussed in Sec. 6.3. Upon entering the secondary arm at hole A, half of the coupled wave will propagate toward the left and half toward the right. Similarly, the coupled wave entering hole B will have leftward and rightward traveling parts. Due to the phase shift entailed in the structure, the rightward traveling wave from B will cancel the rightward traveling wave originated from A. Any residual rightward traveling wave is absorbed by the resistive vane. The overall result is that signals propagate toward the left direction and emerge at the exit port of the secondary arm.

If a signal enters the main line from the left, ideally none of it will show up at the exit port of the secondary arm. In practice, a small amount is present. How well the directional coupler maintains the directive property is given by its directivity. If a directional coupler has a 40 dB coupler, then 0.01% of the power entering the main line in the wrong direction will find its way to the secondary exit port.

6.7 POWER DETECTORS

Both diode detectors and thermistors can detect microwave power, but their operating principles are different. We shall consider the thermistor first.

SEC. 6.7

POWER DIVIDERS

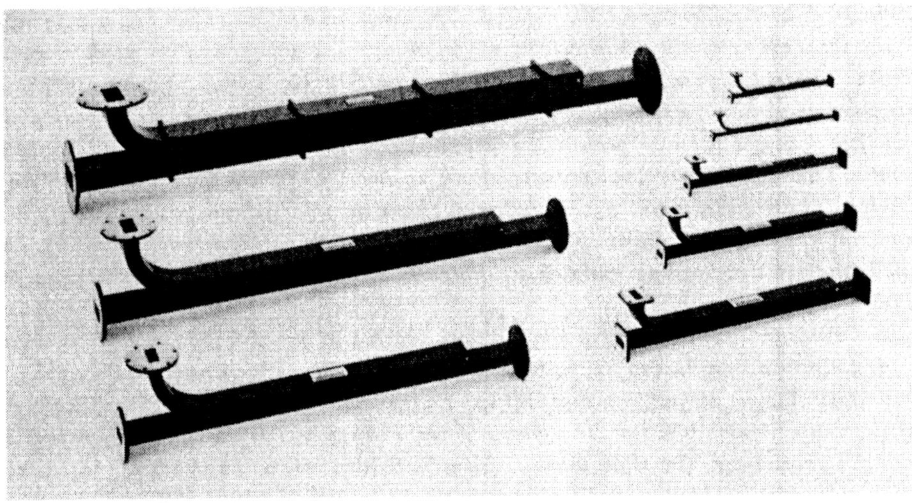

Figure 6.6 Waveguide directional couplers. Courtesy of Hewlett Packard Company.

Thermistor stands for thermal resistor. The resistance value of a thermistor is a function of temperature. Some thermistors have positive coefficients, i.e., their resistance values increase with increasing temperature. The resistance values of thermistors having negative coefficients decrease with increasing temperature. When mounted in the midst of the propagation path, the microwave power warms up the thermistor, which is one leg of a resistance bridge followed by an amplifier and its associated electronics. The change in the thermistor's resistance will unbalance the bridge, and the amplifier's output is proportional to the microwave power imparted on the thermistor. The reading is due to heat which has a slow response time, but the reading is usually independent of the signal frequency. Typically, a thermistor is mounted in a bolometer mount inside of which are temperature sensing elements (connected to the power meter) that will aid the power meter in reducing unwanted temperature drift.

Diode detectors, also known as crystal detectors, are semiconductors point-contacted to thin, 0.003", whisker wires. The rectifying property of a diode gives either a dc or a slowly varying voltage output, depending on whether the signal being detected is steady or varying.

In general, thermistors give *more accurate* readings of power level than crystal detectors. Unfortunately, thermistors tend to be *slow* in reacting to changing signal powers. (Heat-related devices are usually slow, as evidenced from the long time required to warm up a room.)

That a crystal detector does not give accurate power readings is due to its nonlinearity. The response curve of the diode's output voltage *versus* power imparted is shown in Fig. 6.7. Nonlinearity can be mathematically approximated as a series of polynomials, i.e.,

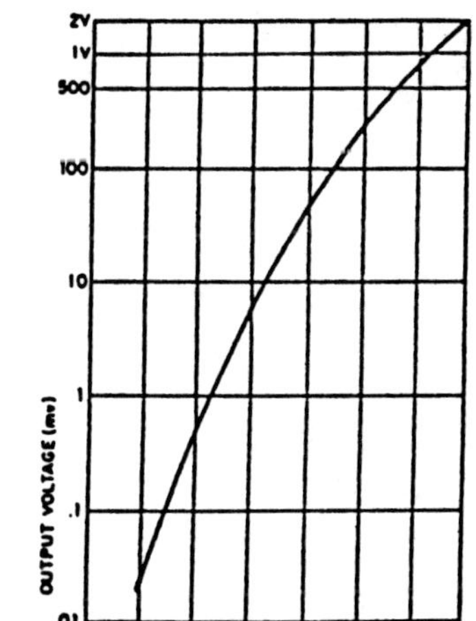

Figure 6.7 The diode output voltage versus applied microwave power. The response is nonlinear.

$$R = aP + bP^2 + cP^3 + \ldots$$

where R is the diode's response, P is the power of the signal to be measured, and a, b, and c are the coefficients of the polynomial. For most purposes, it is sufficient to go up to P^2 only, hence the square-law. The diode response is also dependent on the signal frequency. This is because rectification of an ac signal relies on the electron movements inside the diode. The diode response time and, therefore, the effectiveness of rectification are related to the mobility of electrons.

6.8 MICROWAVE GENERATORS

Microwave generators, or oscillators, generally fall into two categories: vacuum tube and solid-state. Klystrons, traveling wave tube (TWT), magnetron, gyrotron, and cross field amplifiers are tube devices which can be used as oscillators. These tubes will be covered in detail in Ch. 10. Tubes usually yield high microwave power (watts to megawatts) and are not very common on laboratory benches, except for low power klystrons whose outputs vary from milliwatts to watts. The solid-state devices employed as oscillators are Gunn and IMPATT diodes. These solid-state oscillators can generate microwave power from a few milliwatts to several watts. Solid-state devices will be covered in Ch. 11. We shall discuss the Gunn oscillator in more detail because of its popularity in the laboratory.

A Gunn oscillator delivers microwave signals at around 10 GHz. It is an *n*-type gallium arsenide crystal in which an applied dc voltage causes electrons to bunch up due to an uneven distribution of the electric field. The physical length of the chip determines the frequency of the output because the length is related to the transit time, the time it takes for a bunch of electrons to travel across the chip. The output power is limited by the difficulty of removing heat from the small chip.

When operating inside a cavity, fixed or adjustable, the current pulses induce electromagnetic waves at microwave frequencies near 10 GHz. Figure 6.8 shows a Gunn oscillator, which is waveguide mated, and the power supply. Its frequency is tunable from 8.5 to 10.5 GHz with a minimum output of 5mW. The Gunn oscillator can be operated at CW (continuous wave) or in a pulsed (1000 Hz) mode. The operating voltage of the dc supply has an effect on the output power level as well as its frequency.

6.9 METERS

Common laboratory meters for microwave testing include power meters, standing wave ratio (SWR) meters, and frequency meters. The power meter and the SWR meter are active devices, i.e., they require line power to operate. A frequency meter is basically a cavity connected to a waveguide (or coaxial cable) via a small opening. Therefore, a frequency meter requires no line power to operate.

6.9.1 Power Meter

A power meter consists of all the electronics except for the thermistor detector (see Sec. 6.7) with which microwave power can be detected by its heating effect on the thermistor. The thermistor is one leg of a resistance bridge which is balanced without microwave power. Incident power changes the resistance of the thermistor, therefore, upsetting the balance. One common practice is that the bridge is balanced by using feedback to generate low frequency power for balancing the bridge.

Figure 6.9 shows an analog and a digital power meter. A thermistor package must be compatible with the power meter. The lowest scale for the analog meter is −50 dBm full scale, and the highest scale is 10 dBm full scale. One of the five ranges can be selected from the range buttons.

6.9.2 SWR Meter

A SWR meter is a low noise voltage amplifier that derives voltage information about the detected power from the crystal diode. The amplifier is tuned at 1 kHz so that a pulsed microwave signal, i.e., 1000 groups of microwaves (at several GHz) per second, can be rectified by the diode detector. The advantage of the crystal diode/SWR meter combination over the thermistor/power detector is that if the signal power level changes, the former will respond more quickly.

SEC. 6.9

METERS

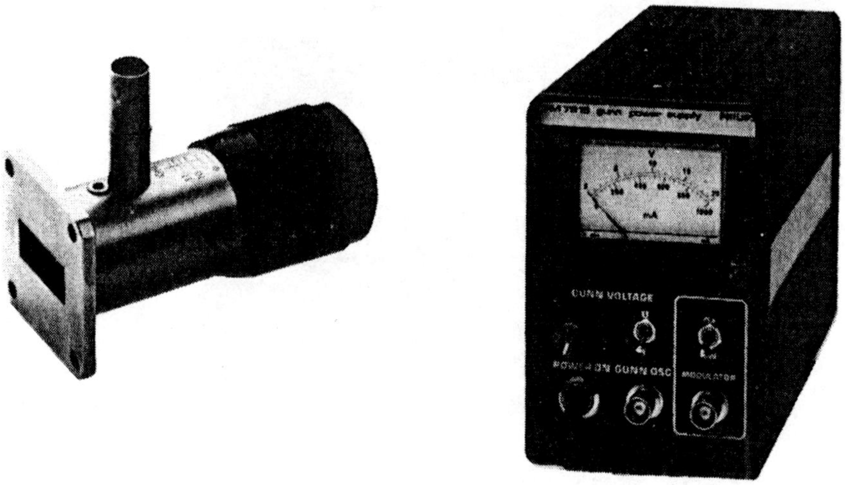

A B

Figure 6.8 A waveguide Gunn oscillator. Courtesy of Silvers Lab.

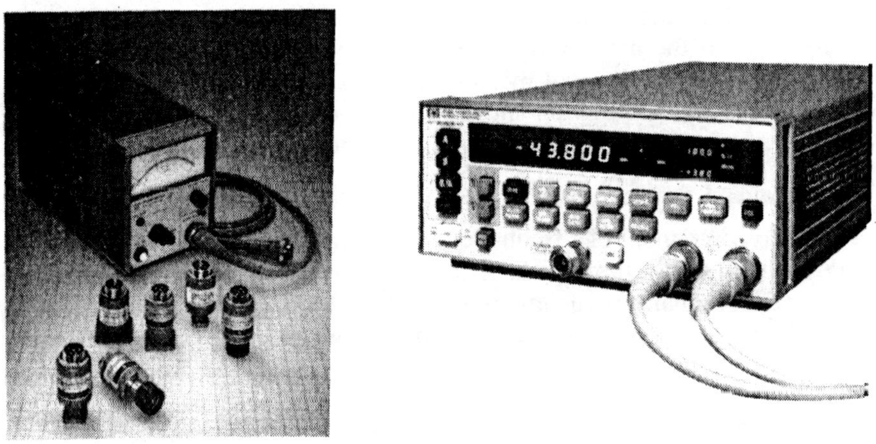

A B

Figure 6.9 (A) Analog power meters. Courtesy of Wavetek Pacific Measurements. (B) Digital power meter. Courtesy of Hewlett Packard Company.

The SWR meter can be used to measure the relative power level of two powers as well as the VSWR of a standing wave. The relative power measurement can be used for experiments on return loss and insertion loss; the absolute values of the powers being compared are not important. Figure 6.10 shows a SWR meter.

Consider the data obtained from the following procedure. By trial and error, it is found that the larger of the two powers to be compared can be adjusted to 0 dB (lower scale on the meter display) at a certain range selection. The detector is then exposed to the lower power and the marker is at 3 dB and the range is unchanged. The lower power, according to the reading, is then 3 dB weaker than the higher power. In a separate instance, if the lower power is at the 3 dB mark on the meter display, but the range is increased from 30 dB to 40 dB, then the lower power is actually 13 dB weaker than the high power.

SEC. 6.9
METERS

Figure 6.10 A standing wave ratio (SWR) meter. Courtesy of Silvers Lab.

The VSWR of a standing wave can be similarly measured. The upper scale on the meter display must be used. The maximum of the standing wave is located by observing the marker movement tendency. The marker position knob can be used next to put the marker at 1.0 and the range is noted. The minimum of the standing wave can be located by, for example, sliding the detector on a slotted line and observing the marker movement. When the minimum of the standing wave is located, the marker's position on the upper scale shows the VSWR.

6.9.3 Frequency Meter

The main component of a frequency meter is a resonant cavity. The dimension of the cavity, usually the length, if the cavity is cylindrical, is the determining factor for the resonant frequency, typically length is equal to half the wavelength. The length of the cavity can be mechanically adjusted by a plunger (electrical short circuit) so that the new length, i.e., the new resonant frequency, matches that of the measured signal.

There are two types of frequency meters (sometimes known as wavemeters): absorption-type and transmission-type. Laboratory frequency meters are usually of the absorption type. An absorption-type frequency meter consists of an adjustable cavity attached to a main transmission line through a small hole (Fig. 6.11). The equivalent circuit of the absorption-type is that of an LCR series connection. In basic electronics, the impedance of an LCR series circuit is minimum at the resonant circuit, the smaller is R, the higher the quality factory Q. The microwave signal going through the main line will regard the attached cavity as an LCR circuit. If the frequency of the signal matches that of the resonant frequency of the cavity, almost all the main line signal will be absorbed by the cavity which stores the energy. A power meter located at the receiving end of the transmission line will register very little power. If the signal frequency does not match the cavity resonant frequency, the signal power will suffer little or no loss.

Figure 6.11 A waveguide frequency meter. Courtesy of Hewlett Packard Company.

Absorption-type frequency meters are available for waveguide and coaxial connections. The user should make sure that the signal frequency falls in the operating range of frequency meter. Typical accuracy is $\pm 0.1\%$. Humidity and temperature affect the dielectric inside the cavity and the physical dimension of the cavity and therefore the meter's accuracy.

Transmission-type frequency meter is analogous to an LCR in parallel. One hole in the structure is connected to the main line where the signal flows. A second hole can be connected to a flange and then monitored by a power meter. If the signal frequency matches that of the cavity resonant frequency, the main line will lose most of its signal power to the cavity, which will then be coupled to be measured by the power meter. Hence, the power meter will register a large reading; the main line still loses most of its power.

6.10 NON-RECIPROCAL DEVICES

Both isolators and circulators are *non-reciprocal* devices employing ferrites biased by static magnetic fields.

An ideal isolator is like an ideal low frequency diode. There is 100% conduction in the forward direction and 0% conduction in the reverse direction. Hence, an isolator placed in front of a signal generator will not affect the forward signal, but will prevent any reflected signal from reaching, and thereby damaging, the generator by absorbing the reflected wave.

Figure 6.12 A coaxial isolator. Courtesy of Eaton Corporation, Microwave Products Division.

How well an isolator functions is given by its forward insertion loss and its isolation. The forward insertion loss should ideally be 1.0 numerically, or 0dB, and practically it should be as small as possible. Loss of power to the isolator in the forward path is inevitable. Isolation is a measure of the attenuation of signals traveling in the reverse direction. Typical isolation values are 30 dB and 40 dB. Isolators are available in the form of waveguides, coaxial cables, and striplines. A coaxial isolator is shown in Fig. 6.12.

The symbol of a circulator is shown in Fig. 6.13 with the three ports marked as A, B, and C for the sake of our discussion. The arrows indicate the direction of signal flow. Hence, a microwave signal entering port A will exit at port B, ideally with 0 dB attenuation; no signal from port A will appear at port C at any magnitude. A circulator is shown in Fig. 6.14.

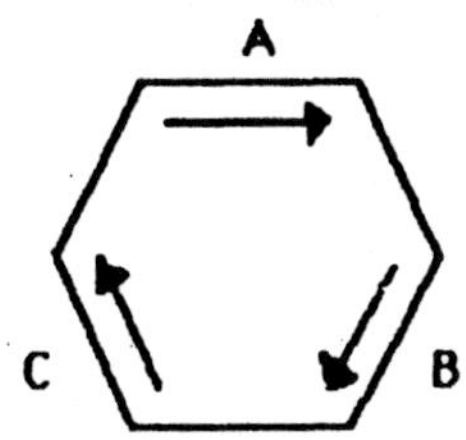

Figure 6.13 Symbol of a circulator.

One application of a circulator is to separate the transmitted signal from the received signal present at an antenna used as both a transmitter and a receiver. This is illustrated in Fig. 6.15.

In choosing an isolator or a circulator, the user should make sure the component will operate at the chosen frequency. Quantities such as (foward) insertion loss, isolation, VSWR, and average and peak power capability are all relevant.

Figure 6.14 A coaxial circulator. Courtesy of Eaton Corporation, Microwave Products Division.

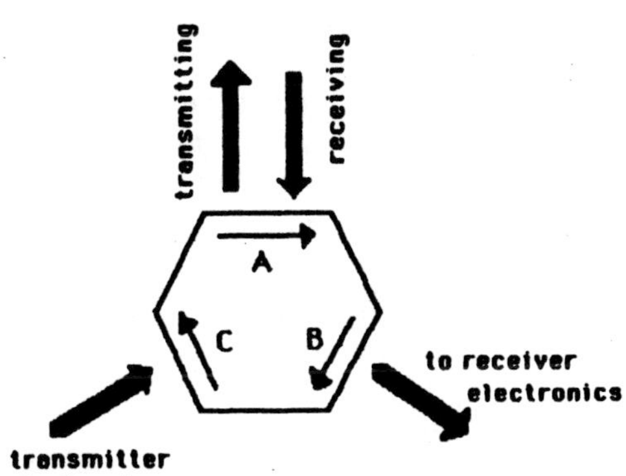

Figure 6.15 One application of the circulator is in the transmission and reception of signals using the same antenna.

SEC. 6.11
SLOTTED LINE

6.11 SLOTTED LINE

The slotted line is a handy component on a microwave laboratory bench. Figure 6.16 shows a slotted line waveguide.

The carriage can be moved longitudinally along the center slot of the main line. The center slot is cut as thin as possible to minimize leakage of microwave signal. A vernier is mounted on the stand to allow accurate location of the carriage movement.

Figure 6.16 A waveguide slotted line. Courtesy of Hewlett Packard Company.

The carriage typically carries a crystal detector which is connected to the SWR meter. This makes the slotted line very useful in measurements such as the standing wave ratio and the guide wavelength.

The slotted line carriage can be modified with a micrometer. The micrometer controls the vertical movement of a plunger (an electrical short circuit). This set-up is useful in the matching exercise on a laboratory bench.

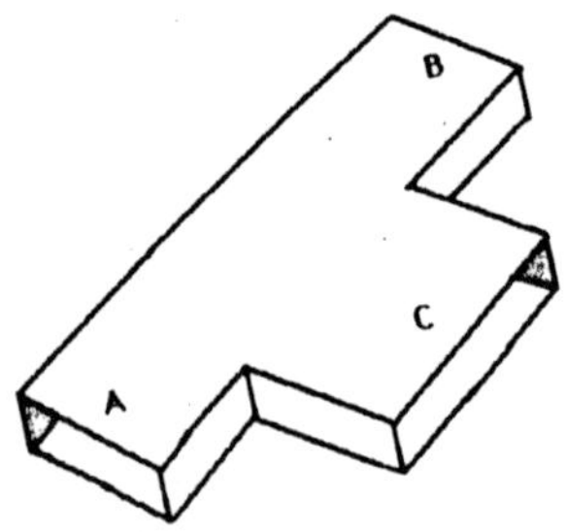

Figure 6.17 A waveguide shunt tee.

6.12 T-SECTIONS

T-sections or *tees*, are waveguides with one or more side-ports. Shunt *tee*, series *tee*, and the hybrid (magic) *tee* will be discussed here.

A shunt *tee* is illustrated in Fig. 6.17. A signal entering port C is split evenly, and exits at port A and port B. Two signals entering port A and port B which are in phase will exit port C with a power level equal to the sum of those of A and B. A series *tee* is illustrated in Fig. 6.18. If a signal enters port D, the output at ports A and B will be equal, but 180° out of phase.

The shunt *tee* and the series *tee* can be combined to form the magic (hybrid) *tee* as illustrated in Fig. 6.19. The operation of a magic *tee* is summarized in Table 6.1. The quantities in parentheses are power units.

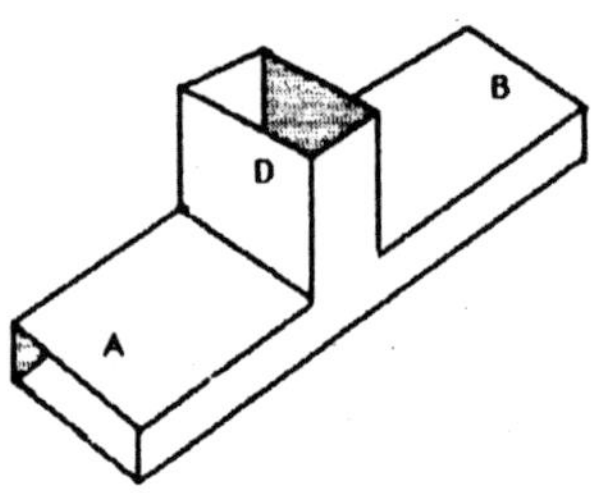

Figure 6.18 A waveguide series tee.

Table 6.1

In	Out
A($\frac{1}{2}$), B($\frac{1}{2}$)	C(1) D(0)
C(1)	A($\frac{1}{2}$), B($\frac{1}{2}$), D(0)
D(1)	A($\frac{1}{2}$), B($\frac{1}{2}$, 180° with A), C(0)

SEC. 6.12
T-SECTIONS

One useful application of the magic *tee* is to act as a transmitter-receiver switch before an antenna. The antenna is used as a transmitter as well as a receiver. Signals received must be routed away from the signals to be transmitted in order to avoid interference. The connection is illustrated in Fig. 6.20.

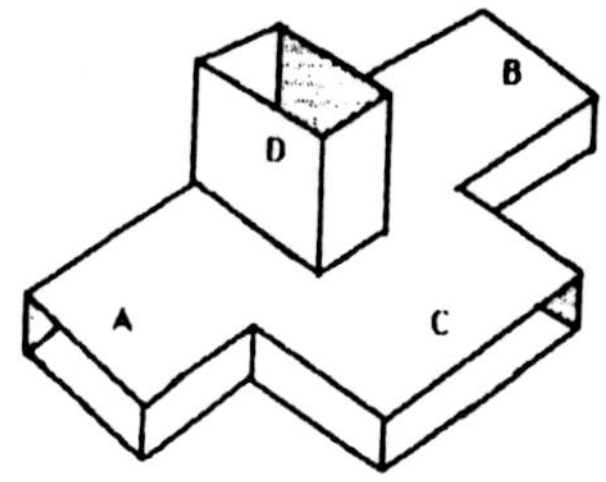

Figure 6.19 A waveguide magic tee is formed by combining a shunt tee and a series tee.

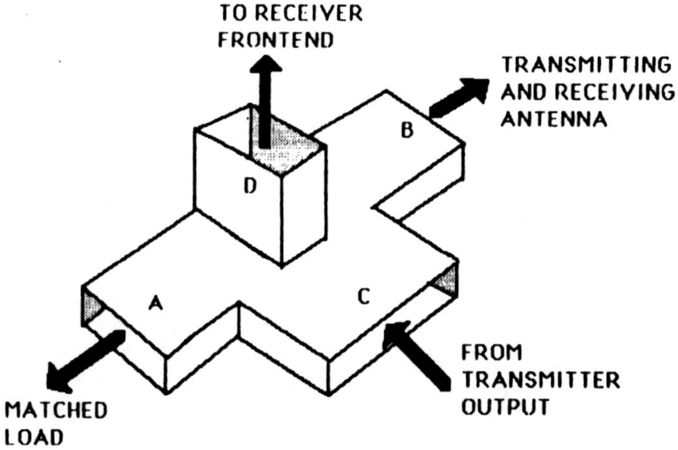

Figure 6.20 One common application of the magic tee is to separate the transmitted signal from the received signal if a common antenna is used.

CHAPTER 7

THE SMITH CHART

Lawrence A. Stark

7.1 INTRODUCTION

In Ch. 4 and Ch. 5 we learned that a component connected to the end of a transmission line causes reflections if Z_L, the impedance of the load, is not equal to Z_0, the characteristic impedance of the transmission line. The reflection coefficient Γ can vary between 0 and 1, depending on the value of Z_L. If Z_L is nearly equal to Z_0, the reflection coefficient is nearly 0. However, if Z_L is substantially different than Z_0, the reflection coefficient will be nearly unity, which represents 100% reflection of the incident signal.

The question naturally arises, what is the exact relationship between the reflection coefficient and the load impedance Z_L of a component connected to a transmission line? Mathematically, the relationship can be expressed quite simply. We have,

$$\Gamma = \frac{Z_0 - Z_L}{Z_0 + Z_l} \tag{7.1.1}$$

This expression appears simple enough, but the difficulty arises because impedance and reflection coefficient are complex numbers which involve the imaginary number

$$j = \sqrt{-1}$$

Calculations involving complex numbers are rather tedious. Thus, the relationship between load impedance Z_L and reflection coefficient is not a simple one to calculate.

This obstacle to understanding the behavior of transmission lines was substantially eliminated in January 1939, when Phillip Smith published the description of his famous chart for performing transmission line calculations. The chart became so widely used that it now bears his name, the *Smith chart.*

In Fig. 7.1, a modern Smith chart is shown. To the untrained eye, it appears to be a confusing mass of curved lines and circles. As we shall see, these lines map out an exact relationship between impedance, expressed as

$$Z = R + jX, \tag{7.1.2}$$

the sum of a resistance and a reactance; and a reflection coefficient expressed as

$$\Gamma = \rho\angle\theta \tag{7.1.3}$$

SEC. 7.1
INTRODUCTION

where

ρ = magnitude of the reflection coefficient,

and

θ = phase angle of the reflection coefficient.

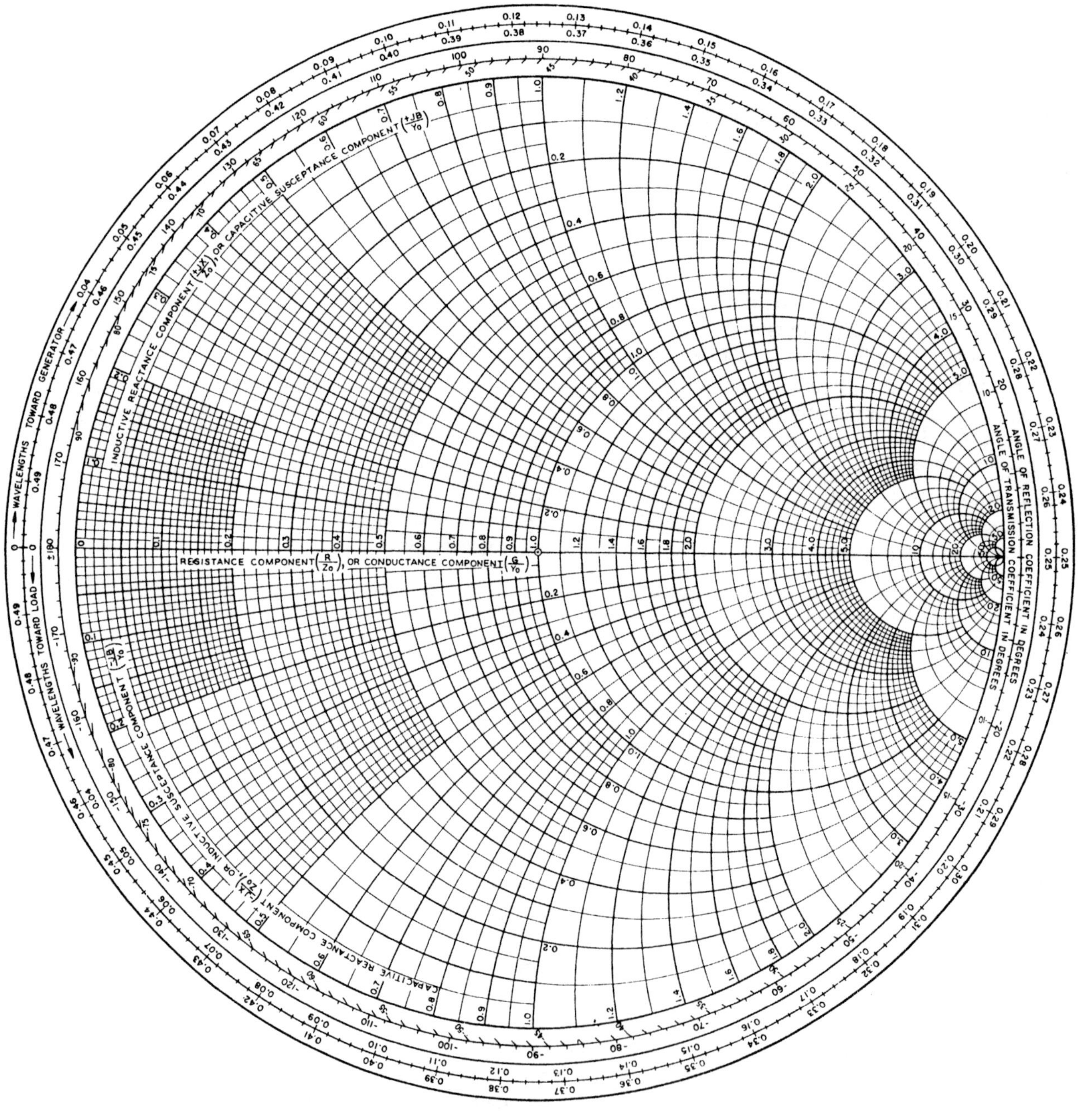

Figure 7.1 The Smith chart.

7.2 REFLECTION COEFFICIENT

The reflection coefficient of an arbitrary load impedance must be expressed as a magnitude and a phase angle. The magnitude varies between 0 and 1, and expresses the ratio of the magnitude of the reflected voltage wave divided by the magnitude of the incident voltage wave. The phase angle of the reflection coefficient is due to the fact that the reflected wave is not necessarily reflected in phase with the incident wave. Therefore, a reflection coefficient of

$$\Gamma = 0.15\angle -70^\circ$$

conveys the information that the reflected voltage wave is 0.15 (or 15%) of the incident wave, and the reflected wave has a phase angle of −70 degrees with respect to the incident wave. The phase angle is taken at the position of the load. As we shall see, at points other than the load, the phase angle is different than the value it has at the load, although the magnitude remains the same as long as the transmission line is lossless.

The first step in understanding the Smith chart is to understand that the term "reflection coefficient" can be applied to describe the ratio between the reflected voltage wave and the incident voltage wave *at any point along the transmission line*. For example, as shown in Fig. 7.2, a load impedance Z_L has a reflection coefficient with a magnitude of 0.35 and a phase angle of 60 degrees. This means that, at the right-hand end of the transmission line, the reflected voltage wave has a magnitude which is 35% of the incident wave, and the phase of the reflected wave leads the incident wave by 60 degrees.

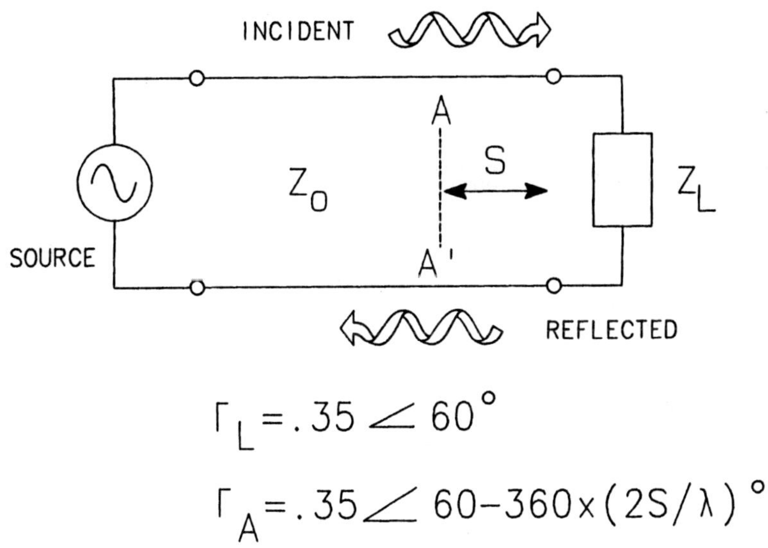

Figure 7.2 The variation of reflection coefficient with position.

Consider the situation at the location labeled *A—A′*. If the transmission line is lossless, the amplitudes of both the reflected and incident waves will be the same as at the end of the line. But the phase of the waves at location A—A′ will be different due to the propagation of the wave along the line. the phase of the incident wave measured at *A—A′* will be advanced by an angle

$$360S/\lambda_g \tag{7.2.1}$$

where λ_g is the guide wavelength of the signal. Because the reflected wave travels in the opposite direction to the incident wave, its phase is not advanced, but is retarded when measured at location *A—A'*. The net result is that the ratio of reflected wave to incident wave, when measured at point *A*, has the same magnitude, but has a phase angle which is less than the phase angle at the load by an amount equal to

$$\theta = 360(2S/\lambda_g) \tag{7.2.2}$$

For example, if the guide wavelength is

$$\lambda_g = 15\text{mm}$$

and the distance from the end of the line is 1.3mm, then the reflection coefficient at the location *A—A'* will be 0.35 at an angle of

$$60 - 360 \times (2 \times 1.3/15)$$

or

$$60 - 62.4 = -2.4°$$

In other words, reflection coefficient is a quantity that can be calculated at any arbitrary point along a transmission line if the magnitudes and phase angles of the incident and reflected voltage waves can be measured. Furthermore, if the reflection coefficient is known at one point along a uniform lossless transmission line, it can be easily calculated at any other point along the line by adding or subtracting a phase angle equal to

$$360(2S/\lambda_g)$$

where S is the distance from the starting point and λ is the guide wavelength of the signal.

Figure 7.3 shows how the reflection coefficient can be plotted on polar graph paper. The distance from the center represents the magnitude of the reflection coefficient, and the angle counterclockwise from the horizontal represents the phase angle. The point *A* represents a reflection coefficient of

$$\Gamma = .60\angle 45°$$

while the point *B* represents the same reflection coefficient at another point on the line. The angle between *A* and *B* is 90 degrees in this case, so the point *B* is one-eighth of a wavelength further from the end of the line than *A*. The usefulness of this polar plot is in easily determining the reflection coefficient of different points on a transmission line by rotating about the center of the plot by an angle given by

$$360(2S/\lambda_g)$$

7.3 IMPEDANCE

At the beginning of this chapter, we stated that the Smith chart makes it easy to find the relationship between load impedance and the reflection coefficient of the load. In fact, the Smith chart is a polar plot of reflection coefficient, as shown in Fig. 7.3, over which graph lines for resistance and reactance have been laid.

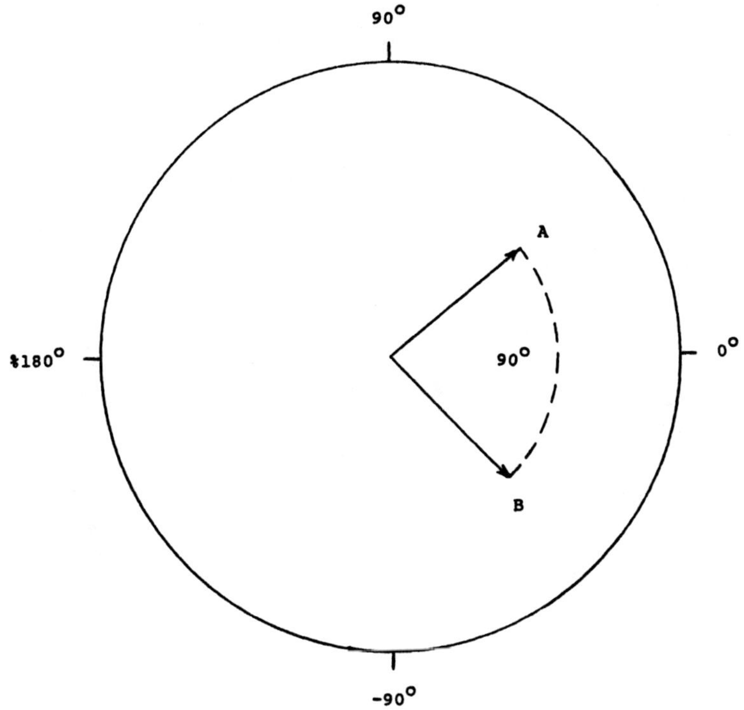

Figure 7.3 A polar plot of reflection coefficient.

We can plot reflection coefficient on the Smith chart in polar form using help features on the chart. A radial scale from 0 to 1 can be found at the bottom of the chart on the left-hand side labeled "REFL. COEFF., E OR I." By using dividers, the distance of any point on the Chart from the center can be determined accurately. The angular displacement of a point can be found by projecting a line from the center through the point and out to the periphery of the chart. Find the scale labeled "ANGLE OF REFLECTION COEFFICIENT IN DEGREES" and use it to determine the angle. See *Example 7.3.1* below for an illustration of this technique.

We shall now examine the lines of resistance and reactance separately. The resistance lines are circles centered on the horizontal axis of the chart as shown in Fig. 7.4, and the reactance lines are arcs which originate at the right-hand edge of the chart as shown in Fig. 7.5.

The resistance circles range in value from zero to infinity. The $R=0$ circle is the largest circle centered on the Smith chart which defines its outside border. Larger values of resistance correspond to smaller circles whose centers are located progressively to the right of the center of the chart.

The positive and negative values of reactance shown in Fig. 7.5 represent inductive and capacitive values of circuit impedance respectively. All reactance values found above the center line are positive and represent inductive reactance, while all values below the line are negative and represent capacitive values. The center line itself is the zero reactance line. Any impedance found on this line is purely resistive because its reactance value is zero.

The resistance and reactance values shown in Figs. 7.4 and 7.5 are not values in ohms. They are values normalized to the characteristic impedance of the transmission line. If $Z_0=50\Omega$, then the circle *$R=2.0$ represents a resistance of*

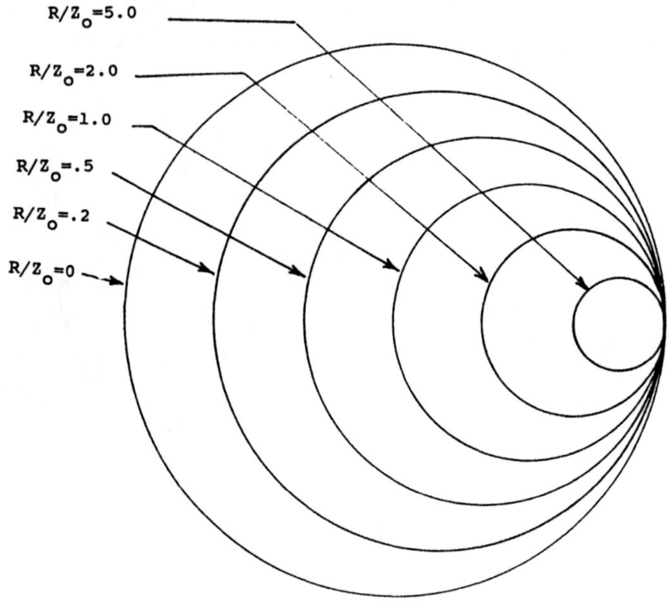

Figure 7.4 The Smith chart; normalized resistance circles.

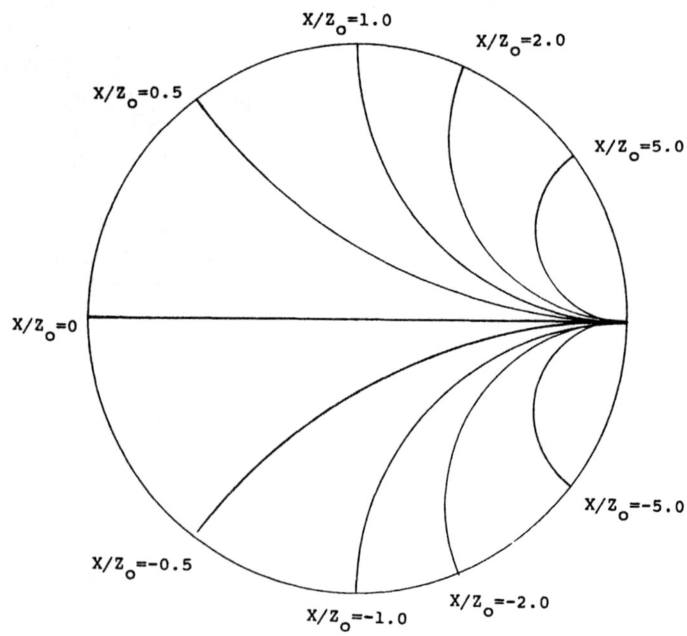

Figure 7.5 The Smith chart; normalized reactance circles.

100 Ω; and the arc labeled $X = -0.5$ represents a reactance of -25Ω. At all times, when using the Smith chart, the reader should remember that the numbers on the R and X contours refer to values of R/Z_0 and X/Z_0.

When the curves in Figs. 7.4 and 7.5 are combined, they yield the Smith chart pattern shown in Fig. 7.1 We can now use the Smith chart to determine the relationship between the impedance of a transmission line load and the resulting reflection coefficient.

Example 7.3.1

Suppose that a transmission line termination with an impedance of $Z_L = 75 - j20\Omega$ is connected to the end of a transmission line which has a $Z_0 = 50\Omega$. What is the resulting magnitude and angle of the load reflection coefficient?

To solve this problem using the Smith chart we make use of the relationship between reflection coefficient plotted in polar fashion, as described in Fig. 7.3, and the resistance and reactance lines which make up the Smith chart. First, locate the impedance value on the Smith chart after normalizing. The normalized value is obtained by dividing both the resistance and the reactance by 50Ω. We obtain $Z_n = 1.5 - j0.4$, where we use Z_n to represent a normalized load impedance. To locate this value on the chart, find the intersection of the $R = 1.5$ circle and the $X = -0.4$ arc, shown as point A in Fig. 7.6. The distance of this point from the center of the Chart as read on the lower scale determines the magnitude, and the angle made by the ray passing through A gives the phase of the reflection coefficient. Thus, the reflection coefficient of the load described above is

$$\Gamma = 0.25\angle -30°$$

The above example illustrates the technique for using the Smith chart to determine the reflection coefficient or impedance of a transmission line termination when one or the other is known.

We saw above that reflection coefficient could be defined anywhere on the transmission line as the ratio of the reflected voltage wave to the incident voltage wave at that point. We can also define impedance in the same way. The impedance at any point on a transmission line is the ratio of the voltage to the current at that point. Here, we mean the total voltage and current, which is the sum of the incident and reflected waves. The total voltage and current will vary with position due to the standing waves described by the VSWR.

Figure 7.7 illustrates this concept. The illustration shows a load impedance Z_L connected to a transmission line. At a distance S from the end of the line, we have a "line impedance," that we will call $Z(S)$, which is the ratio of the voltage at that point to the current as shown. This is the impedance we would see "looking to the right" at point A—A', if the line were broken at that point. Thus, $Z(S)$ is determined entirely by the load impedance Z_L and the length of line with characteristic impedance Z_0.

7.4 THE BASIC APPLICATION OF THE SMITH CHART

The most basic use of the Smith chart is to calculate $Z(S)$ for an arbitrary length of line S as indicated in Fig. 7.7. Adding a length S of lossless line to an arbitrary load changes the phase angle of the reflection coefficient by an amount $360(2S/\lambda_g)$ degrees while the magnitude remains the same. On polar graph paper, this is easily represented, as seen in Fig. 7.3.

We find the line impedance $Z(S)$ by drawing a polar plot of the new reflection coefficient on a Smith chart, and reading off the new impedance value from the resistance and reactance lines. This will be illustrated in the following example.

SEC. 7.4

BASIC APPLICATION

Figure 7.6 Example 7.3.1; determining the reflection coefficient on the Smith chart.

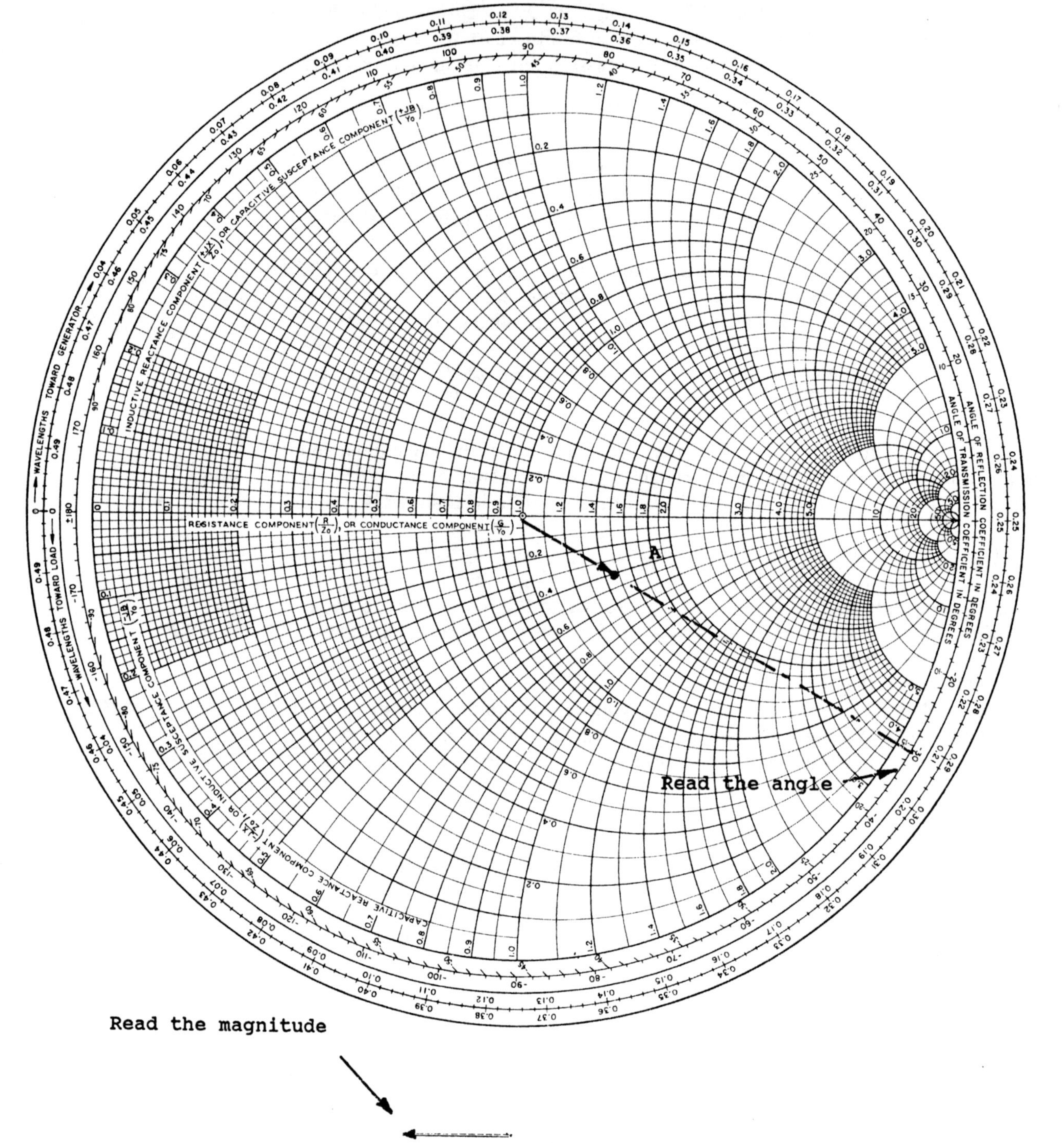

Figure 7.7 The concept of transmission line impedance.

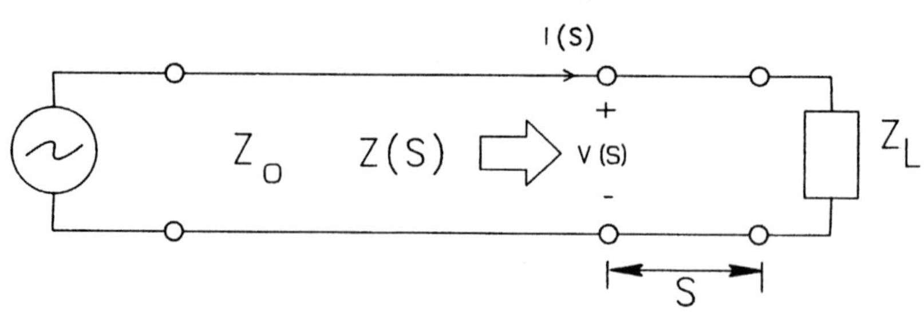

Example 7.4.1

Determine the transmission line impedance at a distance of 1cm from the end of the transmission line in *Example 7.3.1,* if the guide wavelength of the signal is 8cm.

Refer to the Smith chart in Fig. 7.8. The impedance at the end of the line is given as $Z_L = 75 - j20\ \Omega$. Using the chart, we see that the reflection coefficient is

$$\Gamma = 0.25\angle -30°$$

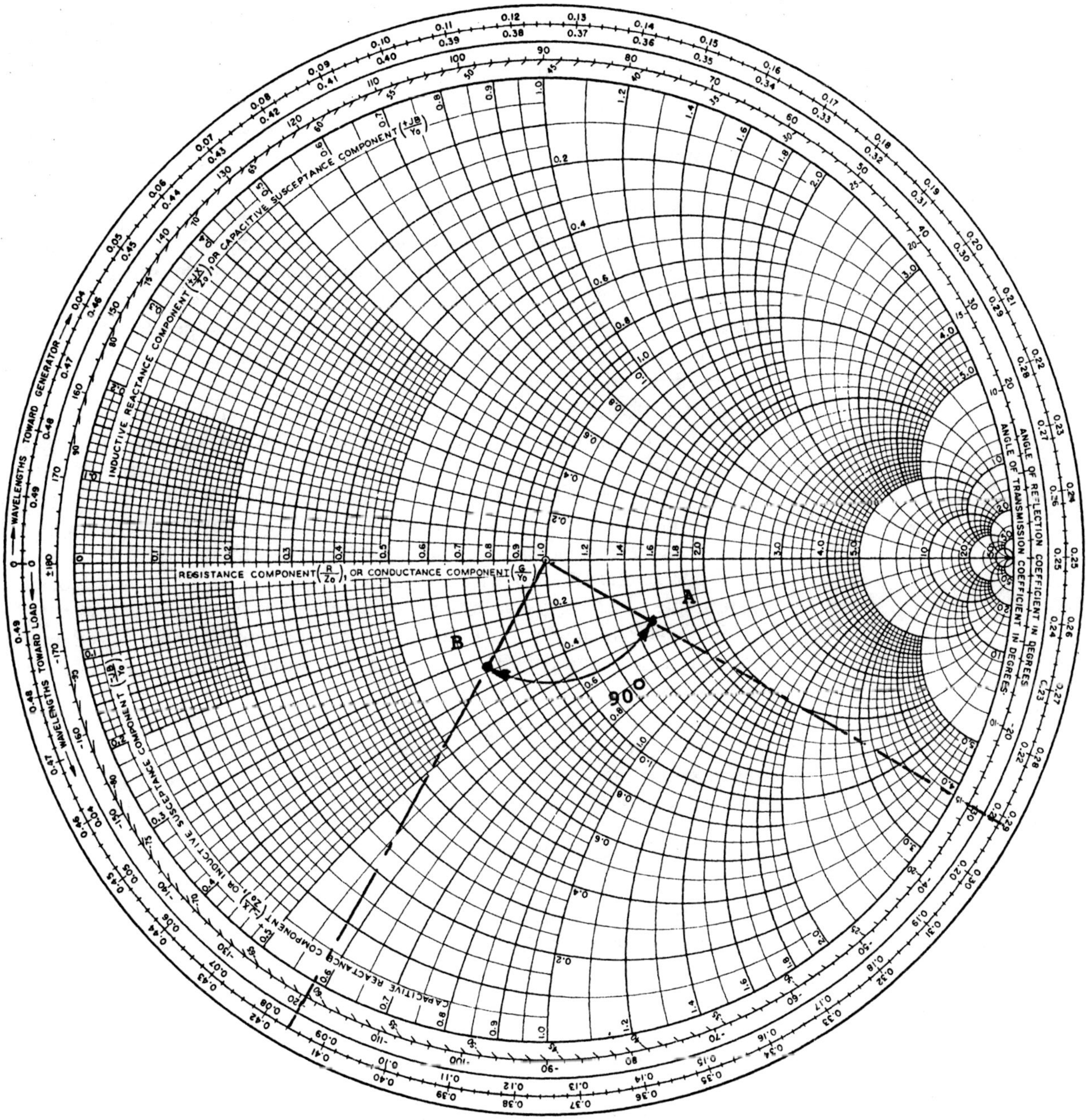

Figure 7.8 Example 7.4.1; impedance change due to line length.

We want to determine the reflection coefficient at the point 1cm from the end of the line. We know that the reflection coefficient at point B will have the same magnitude, but a more negative phase angle as shown in sec. 7.1. above. The change in phase angle will be 90 degrees, calculated from $360 \times (2S/\lambda_g)$ where $S = 1$cm and $\lambda_g = 8$cm. Rotate point A in Fig. 7.8 clockwise about the center of the chart by 90 degrees to represent the reflection coefficient at point B. Now read the new impedance off the chart to obtain

$$Z_B = 1.15 - j0.54$$

We can summarize the basic application of the Smith Chart as follows.

1) The impedance along a transmission line varies due to the standing wave patterns in voltage and current caused by a reflective load.
2) The Smith chart is a tool for calculating the transmission line impedance anywhere on a transmission line, if the impedance is known at a single point. That point may be at the terminating impedance, or it may be at some point in the middle of the line.
3) A single point on the Smith chart represents both reflection coefficient (magnitude and phase angle) in polar coordinates, and impedance (resistance and reactance) in normalized quantities. The parameters of the reflection coefficient are read with the aid of supplementary scales: a radius scale below the chart and an angular scale about the circumference of the chart. The resistance and reactance values are read from the curved axes which define the pattern of the Smith chart itself.
4) The center of the Smith chart represents a purely resistive impedance equal to Z_0, the characteristic impedance of the transmission line. In general, designers try to create microwave components with impedances that lie at or near the center of the chart, so that the reflections which result are small.
5) A length of transmission line S is taken into account by changing the phase of the reflection coefficient by an amount $360 \times (2S/\lambda_g)$ while keeping the magnitude unchanged. Moving away from the load impedance reduces the phase angle, a clockwise rotation on the Chart, while moving towards the load increases the angle, a counterclockwise rotation. The normalized impedance taken from the Smith chart at the new point is the line impedance at the new point.
6) From Eq. (7.2.2), the phase angle of the reflection coefficient advances by 360 degrees, if the line length S is altered by $\lambda_g/2$. Therefore, points on the transmission line which lie $\lambda_g/2$ apart are "identical" in that they have the same value of reflection coefficient and, hence, the same impedance.

7.5 IMPEDANCE-ADMITTANCE TRANSFORMATIONS

So far in our discussion of the Smith chart we have only used the impedance of the line. In many applications, the admittance of the line is more useful for calculations, especially when parallel or shunt circuits are being studied. A useful property of the Smith chart is that the admittance

$$Y = 1/Z \tag{7.5.1}$$

is easily found by rotating the point representing Z by 180 degrees about the center of the chart, as shown in Fig. 7.9. In this example, the impedance is

$$Z_A = 0.4 + j0.4$$

and the corresponding admittance is

$$Y_A = 1.25 - j1.25$$

where both sets of values are *normalized*. Naturally, the equivalence is reciprocal, an admittance can be transformed to the corresponding impedance by rotating the point about the center by 180 degrees. To calculate the actual admittance in siemens, multiply the normalized value of Y, found from the Smith chart, by the characteristic admittance

$$Y_0 = 1/Z_0 \tag{7.5.2}$$

where Z_0 is the characteristic impedance of the transmission lines.

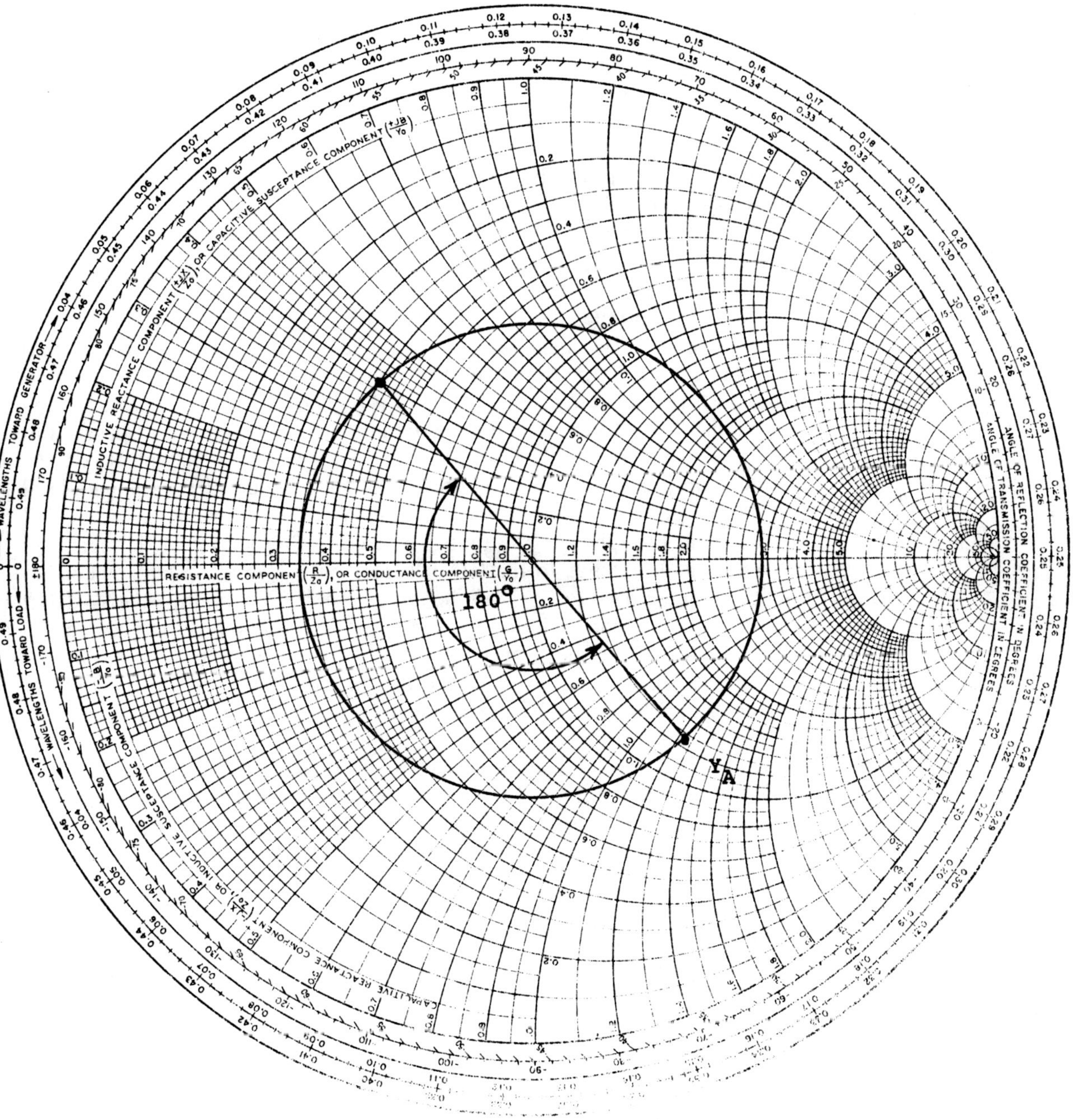

Figure 7.9 Impedance to admittance transformation.

For example, if $Z_0 = 50\Omega$, then

$$Y_0 = 1/50 = .02$$

or

$$Y_0 = 20\text{mS (milli-siemen)}$$

Because a 180 degree rotation about the Smith chart transforms an impedance into its corresponding admittance value, it follows that a length of transmission one-quarter wavelength long will transform a normalized load impedance, Z_L, into the inverse of that value. Thus, a quarter-wave line length transforms a short circuit into an open circuit and *vice versa*. It should also be apparent from examination of the Smith chart, that a quarter-wave line transforms a capacitive load into an inductive load and *vice versa*. This is because of the change of sign of reactance values that occurs when rotating half-way around the chart.

7.6 DETERMINING VSWR

From Ch. 4 on transmission lines, we learned that the VSWR of a transmission line was equal to

$$VSWR = (1+\rho)/(1-\rho) \tag{7.6.1}$$

where ρ = magnitude of the reflection coefficient. In other words, the VSWR depends only on the magnitude of the reflection coefficient of the load termination, but not on its phase angle. The VSWR of a load impedance can be found from the Smith chart, as shown in Fig. 7.10A. The point Z_A represents an arbitrary impedance. Construct a circle centered in the Smith chart which passes through Z_A as shown. This circle is often referred to as the "constant VSWR circle" because all points on this circle have the same value of the magnitude of the reflection coefficient and, hence, the same VSWR. The value of the VSWR can be found in two places. The first is a scale at the bottom of the chart labeled "SWR" which extends to the left of center on the very top line. A compass or dividers can be used to mark off the radius of the circle on the scale to determine the VSWR. A second method is to read the intersection of the constant VSWR circle on the $X=0$ line. The constant VSWR circle intersects this line at two places and we wish to consider the intersection which lies to the right of the center of the chart. The VSWR is equal to the value of normalized resistance at the point of intersection, as shown.

Another important feature concerns the relationship between points on the constant VSWR circle and the standing wave pattern that exists on an actual transmission line. There is a definite relationship between the peaks and nulls of the voltage standing wave pattern and the points on the VSWR curve labeled *A* and *B* in Fig. 7.10B. The figure shows the main features of a standing wave pattern on a transmission line. The impedance at the location of a voltage standing wave minimum corresponds to the point *B* on the VSWR circle, while that at a standing wave maximum corresponds to the point *A*. We can see that the line impedance is purely resistive at these points and that it has its lowest value at the position of the VSWR minimum and its highest value at the VSWR maximum.

The usefulness of the above information lies in interpreting slotted line measurements as discussed below in Sec. 7.8.

7.7 CHANGING CHARACTERISTIC IMPEDANCE

A situation that arises frequently involves circuits with lengths of transmission lines with different values of Z_0. These circuits can be analyzed on the Smith chart using the techniques described above, but handling the change of Z_0 may require some explanation. In the following example, the proper method of accounting for two values of Z_0 is described.

SEC. 7.7

CHANGING CHARACTERISTIC IMPEDANCE

Figure 7.10 Determining the VSWR from the Smith chart.

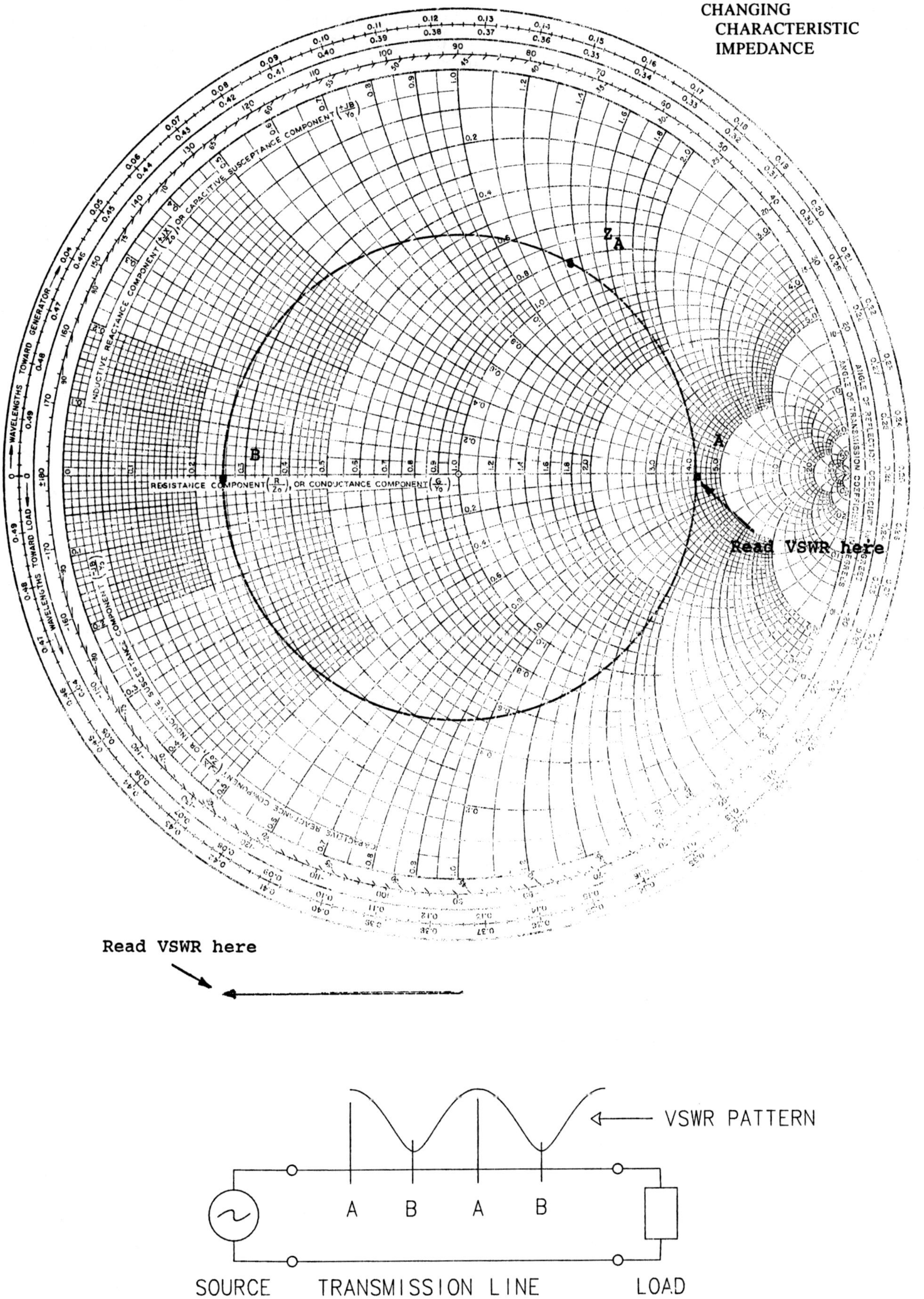

Example 7.7.1

The problem is to find the input impedance to the circuit shown in Fig. 7.11A. The lengths of the lines are given in fractions of a wavelength. In a more practical application, the line lengths would be in centimeters and the wavelength would have to be known.

The problem should be solved in three stages. First, we transform the load impedance from 0–0′ to the point *A*—*A*′ through the $Z_0 = 80\ \Omega$ transmission line. Second, we renormalize the resulting impedance from the 80Ω value to 50Ω. Third, we transform through the $Z_0 = 50\ \Omega$ line to the point *B*—*B*′ to complete the solution.

We wish to use the Smith chart, so we must normalize the load impedance to the value of Z_0, which is initially 80Ω. This yields

$$Z_L = 1.5 + j0.25\ (Z_0 = 80\Omega) \tag{7.7.1}$$

which is plotted as Z_L in Fig. 7.11B. Transforming this value through λ/10 line length results in the point at *A* with the value

$$Z_A = 1.23 - j0.45\ (Z_0 = 80\Omega) \tag{7.7.2}$$

To continue to point *B*, we must renormalize to $Z_0 = 50\ \Omega$. Simply multiply the value in Eq. (7.7.2) by $Z_0 = 80\Omega$ to determine the actual line impedance in ohms, then divide by $Z_0 = 50\Omega$ to re-normalize to the second length of line. This results in

$$Z = 1.97 - j0.72\ (Z_0 = 50\Omega) \tag{7.7.3}$$

which is plotted as Z_{A2} in Fig. 7.11B. This point can now be rotated corresponding to a quarter-wavelength of line which is 180 degrees on the Smith chart. The final point is, therefore, Z_B which has the value

$$Z_B = 0.45 - j0.16 \qquad (B\text{—}B',\ Z_0 = 50\Omega)$$

or, multiplying by $Z_0 = 50\Omega$ to un-normalize,

$$Z_{in} = 22.5 + j8\Omega$$

In general, when the characteristic impedance in a circuit changes from one value to another, the normalized impedance on the Smith chart will be

$$Z_{n2} = Z_{n1}(Z_{01}/Z_{02}) \tag{7.7.4}$$

where Z_{n1} and Z_{n2} are the normalized impedance in transmission lines with characteristic impedances Z_{01} and Z_{02}, respectively.

7.8 SLOTTED LINE MEASUREMENTS

The existence of a standing wave pattern is the most notable feature of transmission line behavior that distinguishes mircowave applications from lower frequency circuits. The slotted line measures the characteristics of the standing wave pattern directly. Using the Smith chart, slotted line measurements can be interpreted to find the impedance of the termination on the line. Although slotted lines are not widely used for microwave impedance measurements nowadays. a knowledge of their application yields valuable insight into the behavior of transmission lines as well as the use and properties of the Smith chart.

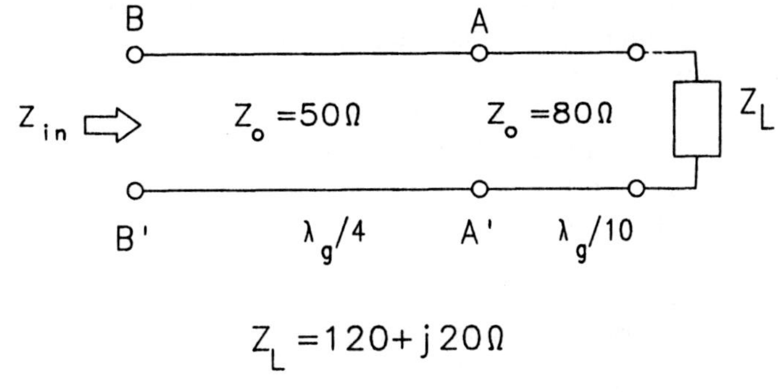

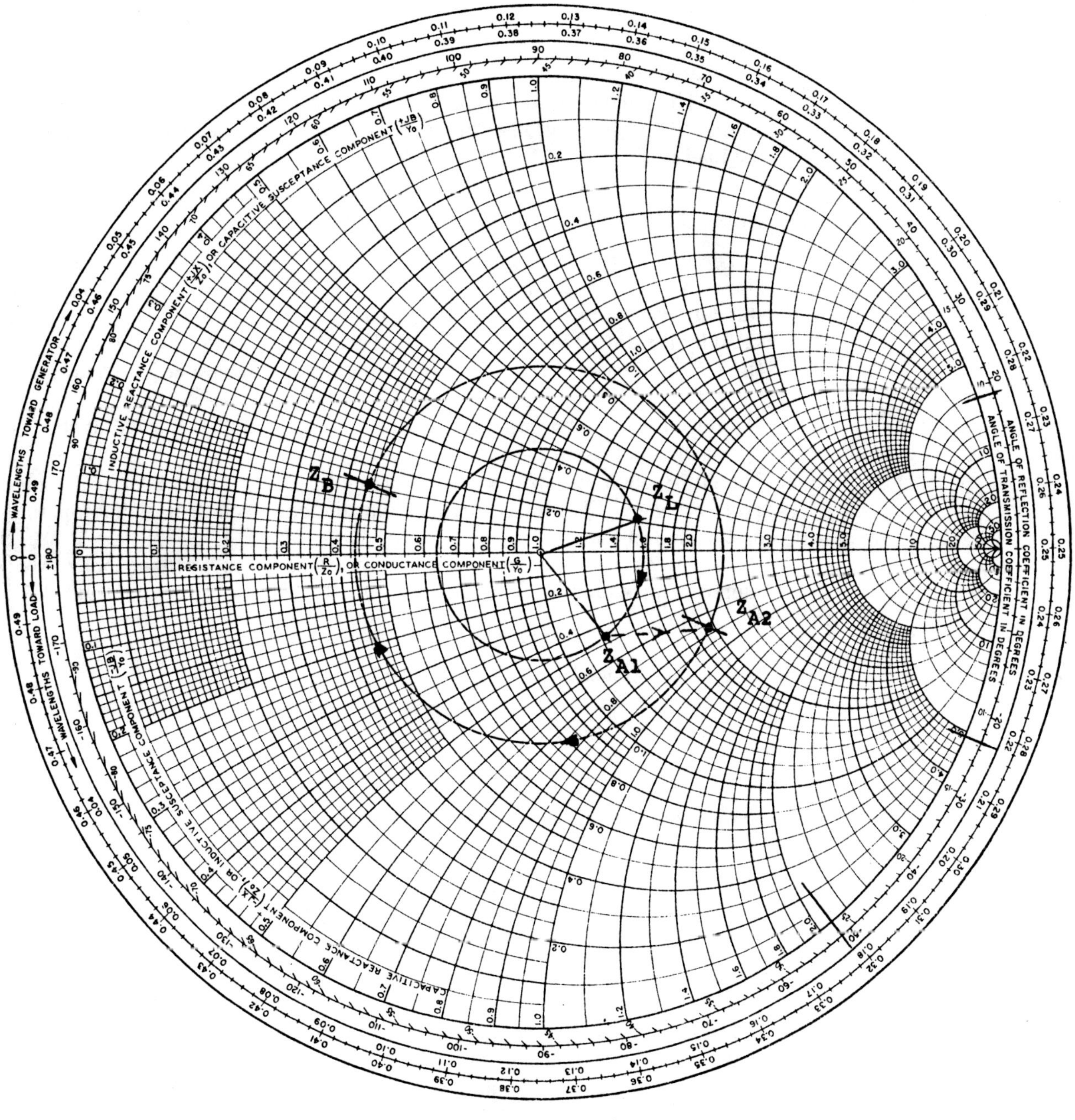

Figure 7.11 Taking account of a change or characteristic impedance.

SEC. 7.8
SLOTTED LINE MEASUREMENTS

Refer to the example in Fig. 7.12. Part A shows the necessary arrangement of instruments for the measurement. A stable source of microwave power is necessary. If the source frequency drifts more than a few MHz, the positions of the standing wave nulls could drift enough to change the position of the calculated impedance by 1 to 2 degrees on the Smith chart. Most stable signal generators will be adequate for this, although some sweepers, intended for broadband use, will drift too much to make accurate measurements.

The SWR meter is used to read out the value of VSWR on the line. The meter measures the voltage from the detector in the slotted line probe. The meter scale is graduated so that the VSWR can be read directly from the meter as the probe is moved back and forth along the line.

A graduated scale, usually with a vernier, is fixed firmly to the slotted line. Using this scale and the SWR meter, the operators can pinpoint the locations of the voltage minima along the transmission line.

Finally, a short circuit is also needed. If the slotted line is in a waveguide, the short circuit can be a metal plate, which can be bolted across the end of the open waveguide. In coaxial systems, shorted connectors are available in all connector types. The purpose of the short circuit is to calibrate the scale on the slotted line so that the scale positions can be related to the distance from the slotted line to the load itself.

The measurement proceeds as follows from *Example 7.8.1.*

Exercise Example 7.8.1: Slotted Line Measurements

The first step is to disconnect the unknown load and replace it with the short circuit, or "reference short," as it is called. This will set up a strong pattern of standing waves on the line with an extremely high VSWR. This pattern is indicated by the dotted line in Fig. 7.12B. Two features of this pattern are of interest:

1) The minima of the standing wave pattern are separated by exactly one-half wavelength.
2) Because the load is a short circuit, a voltage null is located at the position of the load itself.

Therefore, each null of the pattern created by the reference short is located *an exact number of half-wavelengths* from the load itself. This calibrates the scale on the slotted line.

The procedure is to measure the locations of the voltage nulls. We shall suppose, for example, that we have measured three successive null positions, which are

$$d_1 = 4.63\text{cm}$$
$$d_2 = 6.75\text{cm} \qquad \text{(reference short positions)}$$
$$d_3 = 8.87\text{cm}$$

We can calculate the guide wavelength from these values because each null is located one-half wavelength apart. We have

$$\lambda_g/2 = d_2 - d_1 = d_3 - d_2$$

or

$$\lambda_g = 4.24\text{cm}$$

The next step is to reconnect the load and make a second set of measurements, recording the locations of the voltage minima, and also recording the value of the measured VSWR. We shall suppose that the following data were recorded:

SEC. 7.8

SLOTTED LINE MEASUREMENTS

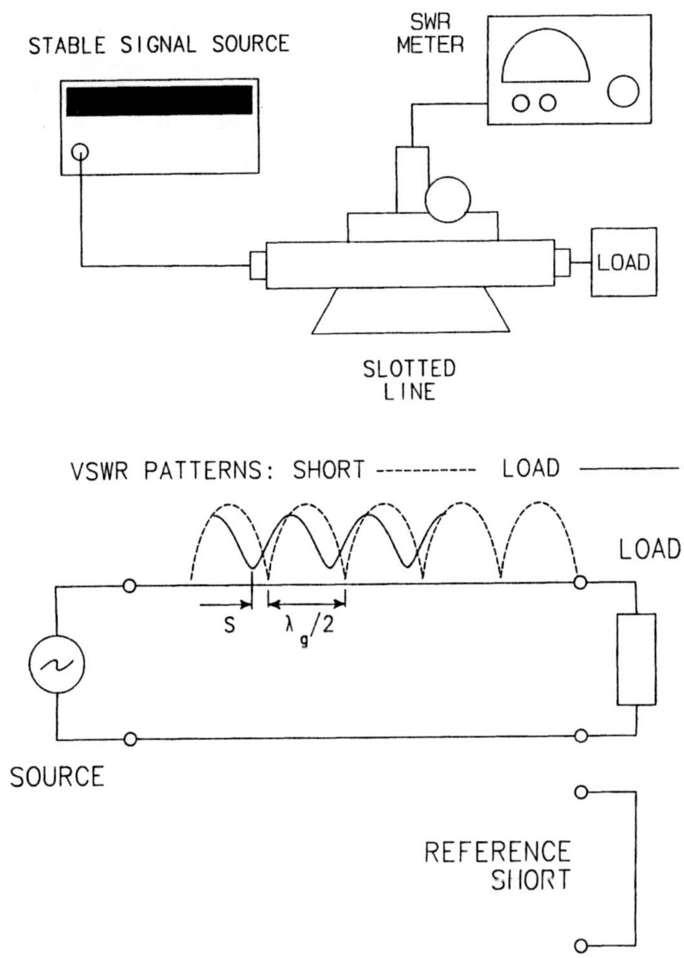

Figure 7.12 The slotted line measurement—experimental arrangement.

$$
\begin{aligned}
VSWR &= 2.6 \\
d_1' &= 5.10\text{cm} \\
d_2' &= 7.22\text{cm} \\
d_3' &= 9.34\text{cm}
\end{aligned}
$$

The measurement scale is arranged so that larger numbers are farthest from the load. We can confirm that the separation between the locations of the minima is still the one-half wavelength measured when the short was connected.

We can now determine the impedance of the unknown load impedance. Given that the VSWR is 2.6, we can construct a circle which corresponds to that value using the information in Sec. 7.6, as shown in Fig. 7.13. The circle is drawn so that it passes through the point $R=2.6$ on the zero reactance lines. Point A can also be identified as the impedance at the points which were labeled d_1', d_2', and d_3' with the load connected, as explained in Fig. 7.10. We can now determine the impedance at the load by referring to our "scale calibrations," the locations of the voltage nulls with the load connected.

The situation is shown schematically in Fig. 7.12B. We know that the impedance at, let us say, d_2' is

$$Z = .385 + j0 \qquad \text{(normalized)}$$

and we want to determine the impedance at the location of one of the reference short positions. Because these reference short nulls are all an exact number of half-wavelengths from the load, the line impedance at these reference positions will be equal to the load impedance, as explained in Sec. 7.4. Therefore, we only need to determine the distance between one of the d' positions (measured with the load connected) and the closest reference short position, and use the Smith chart to transform the impedance Z_A over that distance.

The separation between d_2' and d_2 is

$$s = 7.22 - 6.75 = .47\text{cm}$$

The Smith chart is calibrated in fractions of a wavelength, so we need to determine s/λ_g, which is

$$s/\lambda_g = .47/4.24 = 0.11$$

The reference short positions are located closer to the load than the corresponding VSWR minima from the load measurement. Thus, we must rotate counterclockwise (increasing phase angle of the reflection coefficient) about the chart by 0.11 wavelengths using the scale on the outside of the chart. The final point is the impedance at a reference null and, hence, the impedance of the load. We have

$$Z_L = 0.59 - j0.64$$

as shown in Fig. 7.13. Had the measurement shown that the reference short positions lay towards the load from the VSWR minima, we would have rotated in a clockwise direction to find Z_L.

7.9 THE Z/Y SMITH CHART

In many situations, particularly the design of impedance-matching circuits, as discussed in Ch. 8, we wish to analyze transmission line circuits which involve parallel circuit connections. Parallel connected circuits are easily analyzed using the admittance of the circuit elements.

Figure 7.14 shows a schematic circuit involving two lengths of transmission line with a parallel inductor between them. This can be analyzed using the basic Smith chart operations described above. The impedance

$$Z_L = 0.5 - j0.4 \qquad \text{(normalized)}$$

is transformed away from the load through the length of line S_1/λ_g to the point Z_A. To add the effect of the inductor at this point, we must determine the admittance.

$$Y_A = 1/Z_A$$

by rotating about the Smith chart 180 degrees, as explained in Sec. 7.5. This brings us to the point

$$Y_A = 1.05 - j0.93$$

The parallel element is an inductor with a normalized reactance of

$$X_s = 1.0$$

The admittance of an inductor is given by its susceptance which is

SEC. 7.9
THE Z/Y SMITH CHART

$$B_L = -1/X_L$$

where

$$X_L = \text{normalized reactance of the inductor,}$$

and

$$B_L = \text{normalized susceptance of the inductor,}$$

or

$$B_L = -1.0$$

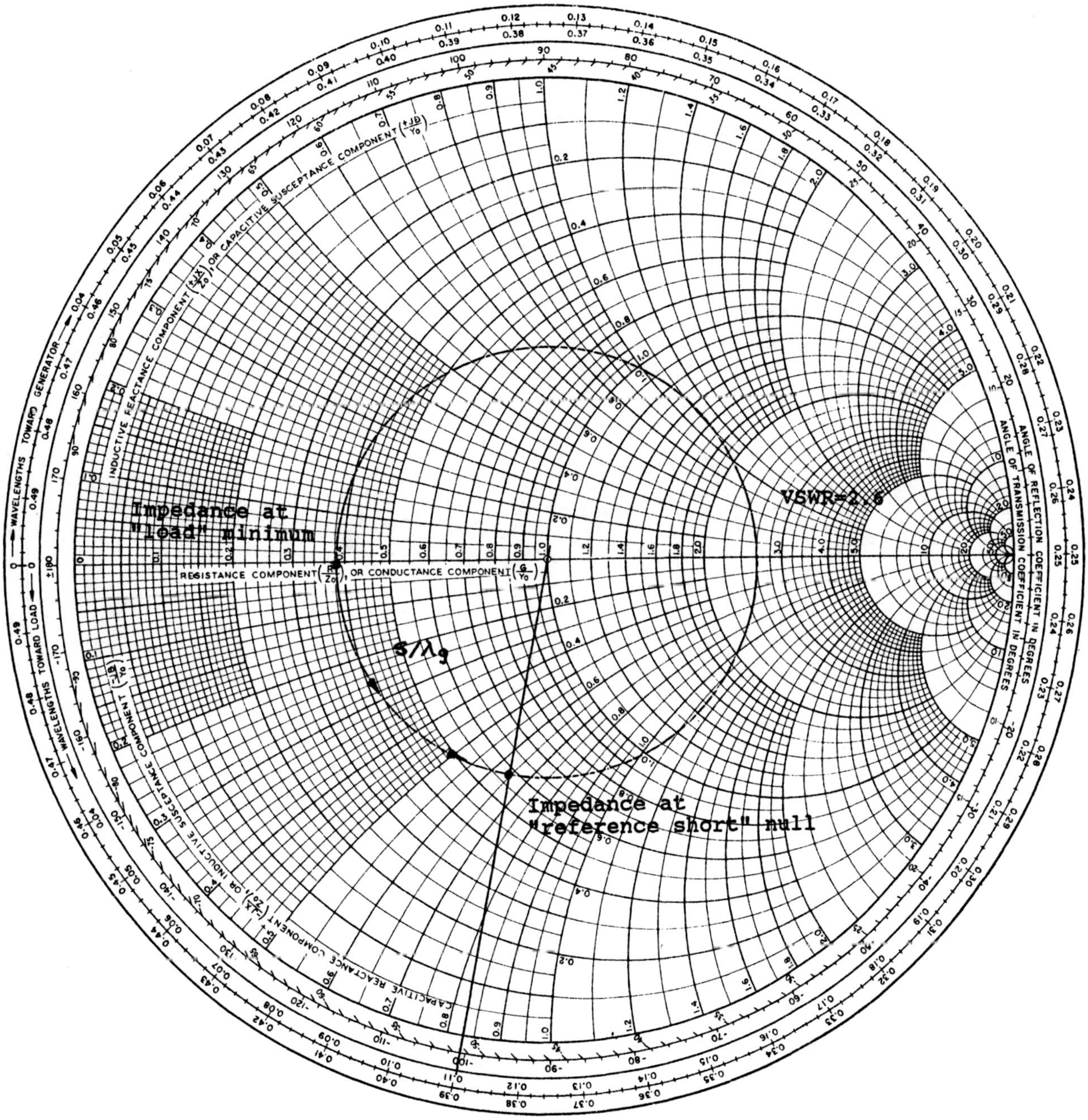

Figure 7.13 The slotted line measurement—use of the Smith chart.

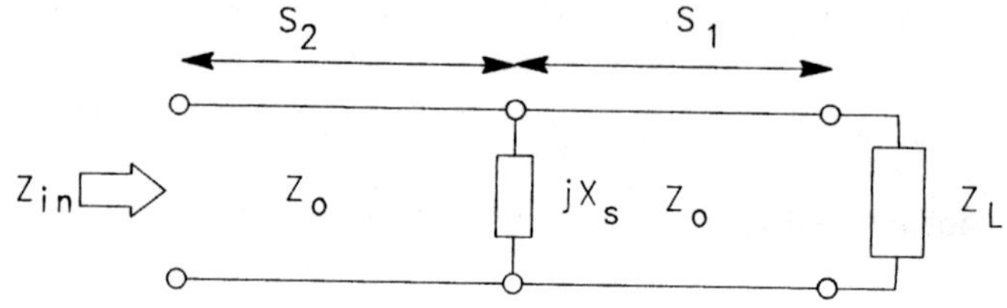

$S_1 = .16\lambda_g$
$S_2 = .09\lambda_g$
$X_s = 1.0$
$Z_L = 0.5 - j0.4$

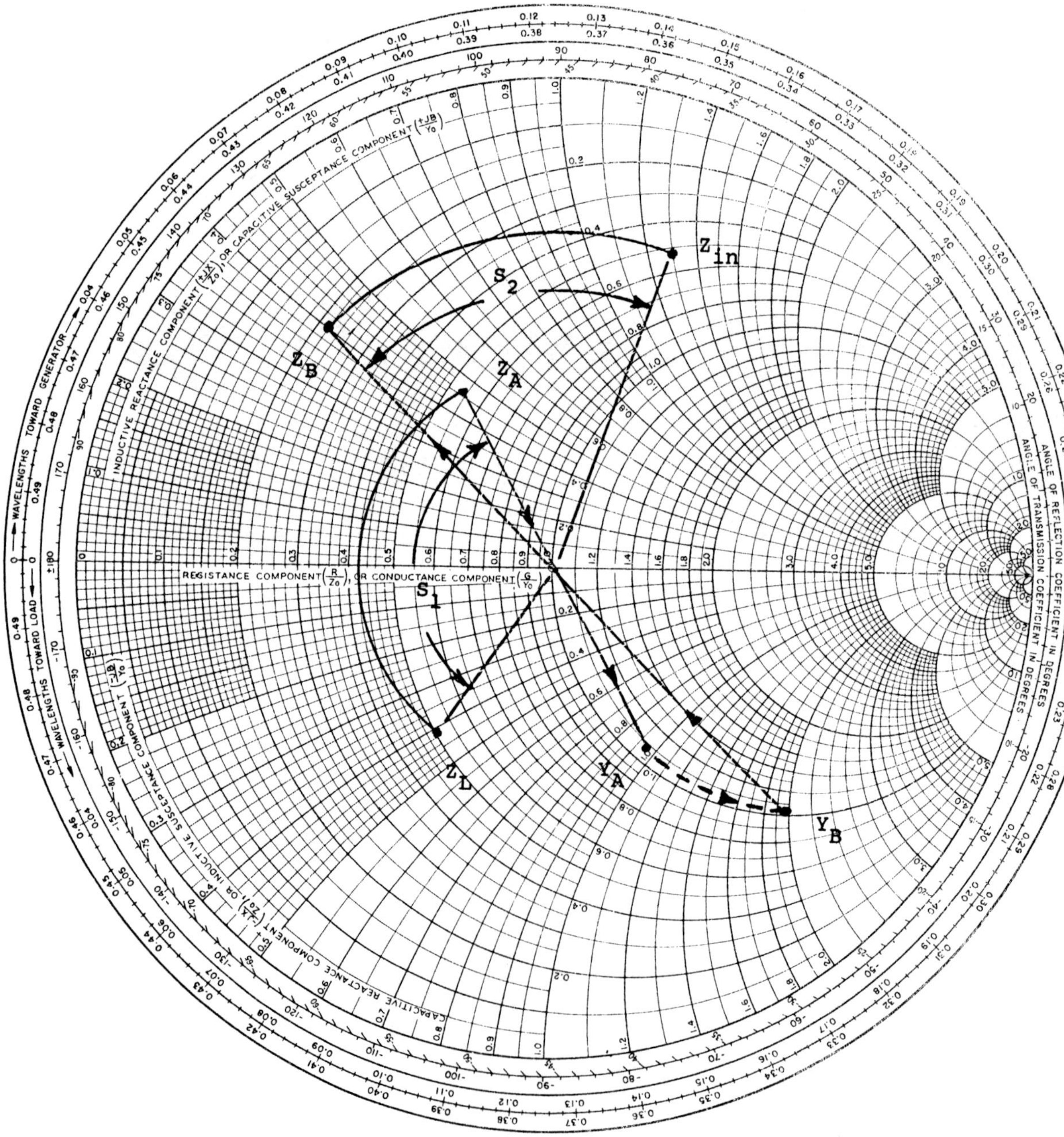

Figure 7.14 Using admittance for connected circuits in parallel.

Therefore, we must subtract a susceptance of 1.0 from the admittance at Y_A as shown and transform back to impedance, which is shown as Z_B. This represents the impedance of the inductor in parallel with the transmission line circuit to the right of the inductor.

Now we can rotate to the input of the circuit through the length $s_2\lambda_g$ to get the input impedance,

$$Z = 0.52 + j1.28$$

The process of converting from Z_A to Y_A and back again is facilitated by the Smith chart, but is still bothersome. This is particularly true if several lengths of transmission line and parallel circuit elements are to be analyzed. A variation of the Smith chart which simplifies the transformation to admittance is called a *Z/Y Smith chart,* shown in Fig. 7.15. To convert impedance to admittance, we

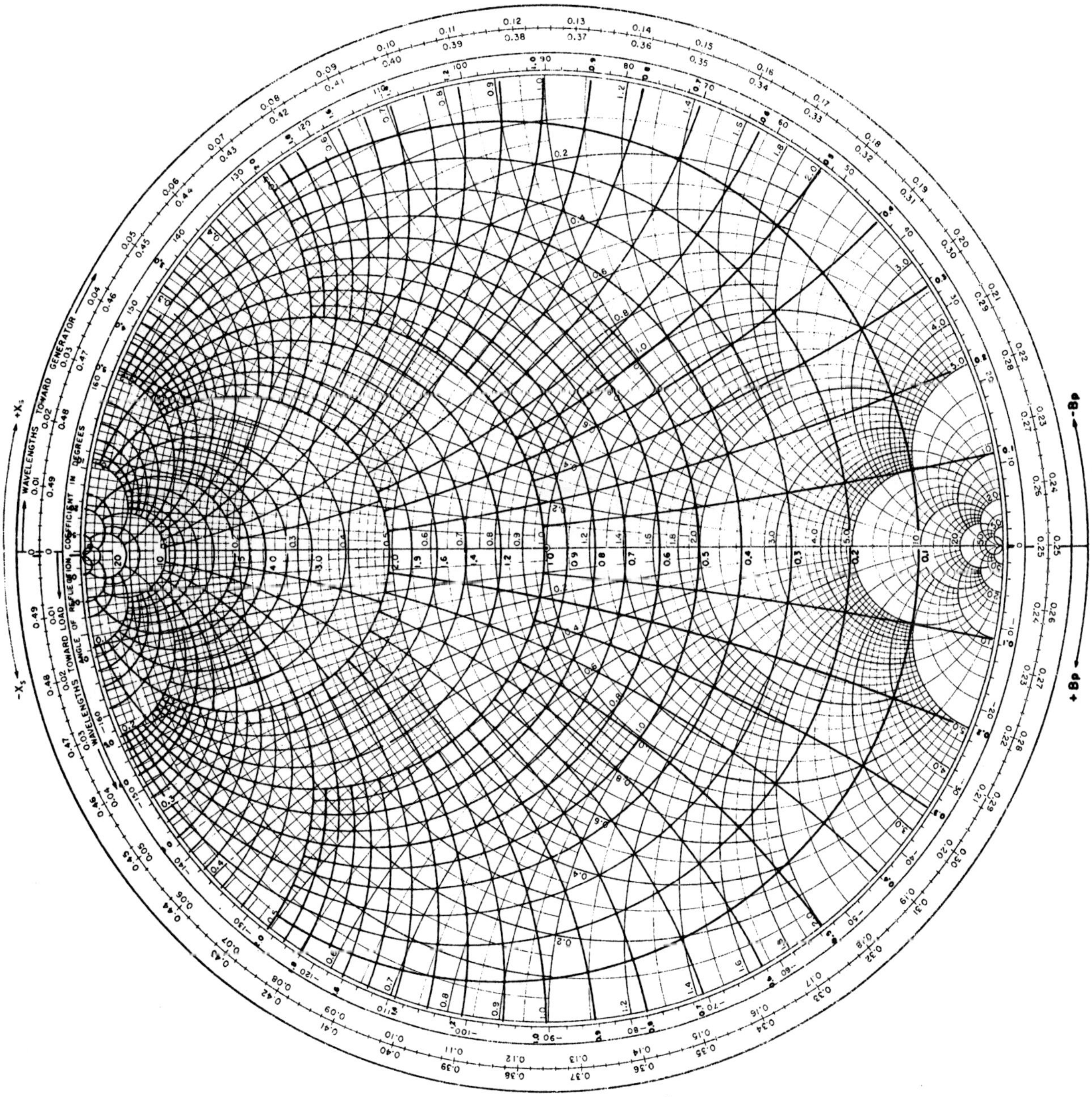

Figure 7.15 The Z-Y chart.

rotate the impedance point 180 degrees, use the new position as admittance, and then rotate back. We could easily imagine a variation in which the pencil is held fixed and the Smith chart is rotated 180 degrees underneath the writing instrument, bringing the correct "admittance" coordinates into play. We would rotate the chart back into "impedance" position, again holding the pencil point motionless.

This procedure can be followed by simply drawing two Smith charts, onto the same sheet of paper, one rotated by 180 degrees, using different colors to separate the two sets of coordinates. The circuit in Fig. 7.14 can now be analyzed more efficiently by transforming from Z_L to Z_A using impedance coordinates (black lines), adding the *negative* inductive susceptance using the admittance coordinates (red lines), and then switching back to impedance coordinates (black lines) for the rest of the solution. With a little practice, an experienced user can easily design transmission line circuits with parallel circuit elements, without cluttering the Smith chart with excessive operations.

CHAPTER 8

IMPEDANCE MATCHING

Lawrence A. Stark

8.1 INTRODUCTION

In an ideal microwave system, all the components would have impedances equal to the characteristic impedance of the transmission lines. This would ensure that no reflected waves would exist at any junction between transmission lines and components. Such a system would be "matched," a shortened version of the phrase "impedance matched."

The importance of a matched system is that it eliminates three potential problems which arise when reflected waves are present. These are:

- Transmission ripple *versus* frequency
- Reduced signal level
- Greater measurement error

Transmission ripple: When reflected waves are present they can reinforce each other or partially cancel each other. Which occurance actually takes place depends on the relative phase of these waves, which in turn depends on the frequency and the path length of the signals. Over a band of frequencies, there will be alternate reinforcement and cancellation resulting in a fluctuation of the transmitted power. This fluctuation is generally called "ripple" and can adversely affect the operation of many types of microwave systems, particularly communications systems.

Reduced signal level: A common problem found in microwave active devices such as diodes and transistors is that they present such a high reflection that the useful signal power in the systems is seriously reduced. This is particularly evident in "active" components such as amplifiers and mixers.

Measurement error: A serious problem in making accurate power measurements on any microwave device is the "mismatch uncertainty" of the measurement. This refers to a potential error caused by the same effect that causes ripple (see above). Basically, reflected waves combine with each other in random ways, obscuring the power in the "original" incident wave that we wish to measure. Thus, the measured power is the desired signal plus an unknown component composed of reflections and re-reflections.

The term "matching" refers to techniques of circuit design or modification that eliminate unwanted reflected signals. Transmission line theory tells us that this is accomplished by setting the impedance of the load equal to Z_0, the characteristic impedance of the transmission line in the system.

A key parameter of a matching circuit is the bandwidth over which the matching is effective because many systems must operate over a broad frequency range. The bandwidth of a circuit is often expressed as a percentage where

$$\text{bandwidth (\%)} = 100 \times (f_u - f_l)/f_0 \tag{8.1.1}$$

where

f_u = the upper frequency of operation,
f_l = the lower frequency of operation,

and

f_0 = the center frequency.

For example, if the operating frequency range lay in the range from 10.7 to 11.2 GHz, then f_0 would be 10.95 GHz and the bandwidth would be

$$\text{bandwidth (\%)} = \frac{0.5}{10.95} \times 100$$

or 4.57% bandwidth.

A variety of matching techniques have been developed, but they fit into two basic categories. The first is called "lossless matching" which gets its name from the use of circuit elements which themselves do not absorb any microwave energy. The advantage of this technique is obvious; namely, no microwave power is unnecessarily lost to the matching circuit. The disadvantage is that these techniques tend to result in circuits which are only matched over a relatively narrow band of frequencies, for example, less than 10%. To achieve wider bandwidths, it is generally necessary to resort to the digital computer and use CAD (*computer aided design*) programs. The power of these programs and their general availability has resulted in the development of ultra-broadband circuits with bandwidths from 2–20 GHz in some instances.

The other method of designing matching circuits is simple enough that it can achieve very broadband results without the need for computer design, but it uses "lossy" matching elements. Thus, power is lost which results in degraded system performance. However, these lossy circuits are usually quite easy to design and can operate over exceptional bandwidths, approaching the theoretical limit of dc to the highest frequency of operation, which is a bandwidth of 200%.

One area where lossy matching is frequently used is in the design of broadband diode detectors. The disadvantage of lower sensitivity is compensated by the exceptionally wide frequency range achieved. Figure 8.1 shows a typical circuit with a detector diode shunted with a 50 Ω matching resistor. At normal operating signal levels, the diode is a high impedance, so the effect of the resistor is to provide a well-controlled 50 Ω input impedance, independent of the impedance of the diode. The capacitor provides an RF short to ground to complete the circuit.

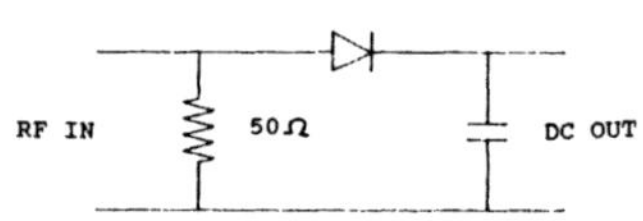

Figure 8.1 Resistive matching of a diode detector.

Matching is not a technique unique to microwave circuits. In fact, impedance matching at lower frequencies using transformers and LC ladder networks can be analyzed quite easily on the Smith chart. The primary difference found at microwave frequencies is the use of transmission lines as part of the matching structure.

8.2 BASIC MATCHING TECHNIQUES

The basic technique for matching circuits is best illustrated with an example. We will match an impedance of $Z = 25 - j75\ \Omega$ to a 50 Ω characteristic impedance transmission line. The first step is to locate the impedance to be matched on a

Smith chart. Normalized to 50 Ω the impedance is

$$Z_L = 0.5 - j1.5 \tag{8.2.1}$$

The objective is to add lossless circuit elements, either in series or parallel, to "move" the load impedance to the center of the Smith chart. Remember that the center of the Smith chart represents an impedance of exactly 50 Ω, which is a reflection coefficient of zero. By "moving" the impedance, we mean that the total impedance of the original load plus the matching components will be at a different location on the Smith chart than the original.

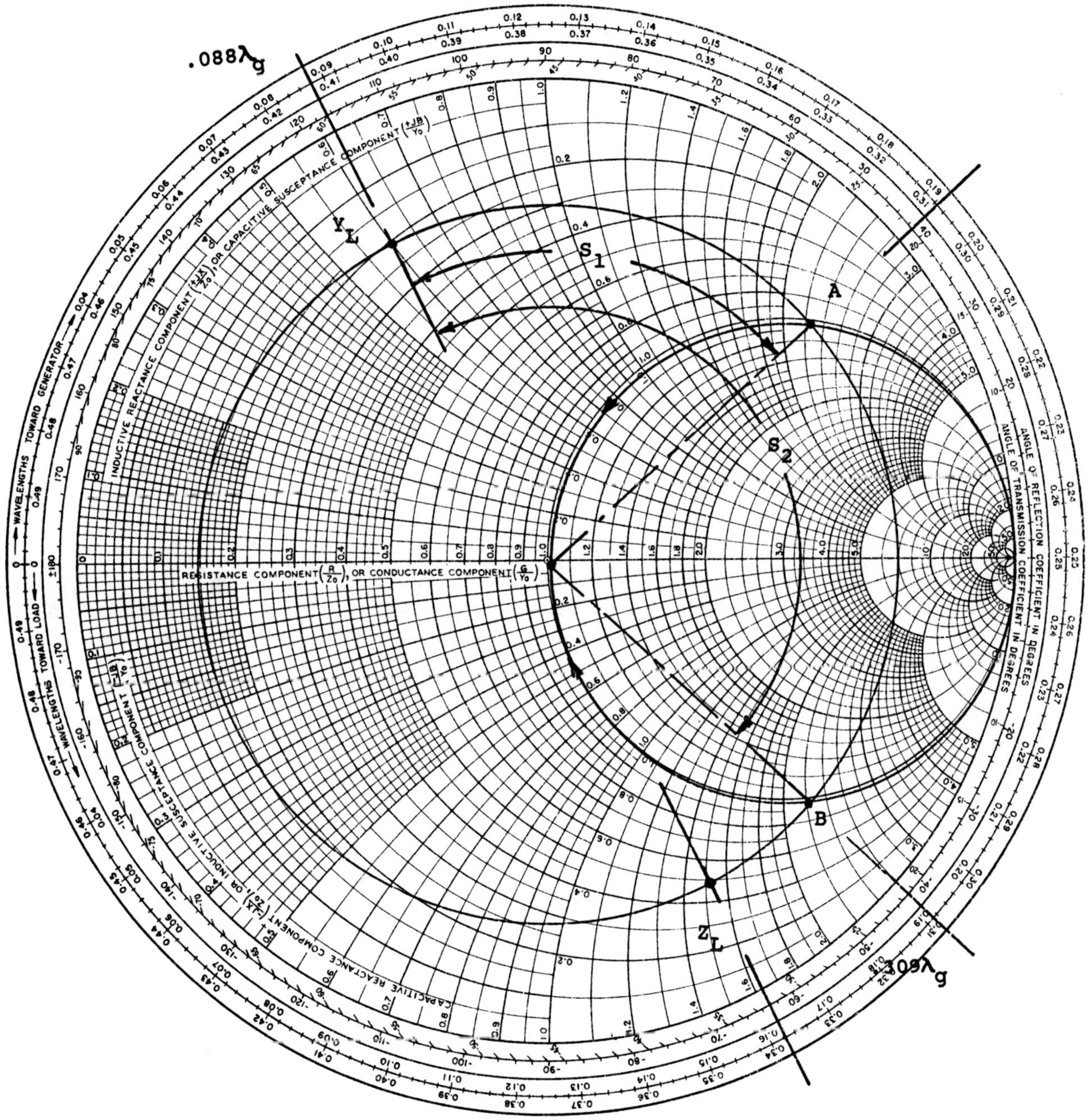

Figure 8.2 Shunt reactive matching of an impedance Z_L. Point A represents the location of a shunt inductive match, while point B represents a shunt capacitive match.

SEC. 8.2
BASIC MATCHING TECHNIQUES

We will see that there are two points on the line where a parallel inductor or capacitor of the right value will transform the load impedance to exactly 50Ω.

Consider the Smith chart in Fig. 8.2. The impedance and the admittance of the load are shown as Z_L and Y_L, respectively. We need to work with the admittance because we will be adding a parallel tuning element across the transmission line to match the load. The complete circle which passes through the points Z_L and Y_L is the constant VSWR circle which describes transmission line admittance at any point on the line. If we move away from the load by a distance S_1 to the point A as shown, we will have

$$Y_A = 1.0 + j2.23 \tag{8.2.2}$$

If we now add a parallel inductor at this point with a susceptance of

$$B_p = -2.23$$

we will cancel the capacitive susceptance of the line admittance at point A and be left with

$$Y_A' = 1.0 + j0 \tag{8.2.3}$$

which is the center of the Smith chart.

Similarly, if we move away from the load by a distance S_2 to the point B on Fig. 8.2, we obtain a line impedance of

$$Y_B = 1.0 - j2.23 \tag{8.2.4}$$

which can be cancelled by adding a parallel capacitive susceptance of

$$B_p = 2.23$$

Either approach yields the same result, which is a matched line. Figure 8.3 shows the two alternatives schematically. The choice of whether to add the tuning element at point A or B usually depends on whether you wish to add a capacitive or inductive susceptance onto the transmission line. In many practical circuits, capacitive tuning elements are easier to fabricate, so the point B is often chosen for practical considerations.

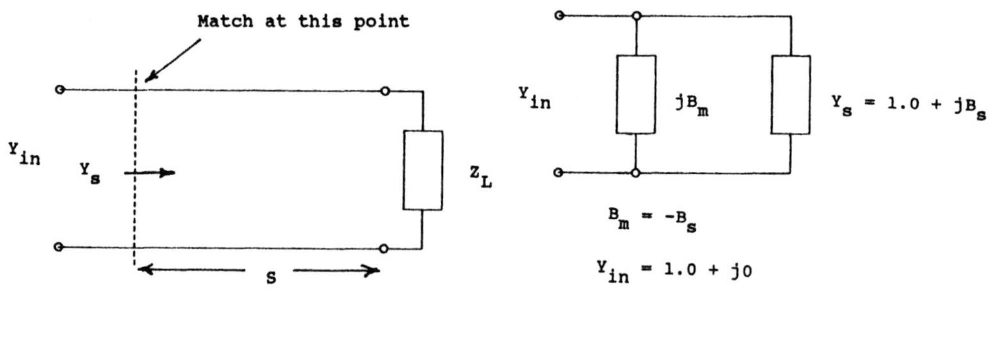

Figure 8.3 The condition for shunt impedance matching. When S is chosen correctly, using the procedure in Fig. 8-2, the line admittance Y_S will have a conductance of 1.0.

Example 8.2.1

Determine the line lengths and tuning element values of the matching example in Fig. 8.2, given that $Z_0 = 50\ \Omega$, $\lambda_g = 3.4$cm and the operating frequency is $f =$ 8.82 GHz.

We see from the Smith chart wavelength scales that the distances S_1 and S_2 are determined in wavelengths as

$$S_1 = .192 - .088 = .104\lambda_g = .354\text{cm}$$

and

$$S_2 = .309 - .088 = .221\lambda_g = .751\text{cm}$$

after multiplying by the guide wavelength. If we elect to match at point A, we need an inductive susceptance of $-2.24Y_0$ where

$$Y_0 = 1/Z_0$$

or

$$B_p(\text{inductive}) = 2.24 \times .02 = 44.8\text{mS}$$

The susceptance of an inductor is

$$B = (1/\omega L)$$

where $\omega = 2\pi f$. Thus,

$$L = (1/\omega B)$$

or

$$L = (1/6.24 \times 8.82 \times 10^9 \times .0448)$$

Calculating this value gives us the required valuc for the tuning inductor, which is

$$L = .406\text{nH}$$

For a capacitive tuning element at point B, we must calculate the value of the capacitor from the formula for capacitive susceptance,

$$B = \omega C$$

Solving for C and substituting the values of B and ω yields

$$C = 2.24/(6.28 \times 8.82 \times 10^9)$$

or

$$C = 40.4\text{pF}$$

To summarize this matching technique, along a mismatched transmission line, the normalized conductance fluctuates above and below its matched value of 1.0. If you add the correct value of parallel susceptance at points where the conductance equals 1.0, you can cancel the line susceptance and leave a total line admittance equal to $Y=1.0+j0$. The Smith chart is invaluable for determining the distance from the load to the proper points for adding these tuning elements, and for determining the required value of the tuning elements themselves.

8.3 PRACTICAL MATCHING TECHNIQUES

The example in the previous section showed the fundamental principle of the simplest type of impedance matching. The advantage of this type of matching is that it is simple and can be used to match practically any impedance value. This primary disadvantage is that this type of matching is inherently narrowband. This means that the frequency range over which the return loss of the circuit is acceptable (let us say, greater than 25 dB) will only be a few percent of the center frequency.

8.3.1 Practical Tuning Elements

Before we consider several commonly used matching techniques in more detail, we should look at common techniques for realizing the capacitive and inductive circuit elements that are commonly used in matching techniques. Briefly, we will use lengths of short-circuited and open-circuited transmission lines when we want inductive and capactive tuning elements, respectively.

Figure 8.4 shows the input impedance of a length of short-circuited transmission line. The input resistance is zero and the reactance is given by

$$X_{in} = Z_0 \tan(360S/\lambda_g) \qquad (8.3.1)$$

where

Z_0 = the characteristic impedance,
S = the line length,
λ_g = the guide wavelength.

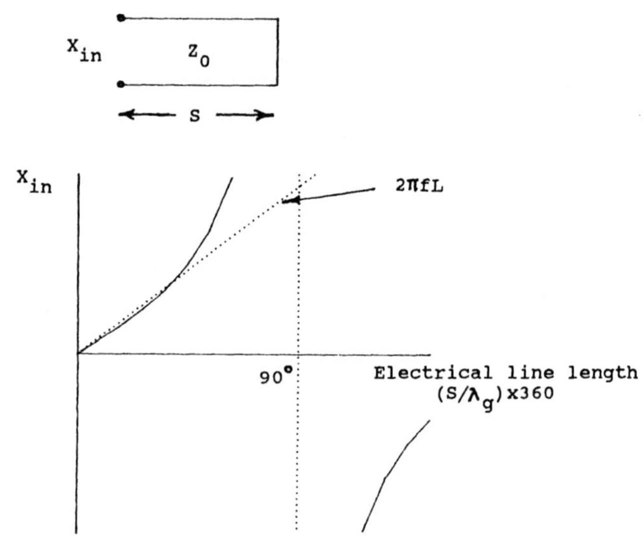

Figure 8.4 The reactance of a short circuited transmission line (solid line) versus electrical line length. At a line length of 90 degrees, the reactance becomes infinite. The reactance of an inductor which approximates the transmission line is shown for comparison (dotted line).

The graph shows that the reactance is positive for line lengths less than 90°, so the shorted line has the properties of an inductor as long as the transmission line is less than a quarter-wavelength long at the operating frequency. This is equivalent to requiring that the frequency satisfy the requirement

$$f < (v_p/4S) \tag{8.3.2}$$

where v_p is the phase velocity on the line and S is the length of the line.

As shown in Fig. 8.5, a similar use can be made of open-circuited lines where the input reactance is negative when the line is less than a quarter-wavelength long. This line looks like a capacitor as long as Eq. (8.3.2) holds. Practical considerations limit the useful frequency range to about half the value shown in Eq. (8.3.2) for both examples.

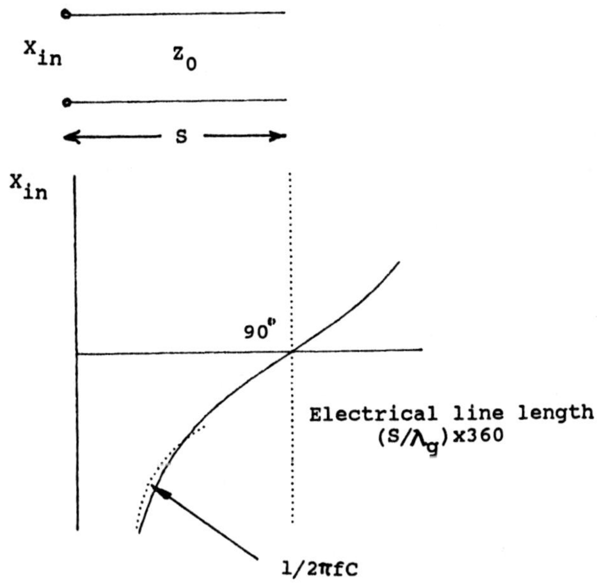

Figure 8.5 The reactance of an open circuited transmission line (solid line) versus electrical line length. At a line length of 90 degrees, the reactance becomes zero. The reactance of a capacitor which approximates the transmission line is shown for comparison (dotted line).

Another frequent application of transmission lines to realize either inductors or capacitors depends on the characteristic impedance value for the line. The characteristic impedance of a transmission line is given by

$$Z_0 = \sqrt{L/C} \tag{8.3.3}$$

where L is the series inductance per unit length on the transmission line and C is the parallel capacitance per unit length. We see that a transmission line with a low characteristic impedance must have a high value of C, whereas a line with a high characteristic impedance must have a high value of L. Thus, high impedance lines find application as series inductors and low impedance lines can be substituted for shunt capacitors.

This is a useful technique provided that the length of the transmission line is less than a quarter wavelength. In fact, a good rule-of-thumb is to keep the line length less than one-eighth wavelength to prevent the transmission line effects from being noticeable.

8.3.2 Lumped Tuning Elements

The reason for finding transmission line replacements for conventional inductors and capacitors is that these components lose their useful properties at several hundred MHz due to self-resonant effects. The inductance of a capacitor, although small, becomes noticeable at these high frequencies. The effect is like a series LC resonant circuit, and the capacitor will become inductive at high frequencies. Similar effects occur in an inductor where the minute capacitance between the turns of the coil eventually resonate with the inductance to form a parallel LC resonant circuit.

The frequencies at which these resonances occur are difficult to predict exactly due to simple variations in the manufacturing process of these circuit elements. Therefore, we must operate safely below the resonant frequencies in order to be able to accurately predict the effects of *L*'s and *C*'s in high frequency circuits.

Some effort has been spent at miniaturizing these elements so that they can be used above 1 GHz. The most success has been achieved with capacitors. Miniature capacitors, called "chip" capacitors, are available that are useful at frequencies as high as 20 GHz. These are made from very low loss dielectric material, only a few thousandths of an inch thick.

Similar efforts with inductors have yielded "printed circuit" inductors which are small enough to push the self-resonant frequency well into the microwave range. These efforts have made it possible to use lumped circuit elements for designing tuning and matching circuits at the operating frequencies of many important microwave systems.

The above efforts notwithstanding, most matching circuits use transmission lines to realize the capacitive and inductive effects which are the foundation of impedance-matching circuits. The principle reason is that such circuit elements are easy to construct as part of microstrip and other forms of hybrid MIC circuits. Using a lumped circuit element requires a manual operation of mounting and connecting the chip capacitor or printed circuit inductor, which raises the price of the component. For the most demanding applications, however, lumped capacitors and inductors will provide an extra margin of performance when the broadest bandwidths are desired.

8.4 COMMERCIALLY AVAILABLE MATCHING CIRCUITS

The situation often arises that a matching circuit cannot be built into the load itself, for example, in an instrumentation set-up where the component to be matched is a complete unit which cannot be disassembled. Then, commercially available tuners are useful. These tuners are available in waveguide and coaxial design.

A waveguide tuner is called a slide-screw tuner, as shown in Fig. 8.6. A post extending into the waveguide through a narrow slot at the top simulates a shunt inductor at that point. By moving the post to the proper location and adjusting the penetration of the post, a wide range of mismatches can be matched. Usually trial and error is sufficient to tune out the undesired reflections.

Another type of waveguide tuner is the E-H tuner, shown in Fig. 8.7. Two mutually perpendicular waveguide arms are fastened together across a rectangular waveguide, and moveable short circuits are installed in the arms. By moving the sliding shorts in the perpendicular arms, we can tune out any reflection from the main arm connected to the load.

In coaxial circuits, there are also two types of matching components. The double-stub tuner, shown in Fig. 8.8, is a useful circuit for matching a wide range of impedances; but it cannot provide a match for all load impedances. To overcome this limitation, a third stub can be added, resulting in the triple-stub tuner, which can match to any load impedance on the Smith chart.

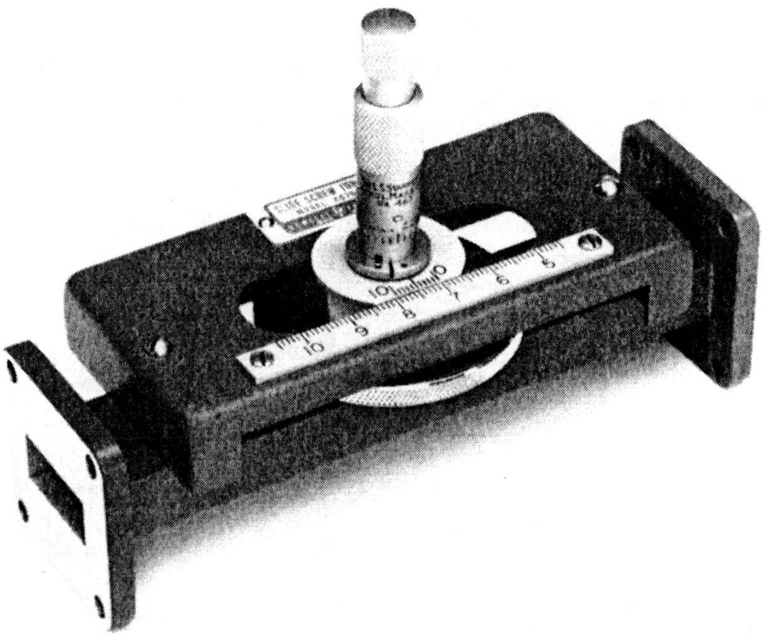

Figure 8.6 *A waveguide slide-screw tuner. The position and the depth of the probe can be varied. Photo courtesy of Hewlett Packard Company.*

Figure 8.7 *A waveguide EH tuner. By varying the positions of the short circuits in the orthogonal arms, any impedance can be matched. Photo courtesy of M/A-Com.*

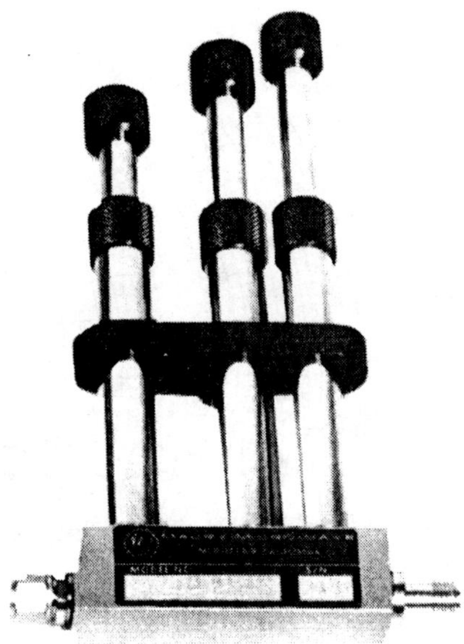

Figure 8.8 *A coaxial double-stub tuner. The length of the stubs can be varied independently and locked into position. Photo courtesy of Maury Microwave.*

The second type of matching circuit is similar to the waveguide slide-screw tuner. A picture of one in Fig. 8.9 shows that two adjustable screws are used to tune for the magnitude of the VSWR to be tuned, and the phase of the VSWR pattern is tuned by moving the probes along the line.

SEC. 8.5
OTHER MATCHING TECHNIQUES

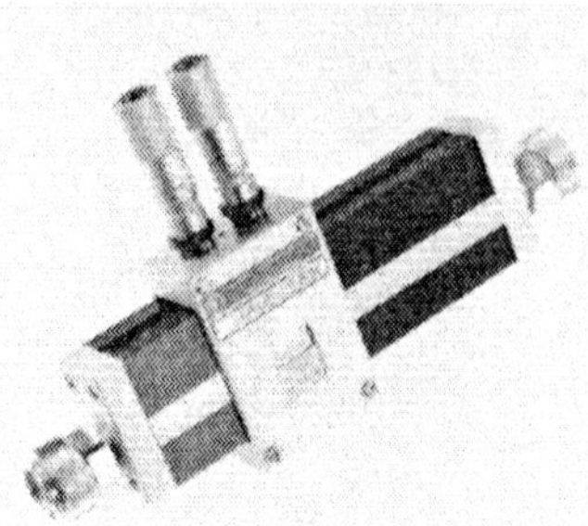

Figure 8.9 A coaxial double slide screw tuner. The depth and relative positions of both probes can be varied. Photo courtesy of Maury Microwave.

8.5 OTHER MATCHING TECHNIQUES

When designers have the option of designing the matching circuit into the circuit itself, they can avail themselves of techniques which achieve matching over a wider bandwidth than the simple technique presented in Sec. 8.2.

8.5.1 The Quarter-Wave Transformer

As the name implies, quarter-wave matching uses the properties of a length of transmission line which is one-quarter wavelength long at the frequency of operation. The characteristic impedance of this section of line must be chosen to match the load impedance.

Figure 8.10A shows how a quarter-wave transformer should be analyzed. As described in Ch. 7, a length of transmission line one-quarter wavelength long

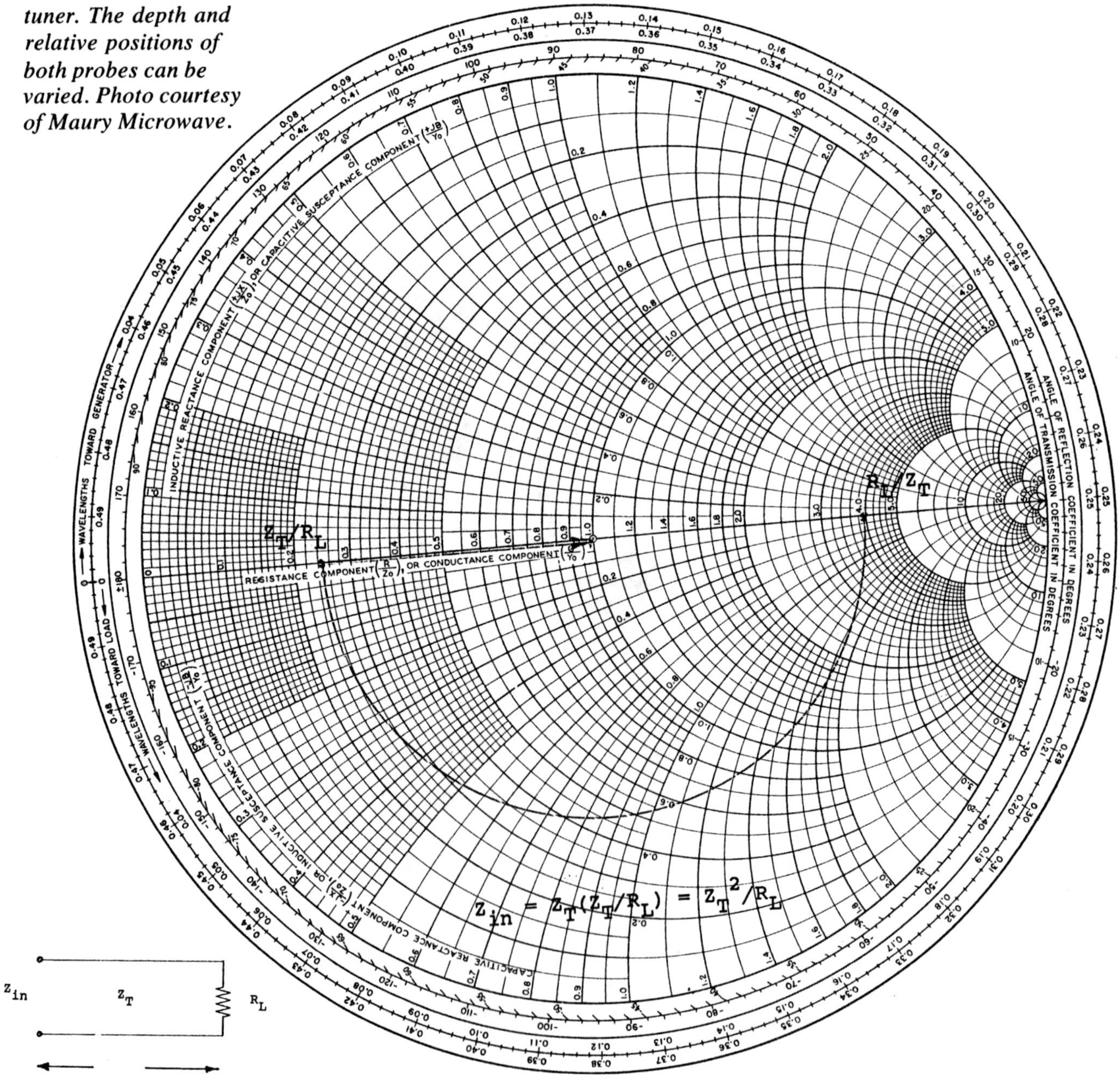

Figure 8.10 Quarter-wave transformer matching for a resistive load impedance.

transforms an impedance Z_A into the impedance $1/Z_A$. These values are, of course, normalized. Therefore, the normalized input impedance to the quarter-wavelength of line in Fig. 8.10A is

$$Z_{in} = Z_T/R_L \qquad \text{(normalized to } Z_t\text{)} \qquad (8.5.1)$$

Multiplying by Z_T to un-normalize, we obtain

$$Z_{in} = Z_T^2/R_L\Omega \qquad (8.5.2)$$

We can choose the value of Z_T to make Z_{in} equal to Z_0 by making

$$Z_T = \sqrt{Z_0 R_L}\ \Omega \qquad (8.5.3)$$

or, expressed in another form

$$Z_T = \sqrt{(R_L/Z_0)} \qquad \text{(normalized to } Z_0\text{)} \qquad (8.5.4)$$

For example, if $R_L = 30\ \Omega$ then we can match the load to a 50 Ω transmission line by choosing

$$Z_T = \sqrt{50 \times 30}$$

or

$$Z_T = 38.7\ \Omega$$

Suppose the load to be matched is not purely resistive. A short length of transmission line can be added before the quarter-wave transformer to make the load resistive, and the impedance of the transformer is chosen as before in Eq. (8.5.3). The proper length of line to be added can be determined from the Smith chart (Fig. 8.11). As shown in the present example, two lengths of line are possible. Choosing a length

$$S_1 = .16\lambda_g$$

results in a value of

$$R_L = .187 \qquad \text{(normalized)}$$

which would be matched with a quarter-wave impedance of

$$Z_T = \sqrt{.187} = .432 \qquad \text{(normalized)} \qquad (8.5.5)$$

If a length of

$$S_2 = .41\lambda_g$$

is added, which is S_1 plus another quarter-wavelength of line, the load impedance will be at point B, which would require a transformer impedance of

$$Z_T = \sqrt{5.346} = 2.31 \qquad \text{(normalized)} \qquad (8.5.6)$$

SEC. 8.5

OTHER MATCHING TECHNIQUES

To find the transformer impedances in ohms, we should multiply the values obtained above by the value of Z_0 for the system. The choice of whether to add a length S_1 or S_2 would rest on practical circuit fabrication issues of whether it was more desirable to use a higher or lower value of the transformer impedance, Z_T.

Also, although it is beyond the scope of this chapter to explain, matching circuits with longer lengths of line (such as S_2) result in smaller operating bandwidths. Still, quarter-wave transformers have bandwidths which are larger than the shunt susceptance types of tuners, although typically still less than 5%.

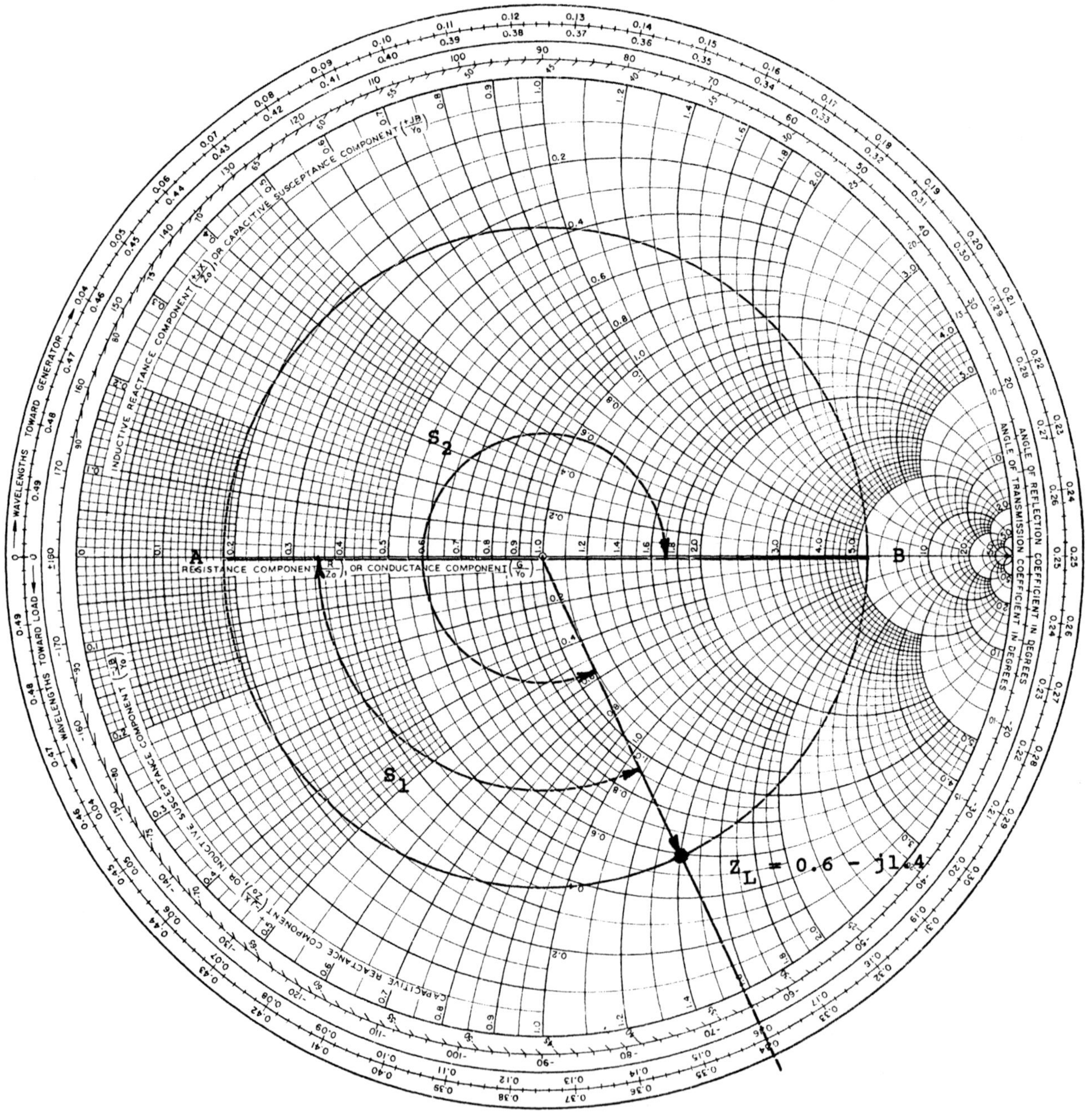

Figure 8.11 Quarter-wave transformer matching for an arbitrary load impedance, Z_L. Add a length of line which will transform Z_L into a resistive load, then apply the quarter-wave transformer.

8.5.2 The Stepped-Impedance Transformer

A method of increasing the bandwidth of the quarter-wave matching transformer is to use a series of quarter-wavelength lines with impedances which are stepped between the value of the load and Z_0. As the number of transformer elements is increased, the impedance values of the individual elements become closer together. Eventually, we would approach another type of matching structure which is the tapered line. A tapered line uses a transmission line matching section which simply varies the characteristic impedance smoothly from Z_0 to R_L over a distance of several guide wavelengths. By making the transition quite gradual, the reflections are kept negligibly small, resulting in a well matched load.

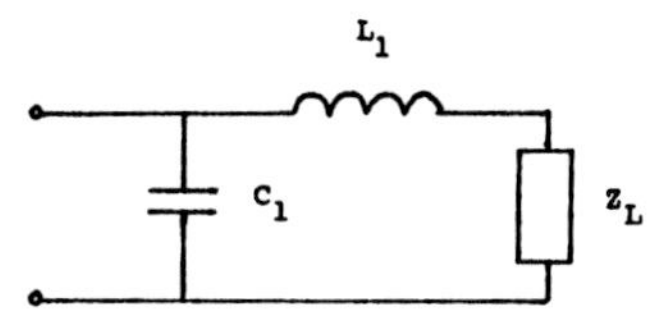

Figure 8.12 A low-pass prototype impedance-matching circuit using lumped circuit elements.

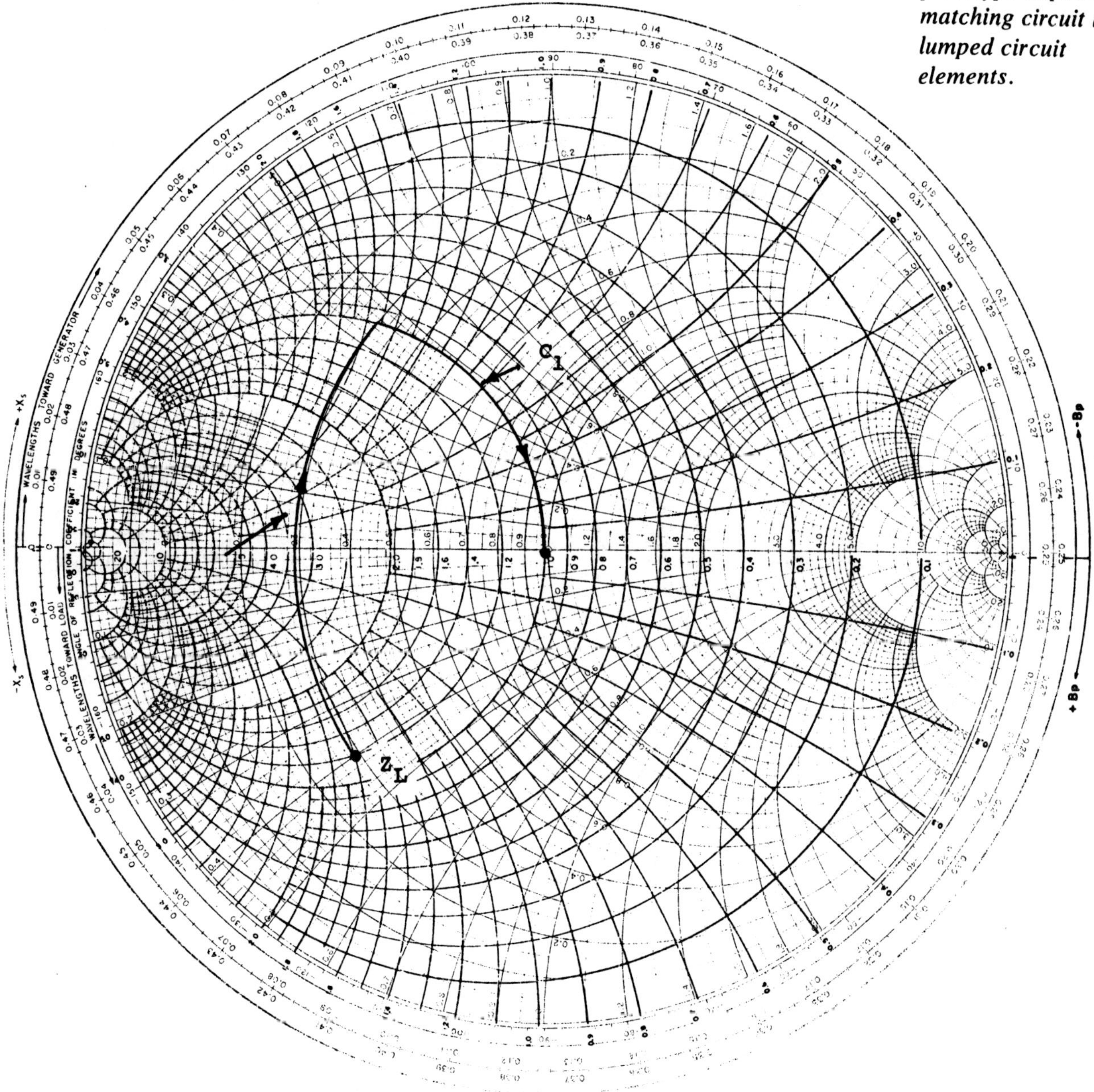

Figure 8.13 A low-pass impedance-matching example; series inductor followed by shunt capacitor.

SEC. 8.5

OTHER MATCHING TECHNIQUES

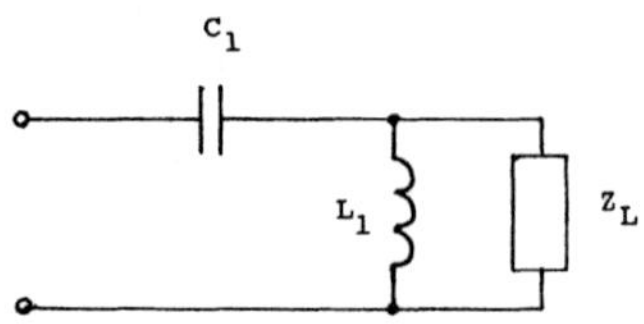

Figure 8.14 A high-pass prototype impedance-matching circuit using lumped circuit elements.

8.5.3 Low-Pass and High-Pass Matching Circuits

For broader band applications, a matching structure which uses the structure of a low-pass filter can be used. The low frequency equivalent circuit of this matching circuit, referred to as the "prototype" circuit, is shown in Fig. 8.12. Circuits of this type are effective for impedance matching over bandwidths as large as 100%. The basic principle of this type of circuit is shown in the impedance/admittance Smith chart in Fig. 8.13. The combination of series inductances and parallel capacitances can effectively transform any impedance to the center of the chart.

Another effective matching circuit is based on the structure of a high-pass filter as shown in Fig. 8.14. The impedance transformation works the same way as the low-pass structure in Fig. 8.12. The alternating series *C*'s and shunt *L*'s successively transform the initial load impedance towards the center of the Smith chart.

Both of these circuits are often applied to the design of GaAs FET amplifiers. A fundamental problem in the design of these amplifier circuits is matching the input of the FET to the ouput of the previous stage. Low-pass and high-pass matching circuits, sometimes combined together, are implemented using all of the techniques described above. Series *L*'s and shunt *C*'s are usually built using high and low impedance transmission lines, respectively. Series capacitors are implemented using chip capacitors, and shunt inductors are built with short circuited lengths of transmission line. Because these circuits are almost always built on microstrip transmission lines, these circuit elements are relatively easy to construct.

8.5.4 Computer Aided Design

No discussion of impedance-matching techniques would be complete without a discussion of *computer aided design,* or CAD. The most difficult impedance-matching challenges arise in the design of broadband multistage amplifiers. Here, the design of a suitable matching circuit becomes too difficult for calculation by hand. The availability of several different software packages, which compute and even optimize the required matching circuits, has made possible broadband amplifier designs which surely would not have been possible otherwise. The programs are not omnipotent! They require the supervision of an informed designer to make intelligent choices at various points along the way. Nevertheless, without these programs, many of the sophisticated microwave circuits which are described regularly in the trade literature would not have been developed.

CHAPTER 9

NOISE

W. Stephen Cheung and Robert Owens

9.1 INTRODUCTION

Noise owes its original description to the audible cracking sound accompanying the expected sound from a speaker. In a broad sense, noise is either disturbance from nature or any undesirable signal that may degrade the performance of a particular system.

Noise falls into two categories: natural and man-made. Lightning, cosmic rays, sun spots, and ambient temperature are examples of natural noise with the *thermal noise* due to the ambient temperature being the most important. Turning an electrical switch on and off, operating rotating machinery (drill, lathe, et cetera), and atomic and nuclear explosions are all examples of man-made noise. The study of noise would not be absolutely essential if its scope were only confined to the above examples. Most of the noise sources mentioned above can be greatly reduced by proper shielding and circuit layout.

Telecommunications involves the transmission of electromagnetic waves through the atmosphere or outer space. The signal at the receiver is usually weak and contaminated by noise of various kinds from the environment in which the receiver is located. The same thing happens to other types of microwave signals, such as radars, intelligence gathering systems, et cetera. In most cases, thermal noise introduced in the receiver is the dominant noise source and places a limit on the resolution of a detector.

The received signal must be amplified, filtered, and processed so that the receiver end can extract the information carried by the wave. Each component in the receiver injects its own noise as it processes the signal. The study of environmental noise and device noise is therefore crucial in telecommunications.

The definition of noise is somewhat subjective. This can be illustrated by the fact that our atmosphere is filled with man-made electromagnetic waves from radio, television, and other communication institutions. These signals all find their way, in various magnitude, to our receiver. If we are only interested in AM station A, waves within the bandwidth of station A comprise our signal and all others outside of the bandwith become noise. Because no filter is perfect, the unwanted waves can never be totally eliminated. If the signal strength of station A is not strong, but only comparable to, or even weaker than, the residual unwanted waves, we will not have "clean" music from the speaker no matter how much we turn up the gain (volume).

This chapter describes a few noise sources and quantities pertinent to the description of noise. These quantities are useful in the following chapters.

9.2 THERMAL NOISE

The energy source of thermal noise is the temperature of the immediate environment in which the object of concern is immersed. The temperature must be expressed in absolute unit, i.e., K (degrees Kelvin), and the conversion between

Celsius and Kelvin is

$$\mathrm{T\ (K)} = \mathrm{T\ (C)} + 273 \tag{9.2.1}$$

Therefore, the absolute zero is 273 C *below* the freezing temperature of water (which is defined as 0 C). At 0 K, all objects cease to move and such "frozen" state has not been realized in the laboratory.

When the absolute temperature of the environment is at some value other than zero, all objects enclosed in the environment will inherit a certain amount of *thermal* energy. The amount of thermal energy is proportional to the absolute temperature of the environment; in other words,

$$\text{Thermal energy} = k\,T \tag{9.2.2}$$

where k is the proportionality constant known as the Boltzman constant. The Boltzman constant k takes on either one of the following values depending on the unit used.

$$k = 1.38 \times 10^{-23} \text{ Watt-second/K} \tag{9.2.3A}$$

$$= 1.38 \times 10^{-11} \text{ pW-second/K} \tag{9.2.3B}$$

The thermal energy of all objects, e.g. electrons, atoms, *et cetera*, is responsible for the functioning of resistors, semiconductors, and many other devices. It is unfortunately also a source of noise which ultimately limits the detection of weak signals.

Thermal energy manifests itself in an object as *random* physical motions such as translation, rotation, and vibration. It also manifests itself as radiation (electromagnetic wave) filled within the environment. The frequency of the radiation theoretically spans across the entire spectrum, from very low to very high frequency, due to the random nature of thermal energy.

A typical illustration of thermal noise is that produced by a 50 ohm resistor. A wire is connected across the resistor to form a complete loop (Fig. 9.1). Because of the thermal energy of the electrons in the resistor, there will be a random motion of electrons through the loop. A sensitive ammeter would register this motion as a random current. Because this current is not a steady flow and, in fact, must average over time to be zero, it is an alternating current. Thus, thermal noise is a weak ac signal of random frequency.

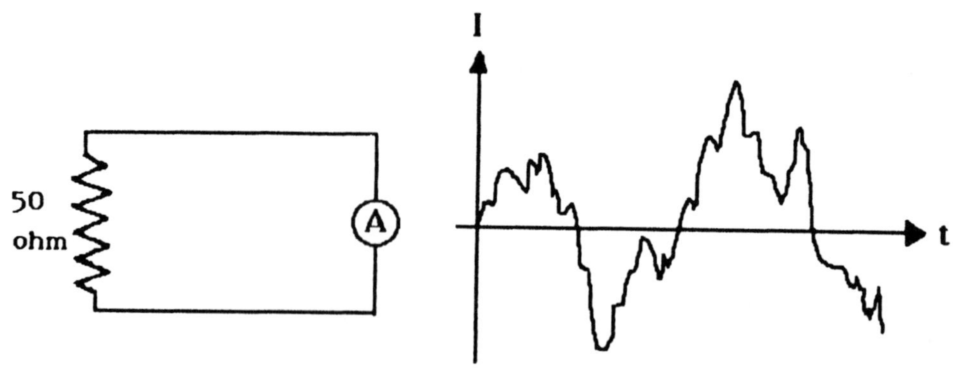

Figure 9.1 The thermal noise current observed when a standard 50 ohm resistor is short circuited.

Now consider a battery in place to drive a direct current through a resistor (Fig. 9.2A). Theoretically, the direction of dc is always one-sided. However, the thermal energy derived from the environment results in an ac on top of dc (Fig. 9.2B).

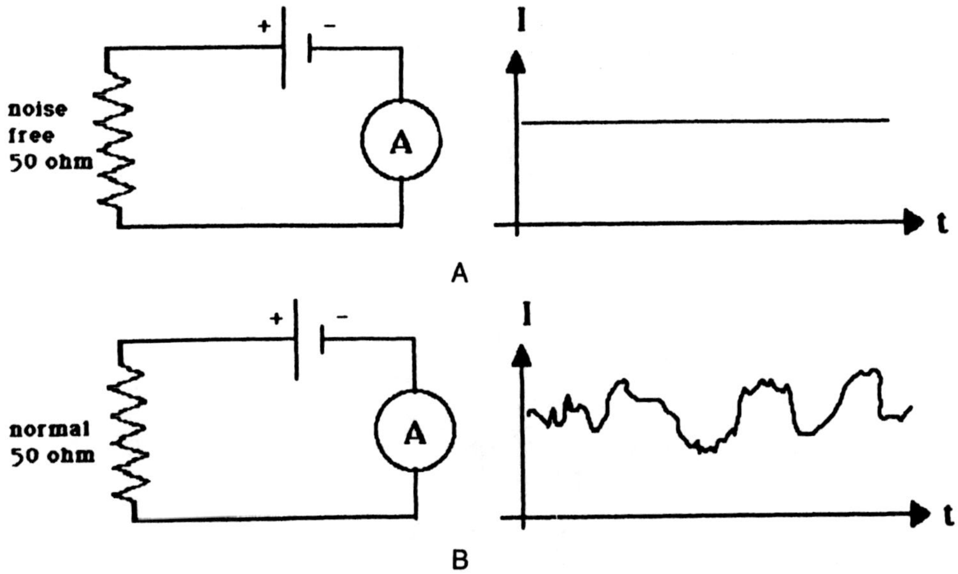

Figure 9.2 When connected across a battery, the current observed is (A) pure dc if the resistor is noise free, (B) ac on top of dc if the resistor is normal.

Similarly, if the signal source is not a battery but an ac generator, the actual current flow is a combination of the theoretical alternate current and the random thermal current (Fig. 9.3A, B). It is easy to see that if the theoretical signal is about the same magnitude as that of the thermal current, the signal current is greatly distorted by the thermal current.

In the above example the thermal current is undesirable, and is therefore termed thermal noise. The magnitude of the thermal noise is directly proportional to the absolute temperature of the device. Another important factor is the bandwidth B, which is explained as follows. As mentioned, the randomness of the thermal energy results in radiation covering the entire frequency spectrum. Usually, an electronic system operates over a range of frequency called bandwidth. As an example, an AM station has a bandwidth of 40 kHz. If the carrier frequency is 1000 kHz, then the station's signal spans from 980 kHz to 1020 kHz. In this case, the thermal noise within this frequency range will contaminate the signal of interest.

In general, the noise power perturbing an electronic system, i.e., the amount of thermal noise energy manifested within a period of time, is given by

$$NP \text{ (Thermal) Noise power } = kTB \tag{9.2.4}$$

For practical applications, noise power is expressed in pW (picowatt) and B in MHz. As an example, room temperature is usually taken as 25 C, or 298 K. The noise power over a bandwith of 1 MHz is

$$\begin{aligned} NP &= 1.38 \times 10^{-11} \times 298 \times 10^{6} \quad \text{pW} \\ &= 4.1 \times 10^{-3} \quad \text{pW} \end{aligned}$$

A popular term known as *noise density* is defined as the noise power divided by the bandwidth. Our chosen unit for noise density *ND* is pW/MHz.

$$\begin{aligned} \text{Noise Density} &= \text{Noise Power/Bandwidth} \\ ND\ (\text{pW/MHz}) &= NP\ (\text{pW})/B(\text{MHz}) \end{aligned} \tag{9.2.5}$$

At room temperature (298 K), the noise density is 4.1×10^{-3} pW/MHz.

Noise power and noise power density can be expressed as dBm and dBm/MHz. Hence at 298 K and 1 MHz bandwidth,

$$\begin{aligned} NP \text{ at } 298\text{ K} &= 4.1 \times 10^{-3} \text{ pW} \\ &= -114 \text{ dBm} \end{aligned}$$

and

$$ND \text{ at } 298\text{ K} = -114 \text{ dBm/MHz}$$

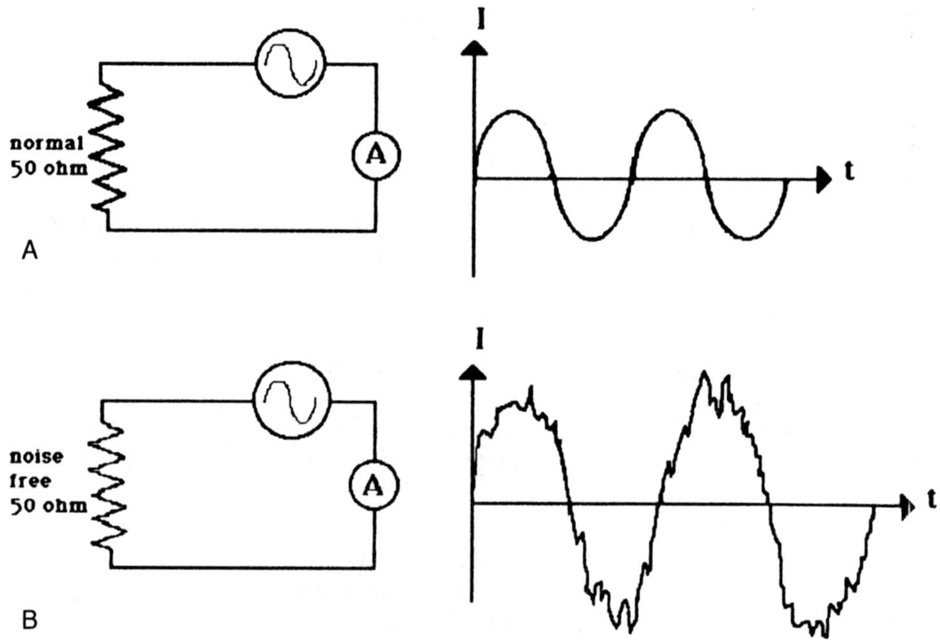

Figure 9.3 When connected across a signal source, the current observed is (A) pure ac if the resistor is noise free, (B) noisy ac if the resistor is normal.

9.3 SHOT NOISE

We shall briefly discuss the origin of shot noise, but mathematical details will not be covered. Shot noise occurs as random fluctuations of electron emissions from the cathodes of vacuum tubes, or across a potential barrier of devices such as diodes and transistors.

Consider the cathode of a vacuum tube where electrons will be released, when heated by high current. These electrons are subsequently accelerated toward the anode which is at a higher potential. The release of electrons from the cathode surface, however, is not uniform. Certainly, the average number of electrons released is directly proportional to the direct current in the cathode because this direct current heats up the cathode, and therefore determines the number of electrons released. However, the actual number of electrons released at an arbitrary moment may be more or less than the average, or expected, number.

Such fluctuations constitute the physical basis of shot noise.

The above description for a vacuum tube can be easily generalized to a diode. Shot noise is found to be totally random and its frequency spans the entire spectrum.

9.4 WHITE NOISE AND PINK NOISE

White noise refers to totally random noise where the noise powers of all frequencies are equal, while the noise power of pink noise decreases as frequency increases. Such terminology arose because white light passing through a prism was discovered to decompose into various colors, i.e., various frequencies. Similarly, pink color has lower frequency, or larger wavelength, than other colors such as yellow, green, and blue.

9.5 SIGNAL-TO-NOISE RATIO (SNR)

Suppose a beam of electromagnetic wave containing information is received by an antenna. Let us assume that the carrier frequency of the wave is 10 GHz with its information spread over a bandwidth of 1 MHz. The strength of this wave is determined by its power which we shall call P_s (for signal power). We have already seen that any environment at absolute temperature T is contaminated with thermal and other noise. Let this power be P_n (for noise power).

The ratio of signal power to noise power over the bandwidth of concern is called the signal to noise ratio, *SNR*, i.e.,

$$SNR \text{ (Signal to Noise Ratio)} = \frac{\text{Signal Power}}{\text{Noise Power}} = \frac{P_s}{P_n} \tag{9.5.1}$$

As an example, suppose the signal power and the noise power are determined to be 1pW and 10^{-3} pW, respectively. Then,

$$SNR = 1\text{pW}/10^{-3}\text{ pW} = 1000$$

In other words, the signal strength is 1000 times stronger than the noise within the bandwidth.

Because *SNR* is the ratio of two power values, it can be expressed in dB.

$$SNR \text{ (dB)} = P_s \text{ (dBm)} - P_n \text{ (dBm)} \tag{9.5.2}$$

The above example, therefore, can be computed as

$$SNR \text{ (dB)} = -90 \text{ dBm} - (-120 \text{ dBm}) = 30 \text{ dB}$$

It is obvious that the higher the *SNR*, the smaller the signal distortion due to noise. As a simple rule, an *SNR* of 100, or 20 dB, is the margin.

9.6 NOISE TEMPERATURE

Our environment is filled with noise. Thermal noise is just one of the many noise sources. Turning a light switch on and off generates electromagnetic radiation, electronic equipment constantly radiates; the examples are numerous.

Although the noise in a receiver may come from many types of noise sources, it is common practice to describe the resulting noise in terms of thermal noise having the equivalent power. To do this, we introduce the term *noise temperature*, or T_n. An example will illustrate this.

Suppose the absolute temperature inside a laboratory is determined to be 298 K and the noise power over a standard 1 MHz bandwidth is measured to be 1pW. The following calculations can be made:

SEC. 9.6

NOISE TEMPERATURE

$$\text{Noise power over 1 MHz due to 298 K alone} = k \times T \times 1 \text{ MHz}$$
$$= 4.1 \times 10^{-3} \text{ pW}$$

Clearly, thermal noise alone cannot account for the measured 1pW noise power. Instead of trying to account for all the noise sources, we can ask an alternate question: What equivalent temperature T_n would result in the thermal noise of 1pW over this 1 MHz bandwidth?

$$k \times T_n \times 10^6 = 1\text{pW}$$
$$T_n = 7.25 \times 10^4 \text{ K!}$$
$$= 72{,}207 \text{ C!}$$

Therefore, if thermal noise were the sole source of the 1pW noise, the environment would be at 7.25×10^4 K. This temperature is known as the noise temperature of the environment, *not* its true temperature.

Noise temperature gives the designer a rough idea whether thermal noise is his ultimate limit. If the noise temperature of an environment is found to be 350 K and the true temperature is 298 K, it is then clear that thermal noise is the dominant source. Measure can then be taken to reduce the noise by reducing the true temperature via cooling. On the other hand, if the noise temperature of an environment is found to be 50,000 K and the true temperature of the environment is 298 K, then thermal noise is not the dominant noise source.

A common phenomenon occurs in a receiving dish antenna. If the antenna is pointed toward the sky and the noise density is measured, the noise is found to be different from the measured value when the antenna is pointed toward Earth (see Fig. 9.4A). We can easily see that the environmental temperature of the antenna is constant in both cases, and is not the source of problems. In fact, the noise detected when the antenna is pointed toward Earth is mostly due to thermal radiation. Comparatively, the antenna is much more receptive to noise from the atmosphere and space when it is pointed toward the sky. Figure 9.4B shows the noise densities as a function of frequency for the two cases. The noise temperature when the antenna is pointed toward the sky can be calculated to be about 11,600 K, or 40 times higher than the ambient temperature.

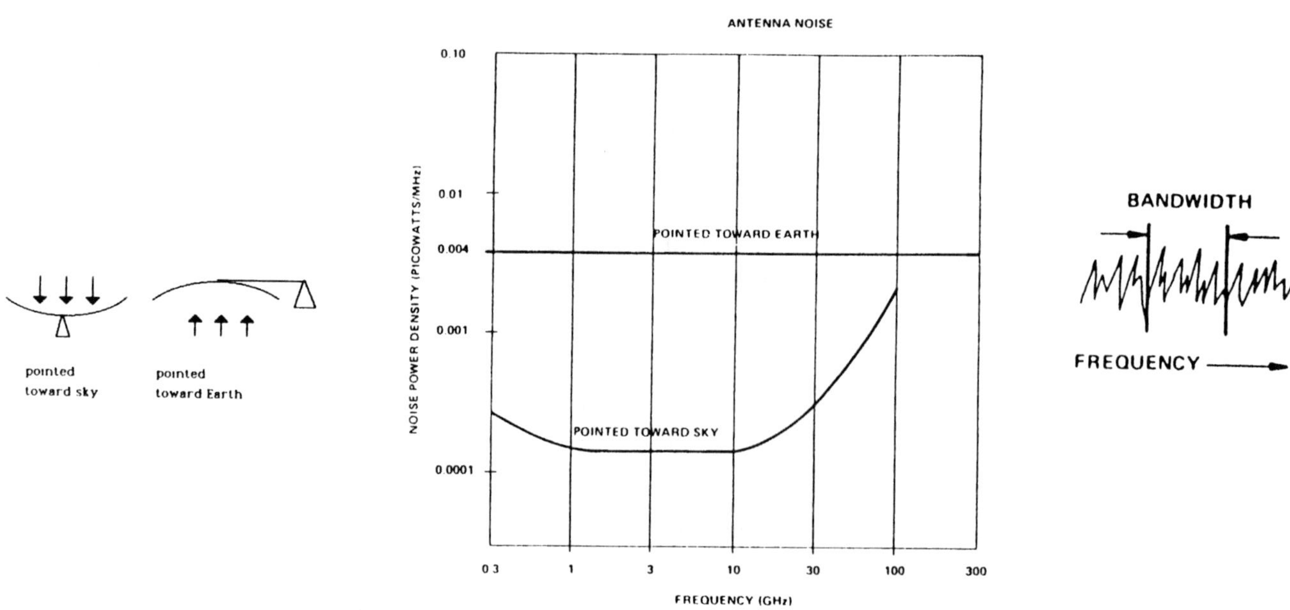

Figure 9.4 A receiving antenna picks up different amounts of noise when it is facing toward the sky and when facing toward Earth.

The utility of the concept of noise temperature can be seen by an example. Figure 9.5 shows an amplifier fed by a source having noise temperature of 290 K.

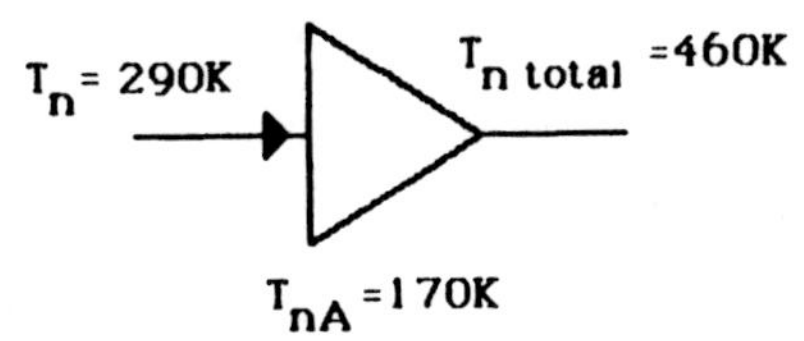

Figure 9.5 When noise equivalent to 290 K is fed to an amplifier of equivalent noise temperature 170 K, the noise output is equivalent to 460 K.

The noise temperature of the amplifier in this example is 170 K. This means that the amplifier adds a noise power (regardless of the noise or signal coming in the input) equivalent to 170 K, or a noise power over a 1 MHz bandwidth equal to $kTB = 1.38 \times 10^{-11} \times 170 \times 10^6$ pW $= 2.35 \times 10^{-3}$ pW, at its input. The total equivalent noise power at the amplifier input can be expressed as a noise temperature by adding the input noise temperature to the amplifier noise temperature: $290 + 170 = 460$ K. Note that this noise temperature, which corresponds to 6.45×10^{-3} pW, is not necessarily all produced at the input but is an equivalent figure, representing noise produced throughout the amplifier. The actual noise output power, on the other hand, can be found by multiplying this equivalent power by the gain of 30 dB, or 6.45pW.

9.7 EFFECT OF AN AMPLIFIER ON SNR

An amplifier (or any other device) cannot distinguish a signal containing desired information from noise as long as both are within its operating bandwidth. Unfortunately, this means that the amplifier cannot selectively amplify the signal and not the noise. For example, an amplifier with gain G will amplify both the signal and the noise by the same factor G. Then, at best, the *SNR* at the output of the amplifier will be no better than the *SNR* at the input.

No device is perfect. The input stage of the device has its own noise. This can be easily seen because, if the input terminals are short-circuited, noise is still generated by the internal resistance of the device. The noise source can be thermal noise, shot noise, et cetera. Therefore, the total input noise is the sum of the noise accompanying the signal and the device noise. During signal processing, the semiconductors, tubes, or other elements involved will inject their own noise. Hence, *the signal-to-noise ratio at the output of the device can only be degraded, not improved.*

9.8 NOISE FIGURE AND EQUIVALENT NOISE TEMPERATURE

From the previous section, we learned that a device degrades the signal to noise ratio by injecting its own noise into the signal processing. A figure of merit can be assigned to the device as a measure of how good the device is as far as noise injection is concerned. Noise figure *NF* is the ratio of *SNR* at the input to the *SNR* at the output of the device.

$$NF \text{ (Noise Figure)} = \frac{SNR\ (in)}{SNR\ (out)} \tag{9.8.1}$$

Expressed in dB,

$$NF \text{ (dB)} = SNR \text{ (in, dB)} - SNR \text{ (out, dB)} \tag{9.8.2}$$

As a numerical example, let the *SNR* at the input of an amplifier be 100 (20 dB)

SEC. 9.8

NOISE FIGURE & EQUIVALENT NOISE TEMPERATURE

and that at the output be 80 (19 dB). Then, the noise figure of the amplifier is

$$\mathrm{NF} = 100/80 = 1.25$$

or

$$NF\ (\mathrm{dB}) = 20\ \mathrm{dB} - 19\ \mathrm{dB} = 1\ \mathrm{dB}$$

The noise figure of an ideal device is 1 or 0 dB. The reader can easily convince himself by putting the input *SNR* equal to the output *SNR*.

In Sec. 9.5, the noise temperature was employed as the equivalent temperature of the environment in order to produce the same amount of thermal noise as measured. This concept can be easily applied to a device. Because an ordinary device injects its own noise to the signal processing, we can ignore the origin of the noise but still find the equivalent temperature of the device. In other words, if thermal noise were the sole source of noise, the device's equivalent temperature is the temperature the device must be at in order to produce the same amount of noise that it injects.

Temperature 290 K is conventionally chosen as the standard operating temperature for most systems. The relationship between the noise figure *NF* and the device equivalent noise temperature T_n is

$$NF = 1 + T_n/290\ \mathrm{K} \tag{9.8.3}$$

or

$$T_n = (NF - 1) \times 290\ \mathrm{K} \tag{9.8.4}$$

If the device were perfect, no noise would be added. Equivalently, the device itself could be regarded as being at absolute zero temperature because absolute zero temperature produces no thermal noise. Hence, the noise figure of the ideal device would be 1.0 and its equivalent noise temperature would be 0 K.

The reader should note that the noise figure *NF* used in Eqs. (9.8.3) and (9.8.4) is in number, *not dB*. We shall continue the example given earlier in this section. The noise figure was calculated to be 1.25. The equivalent noise temperature, according to Eq. (9.8.4) is

$$\begin{aligned} T_n &= (1.25 - 1) \times 290\ \mathrm{K} \\ &= 72.5\ \mathrm{K} \end{aligned}$$

The equivalent amount of thermal noise power generated by the device can also be calculated.

$$P_N = k\ T_n\ B \tag{9.8.5}$$

or

$$P_N = k \times (F - 1) \times 290 \times B \tag{9.8.6}$$

The noise figure of a device, amplifier and otherwise, is usually given by the manufacturer in dB. Noise figure can never be less than 1.0 (0 dB). The closer the noise figure to 1.0, the less noisy (i.e., the better) is the device.

In general, the gain of an amplifier must be traded in order to achieve low noise. The received signal at the detector stage of a system is usually weak and contaminated with noise. It is, however, undesirable to have too high a gain at the front stage because high gain is accompanied by much noise. Therefore, the detector stage must be a low noise amplifier (LNA), i.e., one that has reasonably good gain but still maintains a low noise figure.

In Ch. 4, we learned the input impedance of a receiving load must be equal to, or match, the line's impedance in order to have maximum power transfer. Maximum power transfer to the input of an amplifier means the amplifier will operate at maximum gain. Unfortunately, maximum gain is not accompanied by minimum noise figure. Conversely, minimum noise figure does not usually give maximum gain.

9.9 AMPLIFIERS IN CASCADE

In Sec. 9.8, it was pointed out that the input stage of an amplifier has its own noise. The amplifier regards this noise as part of the received noise and amplifies it. The process of amplification adds more noise.

Referring to Fig. 9.6, two amplifiers are connected in cascade. The input noise P_N *(in)* is the sum of the noise accompanying the signal and the noise of the input stage of amplifier 1. At the output of amplifier 1, the signal output P_s' *(out)* is the signal input P_s *(in)* multiplied by G_1. The noise output P_N' *(out)* is, however, the sum of $G_1 P_N$ *(in)* and the internal noise of the amplifier which is quantified by its noise factor *NF*1. The signal P_s' *(out)* is now the input signal P_s' *(in)* to amplifier 2. However, the noise P_N' *(out)* is only part of P_N' *(in)*, the other part of P_N' *(in)* is due to the input stage of amplifier 2. At the output of amplifier 2, the signal strength P_s *(out)* is P_s' *(in)* multiplied by G_2. The noise output P_N' *(out)* is again

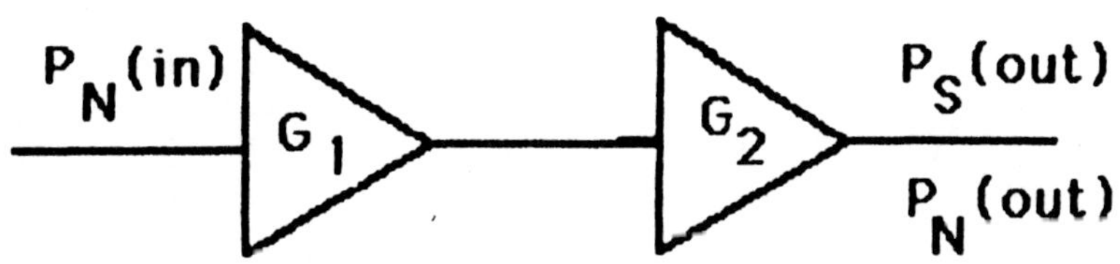

Figure 9.6 When noise is input to one of two amplifiers in cascade, the output noise is a mixture of input noise amplified by the overall gain and the amplifiers' noise.

the sum of P_N' $(in) \times G_2$ and the internal noise of amplifier 2 which is quantified by the noise factor *NF*2.

The overall gain of the amplifier is $G_1 \times G_2$. The signal output is simply P_s $(in) \times G_1 \times G_2$, but the output noise is related to G_1, G_2, *NF*1, and *NF*2. The overall noise factor can be calculated following the argument of the previous paragraph. For two amplifiers,

$$NF \text{ (overall)} = NF1 + \frac{NF2 - 1}{G_1} \tag{9.9.1}$$

Note that the noise figure values in Eq. (9.9.1) are all in numbers, not dB.

Example (9.9.1)

Two amplifiers are connected in cascade. The first amplifier has a noise figure of 10 (10 dB) and a power gain of 50 (17 dB). The second amplifier has a noise figure of 8 (9 dB) and a power gain of 30 (15 dB). Calculate the overall noise figure.

Solution:

$NF1 = 10,\ G_1 = 50$
$NF2 = 8,\ G_2 = 30$

SEC. 9.9
AMPLIFIERS IN CASCADE

The overall noise figure, according to Eq. (9.9.1), is

$$NF\ (\text{overall}) = 10 + \frac{8-1}{50} = 10.14\ (= 10.06\ \text{dB})$$

An observation may be made about Eq. (9.9.1): The gain of the second amplifier appears to be *not* involved. This can be clarified by concentrating on a single stage amplifier.

By definition,

$$NF = \frac{SNR\ (out)}{SNR\ (in)} \tag{9.9.2}$$

$$= \frac{P_s(out)/P_N(out)}{P_s(in)/P_N(in)}$$

$$= \frac{P_s\ (out)}{P_s\ (in)} \cdot \frac{P_N\ (in)}{P_N\ (out)} \tag{9.9.3}$$

Note that the ratio $P_s\ (out)/P_s\ (in)$ is simply the gain of the amplifier. Therefore,

$$NF = G \times P_N(in)/P_N(out) \tag{9.9.4}$$

Hence, the noise figure of an amplifier contains information about its gain.

Returning to Eq. (9.9.1), we can view the overall noise figure as two contributions. The first term on the right-hand side of Eq. (9.9.1) is the contribution from amplifier 1. The second term is contribution from amplifier 2. We see that the second term also contains G_1, the gain of amplifier 1. We can understand this qualitatively whereby the input noise to amplifier 2 is partly related to G_1 and partly due to the input terminals of amplifier 2. If G_1 is large, it means the noise input to amplifier 2 is dominated by the noise coming out of amplifier 1 and not so much by the input terminal of amplifier 2.

A very important design criterion can be inferred from Eq. (9.9.1). If the first amplifier has a relatively low noise figure but a moderately high gain, i.e., a low noise amplifier, the overall noise factor for a long series of devices is mainly due to the first amplifier. The noise figures of the devices following the first amplifier have much less significance. Figure 9.7A and B shows two arrangements of a mixer and an amplifier. In Fig. 9.7A, the overall noise figure is dominated by that of the mixer and is usually high. The amplifier after the mixer has little effect on the overall *SNR*. In Fig. 9.7B, on the other hand, the overall noise figure is dominated by that of the LNA. Even though the mixer has a high noise figure, its contribution is suppressed by the gain of the LNA.

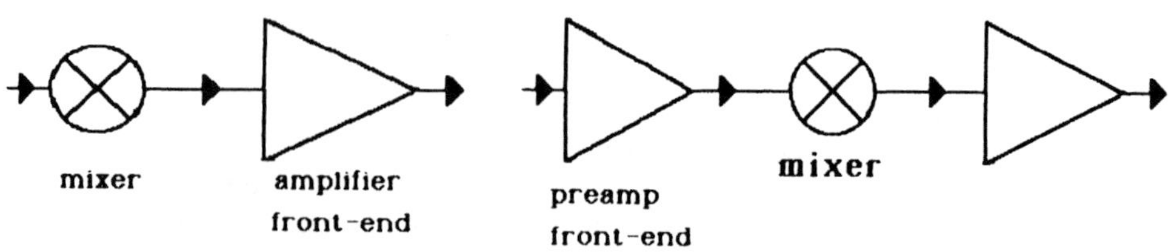

Figure 9.7 When two devices of different noise figures, e.g., an amplifier and a mixer, are in cascade, the overall noise figure depends on whether (A) the mixer is in front, or (B) the amplifier is in front.

The overall noise figure of three or more amplifiers in cascade can be generalized from Eq. (9.9.1) as

$$NF\ (overall) = NF1 + \frac{NF2 - 1}{G_1} + \frac{NF3 - 1}{G_1 G_2} + \ldots \tag{9.9.5}$$

This is known as Friiss' formula.

Friiss' formula can be expressed in terms of equivalent noise temperature. This is easily done by noting the conversion between the noise figure and the equivalent noise temperature as given by Eqs. (9.8.3) and (9.8.4):

$$T_n\ (overall) = T_{n1} + \frac{T_{n2}}{G_1} + \frac{T_{n3}}{G_1 G_2} + \ldots \tag{9.9.6}$$

Again, the equivalent noise temperature of the first stage will dominate if it has high enough gain G_1 to suppress the contributions from the following stages. This intuitively leads to the idea of cooling the first stage, sometimes to cryogenic temperatures (a few degrees K).

9.10 THE SPIRAL CHART

The noise figure and the equivalent noise temperature are related quantities (much like the standing wave ratio and the return loss). The conversion between the two quantities can be quickly read from the spiral chart (Fig. 9.8). Note that the noise figures are in dB.

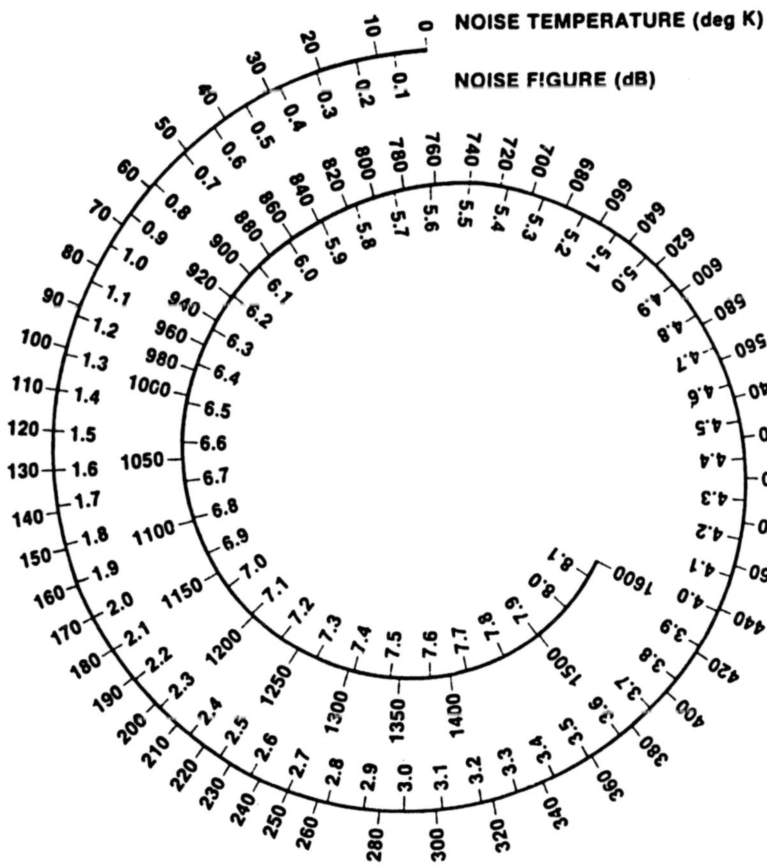

Figure 9.8 A standard spiral chart converting noise temperature and noise figure.

CHAPTER 10

MICROWAVE TUBES

Allan W. Scott and W. Stephen Cheung*

10.1 INTRODUCTION

SEC. 10.1
INTRODUCTION

The major advantage of microwave tubes is their ability to generate large amounts of microwave power. A high density electron beam (generated by various means), when made to pass through a microwave cavity, can deliver a significant portion of its energy to the cavity. External load can then obtain microwave energy from the cavity. The components mentioned above must maintain a vacuum level of at least 10^{-7} torr to avoid excessive collision between the electron beam and air molecules as well as electrical arcing (breakdown) when the cavity has acquired much power.

In order to handle high power, i.e., tens and hundreds of kilowatts, microwave tubes are usually bulky and heavy. Metal parts are employed for appropriate heat conduction, and heavy permanent magnets are often found to be necessary because their effect on the electron beam is usually the principle of operation. Also, much room and weight are occupied by a built-in power supply that can deliver both high voltage (tens of kilovolts) and current (tens of amperes).

Figure 10.1 compares the average power capability of microwave tubes and solid-state devices *versus* frequency. It can be seen that tubes can usually handle at least a thousand times or more power than the solid-state devices at a given frequency. Another important feature to notice is that the power capacity of any device, tube or solid-state, decreases as frequency increases.

As a general rule of thumb, if the power and frequency requirements of a given piece of equipment lie below the solid-state curve, then a solid-state device should be used because it offers the advantage of low weight, small size, low operating voltage, and long life. If, however, the power and frequency requirements lie above the curve, as do most microwave systems, a microwave tube should be used. It is theoretically possible to combine several solid-state sources (IMPATT diodes) to generate the necessary power. Although the combined weight of the solid-state devices is less, extra components must be carefully designed to combine and, therefore, handle the desired power as well as the phase relationships of the devices. The reliability of such an alternative is often questionable.

10.2 TYPES OF TUBES

Several different kinds of microwave tube amplifiers and oscillators are listed by application in Table 10.1. The tube listing in Table 10.1 is grouped according to the type of interaction process, and the interaction process for each class of tube is listed in the right-hand column. This grouping shows the relationship between the types used in amplifiers and oscillators. For example, the cross field

*The contents of this chapter were transcribed and edited by Morris Tischler from the author's (Allan Scott) lecture tapes. Mr. Tischler is with Science Instruments Company, 6122 Reisterstown Road, Baltimore, Maryland 21215. All diagrams and pictures in this chapter are courtesy of the Microwave Training Institute, Mountain View, CA.

SEC. 10.2
TYPES OF TUBES

amplifiers and oscillators, such as the carcinotron, fixed frequency magnetron, coaxial magnetron, and voltage tuned magnetron, all use cross field interaction. The list also shows other types of interaction that are used by amplifiers and oscillators, and how amplifiers and oscillators are related to each other. In design it is necessary to select a tube which may have been produced for use as an amplifier or oscillator.

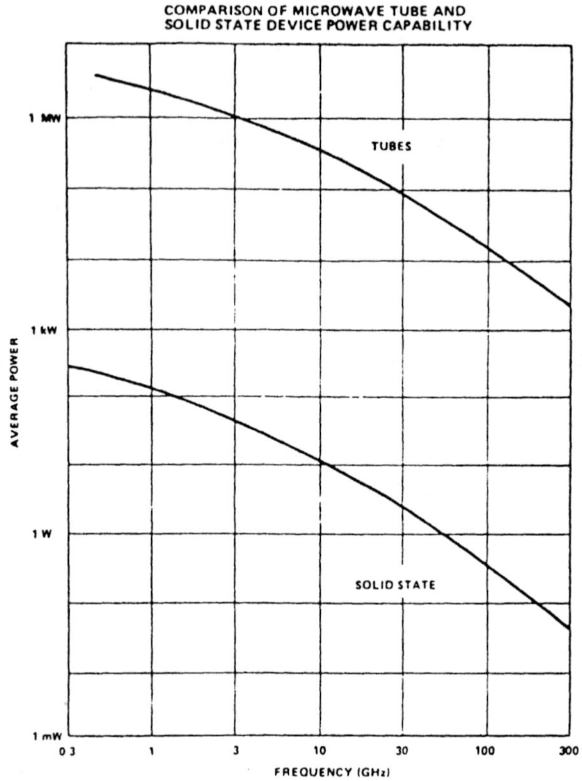

Figure 10.1 Comparison of microwave tube and solid-state device power capability.

Table 10-1
Types of Microwave Tubes

Amplifier	Oscillator	Interaction Process
Gridded Tube	Gridded Tube	Grid Control of Beam Current
Klystron	2-Cavity Oscillator Exended Interaction Oscillator Reflex Klystron	Velocity Modulation with Resonant Cavities
Helix Traveling Wave Tube Coupled Cavity Traveling Wave Tube	Backward Wave Oscillator	Velocity Modulation with Traveling Wave Structure
Crossed-Field Amplifier	Carcinotron Fixed-Frequency Magnetron Coaxial Magnetron Voltage Tuned Magnetron	Crossed Field
Gyrotron	Gyrotron	Spiraling Beam

The average power handling capability of a microwave oscillator tube is considerably less than that of microwave amplifier tubes. In high power equipment, effective results can be achieved by using a low power tube as an oscillator, or using a solid-state oscillator to drive a high power amplifier tube. Furthermore, there is less likelihood of spurious signals being produced when a low power oscillator drives a high power amplifier tube. High power oscillators often generate undesired harmonics. However, if the average power and noise requirements of a high power oscillator tube are adequate, the use of a single oscillator tube will result in a simpler overall system.

10.3 COMPARISON OF DIFFERENT TUBES

The performance characteristics of various types of microwave tubes are now compared. We will find that some tubes are better in one characteristic, while others are better in another characteristic. That is why there are so many different types of microwave tubes available.

First, we will make a comparison of amplifier tubes. Figure 10.2A compares the tubes' average power capabilities, Fig. 10.2B compares their peak power capabilities, and Table 10.2 compares their remaining characteristics. Grid tubes provide both high average power and high peak power, but only at the low frequency end of the microwave band. Klystrons provide the highest peak and average power with over 1 megawatt of average power and up to 100 megawatts of peak power. Above 30 GHz, the newly developed gyrotron provides the highest peak and average power. Coupled-cavity TWTs, helix TWTs, and cross field amplifiers provide lower average and peak power, but other advantages are available. These other advantages, as shown in Table 10.2, include a wider bandwidth, higher efficiency, better gain, lower relative spurious signal level, lower relative operating voltage, and less relative complexity of operation. Although klystrons and gyrotrons provide the highest power, they are limited in bandwidth to only a few percent. In contrast, the helix TWT provides 100 times more bandwidth, but at a lower power efficiency. The coupled-cavity TWT has nearly as good power capability as the klystron and 10 times more bandwidth, but at poorer efficiency. Cross field amplifiers have as good a bandwidth as the coupled-cavity TWTs and the best efficiency of any microwave tube. It is twice as good as any TWT, and even better than klystrons. However, cross field amplifiers have the disadvantage of lower gain, high noise, and greater system complexity.

The performance characteristics of microwave oscillator tubes are compared in Fig. 10.3 and Table 10.3. Most microwave oscillator tubes, with the exception of the magnetron and the gyrotron, operate in the continuous wave (CW) mode. The peak power capabilities of these oscillators must compare with the peak power capability of the microwave amplifier tubes as shown in Fig. 10.2B.

In Fig. 10.3, the average power capability of microwave oscillators is considerably less than that of microwave amplifiers, but well above the solid-state limit curve. The fixed frequency magnetron, the extended interaction klystron, and the gyrotron offer the greatest average power capability, but they are only mechanically tuneable. In contrast, a carcinotron and the voltage tuned magnetron provide lower power but they can be electronically tuned. It should be noted that there is high efficiency with the carcinotron and magnetron types, in some cases, greater than 60%. The two-cavity oscillator is unique in that it provides the lowest noise of any microwave oscillator. The backward wave oscillator and the reflex klystron have power capabilities which are below the solid-state limit curve, and these tubes are being replaced in most new equipment by a solid-state electron oscillator.

All microwave tubes have the basic requirement for a beam of free electrons and high vacuum environment, in which these electrons move inside the cavity or transmission line. The major disadvantage of microwave tubes, as compared to solid-state oscillators and amplifiers, is the need to maintain the vacuum inside

SEC. 10.3

COMPARISON OF DIFFERENT TUBES

the tubes and the need to provide an electron emitting cathode. The tube is always subject to catastrophic failure if the vacuum is lost and the life of the tube is determined by the depletion of the cathode emitting materials. However, a well designed microwave tube in an environment where vacuum integrity is maintained can have an operating life of greater than 100,000 hours. It should also be remembered that microwave tubes can provide four orders of magnitude greater power than the highest powered solid-state devices. In the next section the operation of the tube will be presented. The first tube to be considered is the gridded microwave amplifier tube.

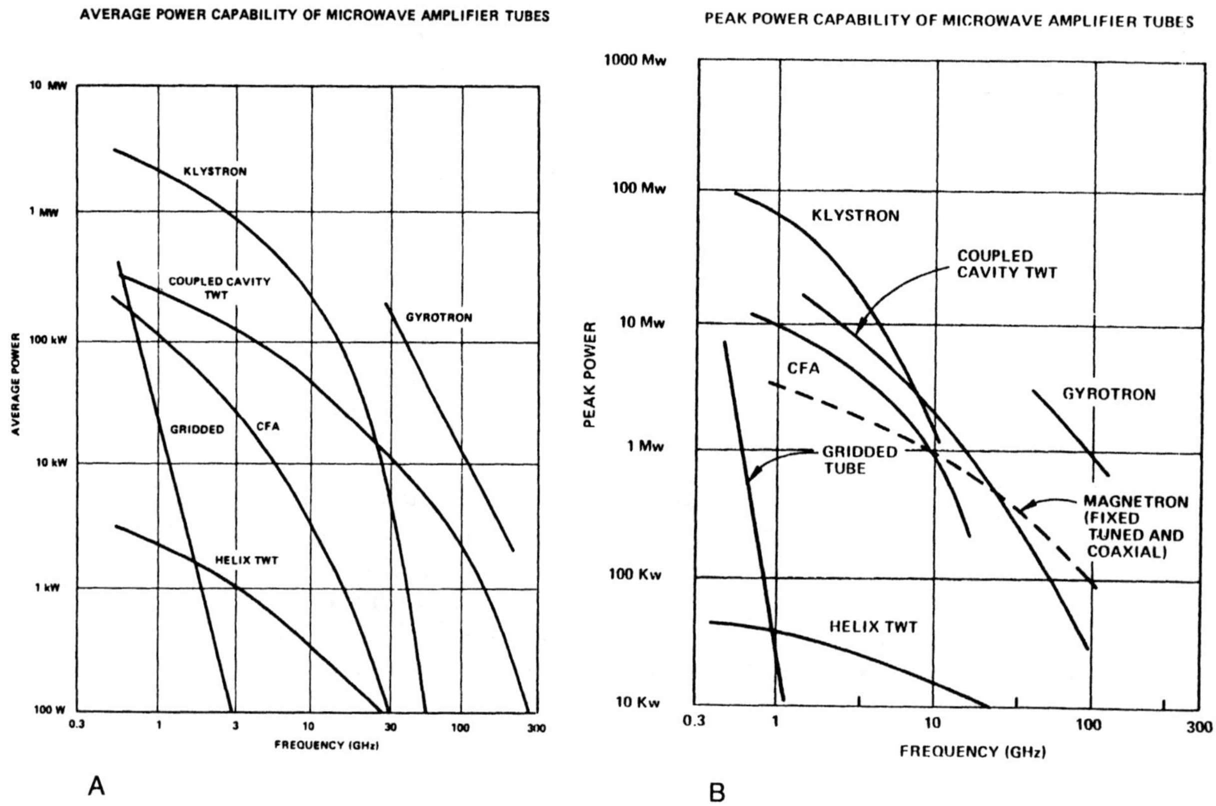

Figure 10.2 Comparison of microwave tube power versus frequency. Average power is shown in (A) and peak power is shown in (B).

Table 10-2
Comparison of Microwave Tubes

Type	Bandwidth (%)	Efficiency (%)	Gain (dB)	Relative Spurious Signal Level*	Relative Operating Voltage	Relative Complexity of Operation*
Gridded Tube	1–10	20–50	6–15	2	low	1
Klystron	1–5	30–70	40–60	1	high	2
Helix TWT	30–120	20–40	30–50	3	high	3
Coupled-Cavity TWT	5–40	20–40	30–50	3	high	3
CFA	5–40	40–80	10–15	5	low	4
Gyrotron	1	20–40	30–40	4	high	5

*1 = best, 5 = poorest

SEC. 10.3

COMPARISON OF DIFFERENT TUBES

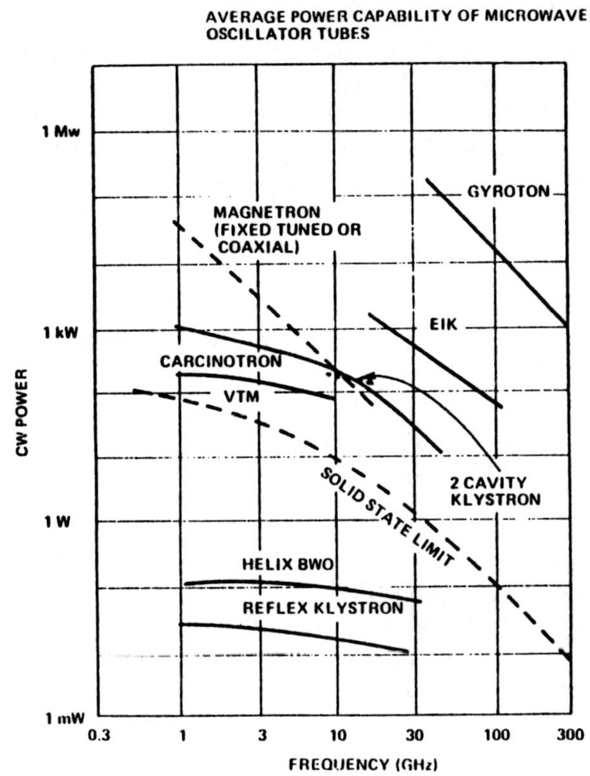

Figure 10.3 Average power capability of microwave oscillator tubes.

Table 10-3
Comparison of Microwave Oscillator Tubes

	Mechanical Tuning Range (%)	Electronic Tuning Range (%)	Efficiency (%)	Relative Spurious Signal Level	Relative Operating Voltage	Relative Complexity of Operation
Gridded Tube	10	—	10–40	2	low	1
Reflex Klystron	10	1	1	3	high	2
2-Cavity Klystron	10	—	10	1	high	3
Extended Interaction Klystron	2	—	10–15	2	high	3
Backward Wave Oscillator	—	30–100	5	4	high	3
Carcinotron	—	30	20–40	5	low	3
Fixed Frequency Magnetron	—	—	40–80	5	low	3
Coaxial Magnetrons	20	—	40–60	5	low	3
Voltage Tuned Magnetron	—	20–30	40–60	6	low	4
Gyrotron	—	—	10–40	2	high	5

SEC. 10.4
GRIDDED TUBE AMPLIFIER

10.4 GRIDDED TUBE AMPLIFIER

In a gridded tube amplifier, the motion of an electron beam (generated by a heated cathode) on its way to the anode is being controlled by an external signal applied to the grid. Consequently, the electron beam delivers a portion of its energy to the surrounding cavity via coupling. A significant term is *transit time*, which is the time required for an electron beam to travel between two points, usually from the cathode to the anode. The transit time ultimately limits the frequency response of the device.

Gridded tubes offer the advantage of high efficiency, small size and weight, low operating voltage, simplicity of operation, and low cost. Their major disadvantage is their limited frequency capability. They are useful only at the low end of the microwave frequency range.

A typical microwave gridded tube is shown in Figures 10.3 A and 10.3 B. The tube consists of two parts, the microwave cavity assembly, and the tube itself. Lead reactance is minimized by mounting the tube inside the microwave cavity. Transit time is minimized by using a very small spacing between the tube electrodes, about 0.2mm. Figure 10.3A shows a cross sectional drawing of the tube and cavity assembly, and Fig. 10.3B shows the tube in the foreground and also the microwave cavity assembly into which the tube fits. The particular tube illustrated is a triode, designed to provide 200 watts of CW power in the 850–870 MHz band, with a gain of 13 dB, i.e., a factor of 20.

The tube is operated with its grid grounded. The input microwave signal to be amplified is brought into the input coaxial cavity through a coaxial connector. The grid and cathode are located at one end of the coaxial input cavity and mounted on concentric metal rings separated by a ceramic cylinder. The inner conductor of the coaxial cavity connects to the cathode by means of one metal ring, and the outer conductor of the coaxial cavity connects to the grid by means of the other metal ring. The input cavity dimensions are chosen so that the coaxial cavity is approximately three-quarters of a wavelength long. When the microwave input power is fed into the cavity and the microwave voltage is applied between the grid and cathode (inside the grid's ceramic envelope), the tube becomes operational. Note that the cathode, grid, catheter grid supports, and ceramic insulator, which forms the vacuum envelope of the tube, also form part of the input cavity.

Most of the microwave cavity is outside the vacuum, however, and need not be replaced each time the tube is replaced, thus reducing overall cost. Heater power is brought in by concentric cylinders located inside the inner conductor of the input coaxial cavity, and so they are shielded from the input microwave signal. The output cavity is a square resonator of approximately one-quarter wavelength long on each side, and is connected between the grounded grid and the tube's plate by the metal rings which support the grid and plate. In the same fashion as the input cavity, grid, plate, grid and plate support, and the ceramic insulator between the grid and plate rings all form part of the microwave output cavity, including two sidewalls which are moveable for tuning. The coupling adjustment between the cavity and external coaxial line, are outside the vacuum envelope of the tube.

Electron current is drawn from the cathode whenever the grid goes positive. However, electrons do not immediately reach the plate because of their finite transit time from the cathode through the grid to the plate. At low frequencies this transit time is an insignificant fraction of an RF cycle, so the plate current is a faithful reproduction of the grid voltage. Consequently, the plate voltage is also a faithful reproduction of the grid voltage. However, at microwave frequencies, the situation is much different. The anode current and the anode voltage, like the grid voltage, are reduced in amplitude and have a distorted waveform. The reasons for these changes at microwave frequencies are characterized by the electron trajectories. Whenever the grid is positive, electrons are drawn from the cathode. The number of electrons drawn from the cathode depends on the magnitude of the grid voltage. Most electrons are drawn when the grid voltage reaches its maximum positive value. When the grid becomes positive, the few electrons from the cathode move slowly from cathode to grid and take almost one-quarter of an RF cycle before they reach the grid. When the grid voltage is decreasing, the electrons are returned to the cathode. This is illustrated by Fig. 10.5.

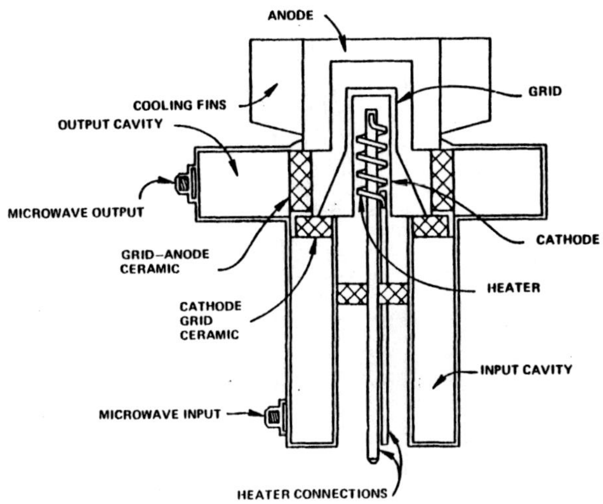

Figure 10.4 (A) The cross section and (B) picture of a gridded tube.

Figure 10.5 indicates that most of the electrons travel from the plate during the part of the microwave cycle when the grid is positive. As this bunch of electrons pass from the grid to the plate, they generate a large microwave field in the output cavity. The generation of this microwave field by the electron beam as it passes through the cavity is illustrated schematically in Fig. 10.6. The sketch on the left-hand side shows a bunch of electrons about to pass through the grid, they induce positive charges on the wall of the cavity near the grid. The center sketch shows the bunch of electrons midway across the cavity, and in this case equal positive charges are induced on both the grid and anode walls of the cavity. Energy of induction comes from the electrons' motion. As the electrons travel across the cavity, electron current flows in the cavity walls to properly balance the positive charges. These induced wall currents generate the microwave field in the cavity. The right-hand sketch of Fig. 10.6 shows the situation as the bunch of electrons are about to strike the plate side of the cavity. In this case, all the positive charges have been drawn to the plate side of the cavity.

The output cavity is designed so that the microwave voltage across the cavity at the maximum of the microwave cycle is almost equal to the anode voltage. This microwave voltage, which is generated by the burst beam of electrons, is decreasing at the time of the microwave cycle when the electron bunch is passing through the cavity. It is clear that the distance the electron beam must travel in the cavity is intimately related to the frequency of the microwave output.

As electrons pass through the cavity, they transfer their energy to the microwave field in the cavity. If all of the electrons could transfer all their energy to the microwave field, the efficiency of the gridded tube would be 100%. However, the electrons cannot do this and still get through the output cavity before the microwave field changes its direction. The typical efficiency of a microwave tube is 30–50%.

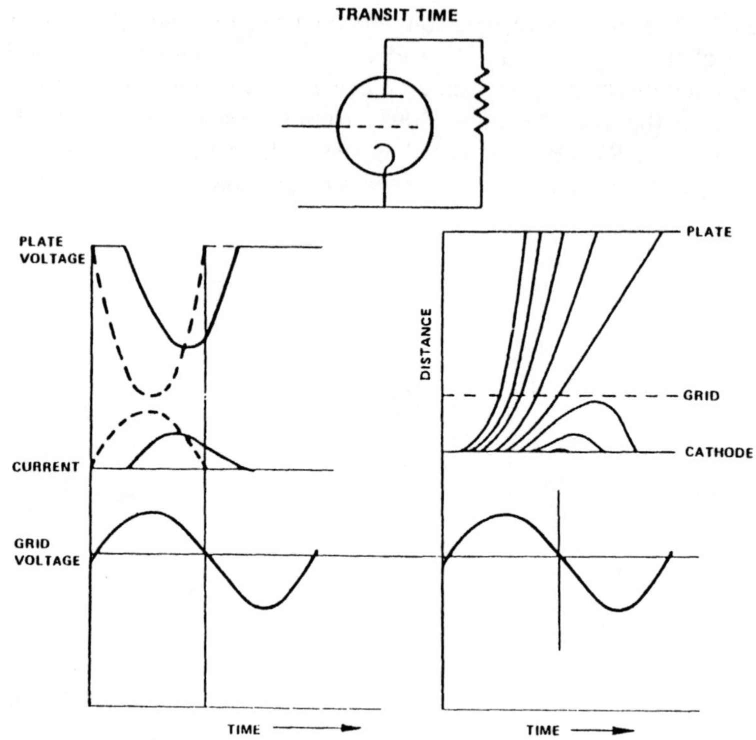

Figure 10.5 The electron transit time of a gridded tube is comparable with the period of a microwave signal. The electron current is distorted.

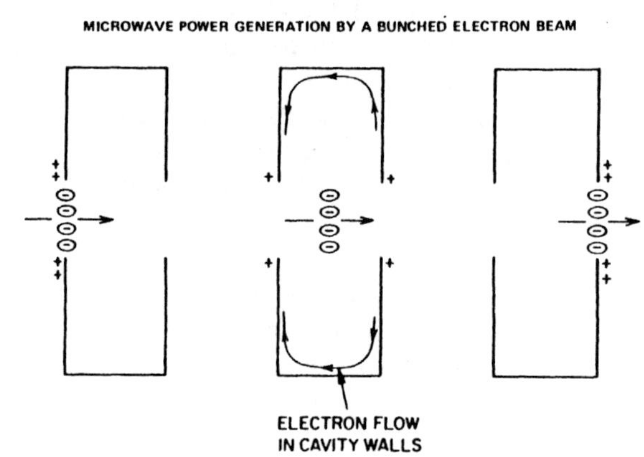

Figure 10.6 When electrons move from one end of a cavity to the opposite end, the induced wall current generates microwave power.

10.5 KLYSTRON AMPLIFIER

The major problem of microwave gridded tubes as we have seen is the transit time. Transit-time effects can be reduced by moving the cathode, grid, and plate closer together, but even with a minimum grid to cathode spacing of a few hundredths of a millimeter (that is, a few thousandths of an inch), microwave gridded tubes are limited in performance to the low frequency end of the microwave band. All other microwave tube types, including klystrons, travelling wave tubes, cross field amplifiers, and gyrotrons, and the various oscillators derived from these amplifiers, all solve the transit-time problem by using velocity modulation. The velocity modulation technique actually uses transit time to generate a burst beam of electrons, rather than trying to minimize transit-time effects.

A schematic drawing of a klystron amplifier is shown in Fig. 10.7. The major elements of a klystron amplifier are an electron gun to form and accelerate a beam of electrons, a focusing magnet to focus the beam of electrons through the interaction cavities where the electron beam power is converted to microwave power, and a collector to collect the electron beam after the microwave power has been generated. There is also an input window, where the small microwave signal to be amplified is introduced into the input cavity. Following the window is a cavity where the electron beam is modulated by the input signal. Next, there are intermediate cavities, where the bunched microwave component of the electron beam is successfully amplified and, finally, an output cavity, where the bunched electron beam generates and amplifies microwave power, is taken out of the tube. The output cavity of a klystron amplifier is similar to the output cavity of the gridded tube. In both types of tubes, the modulated electron beam generates a microwave field as it passes through the output cavity. This microwave field decelerates the electrons causing them to give up their energy to the microwave field. The microwave power is then taken out of the cavity into a transmission line. The major difference between the klystron and the gridded tube is the physical mechanism by which the bunched beam is formed before it enters the output cavity.

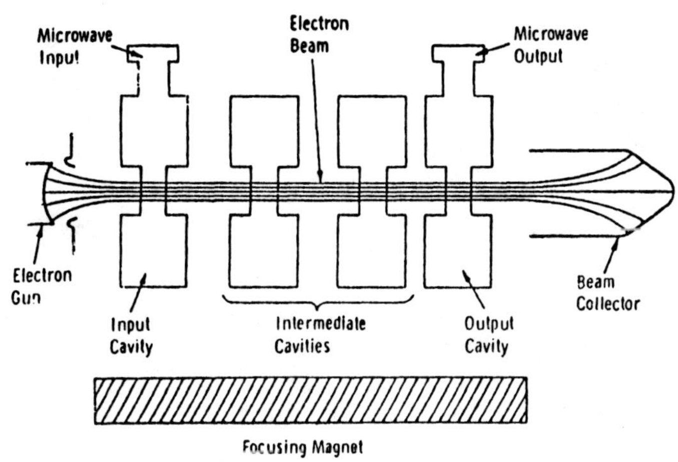

Figure 10.7 The conceptual construction of a klystron.

A typical klystron is shown in Fig. 10.8A, which is a cross sectional drawing of the tube. Figure 10.8B shows the tube itself, and the tube mounted on its focusing magnet.

The velocity modulation process starts with the formation of a beam of electrons which are accelerated from a cathode by a high voltage. The accelerating voltage ranges from a few thousand volts up to several hundred thousand volts, depending on the power level of the microwave tube. The cathode, the accelerating anode, and the beam forming electrode between the cathode and the anode, which shape the electron beam, comprise the electron gun.

In a velocity modulation tube, the electron beam is drawn from the cathode with a constant voltage so that the average number of electrons leaving the electron gun is constant with time and all electrons have the same velocity. This is in contrast to microwave gridded tubes, where both the number of electrons and their velocity depend on when the electrons were drawn from the cathode during the microwave cycle.

SEC. 10.5
KLYSTRON AMPLIFIER

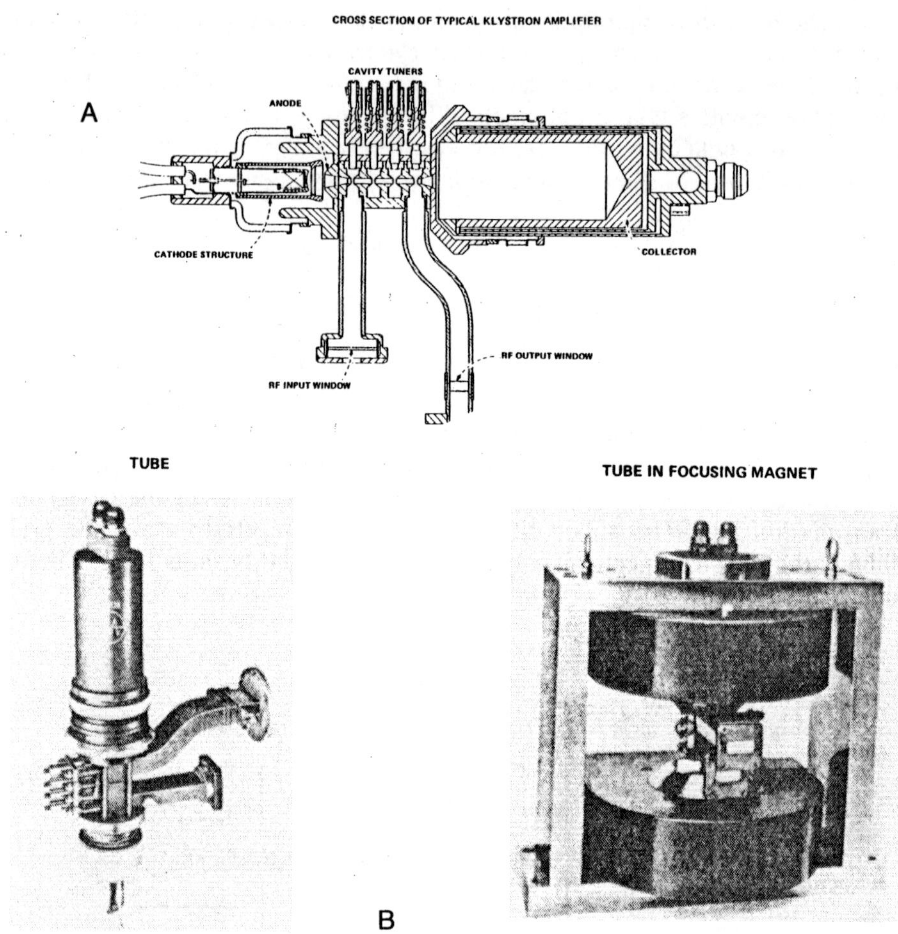

Figure 10.8 (A) The complete cross section of a klystron. (B) The klystron tube itself and inside the focusing magnet.

The electron beam emerging from the electron gun is then shot through the input cavity of the klystron. A small microwave signal to be amplified is applied to the input cavity, and the small electric field set up across the cavity by this input signal interacts with the electron beam.

This interaction process is shown schematically in Fig. 10.9A, where trajectories of the electrons are shown as a function of time during one RF cycle. In the example shown, the electrons are initially accelerated by 10,000 volts, and they reach approximately 20% of the velocity of light as a result of this acceleration process. The microwave voltage across the input cavity is typically only 1 volt, which is very small compared to the accelerating voltage of 10,000 volts, but it speeds up some electrons in the electron beam, and slows down others, depending on the phase of the microwave signal in the input cavity as the individual electrons come through.

In Fig. 10.9A, electrons come through the input cavity at the time in the RF cycle when the microwave field in the cavity is zero. Consequently, the third electron bunch is neither speeded up nor slowed down. It leaves the input cavity and travels down the axis of the klystron with this initial velocity. At times T_A, T_B, T_C, T_D, *et cetera* it reaches distances, *A, B, C, D,* and so forth from the input cavity.

In contrast, the first electron bunch had entered the cavity one-quarter of a cycle earlier, when the microwave field was decelerating, that is, the voltage is negative, and so the first electron bunch takes a longer time to reach a particular point down the tube. However, because the first electron bunch came through the input cavity earlier in time than the third electron bunch, both the first bunch and the third bunch reach point *D* at the same time.

In a similar way, the fifth electron bunch enters the input cavity one-quarter of an RF cycle later than the third electron bunch, when the field is accelerating, and so the fifth electron bunch is speeded up. The fifth electron bunch takes less time to reach point *D* beyond the input cavity because of its higher velocity. It too reaches point *D* at the same time as the first electron bunch and the third electron bunch. In a similar way, the second electron bunch is slightly decelerated, the fourth electron bunch slightly accelerated and, as shown schematically, all of these electrons reach point *D* at the same time.

The basic velocity modulation process is simply that some electrons are speeded up, others are slowed down, depending on when they come through the input cavity during the RF cycle. As the electrons travel farther down the tube beyond the input cavity, the fast electrons catch up with the slow electrons, so as to form a bunched electron beam.

The velocity modulation process is illustrated in another way in Fig. 10.9B, which is a three-dimensional plot showing time during two microwave cycles (horizontally), distance along the tube into the plane of the paper, and the number of electrons (shown vertically).

Note that the same number of electrons arrive at the input cavity at all times during the microwave cycle because the electrons are accelerated from the cathode by a constant voltage. As the electrons pass through the input cavity, some are slowed down and some are speeded up, depending on when during the microwave cycle they passed through the input cavity. At some distance *A* beyond the input cavity, the fast electrons are beginning to catch up with the slow electrons, reaching point *A* at one time during the microwave cycle than at other times.

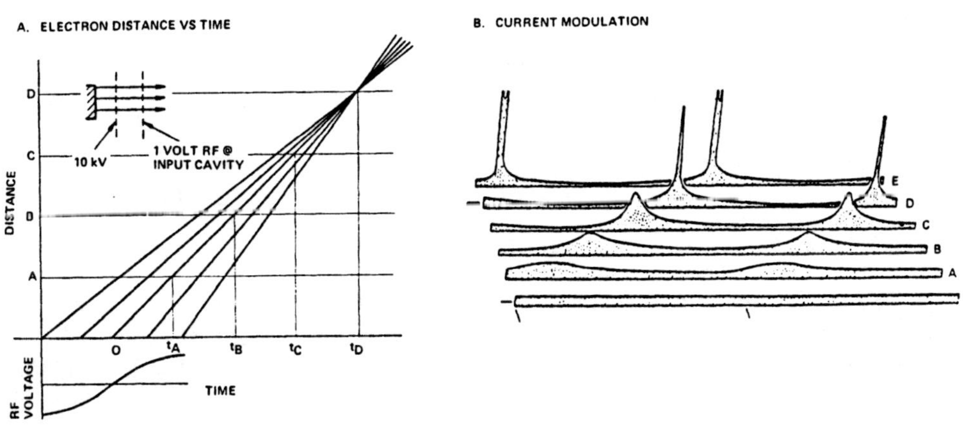

Figure 10.9 The velocity modulation process inside a klystron.

When the electron beam has traveled to distance *B* beyond the input cavity, more of the fast electrons have caught up with more of the slow electrons, and a more tightly bunched beam is formed. The bunching process continues to point *C* and, finally, at point *D*, almost all of the fast electrons have overtaken the slow electrons, and so a tightly bunched beam is formed, with almost all of the electrons reaching point *D* at one instant of time, during the microwave cycle.

One complete cycle later, a second bunch of electrons arrive at point *D*. At point *E*, farther down the tube, the faster electrons have passed the slow electrons, and the bunched beam is beginning to disperse.

SEC. 10.5
KLYSTRON AMPLIFIER

Figure 10.9 suggests that the electron beam could be completely bunched even with a very weak microwave signal if the electron beam were allowed to travel a sufficient distance to give time for the slightly faster electrons to catch up with the slightly slower electrons. This would be true except for space charge forces within the electron beam. As the fast electrons begin to catch the slow electrons and the bunch begins to form, the mutual repulsion of the negatively charged electrons tend to push the bunch apart, and so the velocity modulation process must be enhanced by adding intermediate cavities along the klystron.

The velocity modulation interaction process in a multicavity klystron is illustrated in Fig. 10.10. This figure shows the microwave voltage, the electron velocity, and the microwave current as a function of the distance along the klystron from the gun through the four cavities to the collector. The upper curve shows the voltage applied to the electrons relative to the cathode voltage. The middle curve shows the velocity of the electrons and the lower curve shows the microwave beam current.

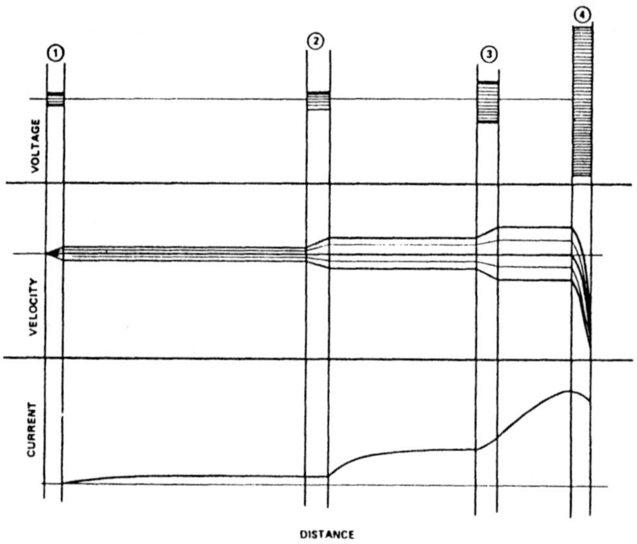

Figure 10.10 The effect on electron velocity and RF current in a multicavity klystron.

The electrons are all accelerated from the cathode through the same voltage, and so they all have the same velocity as they enter the first cavity. As the electrons pass through the input cavity, some see an increased voltage if they pass through the cavity at a time when the microwave signal is accelerating, which adds to the initial accelerating voltage. Other electrons are decelerated if they pass through the input cavity when the phase of the microwave signal is decelerating. As a result of this velocity modulation, some electrons are speeded up as they pass through the input cavity, others are slowed down.

The microwave current as a function of distance is shown in the lower curve of Fig. 10.10. As the partially bunched electron beam travels through the second cavity, the small RF current that exists because of the partial bunching, induces a microwave field into this cavity. This microwave field applies additional velocity modulation to the electron beam. Note that the accelerating and decelerating voltages are larger in the second cavity than they were in the input cavity and, consequently, the faster electrons are speeded up more and the slower electrons are slowed down by a greater amount.

In the input cavity the microwave field was supplied by the external microwave signal to be amplified. In the second cavity, however, the microwave field is generated by the partially bunched electron beam, and the resulting field interacts with the beam to further speed up the fast electrons and slow down the slow electrons.

This greater velocity modulation allows the electrons to further overrun the space charge forces, and the fast electrons come closer to catching the slow electrons as the electrons travel through the third cavity.

The space charge forces, however, still prevent the electrons from becoming completely bunched, and at the point where the maximum current is reached, the third cavity is located. The velocity modulation process is enhanced in the same way in the third cavity, and by the time the electron beam reaches the output cavity most of the fast electrons have caught up with the slow electrons, and the electron bunch looks like that shown in Fig. 10.9B (position *D*).

A large microwave field is generated as the tightly formed beam passes through the output cavity. This microwave power is taken out of the cavity through the output window. The output cavity is designed so that the microwave voltage across the cavity, at the location of the electron beam, is approximately equal to the original accelerating voltage, as is shown in the upper curve of Fig. 10.10. The microwave field in the output cavity is generated by the bunched beam of electrons, and the resulting microwave field is decelerating at the time in the microwave cycle when the electron bunch is passing through the cavity so that they are slowed down, as shown in the middle curve of Fig. 10.10.

All the electrons lose energy as they go through the output cavity, but some electrons lose more energy than others, depending on their velocity as they enter the output cavity, and their particular velocity depends on when during the microwave cycle they entered the input cavity. This reduction of the electron beam energy by the microwave field in the input cavity is the source of power that supplies the output cavity.

Klystrons offer the advantages of high peak power, high average power, good efficiency, high gain, and a low spurious signal level.

Klystrons offer the highest peak and average power of any type of microwave amplifier tube up to about 30 GHz. The only disadvantage of klystrons is their narrow bandwidth, which is only a few percent.

In a typical klystron, the electron beam has 50–70% of its initial energy as it leaves the output cavity. So, typical klystron efficiencies range from 30% to 50%. Note that the velocity modulation process can be enhanced up to its optimum value at the output cavity by adding more intermediate cavities independent of the weakness of the input signal.

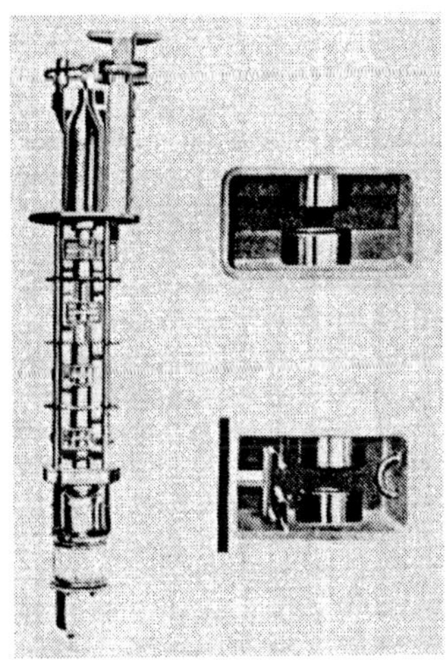

Figure 10.11 The cutaway diagram of a high power klystron and its cavities.

Figure 10.11 shows a cutaway photograph of a high peak power klystron. Also shown is the detail of klystron cavities. The cavities are cylindrical resonators, with nodes projecting from each side in the center of the cavity, to concentrate the electric field at the electron beam. The cavities contain tuning mechanisms so that the center frequency of the klystron can be mechanically adjusted over a 10% range. The instantaneous bandwidth of the klystron is only a few percent. The input microwave signal is introduced into the first cavity from a coaxial transmission line by the coupling loop.

The microwave power generated in the output cavity is coupled from the cavity into a waveguide and out of the tube through a ceramic window, which allows the microwave power to pass into the external waveguide, as can be seen in the output waveguide in the top photograph.

The remaining parts of the klystron include the focusing magnet and the collector. As can be seen from the photograph of a klystron in Fig. 10.8B, the focusing magnet and the collector comprise a major part of the tube's size and weight.

After the electron beam is formed it must be focused through the several klystron cavities from the gun to the collector. Due to mutual repulsion and without the magnetic focusing field, the beam would hit the klystron cavities.

Four methods of obtaining the required focusing field are shown in Fig. 10.12A. An electromagnet yoke is shown schematically in sketch A, and a photograph of such is shown in Fig. 10.8B. The yoke magnet is formed by two electromagnet coils located in an iron yoke which carries the magnetic field from the coils to the cavity region of the klystron. The gun collector fits inside the iron yoke so that it is shielded from the main magnetic focusing field. The advantage of the electromagnet yoke is that the klystron cavities are accessible for tuning. The disadvantages are the large size and weight of the yoke in addition to the 500 to 2500 watts of power required to operate the coils.

The size and weight of the focusing structure can be greatly reduced by using a solenoid instead of a yoke. This is shown in Figure 10.12B. An opening must be cut into the solenoid to provide access for tuning the klystron cavities. The focusing solenoid offers the disadvantage of the high power required. The high peak power klystron shown in Fig. 10.8A is designed for focusing inside a solenoid (solenoid not shown in the figure).

The power requirements of the yoke or solenoid can be eliminated by using a permanent magnet, as shown in Fig. 10.12C. This magnet can be of a barrel or yoke design. The disadvantage of the permanent magnet is the large fringing magnetic field, which can adversely affect surrounding electronic equipment. Another disadvantage is the additional heavy weight of the permanent magnet yoke. The size and weight of the permanent magnet can be drastically reduced by using a periodic permanent magnet for focusing, which is shown schematically in Fig. 12D. This magnet design is used in some klystrons, and is widely used on travelling wave tubes. With periodic permanent magnet focusing, an array of ring magnets with alternating polarities is used. The electron beam is focused with one magnet, allowed to expand, and focused again by the next magnet, and then allowed to expand, focused again by the third magnet, and so on. By alternating the magnets, the external field, and the magnet size and weight are reduced in comparison to the barrel or yoke permanent magnet.

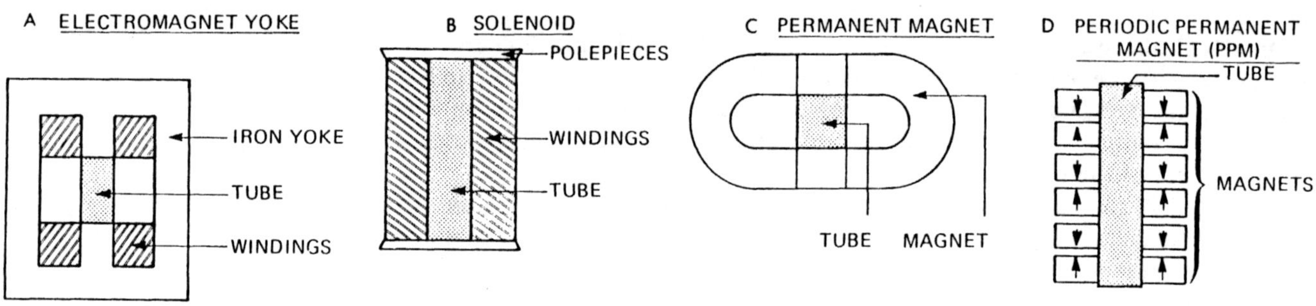

Figure 10.12 Four methods of constructing the klystron's focusing magnet.

The function of the collector is to collect the beam of electrons after it has left the output cavity where the maximum microwave power has been extracted. Because klystron efficiency is in the range of 50–60%, an appreciable amount of beam power is still left in the electron beam as it enters the collector. The electron beam power density and spreading are automatically obtained by stopping the magnetic focusing field at the end of the output cavity so that the beam can spread due to the mutual repulsion of the electrons.

The collector serves no other function except to collect the spent beam of electrons, so it can be made as large as necessary to obtain the required reduction in beam power density at the collector surface. as the electrons strike the collector surface, they transfer their power as heat and this heat must be carried away from the collector by forced air, forced liquid, or evaporation cooling.

SEC. 10.6
TWTA

10.6 TRAVELING WAVE TUBE AMPLIFIER

The main disadvantage of a klystron is its small bandwidth. This disadvantage is eliminated by the travelling wave tube (TWT).

A schematic drawing of a traveling wave tube is shown in Fig. 10.13. The major elements include an electron gun to form and accelerate a beam of electrons, a focusing magnet to focus the beam of electrons through the interaction structure where the electron beam power is converted to microwave power, a collector to collect the electron beam after the microwave power has been generated, an input window where the small microwave signal to be amplified is introduced to the interaction structure, an interaction structure where the electron beam interacts with the microwave signal to be amplified, a microwave output window where the microwave power is taken out of the tube, and an internal attenuator to absorb the power reflected back into the tube from mismatches in the output transmission line.

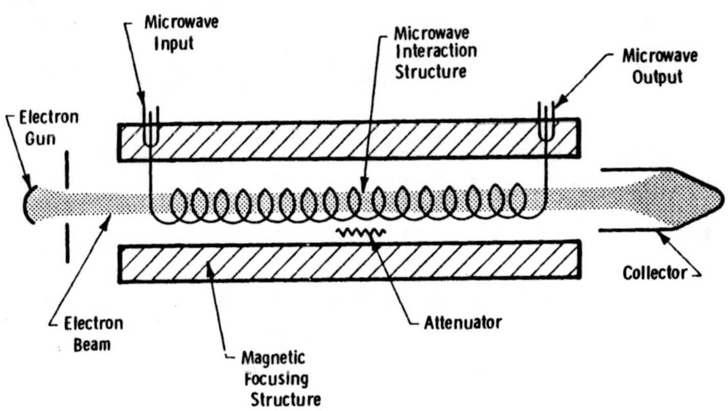

Figure 10.13 A schematic drawing of a traveling wave tube.

The electron gun, focusing magnet, collector, and input and output windows of a TWT are the same as those used in a klystron and serve the identical function. The major difference between a klystron and a TWT is that the input, intermediate, and output cavities of a klystron are replaced by the interaction structure in the TWT.

Figure 10.14 is an exploded photograph of a typical travelling wave tube, and the various parts shown schematically in Fig. 10.13 can be seen in the photograph. The particular tube shown provides 200 watts of continuous wave (CW) power, with 40 dB gain, and an overall efficiency of 25% across the frequency range of 5–10 GHz. The TWT uses the same basic velocity modulation process as the klystron.

Exactly as in a klystron, the process starts with the formation of a beam of electrons by an electron gun. The electron beam is drawn from the cathode. All electron have the same velocity. The electron beam emerging from the electron gun is then shot into the interaction structure. The microwave signal to be amplified is put into the input end of the interaction structure. The key feature of the traveling wave tube is that the electrons in the beam travel at the same velocity as the microwave signal on the interaction structure.

If an electron enters the interaction structure during the positive phase of an RF cycle, it stays in the positive phase of the cycle as it travels with the microwave signal through the interaction structure. It is continuously accelerated. By contrast, if the electron enters the interaction structure during the negative phase of a cycle, it stays in the negative part of the cycle, and is decelerated continuously as it travels through the interaction structure.

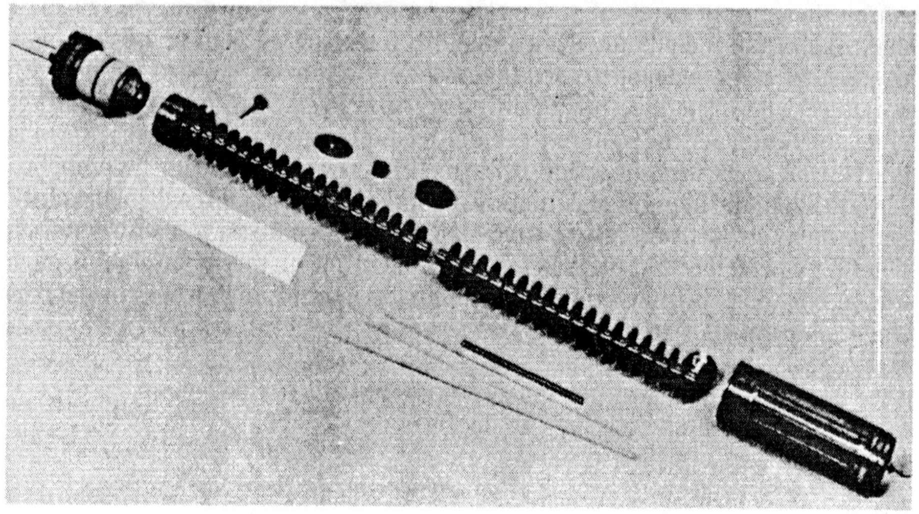

Figure 10.14 The exploded picture of a traveling wave tube.

There, depending on the phase at which they enter the interaction structure, some electrons are speeded up and others are slowed down. As the electrons progress along the length of the TWT, the fast electrons begin to catch the slower electrons, and a bunch beam is formed which excites an increasing microwave signal on the interaction structure. The interaction process described above is shown schematically in Figs. 10.15 and 10.16.

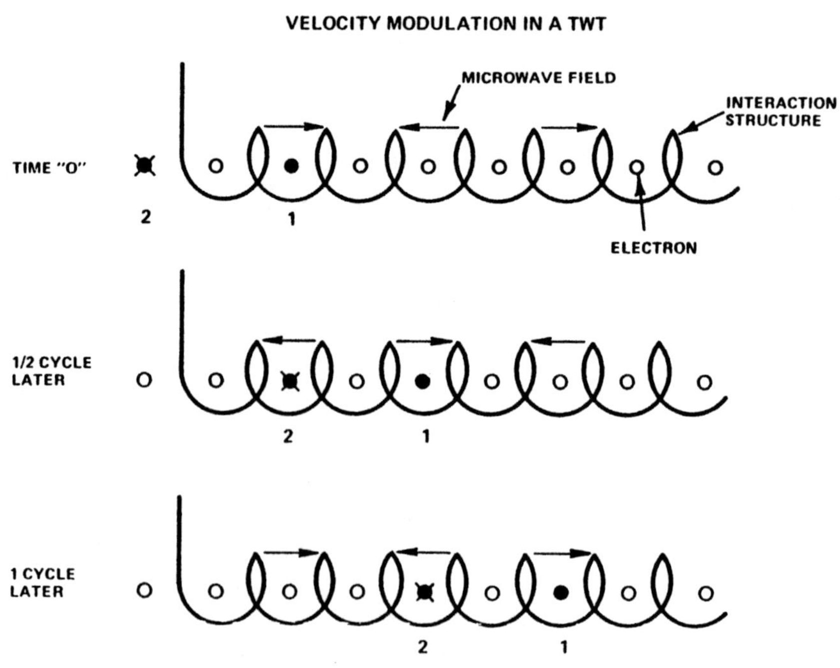

Figure 10.15 The velocity modulation process in a traveling wave tube.

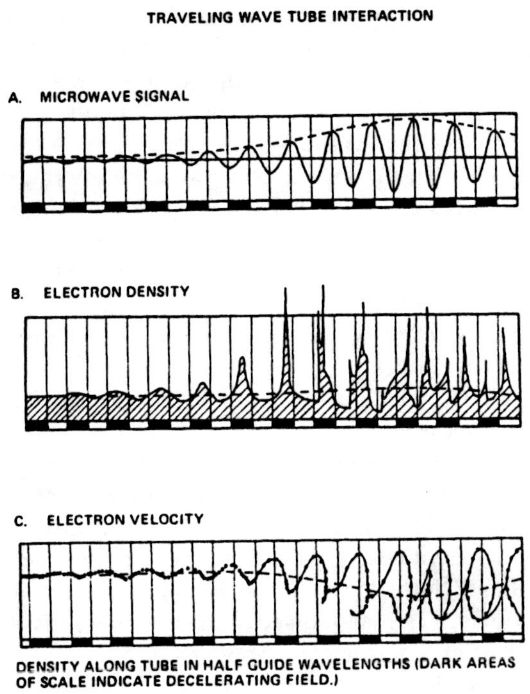

Figure 10.16 The growth of the microwave signal, electron density, and electron velocity along a traveling wave tube.

Figure 10.15 shows the electric field of a TWT interaction structure at three instances of time, and the location of two representative electrons. At time 0, the first electron, which had entered the interaction structure earlier, is in a position where the electric field is accelerating. The second electron has not yet entered the interaction structure. At a time one-half of an RF cycle later, as shown in the middle sketch of the figure, the first electron has travelled farther down the interaction structure. The microwave signal which was fed into the input end of the interaction structure has also traveled down the interaction structure, and the microwave signal has moved down the acceleration structure at the same velocity as the electron, so the first electron is still in an accelerating field. The second electron by this time has entered the interaction structure, and is in the same position that the first electron was in at time 0. Note, however, that the electric field at this same position is now decelerating, so the second electron is being slowed down. At a time one cycle later, as shown in the bottom sketch, the first and second electrons have moved farther down the interaction structure at the same velocity, so the electric field is still accelerating the first electron and still decelerating the second electron.

The TWT interaction process is shown schematically in Fig. 10.16. The electron density of the microwave signal indicated in sketch A is shown in sketch B and its electron velocity is shown in sketch C. All sketches show the interaction effects as a function of distance along the TWT.

The dark areas of the scale represent the decelerating phase and the light areas represent the accelerating phase. At the input end of the tube, on the left-hand side of the figure, the microwave signal is small, the electron density in the electron beam is uniform, and the electron velocity is constant. As the electrons and the microwave signal progress through the interaction structure, some electrons are slowed down, some are speeded up, and bunches of current begin to form. These bunches excite an increasing microwave signal onto the interaction structure. Near the output end of the TWT, as shown in the right-hand side of Fig. 10.16, the microwave power on the interaction structure grows to be about 20% of the beam power. At this point, the beam is tightly bunched with most of the electrons grouped together in the decelerating phase of the RF cycle. Almost all of the electrons have been slowed down and the energy has been extracted from the beam. The electrons have been slowed down to the point that they no longer travel at the same velocity as the microwave signal, and traveling farther through the interaction structure would result in getting into the accelerating field and absorbing power from the microwave field rather than giving it off. Therefore, the microwave signal is taken off of the interaction structure at the point where the maximum signal level occurs.

Note from sketch C, which shows the electron velocity distribution, that although most electrons are slowed down as they give off energy to the input microwave, some are actually speeded up as they have gotten off into the wrong phase. As the electrons leave the interaction structure to enter the collector, they have a large range of velocities. Some have an even greater velocity than when they were initially accelerated from the cathode. The TWT offers the following advantages: wide bandwidth, from 30–120%; high gain; moderate peak power; moderate average power. The main disadvantage of TWTs is their low efficiency, which is the lowest of all microwave amplifier tubes, but in many system applications the wide bandwidth greatly outweighs the low efficiency.

10.7 COMPARISON BETWEEN KLYSTRON AND TRAVELING WAVE TUBE TUBE AMPLIFIERS

The klystron achieves high gain per unit length and high efficiency at the expense of bandwidth by using resonant cavities which provide a high microwave interaction field. The traveling wave tube, because it does not use resonant cavities, can provide extremely wide bandwidth, but the microwave interaction fields are much weaker. However, because the interaction is allowed to build up continuously, high gain and high power can be achieved.

All of the power of a klystron is extracted from the output cavity. The beam can thus be tightly bunched and remain so as it passes through the cavity. In contrast, the build-up of power in the traveling wave tube occurs over a considerable length of the interaction structure, as illustrated in Fig. 10.15, and the electrons go out of phase. Therefore, the efficiency of the traveling wave tube process is limited to from 10% to 30%, as compared to an efficiency from 30% to 50% in the klystron.

All klystrons use the same basic types of cavities. By contrast, two different types of interaction structure are used for traveling wave tubes. These structures are illustrated in Fig. 10.17. One type, the helix, is shown in Fig. 10.17A. This type of interaction structure was shown in the TWT schematic of Fig. 10.13 and was also indicated in the photograph of the typical TWT in Fig. 10.14.

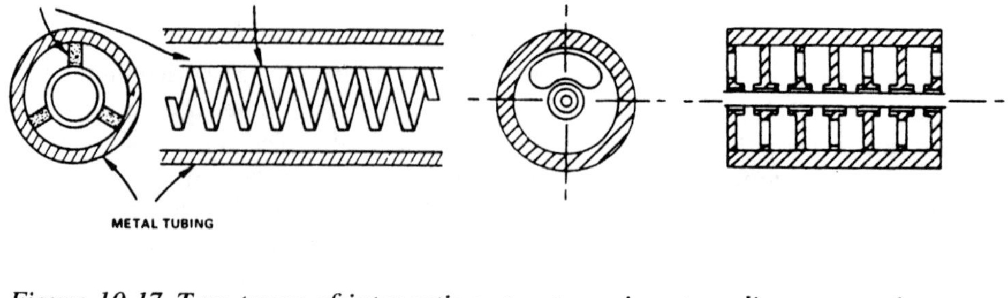

Figure 10.17 Two types of interaction structures in a traveling wave tube: (A) the helix and (B) the coupled cavity.

SEC. 10.7
COMPARISON BETWEEN KLYSTRON & TWTAs

The helix is made of tungsten wire and is supported on ceramic rods inside a metal tubing which forms the vacuum envelope of the tube. The advantage of the helix is that it can cover a wide bandwidth. For example, the entire frequency range of 2–8 GHz can be covered with a single TWT using a helix interaction structure. The disadvantage of the helix is its low average power capability. The average power is limited because of the fragile nature of the helix structure.

The coupled cavity interaction structure shown in Fig. 10.17B solves the average power limitations of the helix, but at the expense of bandwidth. This interaction structure consists of an array of klystron-like cavities, each coupled to successive cavities through a kidney-shaped opening in the coupled-cavity wall. Because the interaction structure is made entirely of metal, heat can be easily conducted from the region surrounding the electron beam out to the outer walls of the cavities from the vacuum envelope.

The average power capability of the coupled-cavity interaction structure is practically as good as that of the klystron, and its bandwidth can be any value from 5% up to 40%, depending on the size of the coupling hole. The coupled-cavity TWT is designed to provide 10kW of peak power, and 1kW of average power, over a 20% frequency range of about 10 GHz.

The electron gun and focusing magnets of a TWT serve the same functions as those elements in a klystron. Most traveling wave tubes, like the one shown in Fig. 10.14 use periodic permanent magnetic focusing.

As in a klystron, the collector of a TWT collects the electron beam after the power has been extracted from the beam in the interaction structure and converted into microwave power. In the klystron, about 50% of the electron beam is converted into microwaves. By contrast, in a TWT only about 15% of the power in the beam is converted to microwave power. The electron beam, therefore, still contains 85% of its original power when it enters the collector.

The overall efficiency of the TWT is normally increased by depressing the collector, that is, operating the collector at a voltage of approximately one-half the interaction structure voltage. A block diagram of a TWT with a depressed collector, which illustrates the relationships between the tube and tube power supplies, is shown in Fig. 10.18. The electron beam is accelerated from the cathode to the interaction structure by a 10,000 volt interaction structure supply. As electrons pass through the interaction structure, the electrons are slowed down as they convert 15% of their power into microwave power.

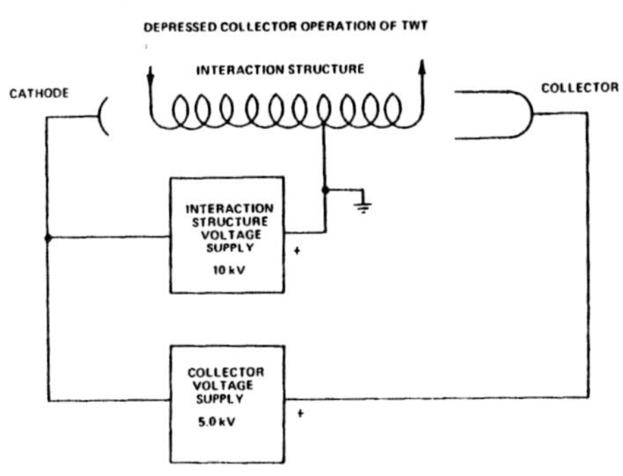

Figure 10.18 Block diagram describing the depressed collector operation of a TWT.

Very few of the electrons strike the interaction structure and the electrons have most of their original power as they leave the interaction structure and enter the collector. This power can be regained through slowing down the electrons by operating the collector at a voltage midway between the cathode and interaction structure. The electron beam is ultimately collected at one-half the original accelerating voltage. In this manner, the efficiency of the TWT can be approximately doubled from 15% to 30%.

10.8 CROSSED FIELD AMPLIFIER

The low efficiency problem of a traveling wave tube can be solved, while retaining its good bandwidth capability, by the crossed field amplifier. The crossed field amplifier, commonly called a CFA, offers the advantages of high efficiency, high peak power, and wide bandwidth. The cross field amplifier has the highest efficiency of any microwave tube, which is as high as 80%. The disadvantages of the cross field amplifier are low gain, high noise output, and low average power capability.

A schematic drawing of a crossed field amplifier is shown in Fig. 10.19A. Just as the schematic drawing of a TWT was related to that of a klystron, the schematic drawing of the crossed field amplifier is related to that of a TWT. The major elements of a crossed field amplifier are an electron gun to form and accelerate a beam of electrons, an interaction structure where the electron beam interacts with the microwave signal to be amplified is introduced into the interaction structure, a microwave output window where the microwave power is taken out of the tube, a collector to collect the electron beam after the microwave power has been generated, a sole electrode, and an external magnet to provide a magnetic field in the interaction region.

The crossed field amplifier has the same basic parts as the traveling wave tube, but a different type of velocity modulation interaction is used to overcome the major limitation of the TWT, its low efficiency. The velocity modulation process in the crossed field amplifier occurs in the region of crossed electric and magnetic fields, and a sketch of the interaction region in a crossed field amplifier is shown in Fig. 10.19B.

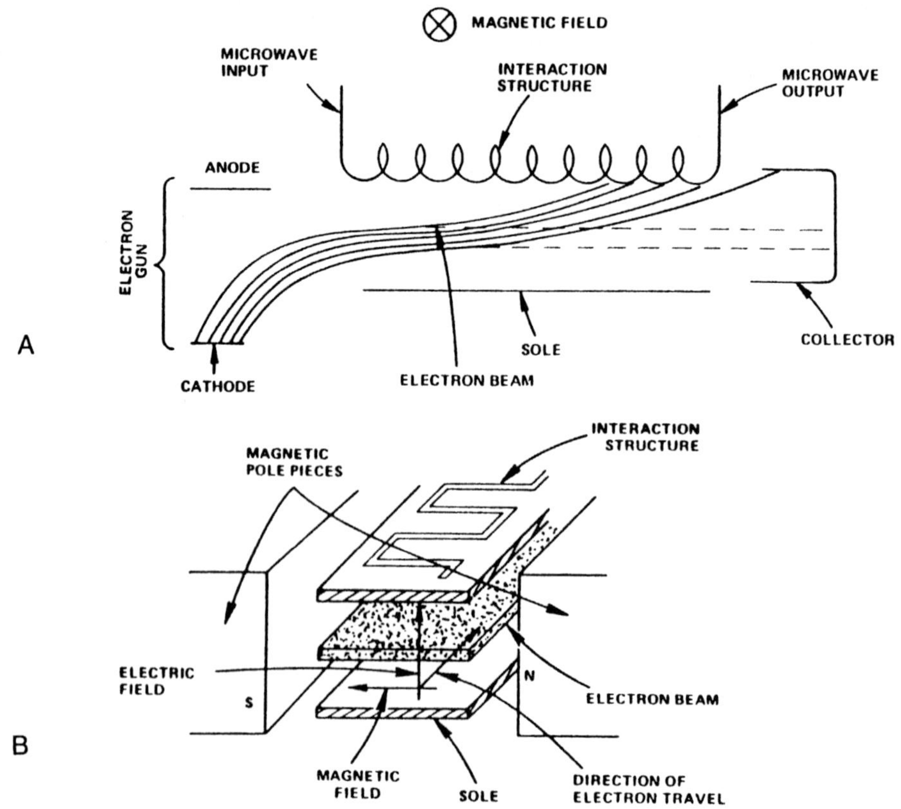

Figure 10.19 (A) The schematic drawing of a crossed field amplifier. (B) The interaction region of a crossed field amplifier.

SEC. 10.8

CFA

This sketch shows the interaction structure, and is taken at right angles to the schematic drawing shown in Fig. 10.19A. The electron beam in a crossed field amplifier is a strip beam which travels in the region between the interaction structure and the negative electrode called the *sole*. In Fig. 10.19B, the electron beam moves into the plane of the paper. The sole is used to provide an electric field at right angles to the direction of electron beam travel. The electric field forces the electron towards the interaction structure. The magnetic field in a crossed field amplifier is at a right angle to the direction that the electron travels and also at a right angle to the direction of the electric field.

The combined action of the moving electron beam and the magnetic field provides a force, which counteracts the electrical force developed by the sole, and pushes the electron away from the interaction structure. The voltage on the sole and the strength of the magnetic field are adjusted so that the two forces exactly counteract each other. When no microwave input is applied to the tube, the electrons move in a straight path along the axis of the tube, as shown by the dotted curve representing the electron beam in Fig. 10.19A. The electric field between the sole and the interaction strucure, the magnetic field, and the electron velocity, are all crossed, that is, they are at right angles to one another. This is the key feature of the interaction process which gives the crossed field amplifier both its name and its high efficiency.

It should be noted that the magnetic field in a klystron or traveling wave tube is only used to focus the electron beam through the cavities or the interaction structure. These tubes would still work without magnetic focusing, for example, if the beam were electrostatically focused through the tube. By contrast, in the crossed field amplifier, the magnetic field not only confines the beam, it is an essential part of the velocity modulation process as well. The crossed field interaction process is illustrated schematically in Fig. 10.20. Sketch A shows the microwave field in the interaction region in the bunching forces on the electrons. If there was no microwave field present, the electrical force from the electric field between the sole and interaction structure would be exactly balanced by the magnetic force of the magnetic field and the moving electron beam. Therefore, the electron would move, as shown by the dotted line, in a straight line along the axis of the tube.

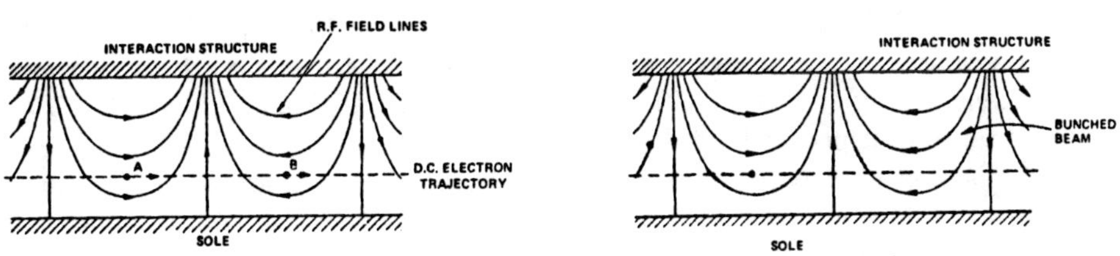

Figure 10.20 The interaction process of a crossed field amplifier: (A) describes the microwave field and forces on the electrons and (B) shows the electron beam bunching.

When a microwave field is applied to the interaction structure the electrons at position *A* are speeded up, and the electrons at position *B* are slowed down. Just as in a traveling wave tube, the velocity of the microwave signal along the axis of the tube is made equal to the velocity of the electrons. An electron in position *A*, therefore, will see the same field as it moves along the tube. As electron *A* is speeded up, the magnetic force which depends on the strength of the magnetic field and the magnitude of electron velocity increases, and becomes greater than the electrical force created by the negative sole. So, the electron, rather than moving in the axial direction, moves away from the circuit. In the same way, the electron at position *B* slows down, and the electrical force from the sole becomes greater than the magnetic force, and the electron moves toward the interaction structure.

This differential movement, depending on the position of the electron during the phase of the microwave field, causes a velocity modulation bunching to occur. The resulting bunched beam is shown in Fig. 10.20B. Because of the crossed electric and magnetic fields, the electrons do not bunch axially as the beam moves along the interaction structure, but rather bunch closer to, or farther from, the interaction structure. Just as in a traveling wave tube, as the bunches form they interact with the interaction structure to generate an increasing microwave field, which further bunches the beam. Unlike a traveling wave tube, however, the electron velocity along the axis of the tube remains the same. The electrons do not slow down as power is transferred from the electron beam to the microwave field. Their power comes from potential energy because the interaction structure is at a higher voltage than the sole, not from the kinetic energy of the electron beam. Because the electrons do not slow down as they travel along the interaction structure, they do not lose synchronism. The interaction process can be continued until almost all the power is transferred from the beam to the interaction structure. Consequently, the crossed field amplifier has extremely high efficiency.

This basic crossed field interaction process that provides the high efficiency also causes the crossed field amplifier's limitations of low gain, high noise, and only moderate average power capability. The beam moving in the crossed electric and magnetic field region is basically unstable, that is, any small noise modulation will enhance itself. Consequently, if the crossed field amplifier's interaction region is made very long, the noise that is always present in the beam due to non-uniform emission of electrons from the cathode will be amplified and become greater than the desired signal.

The interaction region in a klystron or TWT can be as long as necessary to achieve the required gain, but the length of a crossed field interaction region must be limited so that noise will not build up. Consequently, crossed field amplifiers can have no more than 10–15 dB gain, whereas klystrons and TWTs can have gains of 50–60 dB. A crossed field amplifier must have a chain of crossed field amplifiers or a TWT to drive it, whereas a klystron or TWT can be driven from a low level solid-state source. Even at the low gain levels of 10–15 dB, the crossed field amplifier still has a noisier output than a high gain klystron or TWT at the same output power level.

For highest efficiency, the electron beam in a crossed field amplifier must move very close to the interaction structure, and in most crossed field amplifiers the beam actually strikes the structure. This is illustrated schematically in Fig. 10.19A.

Most of the power has been extracted from the beam, but still some remains, and this is converted to heat which must be absorbed by the interaction structure. By contrast, in a klystron or TWT, only a small fraction of the beam strikes the interaction structure. Consequently, the average power of a crossed field amplifier is much less than a klystron or TWT, which collect the unused electron beam in a separate collection region, rather than on their interaction structure.

As with a TWT, either a helix or a coupled-cavity interaction structure can be used for a crossed field amplifier. The helix structure offers wide bandwidth, up to about 60%, but at limited peak power levels up to several megawatts, with average power levels of several kilowatts, and bandwidth limited to about 20%.

In the crossed field amplifier, which is shown schematically in Fig. 10.19A, as electrons are emitted from the cathode, they are initially accelerated to the anode.

Note that the cathode is approximately at the same voltage as the sole, and the anode is approximately at the same voltage as the interaction structure. However, as electrons begin to move away from the cathode, they are acted upon by the magnetic field, and their trajectories are bent, as shown in Fig. 10.19A, so that the electron beam enters the interaction region and moves at right angles to the cathode.

Crossed field amplifiers are not actually made as shown in Fig. 10.19, with the electron beam injected into the interaction structure. This figure was mainly used to compare the crossed field amplifier to the klystron and TWT. Most crossed field amplifiers, however, use an emitting sole-type of electron gun, which is shown schematically in Fig. 10.21. In sketch A, the emitting sole design, the sole is made to be the cathode. Electrons are emitted from the sole, and because of the action of the magnetic field, they are bent to move down the axis of the tube. Electrons are injected in the tube. The electron beam current is small at the input end, and becomes larger and larger towards the output end of the tube. The emitting sole can provide much higher current than the gun of a klystron or TWT and, consequently, for a given power level, the crossed field amplifier can operate at a much lower voltage than other tubes. For example, a klystron or TWT providing one megawatt of peak power requires a beam voltage of 100,000 volts, whereas a crossed field amplifier providing a megawatt of peak output power operates at a beam voltage of only 40,000 volts.

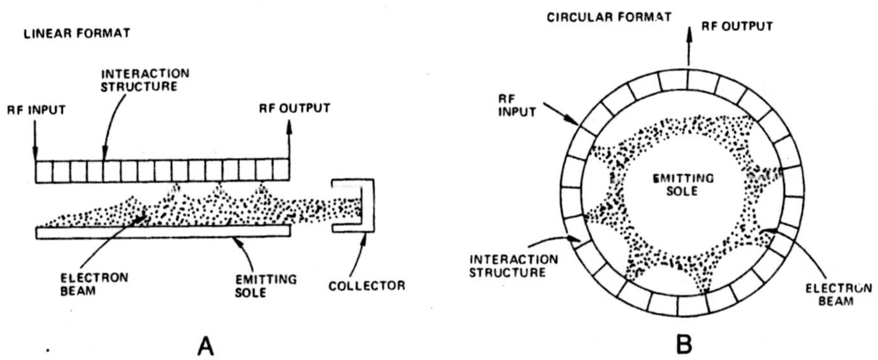

Figure 10.21 The linear and the circular format of emitting sole CFA.

For practical construction reasons, the emitting sole crossed field amplifier is usually made in a circular format, as shown in Fig. 10.21B. The emitting sole becomes a cylinder, which is easier to heat in order to obtain electron emission. The interaction structure in a circular format is more compact, and it is much easier to apply the crossed magnetic field. In both sketches of Fig. 10.21 the magnetic field is applied into the plane of the paper. The collector in the circular format crossed field amplifier is eliminated because most of the current strikes the interaction structure anyway. A drawing of the typical crossed field amplifier using an emitting sole in circular format with a coupled-cavity type of interaction structure is shown in Fig. 10.22. This tube provides one megawatt of peak output power over the 5.4 GHz to 5.9 GHz frequency band, with 13 dB gain and 60% efficiency.

10.9 GYROTRON AMPLIFIER

The gyrotron offers the advantages of very high peak and average powers in the millimeter frequency range of the microwave spectrum. The gyrotron is capable of providing two orders of magnitude more power than any type of microwave tube amplifier. As the frequency of klystrons, or traveling wave tubes, or crossed field amplifiers is increased, the dimensions of the interaction structure get smaller and smaller. The size of the interaction structure of these tubes is inversely proportional to frequency, and this requirement is necessary to make the microwave fields in the cavity or interaction structure couple to the electron beam.

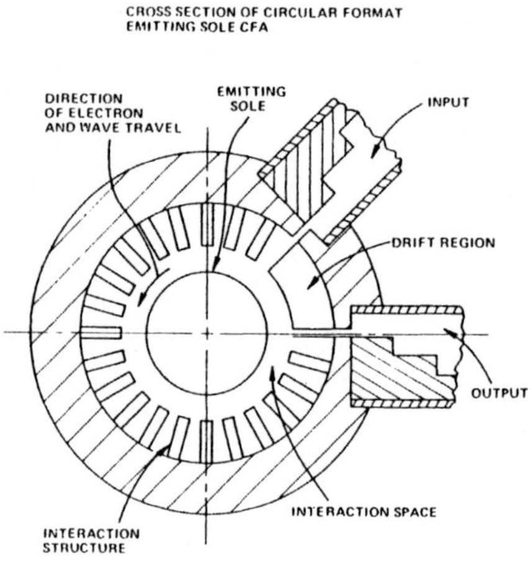

Figure 10.22 The connecting diagram of a circular format emitting sole CFA.

At frequencies in the millimeter part of the spectrum, it is difficult to fabricate the complicated interaction structures that allow the microwaves to travel at the electron beam velocity. Even if they could be fabricated, these small structures cannot handle high average power. Consequently, as shown in Fig. 10.1A the power capability of klystrons, traveling wave tubes, and crossed field amplifiers fall off very rapidly above 30 GHz.

The principle of gyrotron operation is based on the interaction between a static magnetic field and a moving electron. Magnets employed in klystrons and TWTs are for focusing, i.e., confining the electron beam to a small diameter. The magnetic action here is that the magnetic field exerts a force on a moving electron, which, as a result, describes a spiraling path as it advances, i.e., the path of a helix. The spiraling frquency, called cyclotron frequency, can be calculated. If the helix motion of the electron occurs in a cavity there will be coupling of the electron's energy to the cavity. The cyclotron frequency depends on the strength of the magnetic field, the mass of the electron, and the velocity component perpendicular to the magnetic field. We may argue that different electrons have different velocity components perpendicular to the magnetic field, so the cyclotron frequency is not unique. However, special relativity suggests that a fast moving object has more energy and, therefore, more mass, causing the cyclotron frequency to decrease. Also, the energy coupling causes the electron to lose energy, thereby decreasing its mass. Consequently, electron bunching as in klystrons and TWTs is observed.

A schematic drawing of a gyrotron is shown in Fig. 10.23. The gyrotron has the same elements as a klystron. The gun and the magnet are designed so that the electrons take a gyrating path when traveling through the cavities. The cavities are designed as described above, with the beam opening 10 times larger than the beam opening in an equivalent klystron cavity in order to accommodate the large diameter hollow beam.

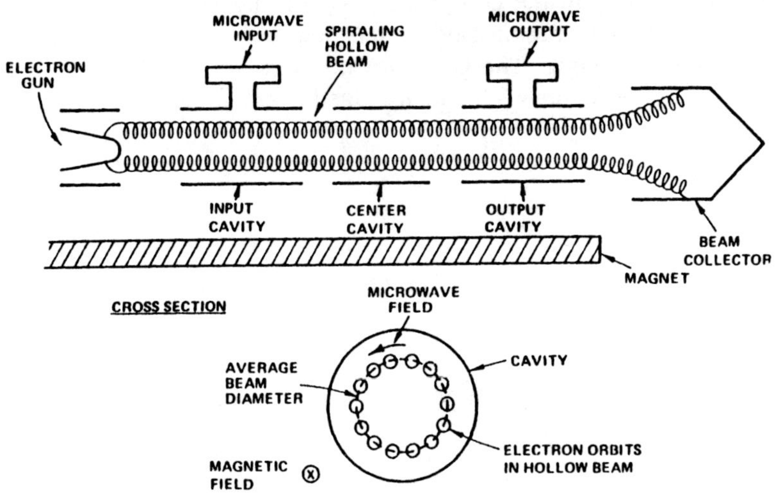

Figure 10.23 The schematic drawing of gyrotron.

10.10 GRIDDED TUBE OSCILLATOR

The gridded tube amplifier previously described can be made into an oscillator by coupling a part of the output from the output cavity back into the input. The resulting gridded tube can be mechanically tuned over a 10% frequency band by tuning both the input and output cavities.

The efficiency of a gridded tube microwave oscillator is good, ranging from 10% to 40%. Their peak and average power capability are also good, but, of course, gridded tubes can only work down at the low end of the microwave band, below 1 GHz.

10.11 KLYSTRON OSCILLATORS

Three types of klystron oscillators will be considered here. The three types of klystron oscillators that can be used are the two-cavity oscillator, the extended interaction oscillator, commonly called the EIO, and the reflex klystron. Schematic drawings of these three oscillators, and photographs of a representative model of each, are shown in Fig. 10.24 A, B, C. Sketch A shows a two-cavity amplifier, with part of the output power coupled back into the input to make the amplifier into an oscillator. This is the same way that the grid tube was made into an oscillator.

The two-cavity klystron oscillator has the lowest FM noise of any microwave oscillator tube or solid-state source, and the lowest AM noise of any microwave oscillator except the transferred electron device. The two-cavity oscillator shown in Fig. 10.24 can provide 10W of output power at 10 GHz. It requires no focusing magnet because the electron beam is designed so that it does not spread enough to hit the cavities in passing them.

The extended interaction oscillator, shown in Fig. 10.24B, consists of a gun, focusing magnet and collector, like any klystron amplifier, and a composite interaction cavity consisting of several individual cavities coupled together. The interaction process is like that of a coupled-cavity traveling wave tube. The microwave signal travels down the composite cavity with the electrons. In addition, the microwave signal is also coupled in the reverse direction so that the

SEC. 10.11
KLYSTRON OSCILLATOR

device becomes an oscillator. The oscillation frequency of the EIO is tuned, not by tuning each cavity in the composite way, but by coupling all the individual cavities into a common tuning cavity, and then tuning this cavity mechanically with an adjustable plunger. The EIO shown in Fig. 10.24B provides 75W of CW power at 50 GHz. It can be mechanically tuned over a 2% band, with an efficiency of about 30%.

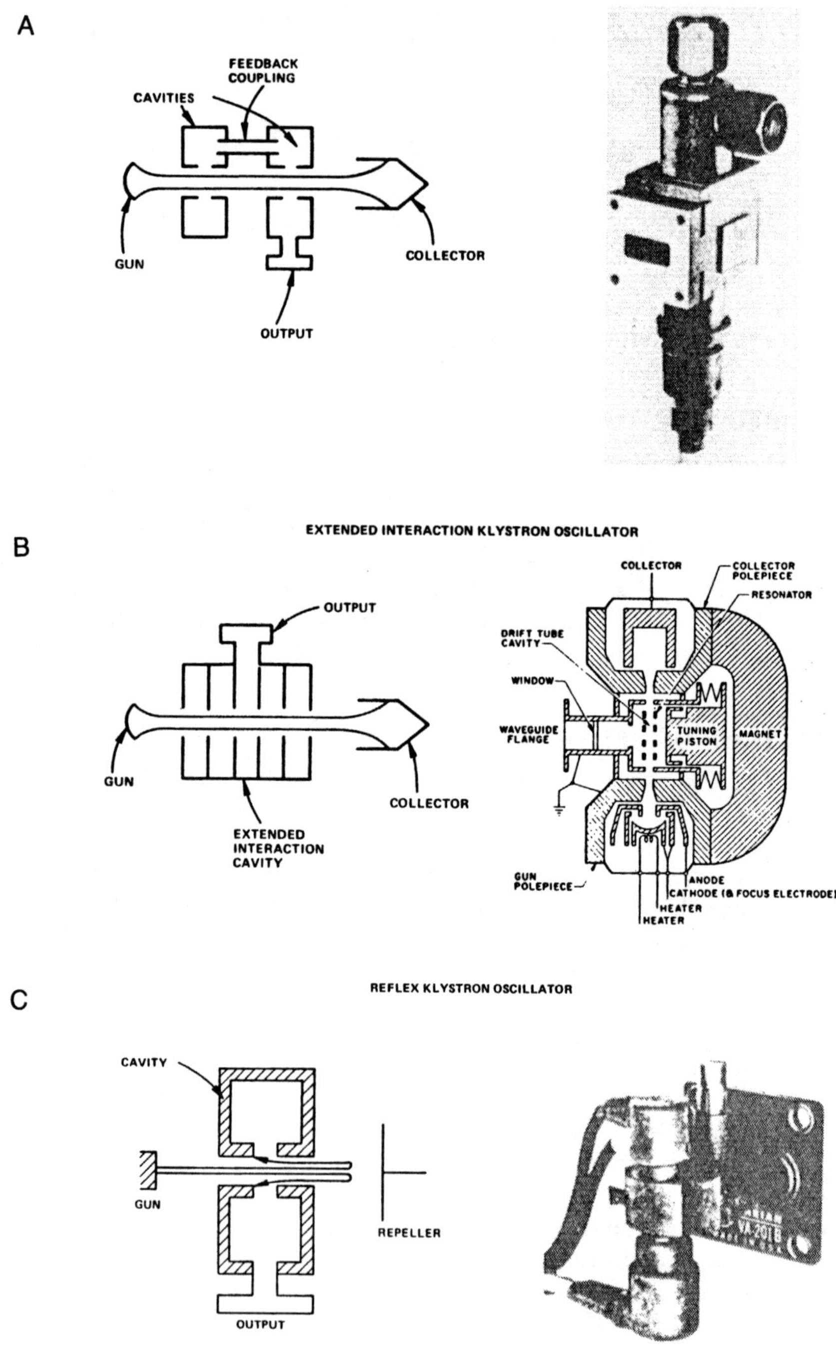

Figure 10.24 Three types klystron oscillators: (A) two-cavity; (B) extended interaction; and (C) reflex klystron oscillators.

As was known in Fig. 10.2, the extended interaction klystron oscillator can provide more power than any other oscillator except for the gyrotron, in the frequency range from 20 GHz to several hundred GHz.

The reflex klystron, shown in Fig. 10.24C consists of an electron gun, a single resonant cavity, and a repeller plate which is connected to a negative voltage supply. The electron beam travels through the cavity, then returns to pass through the cavity a second time by the negative repeller plate. This provides the necessary feedback for oscillation. The reflex klystron is a low power device, providing 50mW of power at 10 GHz with 1% efficiency. As shown in Fig. 10.2, the power output of the reflex klystron is well below the solid-state limit curve, and reflex klystrons are now being replaced by solid-state oscillators in most new systems.

10.12 BACKWARD WAVE OSCILLATORS

A backward wave oscillator, commonly called a BWO, is shown in Fig. 10.25. The unique feature of the BWO is that it can be electronically tuned over extremely wide bandwidths. The backward wave oscillator consists of an electron gun, focusing magnet, collector, and a helix interaction structure, just like a traveling wave tube.

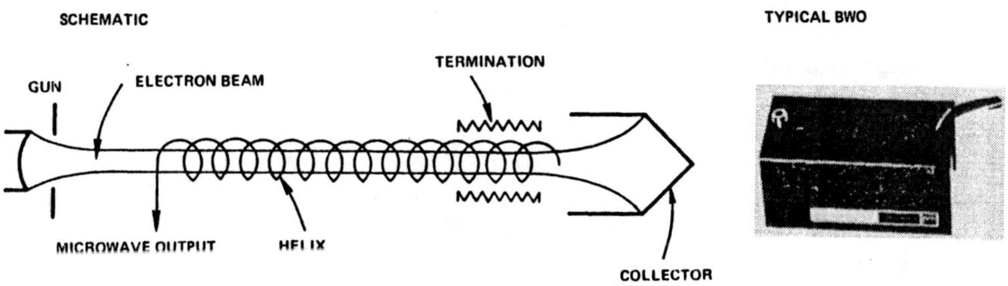

Figure 10.25 The construction of a backward wave oscillator.

The only difference between the BWO and a TWT is that the helix is used in a higher order of backward wave mode. In the BWO, the power is directed along the helix from the collector to the gun end. This is the opposite direction of that used in a TWT amplifier, and power is taken out the gun end of the tube. Although the electron is traveling in the opposite direction, it sees the same phase of the microwave field as it passes each turn of the helix. For example, if the electron enters at such time as it sees an acclerating field, it will continue to see an accelerating field at each successive turn of the helix. If the electron enters to see a decelerating field at the first turn of the helix, it will continue to see a decelerating field as it continues down the tube.

The electron does not travel with the microwave signal, but sees the correct phase at each helix turn from the microwave signal coming down the helix in a backward direction. Because the electrons see the same phase at each turn of the helix, just as in a TWT, the electron beam is progressively bunched, and the bunched electron beam interacts to generate increasing power on the helix. Because the power is traveling in a backward direction, the signal for amplification is always supplied by the tube itself and, consequently, oscillation occurs.

The frequency at which the oscillation occurs is related to the velocity of the electron beam, which is determined by the voltage of the electron beam, which is in turn determined by the voltage that is applied between the cathode and the helix. This allows the backward wave oscillator to be electronically tuned.

The typical backward wave oscillator, shown in Fig. 10.26B, can be electroncally tuned over the frequency range from 40 to 80 GHz, by varying the cathode

SEC. 10.13
CARCINOTRONS

to helix voltage from 200 to 1400V. The tube shown provides a minimum of 50mW of output power across this octave frequency band. The power capability of the backward wave oscillator is well below the solid-state limit curve. Consequently, backward wave oscillators are being replaced in most new equipment by yig-tuned transferred electron oscillators.

10.13 CARCINOTRONS

A carcinotron is a crossed field version of the backward wave oscillator. The backward wave oscillation interaction occurs in a crossed electric and magnetic field. Consequently, the carcinotron provides both electronic tuning and higher efficiency. By using a coupled-cavity type of interaction structure, rather than a helix, the carcinotron can provide much greater average power, but over a narrower bandwidth, than the backward wave oscillator.

The carcinotron is made in a circular format but uses an injected beam gun rather than an emitting sole. The tube provides 150W of CW output power from 8.5 to 11 GHz, with 25% efficiency, as the voltage is varied from 2.3 to 4.0kV.

10.14 MAGNETRONS

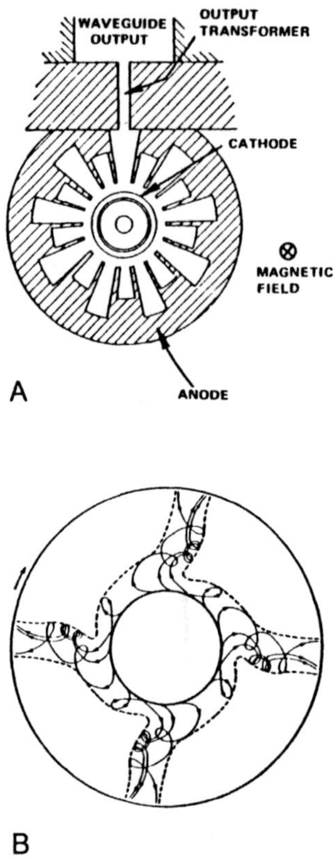

Figure 10.26 Magnetrons: (A) schematic of con struction and (B) the electron motion.

The magnetron is another type of tube oscillator and it is one of the most important microwave tubes. They were the first practical high power microwave tubes ever developed, and are currently produced in larger quantities than any other type of microwave tube. Commercial microwave ovens use magnetrons.

Magnetrons offer the advantages of high efficiency, high peak power, moderate average power, and low cost. Magnetrons have the highest efficiency of any microwave tube. In some magnetrons, an efficiency of over 80% is achieved. The disadvantage of magnetrons is their high noise output. The operation of magnetrons can best be understood by thinking of the magnetron as a crossed field amplifier with its output connected directly to its input.

Figure 10.26A shows a cross sectional drawing of a magnetron. The major element is a cylindrical cathode where electrons are emitted. In a crossed field amplifier this element is commonly called the cathode. The second element of the magnetron is the anode, which is where the microwave signal is propagated. In a crossed field amplifier this element is called the interaction structure.

The third element of the magnetron is the output where the microwave signal generated in the magnetron is taken out into an external transmission line, and the fourth element of the magnetron is the magnet which provides the magnetic field needed for the crossed field interaction.

As Fig. 10.26A shows, the magnetron is similar to the emitting sole crossed field amplifier. The magnetron has no input because it is an oscillator, not an amplifier. The interaction process is the same in the magnetron as in the crossed field amplifier. The microwave signal travels along the anode which is a coupled-cavity type of interaction structure with the vanes forming the cavity. Electrons are emitted from the cathode, but instead of being accelerated to the anode, they are bent by the combined effect of the electric and magnetic fields in order to move around the cathode.

As the electrons bunch under the influence of the microwave field on the anode, the electrons move towards the anode, exchanging potential energy to build up the microwave field, just as in crossed field amplifiers. Finally, the electrons strike the anode.

Figure 10.26B shows the complex electron motion in a magnetron. The output of the magnetron is fed directly to its input. In fact, the interaction region is continuous. Oscillation therefore occurs, and power is coupled from the anode circuit to an output transmission line. A magnetron can be fixed tuned, mechanically tuned, or electronically tuned. A typical fixed-tuned magnetron is shown in Fig. 10.27. This magnetron provides 600W of CW output power at 2450 MHz and is designed for use in microwave ovens. Magnetrons with this design are made in quantity levels of several million per year, and the unit cost for such a cooker magnetron is only $20.00.

Figure 10.28 shows a mechanically tuned magnetron. The magnetron anode is surrounded by a coaxial cavity. For this reason, the magnetron is called a coaxial magnetron. Each cavity formed by the vanes of the anode structure is coupled to the coaxial cavity, and the frequency of oscillation is then determined by the

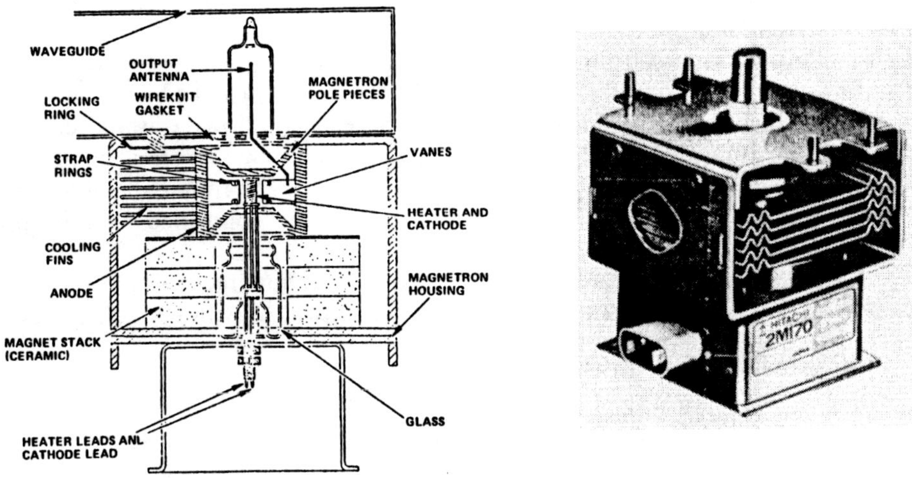

Figure 10.27 The fixed-tuned magnetron.

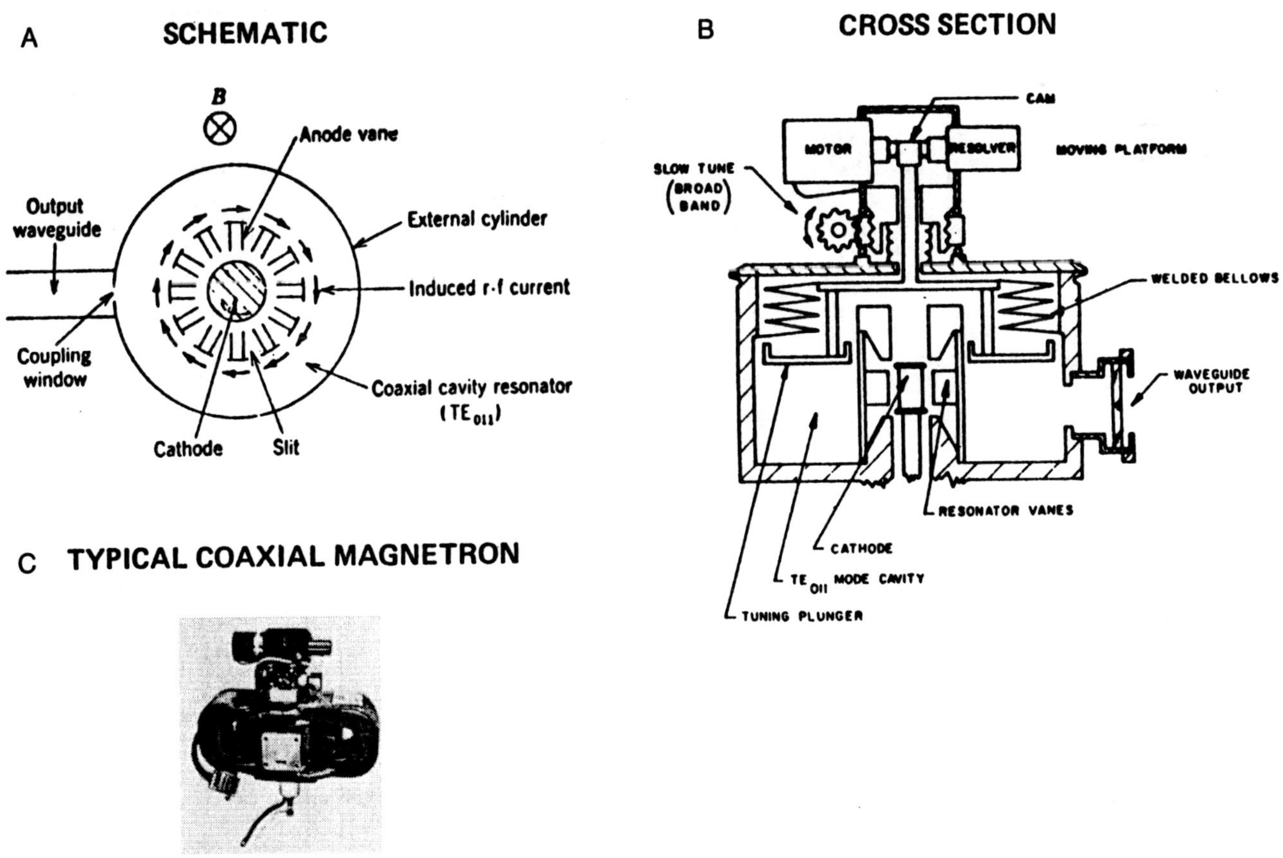

Figure 10.28 The construction of a mechanically tuned, or coaxial cavity, magnetron.

resonant frequency of the composite cavity. The composite cavity is tuned with a tuning plunger at the end of the coaxial cavity. The coaxial magnetron shown in Fig. 10.28C provides 200 kW of peak power, and 200W of average power. It operates at a voltage of 30kV, and a peak current of 32 amps. It is mechanically tuneable over the 8.6 to 9.6 GHz frequency range. The motor mounted on top of the magnetron controls the tuning plunger in the coaxial cavity.

A voltage-tuned magnetron, commonly called a VTM, is shown schematically in Fig. 10.29. The anode of the VTM consists of an array of inner digital fingers which are connected to an external cavity outside the tube's vacuum envelope by an anode ring. The sole is non-emitting, and the electron beam is injected into the sole anode space from one end. Electron flow in a voltage-tuned magnetron is in a spiral path as illustrated in sketch B. In contrast, as shown in a fixed-tuned or mechanically tuned magnetron, the electron flow is in a circular path around the emitting sole. The length of the spiral path in a VTM is controlled by the voltage between the ring cathode and the anode. The frequency of oscillation is determined by this length, and can therefore be controlled by the applied voltage.

Voltage-tuned magnetrons provide up to several hundred watts of CW power, and can be electronically tuned over a 20% bandwidth, with 60% efficiency.

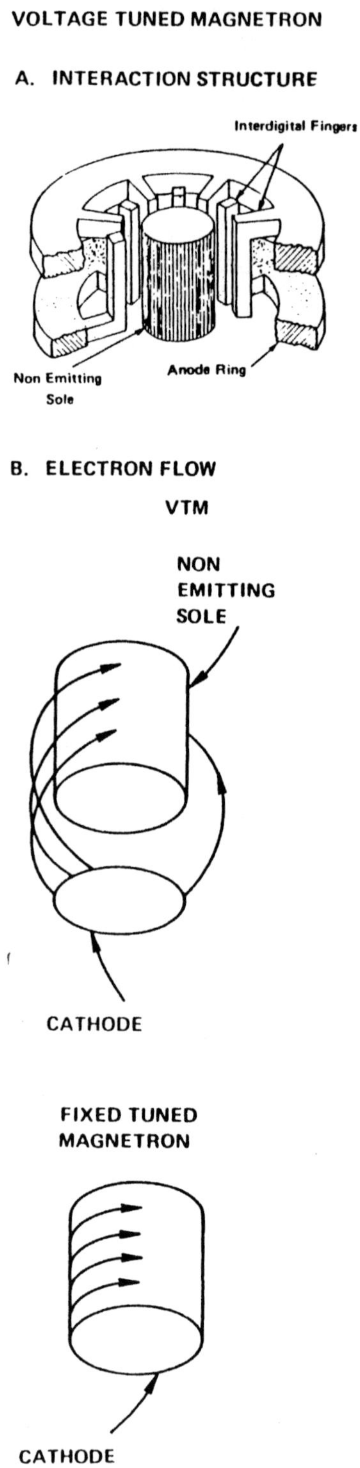

Figure 10.29 A voltage-tuned magnetron.

10.15 GYROTRON OSCILLATOR

The highest power microwave oscillator tubes are the gyrotrons. The gyrotron oscillator is similar to the gyrotron amplifier discussed previously, with internal feedback added so that the tube becomes an oscillator. The gyrotron oscillators have produced over 10kW of CW power, at 100 GHz, and over 1kW of CW power at 300 GHz. Both of these high frequency gyrotron oscillators require such high magnetic fields that cryogenically cooled solenoids must be used.

10.16 SUMMARY

Microwve tubes have been shown in this chapter to be capable of providing extremely high peak power, average power, high efficiency, and many other desirable characteristics such as broad bandwidth, high gain, and low noise. In view of these important characteristics, microwave tubes are, and will continue to be, a most important component in microwave systems.

CHAPTER 11

MICROWAVE SOLID-STATE DEVICES

PART I: DEVICE PHYSICS

W. Stephen Cheung

11.1 INTRODUCTION

This part gives a brief introduction to high frequency solid-state devices. A few low frequency solid-state devices are reviewed first. Microwave devices such as Gunn diodes, IMPATT (impact avalanche transit time) diodes, tunnel diodes, and varactor diodes are then discussed. This part, however, does not intend to cover every solid-state device in the microwave field.

The reader is assumed to have some basic understanding of atomic structure such as the meaning of orbital and valence electrons. They should consult the many excellent solid-state electronics books on the market for a more thorough treatment of atomic structure and low frequency solid-state electronics.

11.1.1 Semiconductor

Electrically, all substances can be classified under three categories: insulator, semiconductor, and conductor. All the atomic electrons of an ideal insulator are so tightly bound (by electrical force) to each nucleus that a tremendously large amount of energy is required to remove them from their orbits. In the other extreme, the valence electrons (electrons in the outermost orbit) of an ideal conductor require virtually no energy to be removed. The valence electrons of a semiconductor can be removed with a moderate amount of external energy.

Upon removal from its valence orbit, an electron can move around the inter-atomic spacing freely, this is the concept of conduction. From the energy standpoint, the valence band is the range of energies of those electrons occupying the valence orbits while the conduction band is the range of energies of freely moving electrons. Hence, an insulator is one in which the conduction band is separated from the valence band by a wide margin as illustrated by Fig. 11.1A. Similarly, the gap between the conduction band and the valence band is moderate for a semiconductor (Fig. 11.1B) and overlapping for an conductor (Fig. 11.1C).

If voltage (electric field) is applied, the electrons in the valence band may have a chance to move into the conduction band. Hence, a moderate voltage can cause current to conduct inside a semiconductor and very low voltage is necessary for a conductor. However, an extremely high voltage is required to overcome the wide energy gap between the valence band and the conduction band in an insulator.

A semiconductor can be "doped" with foreign atoms known as impurities so that the energy gap between the conduction and the valence band is narrowed further. The atoms of an *n*-type semiconductor have more loosely bound electrons than before, while those of a *p*-type have more deficiency in electrons than before. A deficiency in electrons means there are unfilled vacancies in the electron orbits.

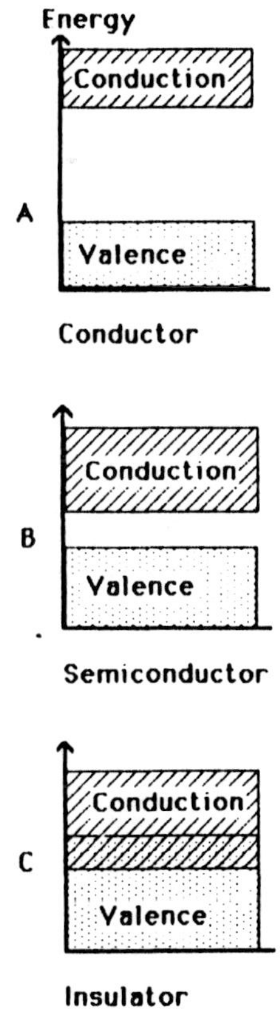

Figure 11.1 The band-gap approach to explain (A) insulator, (B) semi conductor, and (C) conductor.

A deficiency in one electron is called a *hole* and is equivalent to a positive charge. Also, a vacancy can be filled by a nearby orbital electron creating a new vacancy in the nearby atom. Thus, the hole in the nearby atom vacancy can "move" and the motion is equivalent to the flow of a positive charge.

A conduction electron is negatively charged and relatively light, while a hole is positively charged and heavy. The actual mass of an electron, m_e, is 9.1×10^{-31} kg. However, each valence electron is shared by several other atoms constituting the crystal structure. A convenient way to deal with this effect is to replace the electron's actual mass by an *effective mass* m^* which depends on the atomic structure. For silicon, the effective mass of an electron is $0.26 \times m_e = 2.4 \times 10^{-31}$ kg.

In an n-type semiconductor, the electrons are the dominant constituents of the electric current. Holes still exist but are relatively immobile. Hence, the electrons are the *majority* carriers and the holes are the *minority* carriers in an n-type semiconductor. On the other hand, the majority carriers of a p-type are the holes and the minority carriers are the electrons.

A semiconductor can be lightly or heavily doped. Symbollically, a heavily doped n-type semiconductor as n^+ and a heavily doped p-type semiconductor as p^+. Note that the superscript + is not a representative of electric charge.

Typical semiconductors include silicon (Si), germanium (Ge), gallium arsenide (GaAs), and indium phosphide (InP). The latter two are known as compound semiconductors. It turns out that simple semiconductors such as silicon and germanium are suitable for low frequency electronics while compound semiconductors work well for high frequency work.

11.1.2 ELECTRONIC MOBILITY

Figure 11.2A describes a free electron moving randomly between atoms (black dots). This free electron may acquire thermal energy from the ambient temperature to overcome the atomic attraction. Under no other influence, the electron's motion is totally random, i.e., it has equal probability of going in any direction three-dimensionally. The electron encounters and interacts with an atom, and then reflects in an arbitrary direction. Hence, the free electron has no average displacement in any preferred direction over a long period of observation time. The average distance that an electron travels before striking an atom is called the *mean free path*. The typical value of the mean free path is between 10^{-8} m and 10^{-6} m as compared to the typical atomic dimension of 10^{-11} m.

A quick estimate of the velocity of a free electron will be enlightening. At room temperature 300 K (27 C), the thermal energy is approximately equal to kT, where k is the Boltzmann's constant and is equal to 1.3×10^{-23} J/K and T is the absolute temperature of the ambient in K. (A more accurate expression for thermal energy is $\frac{3}{2} kT$, but we shall limit ourselves to using kT. The reader is referred to Ch. 9 on thermal energy and thermal noise.)

By approximating an electron's kinetic energy $\frac{1}{2} m^* v_t^2$ to the thermal energy kT, the thermal velocity v_t is found to be 1.8×10^5 m/s. Hence, the electron moves around randomly at 1.8×10^5 m/s. The time needed to cover a distance equal to the typical separation of two atoms, i.e., the mean free path of, let us say, 10^{-7} m is approximately 5.6×10^{-12} s, i.e., a few picoseconds. However, the average velocity in any preferred direction over a long period of time is still zero due to the randomness of the electron's motion.

Under the influence of an external electric field, e.g., the sample is sandwiched between two metal plates across which a battery is connected, the electron now has a preferred direction of motion. Although it will still have some randomness in other directions, the electron will tend to "drift" in a direction parallel to the electric field with a drift velocity v_d (Fig. 11.2B). The drift velocity obviously depends on the strength of the electric field E and the (effective) mass of the electron; it is usually less than the thermal velocity v_t. Another important factor

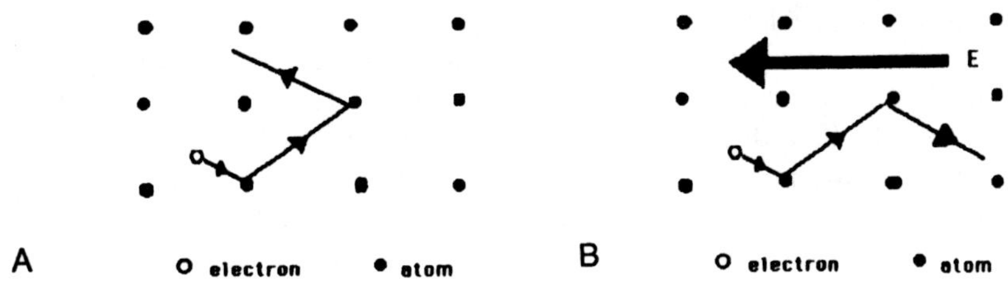

Figure 11.2 (A) The random movement of a free electron in between atoms. The average displacement three-dimensionally is zero. (B) the application of an electric field causes a net displacement of the electron.

in determining the drift velocity is the mean free path which reflects the sample's inter-atomic structure as well as the amount of impurities or dopants. Hence, if the electric field strength and the ambient temperature are kept constant, the drift velocity of an electron in one sample will be different from that in a different sample.

In a given sample, the ratio of the electron drift velocity to the applied electric field is called the *mobility*, i.e.,

$$\mu[(\text{electron) mobility}] = v_d/E \tag{11.1.2.1}$$

If the drift velocity is in m/s and the electric field is in V/m, then the SI unit for mobility is meter^2/volt.second. However, the common unit for mobility is in cm^2/volt.second. The conversion factor is to multiply m^2/V.s by 1000 to obtain cm^2/V.s.

The drift velocity, and therefore the mobility, is related to the mass of the electron, the mean free path of the sample's inter-atomic structure, and the ambient temperature. The concept of mobility also applies to holes. A hole is an electron vacancy site, and can be regarded as a positively charged but considerably more massive particle. Therefore, the hole's drift velocity is correspondingly smaller. Consequently, the mobility of a hole is smaller than that of an electron in the sample.

For comparison purposes, the approximated values of electron and hole mobility (in cm^2/V.s) of several semiconductors at room temperature (300 K) are given as follows:

Material	*Electron mobility*	*Hole mobility*
silicon (Si)	1500	600
germanium (Ge)	3900	2000
galium arsenide (GaAs)	8500	400
galium antimonide (GaSb)	4000	1400
indium phosphide (InP)	4600	150
indium arsenide (InAs)	33,000	460
indium antimonide (InSb)	78,000	750

The higher electronic mobility of GaAs and InP compared to Si and Ge is one reason that *n*-type GaAs and InP are suitable for high frequency work. Note also the tremendous potential of indium compound semiconductors in high frequency work.

11.1.3 p-n DIODE

A diode is formed by joining an *n*-type to a *p*-type as shown in Fig. 11.3A. Note that the two pieces of semiconductor are electrically neutral before they join. When the diode is not biased, the loosely bound electrons in the *n*-type are electrostatically attracted to the holes so that some electrons drift across the junction. Consequently, there is a build-up in electrons in the *p*-type and a depletion of electrons (or a build-up of holes) in the *n*-type. This mutual build-up prevents any further drifting of electrons, and a barrier called a depletion region is formed at the junction.

When forward biased by a battery, as shown in Fig. 11.3B, the electric field set up by the battery causes the electrons and the holes to move toward and overcome the barrier. The electron flow and the hole flow together constitute the overall current flow, which is the reason that the device is *bipolar*. The diode, therefore, conducts heavily when forward biased. The depletion region at the junction is negligibly small.

When reverse biased, as shown in Fig. 11.3C, the holes and the electrons are separately "held back" and no conduction will occur. The barrier is now considerably wider than the unbiased condition. The width of the depletion region depends on the reverse biased voltage. In practice, a small amount of reverse biased current is observed as illustrated by the *I-V* curve shown in Fig. 11.3D.

With reference to Fig. 11.3D, the reverse biased current will suddenly increase by many folds as the reverse biased voltage increases to a threshold value. This is due to the so-called secondary collisions between the electrons under the influence of a strong electric field, and the local atomic electrons resulting in a large release of electrons and holes. The voltage at the knee of the *I-V* curve is called the *breakdown* voltage and is extremely useful in the cases of the zener diode and the microwave IMPATT diode.

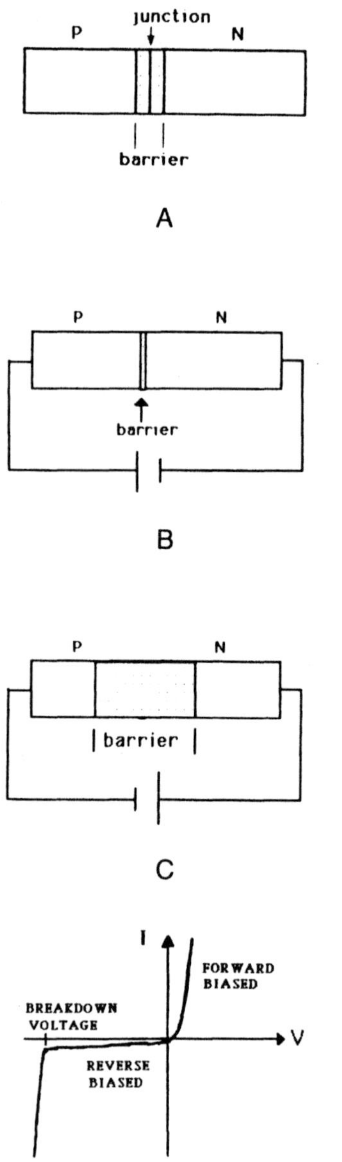

Figure 11.3 The construction of a pn diode and its behavior under voltage. (A) Joining a p-type and n-type forms a diode and a barrier at the junction. (B) Forward biasing the diode makes the barrier very small. (C) Reverse biasing the diode widens the barrier. (D) The I/V characteristics of a pn

11.1.4 BIPOLAR TRANSISTORS

Only the *npn* transistor will be discussed here, and we are also limited to an extremely simplified model. The actual construction and its simplified model are shown in Fig. 11.4A and 11.4B. The symbol of an *npn* transistor is shown in Fig. 11.4C.

The process described above is also valid for the holes, except that the direction of movement is opposite. Both the holes and the electrons are involved in this *transistor action*. The transistor action happens only when the two junctions are sufficiently close together.

In a loose sense, the *npn* transistor can be viewed as two diodes connected back to back. When connected as shown in Fig. 11.4D, the base-emitter junction is forward biased, while the base-collector junction is reverse biased. The important feature here is that the base-collector is reversed biased by a voltage very near the breakdown value (Fig. 11.3D). The base-collector biasing voltage helps push the base-collector into avalanche (it can be viewed as a controlled avalanche).

A large amount of electrons will accelerate from the emitter toward the collector. The base is very thin so that the majority (99%) of the electrons reach the collector; only 1% of the electrons exit at the base.

The controlling factor is the size of the (conventional) current into the base, which is the input terminal. The more the base current, the larger is the (conventional) current drawn from a battery connected to the collector. An operating point should be chosen so that as the input base current varies the collector current will swing between zero and the saturated value. Beyond the saturation point, further increase in base current will result in no more than the already saturated collector current.

11.1.5 JUNCTION FIELD EFFECT TRANSISTOR (JFET)

A low frequency *n*-channel JFET will be considered here. We shall start with a piece of *n*-type semiconductor (which is rich in electrons), such as silicon doped with phosphorous with metal electrodes on both ends. A battery B_1 is connected as shown in Fig. 11.5A. Electrons flow from one end called the source S to the

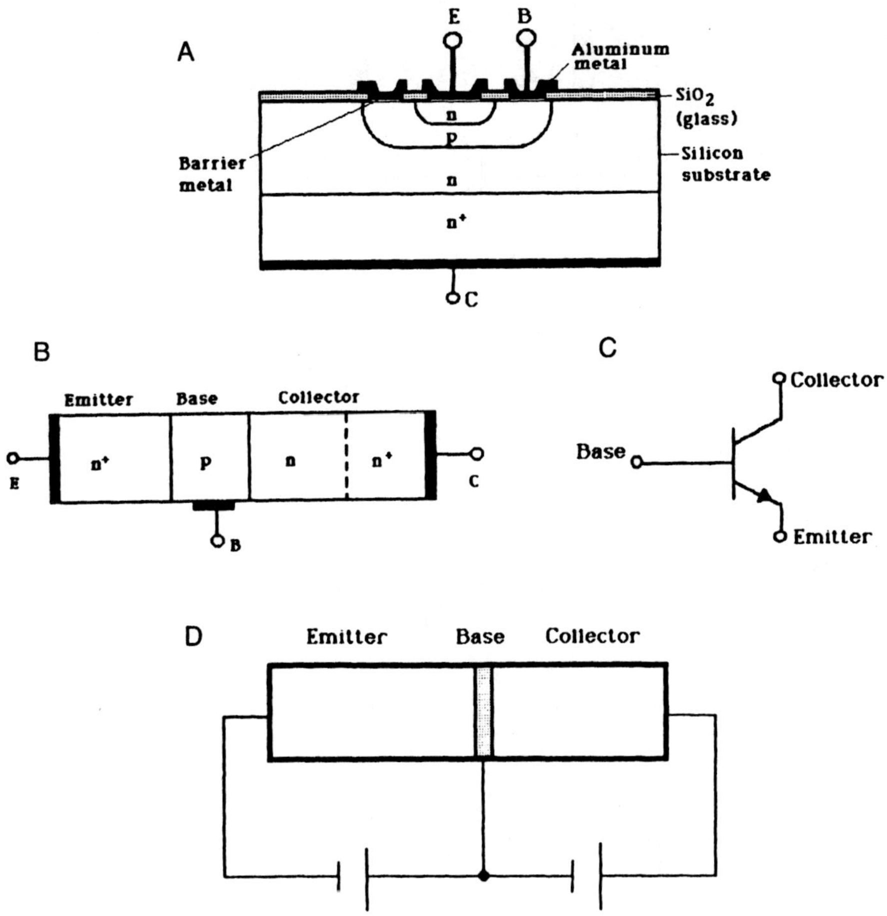

Figure 11.4 The construction and operation of a npn transistor. (A) actual construction; (B) simplified construction; (C) circuit symbol, (D) biasing of the transistor.

opposite end called the drain D. The magnitude of the electron flow is proportional to the battery voltage B_1.

Figure 11.5B shows that p-type material is deposited around the center n-type semiconductor. The electrode to the p-type is called the gate G, which is reverse biased relative to the drain D by battery B_2. The source is conveniently chosen as ground potential so that the drain is above and the gate is below ground potential.

Because the gate is reverse biased relative to the drain, the p-n junction is reverse biased and a depletion region is formed. The width of the depletion region is proportional to the battery voltage B_2. The depletion region cuts down the amount of electron flow. As the voltage of battery B_2 increases, the depletion region will grow and eventually cut off the electron flow. The threshold value of B_2 is called the *pinch-off voltage* at which the electron current is zero and beyond which further increase in B_2 will not bring the current back. Hence, the source to drain current starts out from a maximum value called the *saturated drain to source current*, I_{DSS}, at $V_g=0$ and then decreases to zero as the gate voltage increases in the negative direction and reaches the pinch-off voltage. This is illustrated by in Fig. 11.5C. The operating point must be chosen so that the output source to drain current swing should be contained between the saturated and the pinch-off value as a result of the input signal change.

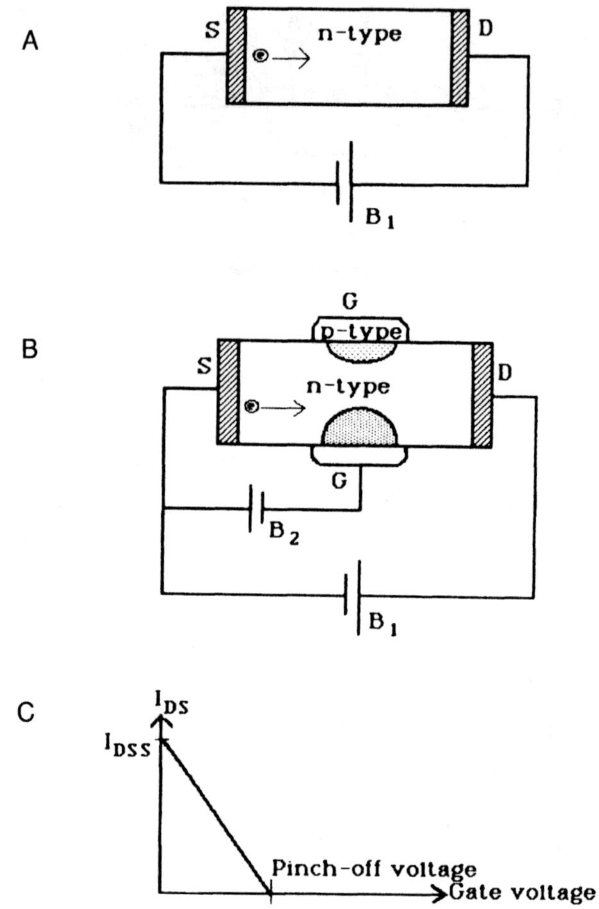

Figure 11.5 The conceptual construction of an n-channel JFET. (A) Starting with a bulk n-type semiconductor, (B) deposit p-type around the n-type, therefore forming a depletion region to limit the current flow. (C) The I/V behavior of the JFET.

The operation of a JFET can be visualized as follows. Consider a garden hose under water pressure. The amount of water flow is, of course, proportional to the pressure. Now imagine the hose is gripped by hand. The water flow will decrease due to the cut down in cross sectional area and eventually stops as the gripping pressure reaches a threshold.

The JFET described above is an *n*-channel because the center semiconductor is *n*-type and electrons are the majority current carriers. A *p*-channel JFET can be formed by using a *p*-type semiconductor as the center piece and *n*-type as the gate, the majority current carriers are now the holes. Because a JFET relies on one kind of carriers only (electrons for the *n*-type and holes for the *p*-type), the transistor is sometimes called *unipolar.*

In the case of the bipolar transistor, the collector current increases with increasing base current. As the base current increases beyond the linear range, the transistor is saturated and the collector current will reach a plateau. The source to drain current in a FET, on the other hand, decreases and finally stops as the gate voltage falls more and more below ground. Also, a bipolar transistor conducts by using both electrons and holes while a JFET relies on only one kind of carriers: electrons for a *n*-channel and holes for a *p*-channel.

The source to drain current is controlled by the size of the depletion region which is created by the gate to drain voltage. Because only the gate voltage but no current is required, the ideal input resistance is infinite, and a JFET is therefore an ideal voltage device. In practice, a small amount of leakage current in the order of nA is present, the practical input resistance is in the order of GΩ. As a quick note, the large input resistance of JFETs makes fabrication of operational amplifiers with input resistance more than 10^{13} Ω possible as compared to the MΩ value for operational amplifiers made of bipolar transitors.

11.1.6 MESFET

A *metal-semiconductor field-effect transistor* (MESFET), which is also known as a metallic Schottky barrier field-effect transistor, can operate at higher frequencies than the JFET. Gallium arsenide is used due to its higher electron mobility than that of silicon. The *p-n* junction at the gate is replaced by a *Schottky barrier* which is basically metal deposited on an *n*-type semiconductor and works on only one type of carrier, the electrons. Its operating frequency has been pushed up to tens of GHz, and is therefore suitable for microwave work. Interested readers are referred to books on semiconductor physics for more detailed discussions on various types of FETs.

11.1.7 TUNNEL DIODE

The tunneling effect will be discussed first. It is a quantum mechanical phenomenon with no everyday analogy. For example, a man is confined in a deep valley and has not enough energy to climb the cliff. Our everyday experience suggests that the man will be confined in the valley forever. According to quantum mechanics, there is a finite probability that the man will "appear" on the other side of the cliff despite his lack of energy, provided that the cliff is not infinitely high. This phenomenon is prohibited in any of the classical sciences.

The above example can be used to explain the tunneling effect of a *p-n* diode. If not biased the holes and the electrons drift across the junction until an electrostatic barrier, let us say, 0.3V, is formed to prevent further drifting. This barrier is analogous to the cliff in the above example. Normally, the diode must be forward biased by a voltage (electric field) higher than that of the barrier in order to overcome this barrier. According to the tunneling phenomenon, however, there is always a finite chance that an electron will "tunnel" through the barrier from the *n*-type to the *p*-type, or through a hole from the *p*-type to the *n*-type. In either case, a tunneling current is formed. The probability will increase if both types are heavily doped (about 100 to 1,000 times more than the normal 10^{17} impurity atoms per cubic centimeter) and the junction is narrow (about 100 to 1000 times narrower than the normal 1/1000in). Another factor that can increase the probability is the application of an external voltage, i.e., an electric field. Note that if the barrier is 0.3V as assumed, an external voltage from 0 to 0.3V will correspondingly increase the probability.

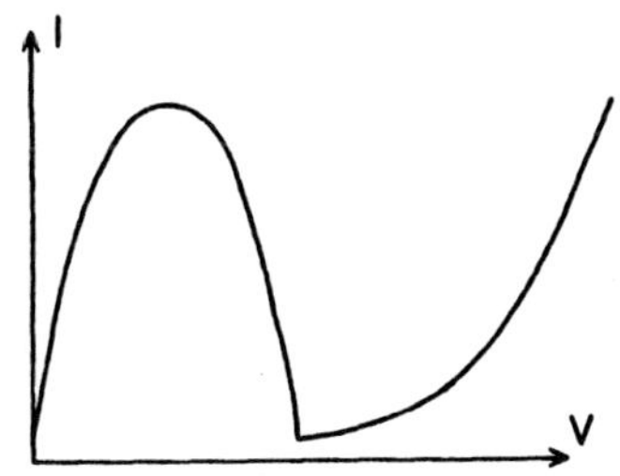

Figure 11.6 The I/V characteristics of a tunnel diode. Note the negative resistance region.

As the forward biased voltage increases, the current increases, followed by a region where the current drops as the voltage increases, whereupon the diode resumes its normal behavior. The *I-V* curve is shown in Fig. 11.6. The *negative resistance* region is where normal forward current start cutting down the tunneling current thereby decreasing the overall current as the voltage increases. The significance of negative resistance is extremely useful for oscillators and will be discussed in Part IV.

When the tunnel diode is reverse biased, the normal (reverse biased) barrier of a diode is overcome by the tunneling current due to the high concentration of dopants. The reverse biased current is therefore as high as the forward biased current, as shown in Fig. 11.6.

11.1.8 GUNN DIODE

A Gunn diode is typically made of an *n*-type compound semiconductor such as gallium arsenide or indium phosphide. Two modes of operation are common: bulk (transit-time) and limited space-charge accumulation (LSA). Using Gunn diodes to generate microwaves is very popular in the solid-state field.

Consider a piece of GaAs semiconductor biased by a battery, as shown in Fig. 11.7A. No junction effect or barrier occurs because the semiconductor is one piece, i.e., bulk instead of two pieces joined together. The semiconductor simply functions like a bulk resistor. Unlike semiconductors such as silicon, the atomic structure of the compound semiconductor causes an uneven and highly localized distribution of the external electric field strength. An electron moving from the cathode will undergo an acceleration and then a deceleration as it sees different electric field strength. Hence, the electron's motion is not uniform but bunched. Domains of electrons are said to have formed.

The bunching phenomenon is a bulk effect due to the slowing down of the electrons by the uneven distribution of local electric field. Atomically, two conduction regions exist and a valence electron starts out at the low conduction region and then switches over to the higher conduction region. Such an effect can be characterized by a decrease in electron mobility, which in turn is equivalent to negative resistance. Figure 11.7B shows the electron velocity *versus* the applied electric field strength. As the electric field increases, the electron velocity increases as expected from any normal semiconductor. Beyond a threshold electric field, the velocity suddenly slows down due to the switch-over to a higher conduction band and a negative resistance region is formed.

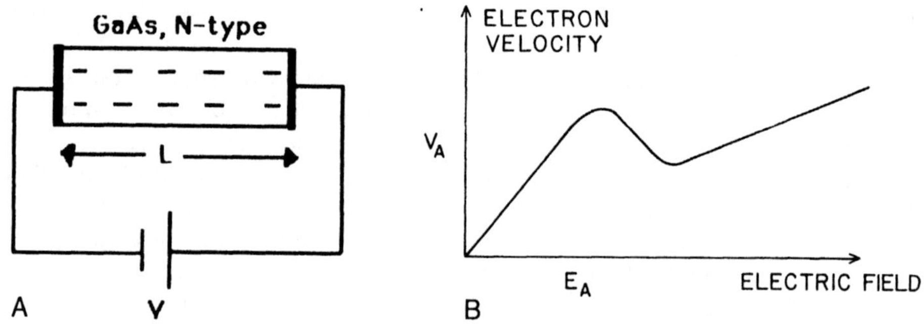

Figure 11.7 (A) The bulk mode of a Gunn diode. (B) The electron velocity versus applied electric field relationship. Note the negative resistance region.

The electron domains burst across the semiconductor at a frequency dependent upon the width of the semiconductor layer, which is typically about 10μm (10^{-5} m) thick. The typical drift velocity with which the electrons travel is about 10^5m/s making the transit time $\tau = 10^{-5}\text{m}/10^5\text{m/s} = 10^{-10}$s. The electron burst, therefore, repeats itself at approximately 10^{10} times per second, i.e., the frequency is 10 GHz. Hence, a Gunn diode biased by a dc voltage inside a cavity of the right dimensions, can result in CW microwave generation. The frequency is largely determined by the width of the semiconductor's active layer. Hence, high frequency CW signals can be formed with a very narrow active layer, which, however, results in limited power capability of the device.

The operation of the bulk or transit time mode is due to the buildup of domains which then burst across the semiconductor. The slowing down of an electron to "wait" for the upcoming one accumulates space-charges within the domain. The limited space-charge accumulation (LSA) mode eliminates the "waiting" mode for the domain to build up and operate directly on the negative resistance region of the Gunn device. The compound semiconductor is biased near the threshold of the negative resistance region and an RF signal is then applied. The RF signal periodically extends into the negative resistance region and signal amplification results.

11.1.9 IMPATT DIODE

When an ordinary *p-n* diode is reverse biased, very little current flows across the diode. A depletion region is said to have formed. As the reverse biased voltage increases further, a breakdown voltage, or avalanche voltage, will be reached. Beyond the avalanche, the reverse biased current will suddenly increase tremendously. A simplified explanation of the avalanche is due to the attraction of an electron in the *p*-type by the high voltage of the battery's positive electrode. The acceleration is so large that the electron collides violently with the local atoms releasing more electron-hole pairs in a multiplying manner. Consequently, the energetic electrons cross the depletion region and a large current is formed. The same process essentially applies to the holes.

An IMPATT diode is a *p-n* junction as shown in Fig. 11.8. When reverse biased the depletion region (drift zone) is chosen to extend into the heavily doped p^+ and n^+ sections. The end result of reverse biasing the diode is that bursts of electrons will cross the drift zone at microwave frequencies. Microwave energy can be extracted by operating the IMPATT diode in a cavity. The microwave frequency is related to the electron's transit time which is determined by the width of the drift zone.

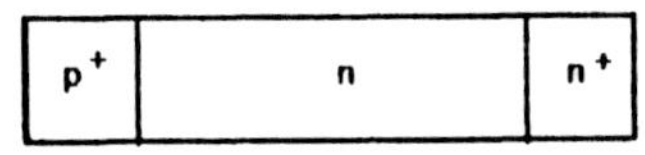

Figure 11.8 The construction of an IMPATT diode.

The IMPATT diode also exhibits negative dynamic resistance. When the diode is pulsed by a RF signal in and out of the breakdown threshold, the current burst turns out to be 180° out of phase with respect to applied voltage. Hence, the diode can also be employed in an amplifier circuit similar to that the tunnel diode. This will be discussed further in Part IV.

IMPATT diodes are compact, lightweight, and relatively easy to use. Microwaves of several hundred GHz and several watts are obtainable. The noise performance of the IMPATT is, however, greater than that of a Gunn diode.

11.1.10 VARACTOR DIODE

When an ordinary *p-n* diode is reverse biased, a wide depletion region is formed. The holes are held back in the *p*-type and the electrons in the *n*-type. The depletion region is electronically equivalent to a capacitor.

The actual width depends on the reverse biased voltage. If the reverse biased voltage varies, the width of the depletion region and, therefore, its capacitance will vary correspondingly. This is the principle of the varactor diode. A reverse biased diode, as will be discussed further in Part IV, can serve as (1) a matching element, (2) a variable capacitance of an LC or an RC voltage-controlled oscillator (VCO) circuit, and (3) a frequency multiplier owing to its nonlinearity.

PART II: BIPOLAR AMPLIFIERS

Salvatore Algeri

11.2 INTRODUCTION

11.2.1 Bipolar Transistor Amplifiers

Bipolar transistors are used to amplify signals in the low end of the microwave frequency range. Practical amplifiers are currently limited to 4 GHz, but advances in semiconductor fabrication techniques constantly challenge the upper frequency limits. Bipolar transistor amplifiers in this chapter will concentrate on RF circuit matching and assumes that the reader has a reasonable understanding of dc biasing. A new type of two-port network parameters, called scattering parameters, shall be introduced. It will be shown how the four scattering parameters define a two-port network. Once these scattering parameters are known, a specialized Smith chart called a *Z-Y* chart will be used to greatly simplify discrete inductor (*L*) and capacitor (*C*) matching. Once matching circuits have been developed, both the RF and dc circuits will be integrated to make a complete bipolar transistor amplifier.

11.2.2 DC BIAS CIRCUITS

Before concentrating on the RF circuits, a brief review of some common dc circuits is presented. Figure 11.9 shows a common resistor biasing scheme using two power supplies, V_{CC} in the collector and V_{BB} in the base. Remember that resistor R_B is selected to set the base current I_B.

$$R_B = (V_{BB} - V_{BE})/I_B \tag{11.2.1}$$

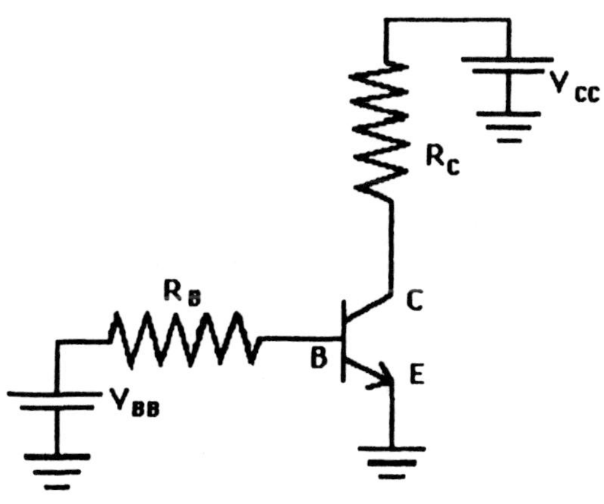

Figure 11.9 A typical low frequency biasing circuit of a npn transistor using two batteries.

Once the base current has been set, the collector current will be predetermined from the *I-V* curve of the transistor. the collector to emitter voltage V_{CE} is the difference between the power supply V_{CC}, and the voltage drop across the collector resistor R_C. Using Ohm's law, the value of R_C is the desired voltage drop across it divided by the collector current I_C through it.

$$R_C = (V_{CC} - V_{CE})/I_C \tag{11.2.2}$$

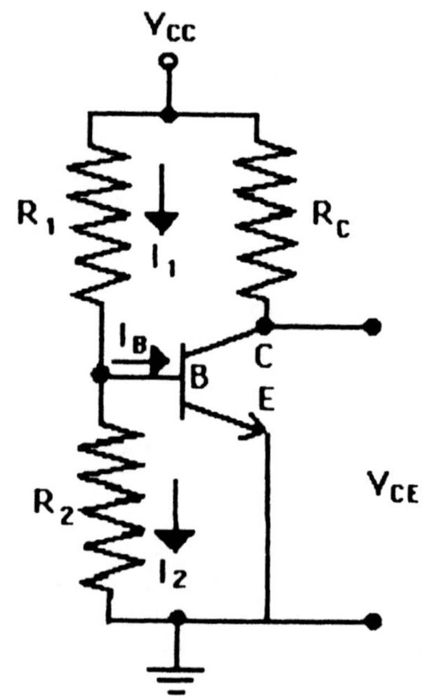

Figure 11.10 Using one battery to bias an npn transistor.

The collector current I_C and collector to emitter voltage V_{CE} determine the operating point of the bipolar transistor.

Another common type of dc bias circuit uses only one power supply as shown in Fig. 11.10. Resistors R_1 and R_2 constitute a voltage divider for the sole purpose of setting the base current I_B.

$$I_2 = V_{BE}/R_2 \tag{11.2.3}$$

$$I_1 = (V_{CC} - V_{BE})/R_1 \tag{11.2.4}$$

and

$$I_B = I_1 - I_2 \tag{11.2.5}$$

Once the base current has been set, the collector current will be predetermined from the *I-V* curve of the transistor. The collector to emitter voltage V_{CE} is the difference between the power supply V_{CC} and the voltage drop across the collector resistor R_C. By applying Ohm's law, the value of R_C is the desired voltage drop across it divided by the collector current I_C through it, i.e.,

$$R_C = (V_{CC} - V_{CE})/I_C \tag{11.2.6}$$

Both the collector current I_C and collector to emitter voltage, V_{CE} constitute the operating point of the bipolar transistor.

11.2.3 LOW FREQUENCY CIRCUIT

The common emitter configuration, or grounded emitter, is used for small-signal class A amplifiers. The low frequency equivalent RF circuit is shown in Fig. 11.11 for a bipolar transistor. Resistor, R_i, is the input resistance. Input signal, V_{in}, is across the base and emitter. The input signal current is identical to base signal current. Beta, β, is the current amplification factor from base to collector. Output signal voltage V_{CE} and signal current i_C are across load resistor R_L. The current gain is ouput signal current i_C divided by base signal current i_B

$$\text{Current Gain} = \frac{i_C}{i_B} = \frac{\beta i_B}{i_B} = \beta \qquad (11.2.7)$$

As can be seen from Eq. (11.2.7), the current gain is a constant β of the transistor. Most modern microwave bipolar transistors have β in the range of 100.

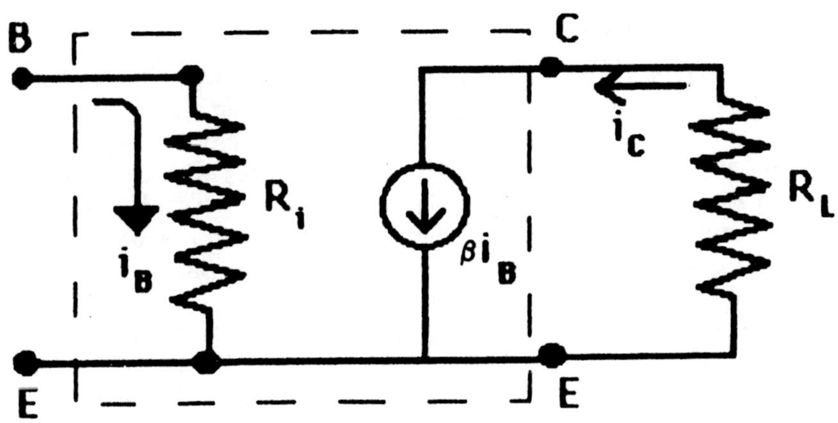

Figure 11.11 The low frequency equivalent circuit of a bipolar transistor.

The power gain is slightly more complex, and is defined as the output signal power divided by the input signal power.

$$\text{Power gain} = \frac{P_{out}}{P_{in}} = \frac{i_C^2 R_L}{i_B^2 R_i} \qquad (11.2.8)$$

Substituting βi_B for i_C,

$$\text{Power Gain} = \frac{(\beta i_B)^2 R_L}{i_B^2 R_i} = \frac{\beta^2 i_B^2 R_L}{i_B^2 R_i} \qquad (11.2.9)$$

$$\text{Power Gain} = \beta^2 R_L / R_i \qquad (11.2.10)$$

The power gain in the low frequency equivalent circuit is dependent upon the current gain β, and the ratio of resistors R_L and R_i of the transistor.

Exercise (11.2.3.1)

Let $\beta = 100$
$R_L = 50$ ohm
$R_i = 200$ ohm,

then,

$$\text{Power Gain} = \frac{(100)^2 \times 50}{200} = \frac{10{,}000 \times 50}{200}$$

$$\text{Power Gain} = 2500$$

$$\begin{aligned}\text{Power Gain (dB)} &= 10 \log (2500)\\ &= 10 \times 3.4 \text{ dB}\\ &= 34 \text{ dB}\end{aligned}$$

A bipolar transistor is a current amplifier. The output signal current is beta, β, times the input signal current. Power gain is dependent upon the current amplification of the device and the ratio of R_c and R_i. Now that the concept of gain has been reinforced, we can examine the complete equivalent circuit of a bipolar transistor operating in the lower microwave frequency range.

11.2.4 MICROWAVE CIRCUIT

Figure 11.12 is a schematic of the complete equivalent circuit at high frequency. Note the similarities to the low frequency equivalent circuit of Fig. 11.11. The controlled current source βi_B is the same. Note the differences between the two schematics. The input impedance in Fig. 11.12 changes with frequency due to the parallel RC combination of C_{BE} and R_{BE}. Signal output current i_C is divided between an output load resistor and the output resistance R_{CE} of the transistor. The most important connection is between the output (collector) and the input (base) of the device through a parallel resistor-capacitor combination. Resistor R_{BC} produces negative feedback, thereby reducing the overall gain of the device. Capacitor C_{BC} also results in negative feedback, but it is frequency dependent. As the frequency increases, the reactance of capacitor C_{BC} decreases. This decreasing reactance results in negative feedback, thereby decreasing gain. A typical gain response of a bipolar transistor operating in the lower microwave frequency range is shown in Fig. 11.13. Capacitor C_{BC} is the primary element in a bipolar transistor limiting the upper frequency use of these devices.

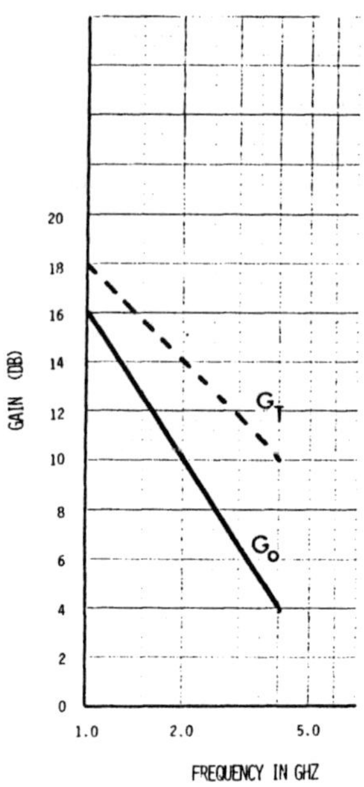

Figure 11.13 Gain response, G_O, of a bipolar transistor at microwave frequency due to high frequency capacitance limitation. G_T is the gain if the input and output circuits are matched.

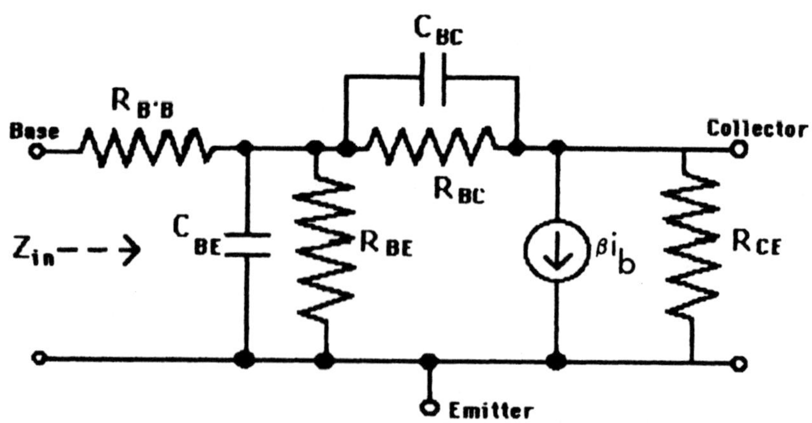

Figure 11.12 The high frequency equivalent circuit of a bipolar transistor.

11.2.5 Scattering Parameters

Due to the extreme complexity of the complete equivalent circuit, it is very difficult to analyze problems using conventional ac circuit theory. Therefore, another approach is used to help simplify the analysis and design of microwave bipolar transistor amplifiers. The transistor is treated as a two-port device with an input port and an output port. The technique utilizes a power flow concept for the two-port network, which corresponds to the input and output ports of a transistor. Before matching networks can be developed, our two-port transistor network must be defined in terms of four scattering parameters.

Scattering parameters, like impedance parameters, are used to define a two-port network. Impedance or Z-parameters use signal voltages and currents at both the input and the ouput of the two-port network. Scattering parameters, or S-parameters, use reflected and transistor signal voltages at both the input and the output of the two-port network. These four S-parameters can be defined by how they are measured (see Fig. 11.14A):

V_{1i} = the incident signal voltage entering the input $(1-1')$,
V_{1r} = the reflected signal voltage at the input $(1-1')$,
V_{1t} = the transmitted signal voltage at the output $(2-2')$ due to V_{1i},
V_{2i} = the incident signal voltage entering the output $(2-2')$,
V_{2r} = the reflected signal voltage at the output $(2-2')$,
V_{2t} = the transmitted signal voltage at the input $(1-1')$ due to V_{2i}.

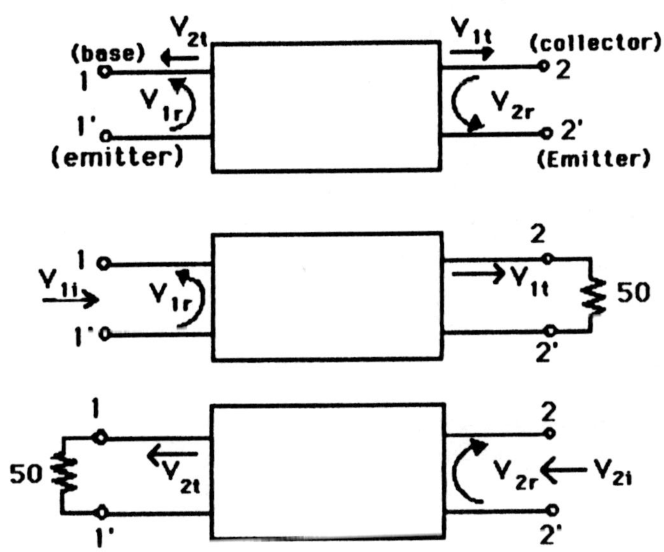

Figure 11.14 The definition of the S-parameters.

Consider Fig. 11.14B, where the signal voltage is incident at the input $(1-1')$ and the output is terminated by 50 ohm with a measuring device. Then

$$S_{11} = \text{input reflection coefficient} = V_{1r}/V_{1i}$$

which is simply the input reflected voltage ratio commonly known as ρ. Knowing ρ, the return loss can be calculated, as shown in Ch. 5. Also

$$S_{21} = \text{forward transmission coefficient} = V_{1t}/V_{1i}$$

which is simply the voltage gain or loss, depending on whether V_{1t} is larger or smaller than V_{1i}.

Consider Fig. 11.14C, where the signal voltage is incident at the output $(2-2')$ and the input is terminated by 50 ohm with a measuring device. Then,

$$S_{22} = \text{output reflection coefficient} = V_{2r}/V_{2i}$$

which is the output reflected voltage ratio identical to ρ except at the output terminals. Also

$$S_{12} = \text{reverse transmission coefficient} = V_{2r}/V_{2i}$$

which is the voltage gain or loss in the reverse direction. For an amplifier this term is commonly called the isolation as discussed in Ch. 5. Therefore, by making four simple microwave measurements the contents of a two-port network can be defined. The proof of this statement can best be illustrated by an example.

Exercise (11.2.5.1)

Four S-parameters were measured at 1 GHz as follows for a two-port network: $S_{11}=0.5$, $S_{21}=10$, $S_{22}=0.333$, $S_{12}=0.1$. Analyze what the two-port network is doing to the 1 GHz signal, and then define the network.

Solution: If S_{11} is 0.5, this means that the input reflection coefficient is 0.5. The input return loss, according to Ch. 5, is

$$\begin{aligned} RL\ \text{(input)} &= 20 \log (1/|S_{11}|) \\ &= 20 \log (1/0.5) \\ &= 6 \text{ dB} \end{aligned}$$

The input *VSWR* is

$$VSWR\ \text{(input)} = \frac{1+|S_{11}|}{1-|S_{11}|} = \frac{1+0.5}{1-0.5}$$

$$VSWR\ \text{(input)} = 3$$

Since S_{21} is greater than 1, the two-port network has a voltage gain* which is

$$\begin{aligned} \text{Gain (Voltage)} &= 20 \log|S_{21}| = 20 \log 10 \\ \text{Gain (Voltage)} &= 20 \text{ dB} \end{aligned}$$

The output return loss is

$$\begin{aligned} RL\ \text{(output)} &= 20 \log (1/|S_{22}|) \\ &= 20 \log (1/0.333) \\ &= 9.6 \text{ dB} \end{aligned}$$

The output *VSWR* is

$$VSWR\ \text{(output)} = \frac{1+|S_{22}|}{1-|S_{22}|} = \frac{1+0.333}{1-0.333}$$

$$= 2.0$$

Since S_{12} is less than 1, the two-port network has an isolation equal to

$$\begin{aligned} \text{Isolation} &= 20 \log (1/|S_{12}|) \\ &= 20 \log (1/0.1) \\ &= 20 \text{ dB} \end{aligned}$$

* *Editor's note:* Power gain in dB is 10 times the log value of the power ratio, but voltage gain is 20 times the log value of the voltage ratio; see Ch. 2.

Therefore, this two-port network is an amplifier with 20 dB gain which has an input *VSWR* of 3.0 and output *VSWR* of 2.0.

Scattering parameters are usually measured with the aid of an automatic network analyzer which is computer controlled. This equipment can measure the four S-parameters at discrete frequencies very rapidly, and produces a printout as shown in Fig. 11.15. These S-parameters can then be plotted on a Smith chart so that matching circuits can be developed.

COMMON EMITTER SCATTERING PARAMETERS

S-MAGN AND ANGLES

V_{CE}=8V, I_C=20mA

FREQUENCY (MHz)	S11		S21		S12		S22	
100	0.55	-75	32.3	145	0.019	60	0.81	-34
500	0.67	-156	11.2	102	0.038	41	0.32	-78
1000	0.68	-174	5.5	87	0.049	53	0.20	-89
2000	0.69	172	2.8	77	0.074	68	0.20	-97
2500	0.67	169	2.3	70	0.090	71	0.24	-97
3000	0.67	167	1.9	66	0.104	73	0.28	-100
3500	0.66	163	1.6	60	0.117	73	0.31	-99
4000	0.67	158	1.5	55	0.130	74	0.36	-99
4500	0.65	152	1.3	51	0.139	73	0.37	-99
5000	0.65	147	1.2	45	0.145	73	0.38	-100
5500	0.65	141	1.0	44	0.150	74	0.37	-103
6000	0.64	137	1.0	40	0.162	76	0.41	-106

Figure 11.15 A computer printout of the S-parameters for a typical bipolar transistor.

11.2.6 CONCEPT OF TOTAL GAIN

Let us assume that the S-parameters for a bipolar transistor have been measured and that both S_{11} and S_{22} are not equal to zero. This means that the transistor has some input and output reflected signal. Then, the term *total gain* is defined as

$$G_{Total}(\text{dB}) = G_0(\text{dB}) + G_1(\text{dB}) + G_2(\text{dB}) \tag{11.2.11}$$

where it is important to remember that all the gain values must be in dB.

The G_0 term is called the 50 ohm gain and is simply the gain of the device when measured in a 50 ohm system. This translates to the gain of S_{21}, or the forward transmission coefficient. The G_1 contribution to total gain is the additional gain measured for the transistor by matching the input to a 50 ohm generator using a lossless network. Also, the G_2 contribution to the total gain is the additional gain measured for the transistor by matching the output to a 50 ohm load using a lossless network.

When a lossless matching network is placed between the input to the transistor and the 50 ohm generator, no reflected signal will be produced, meaning that S_{11} will be zero. If S_{11} is zero, then the signal which was previously reflected before the insertion of the lossless matching network must have entered the amplifier. When the input signal to the transistor increases, then the output signal increase will be the gain times the input signal increase. This has the effect of increasing the gain. Therefore, G_1 is the additional gain measured when we make S_{11} zero at the input, which is called the matched condition. The G_2 contribution to total gain is the same as G_1 except at the output of the device.

A bipolar transistor may only have a 50 ohm gain of 4 dB at 4 GHz, which is the G_0 contribution (see Fig. 11.13). By matching the input the gain may increase by 4 dB which is the G_1 contribution. Also, by matching the output the gain may further increase by 2 dB which is the G_2 contribution to total gain. As a result of matching both input and output, the gain could be increased from a poor 4 dB of 4 GHz to a more reasonable 10 dB, as shown in Fig 11.16.

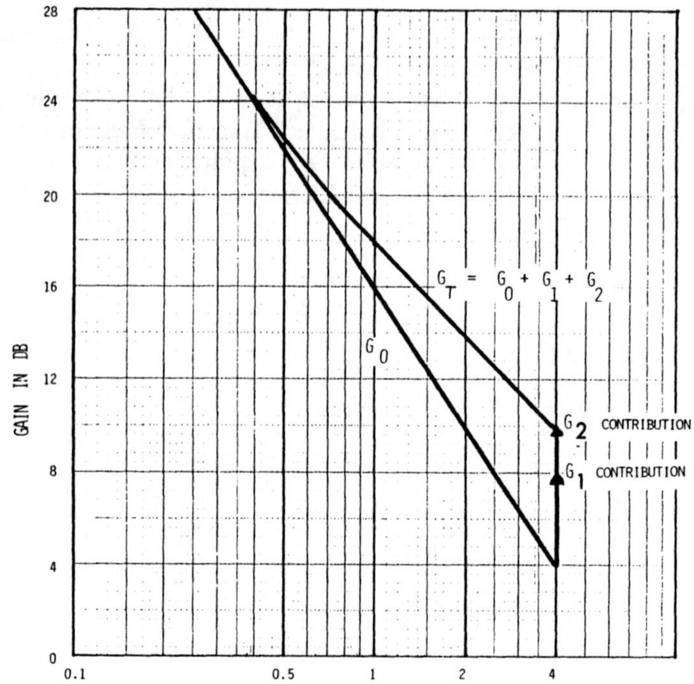

Figure 11.16 By matching the input and the ouput of the bipolar transistor, the total gain of the network can be maximized. Matching the input and the output result in contributions G_1 and G_2, respectively.

11.2.7 THE Z-Y CHART

By matching both the input impedance and output impedance of a microwave bipolar transistor amplifier, the designer can optimize the gain. This is accomplished by using lossless matching elements like capacitors and inductors. Resistors could be used for matching, but they would absorb energy rather than direct all of it to the transistor. A Smith chart could be used to develop matching circuits as discussed in the previous chapters. The primary disadvantage of a Smith chart is that it only shows impedance or admittance at any one time. We must rotate by 180 degrees about the center of the chart to change from impedance to admittance or from admittance to impedance. A *Z-Y* chart has been developed to eliminate the inconvenience of the 180 degree rotation.

A *Z-Y* chart shown in Fig. 11.17 consists of two Smith charts, where one is the mirror image of the other. The black chart is a normal Smith chart and represents the (normalized) impedance plane. The red chart is a mirror image of the black chart and represents the (normalized) admittance plane. To become more familiar with this chart some examples shall be used.

Exercise (11.2.7.1)

Plot $Z_1 = 1 + j1$ and determine its VSWR. Also determine the admittance $Y_1 = 1/Z_1$.

Solution: To plot Z_1 we must use the black chart because Z_1 is a normalized impedance. Find the black 1.0 circle and $+j1$ reactance curve. Where the circle and the curve meet is the point $1 + j1$ (see Fig. 11.18). Note that the key at the left side of the chart (black) shows which are the positive and negative reactive parts of the chart. The *VSWR* can be obtained by plotting a circle about the center through the point Z_1. Where this circle crosses the real axis between 1 and infinity is the *VSWR*. In this case it is 2.6 using the red graph.

To plot the admittance given an impedance using a Smith chart, we must rotate 180 degrees about the center. In this case all we need to do is read the black

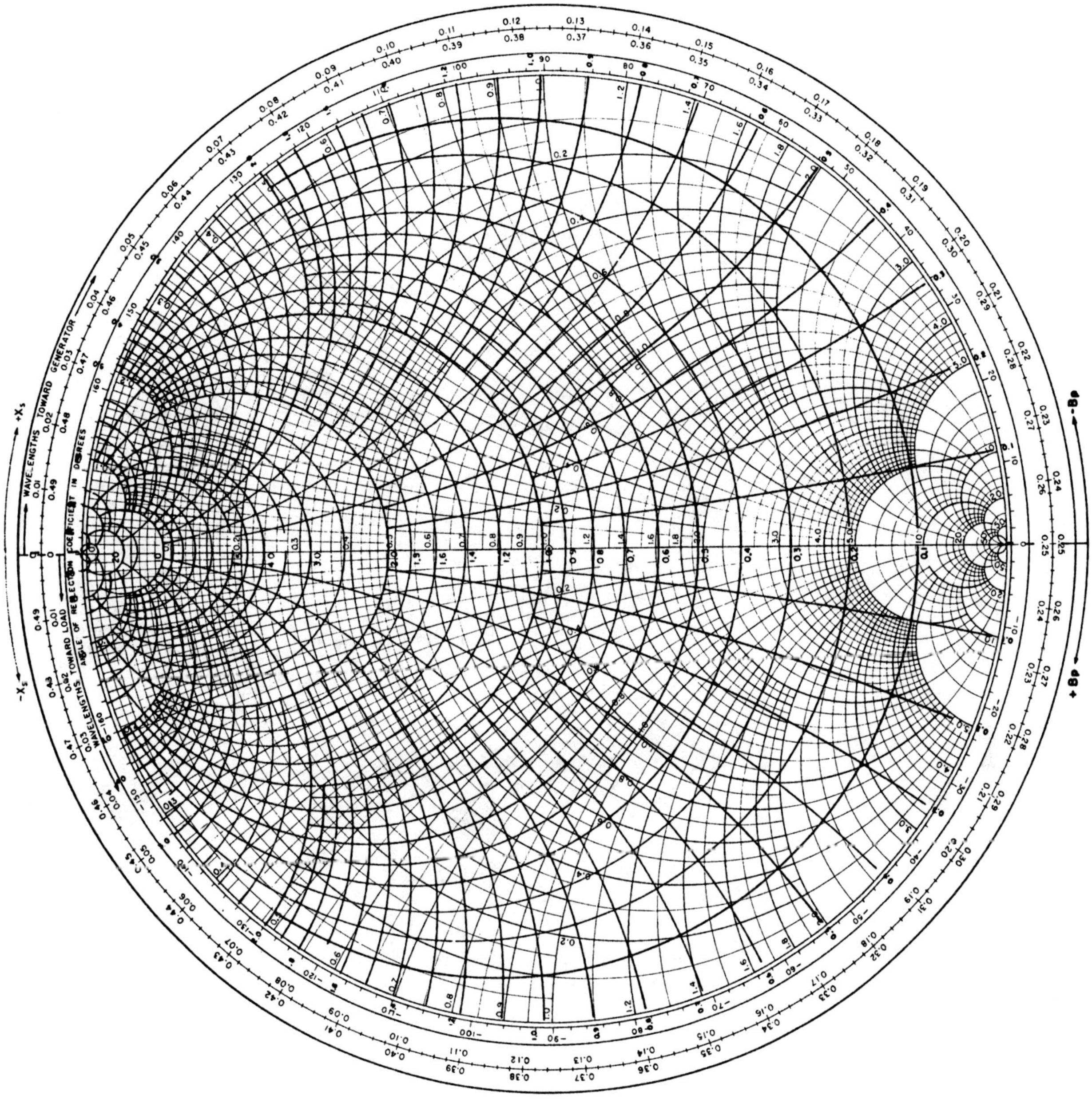

Figure 11.17 A Z-Y chart. Impedance coordinates are in black and admittance coordinates are in red.

chart at the same point. Note that the key to the right (red) tells us that the upper half of the red chart is negative susceptance. Therefore, the point lies on the red 0.5 circle and the red $-j0.5$ susceptance curve. The admittance Y_1 is $0.5-j0.5$.

How to Use the Z-Y Chart for Matching

Directions of movement on either constant resistance (black) circles or constant conductance (red) circles defines either series or shunt matching elements. These movements define capacitors or inductors which are lossless elements.

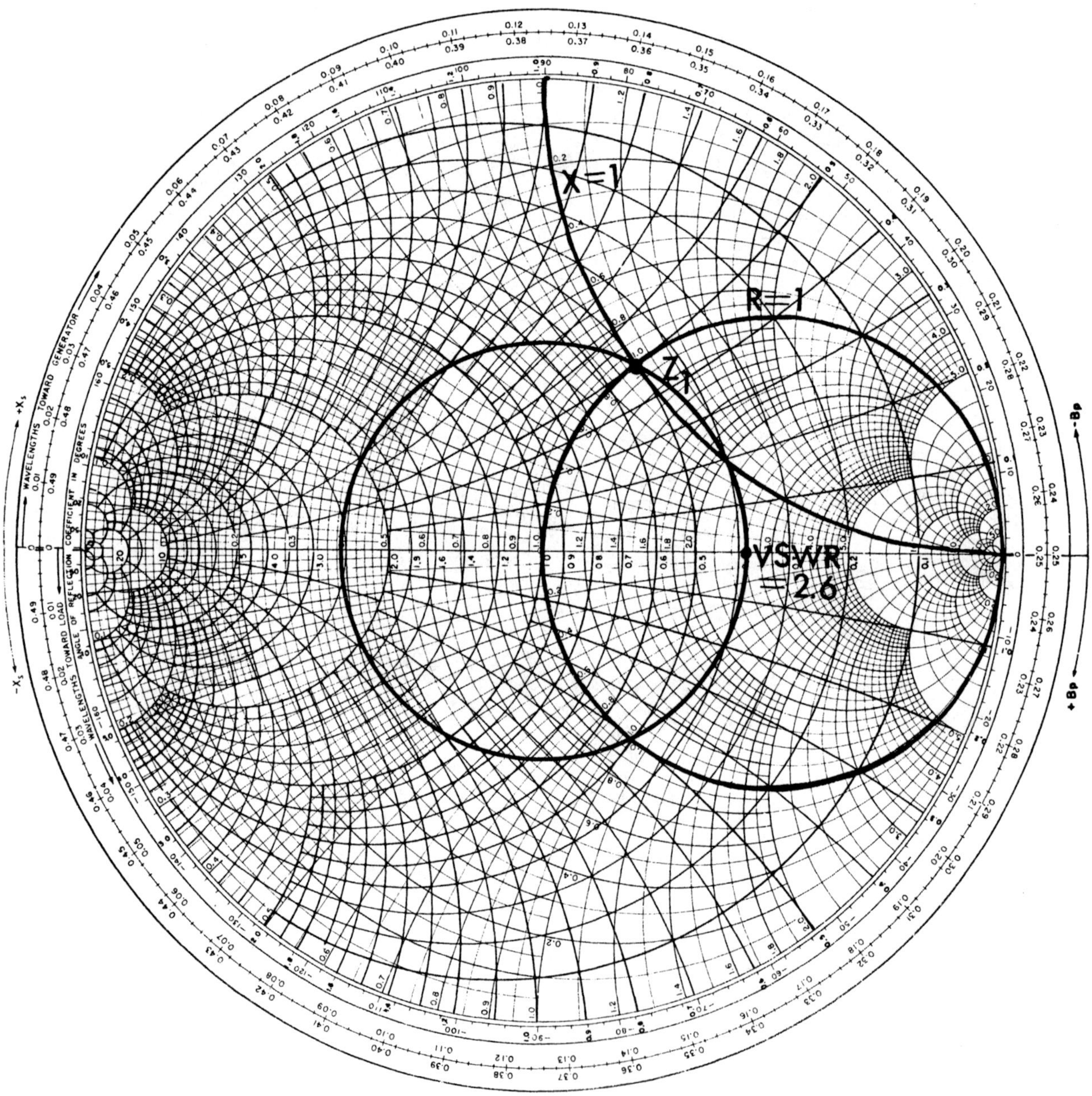

Figure 11.18 *Use the Z-Y chart to plot* $Z_1 = 1 + j1$ *on the black coordinates and find the corresponding Y value in red. The VSWR can also be found.*

Moving along constant reactance curves (black) or constant susceptance curves (red) defines series or shunt resistors, which are lossy elements and not generally used for matching.

Four simple rules to follow and their equivalent circuits are shown in Fig. 11.19:

A. Black Circle Movement (constant resistance):
   1. Clockwise direction means series inductor (Fig. 11.19A);
   2. Counterclockwise direction means series capacitor (Fig. 11.19B).

B. Red Circle Movement (constant conductance):
   1. Clockwise direction means shunt capacitor (Fig. 11.19C);
   2. Counterclockwise direction means shunt inductor (Fig. 11.19D).

By following these four simple rules, we can develop matching circuit topologies at any one frequency more easily than using a standard Smith chart.

Once the S-parameters of a bipolar transistor are measured, the input reflection coefficient S_{11} and the output reflection coefficient S_{22} can be plotted on a *Z-Y* chart. If S_{11} is not at the center of the chart, then a mismatch condition exists at the transistor input. Remember, to increase the total gain G_T by the G_1 contribution, we must match the input to the 50 ohm generator using a lossless network. The *Z-Y* chart is used as a tool to help develop a matching circuit. First, the *Z-Y* chart will be used to develop a matching circuit topology, and later it will be used to calculate the values of the circuit elements. To illustrate this, let us try a couple of examples.

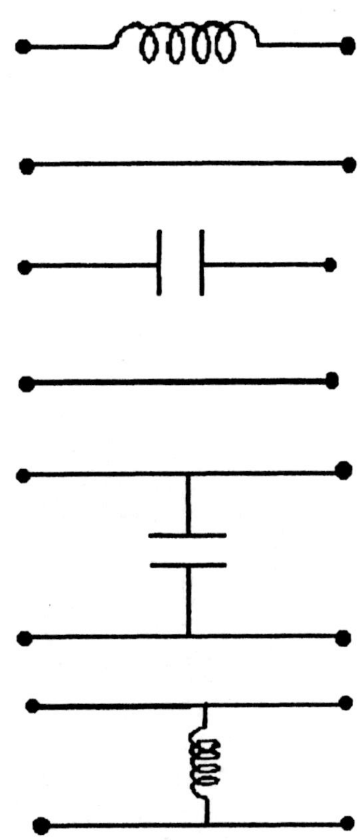

Figure 11.19 Equivalent circuits for movements along black and red coordinates on the Z-Y chart.

Exercise (11.2.7.2)

Let us assume that the S_{11} of a bipolar transistor at 1 GHz is the normalized value of $2-j1$. Develop two matching circuit topologies which would produce no reflection when placed between the input of the transistor and a 50 ohm generator.

Solution: First, we must plot $S_{11}=2-j1$ using the black chart (see Fig. 11.20). If no reflection is to be produced, we must move from S_{11} on the chart to the center of the *Z-Y* chart. There are two ways in which this can be accomplished:

a) Note that S_{11} conveniently lies on the red 0.4 circle. Follow the red 0.4 circle clockwise and stop at the unity black 1.0 circle. This first element (1) maps out a shunt capacitor. Next, we follow the unity black circle clockwise to the center of the chart. This second element (2) maps out a series inductor. Therefore, one possible schematic would be a series inductor L_2 followed by a shunt capacitor C_1 as shown in Fig. 11.21A. If the values of C_1 and L_2 are properly selected, then the input impedance Z_{in} would be $1.0+j0$, i.e., zero reflection or perfect match.

b) Another solution would be to follow the red 0.4 circle in a counterclockwise direction and stop at the unity black circle on the upper half of the chart. This maps out a shunt inductor (3). Now we follow the unity black circle in a counter-clockwise direction to the center of the chart. This maps out a series capacitor (4). Therefore, our schematic would be a series capacitor C_4 followed by a shunt inductance L_3 as shown in Fig. 11.21B. If the values of L_3 and C_4 are selected properly, then the impedance Z_{in} would be $1.0+j0$, i.e., zero reflection or perfect match.

Example (11.2.7.3) The S_{22} of a bipolar transistor operating at 1 GHz is $0.4-j1.0$. Develop two matching circuit topologies which would produce no reflection when placed between the output of the transistor and a 50 ohm load.

Solution: First, plot $S_{22}=0.4-j1.0$ using the black chart (see Fig. 11.22). If no reflection is to be produced, we must move from S_{22} on the chart to the center of the *Z-Y* chart.

a) For the first element, follow the black 0.4 circle clockwise and stop at the unity red circle. This first element (1) maps out a series inductor. Next, we follow the red 1.0 circle counterclockwise to the center of the chart. This second element (2) maps out a shunt inductor. Therefore, one possible schematic is a series inductor L_1 followed by a shunt inductor L_2 as shown in Fig. 11.23A. If the values of L_1 and L_2 are properly selected, then the impedance Z_{out} would be $1.0+j0$. This circuit could now be connected to a 50 ohm load and there would be zero reflection.

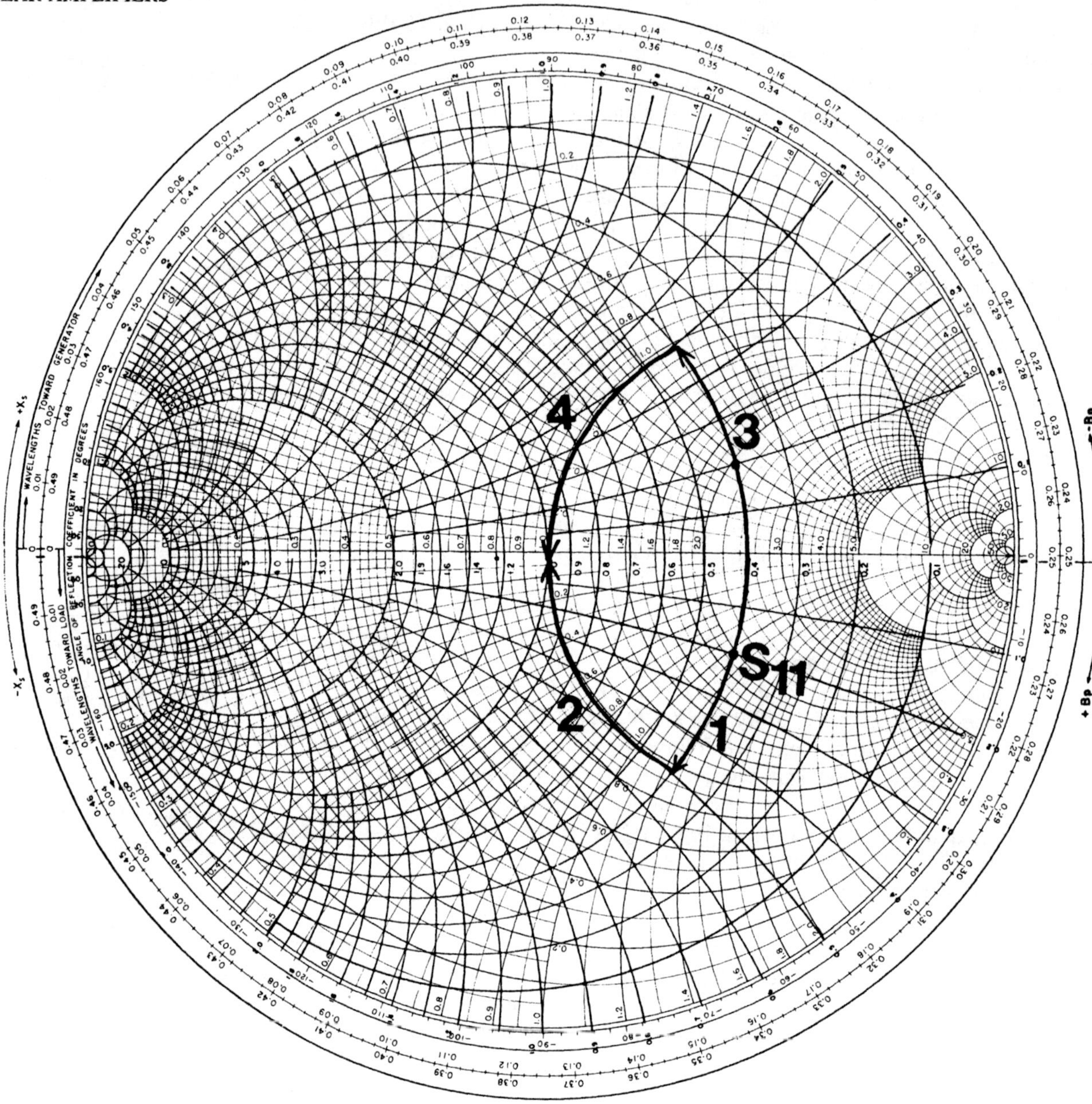

Figure 11.20 The S-parameter S_{11} is given to be $2 - j1$. Shown are two ways (paths 1-2 and 3-4) to go to the center of the chart.

b) A second solution would be to follow a red circle counterclockwise, but our S_{22} does not lie on a red circle. It lies between the 0.3 and 0.4 circles, and appears to be a conductance of 0.34. Because the 0.34 red circle is not plotted on our graph, we should draw in the 0.34 red circle and follow it counterclockwise stopping at the unity black circle. This maps out a shunt inductor (3). Now follow the black unity circle in a clockwise direction to the center of the chart. This maps out a series inductor (4). Therefore, our schematic is a shunt inductor L_3 followed by a series inductor L_4 as shown in Fig. 11.23B. If the values of L_3 and L_4 are properly selected, then the impedance Z_{out} would be $1.0 + j0$. This circuit could now be connected to a 50 ohm load and there would be zero reflection.

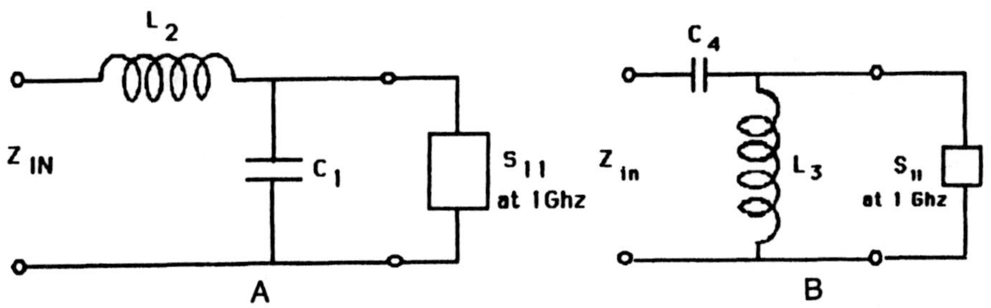

Figure 11.21 (A) Using a series inductor L_2 followed by a shunt capacitor C_1 to match the input of the bipolar transistor circuit at 1 GHz. (B) An alternative is to use a series capacitor C_4 followed by a shunt inductor L_3.

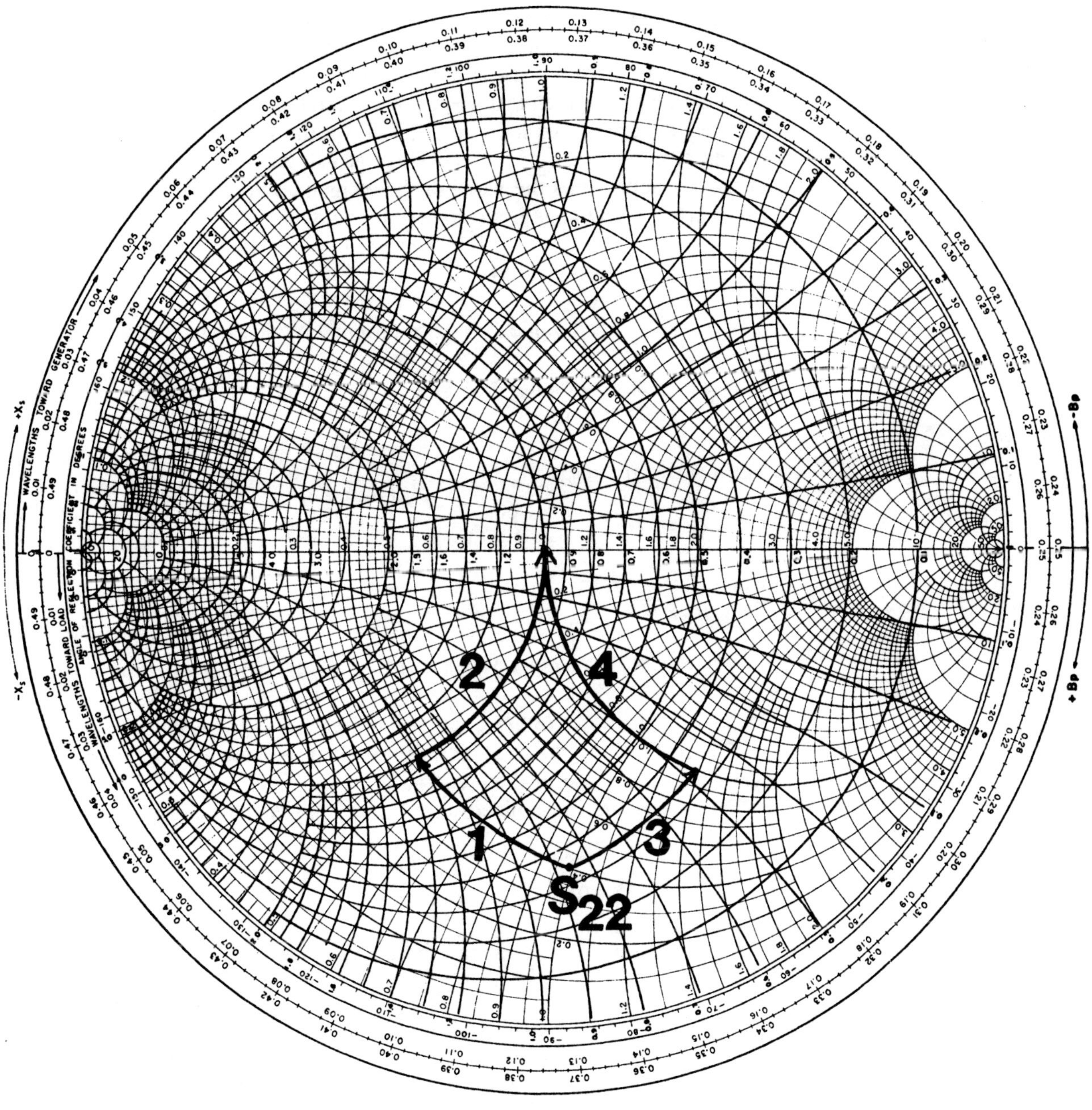

Figure 11.22 The S-parameters S_{22} is given to be $0.4 - j1.0$. Shown are two ways paths 1-2 and 3-4) to go to the center of the chart.

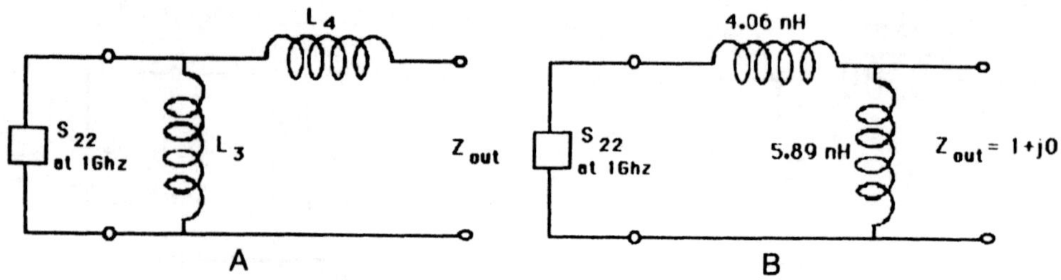

Figure 11.23 (A) *Using a series inductor* L_1 *followed by shunt inductor* L_2 *to match the output of the bipolar transistor circuit at 1 GHz. (B) An alternative is to use a shunt inductor* L_3 *followed by a series inductor* L_4.

How to Calculate Matching Circuit Element Value

The technique for calculating the values of the matching elements can best be illustrated by an example. Let us use solution a) from *Example (11.2.7.3)*. This matching circuit topology is a series inductor L_1 and shunt inductor L_2. First, we must plot S_{22} which is $0.4-j1.0$ and sketch in our mappings (see Fig. 11.24A). Notice that our series inductor L_1 follows the 0.4 constant resistance circle. It begins at a reactance of 1.0 and stops at 0.49 at the red unity circle. The length of the line represents the value of the element. Therefore, the normalized impedance, z_1, is the change in normalized reactance,

$$z_1 = 1.0-0.49=0.51$$

The un-normalized value Z_1 is z_1 multiplied by 50 ohms. Therefore,

$$Z_1 = z_1 \times 50 \text{ ohm} = 0.51 \times 50 \text{ ohm}$$
$$Z_1 = 25.5 \text{ ohm}$$

For a series inductor,

$$Z_1 = 2\pi f L_1$$

where f = 1 GHz.
Therefore,

$$L_1 = Z_1/2\pi f = (25.5/2\pi \times 1 \times 10^9) \text{ henries}$$

Shunt inductor, L_2, follows the red unity circle. It begins at a susceptance of 1.35 and stops at the real axis or zero susceptance. The length of the line represents the value of the element.

Therefore, the normalized admittance, y_2, is the change in normalized susceptance,

$$y_2 = 1.35-0=1.35$$

The un-normalized value Y_2 in mho is y_2 divided by 50.
Therefore,

$$Y_2 = y_2/50 = 1.35/50 = 0.027 \text{ mho}$$

For a shunt inductor $Y_2 = 1/2\pi f L_2$
Therefore,

$$L_2 = \frac{1}{2\pi f Y_2} = \frac{1}{2\pi \times 10^9 \text{Hz} \times .027 \text{ mho}}$$

$$L_2 = 5.89 \text{ nH}$$

The completed matching schematic is shown in Fig. 11.24B.

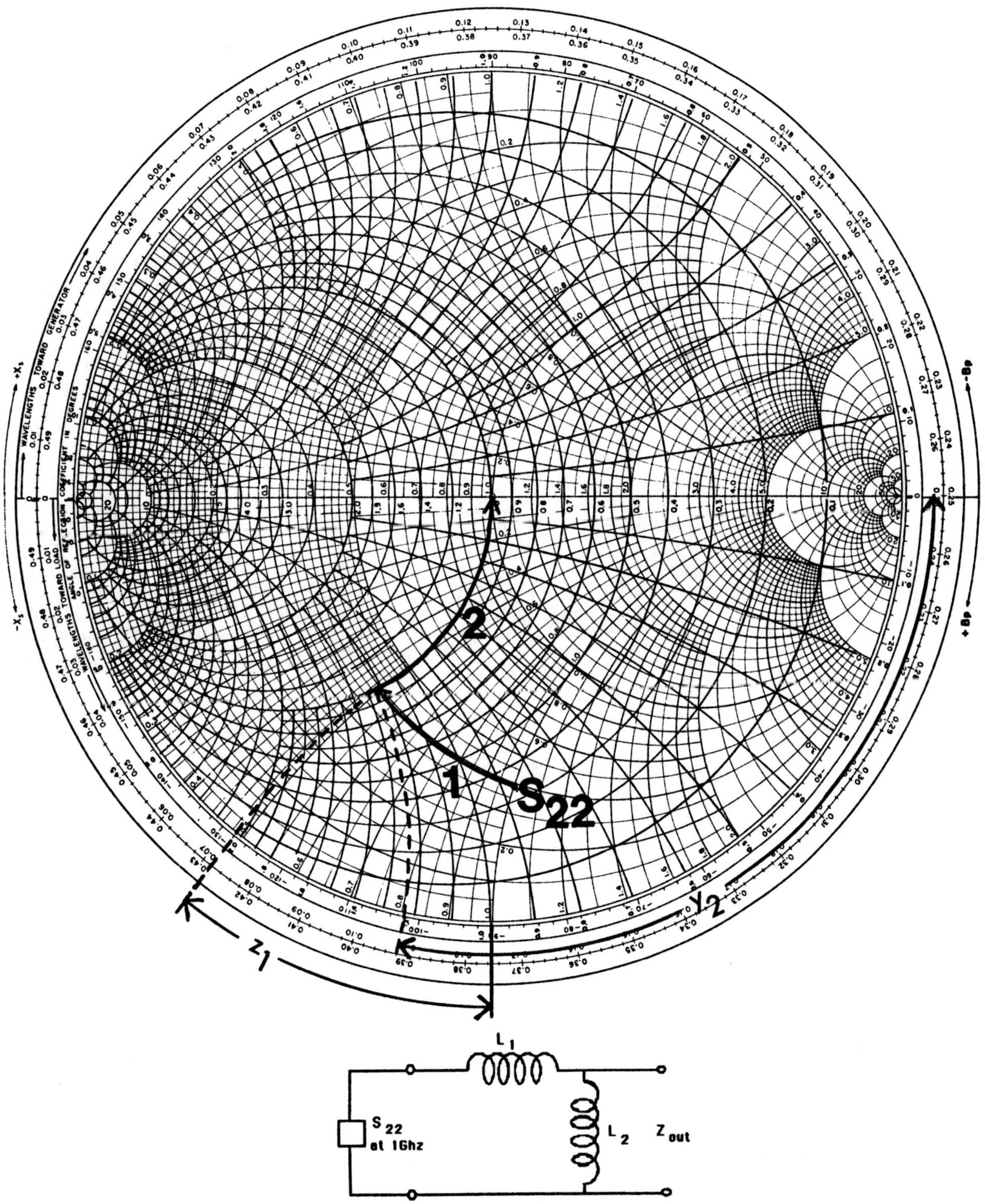

Figure 11.24 (A) Detailed calculation of the matching elements z_1 and y_2. (B) The output matching circuit with calculated values given.

11.2.8 Integrating Both the DC and RF Circuits

Both the dc bias and RF circuits have been discussed separately. Integrating these circuits is not necessarily as simple as we may think. Functionally, the dc circuit must provide bias to the transistor so that it can amplify. Likewise, the RF circuits must match the input and output reflecting coefficients to achieve the desired gain level. Each of these circuits must be integrated without one upsetting the other.

Figure 11.25A shows a typical RF circuit. The entire RF circuit consists of a 50 ohm generator, input match, transistor, output match and load. Values for the matching elements may be as shown. A typical dc circuit is shown in Fig. 11.25B where the dc conditions are $I_B = 0.1\text{mA}$, $I_C = 10\text{mA}$, and $V_{CE} = 10\text{V}$.

Before we integrate these two circuits we must define two elements. An RF choke is a large inductor which passes dc signals but blocks RF signals from passing through it. A general rule is that the RF choke (RFC) should be about ten times the largest inductor value in the matching circuit. A coupling or bypass capacitor is a large capacitor which passes RF but blocks dc signals. Coupling capacitors couple signals either into or out of a circuit. Bypass capacitors usually connect a point in a circuit to ground so that only the RF signal is shunted to a ground without disturbing the dc circuit, this is generally called the ac ground. Again, coupling or bypass capacitors are generally ten times the largest matching element value in the circuit.

The integration of the RF circuit and the dc circuit is shown in Fig. 11.25C. Note that the input matching circuit, C_1 and L_2 as well as the output matching circuit, L_3 and C_4 are connected as in the RF circuit of Fig. 11.25A. The base

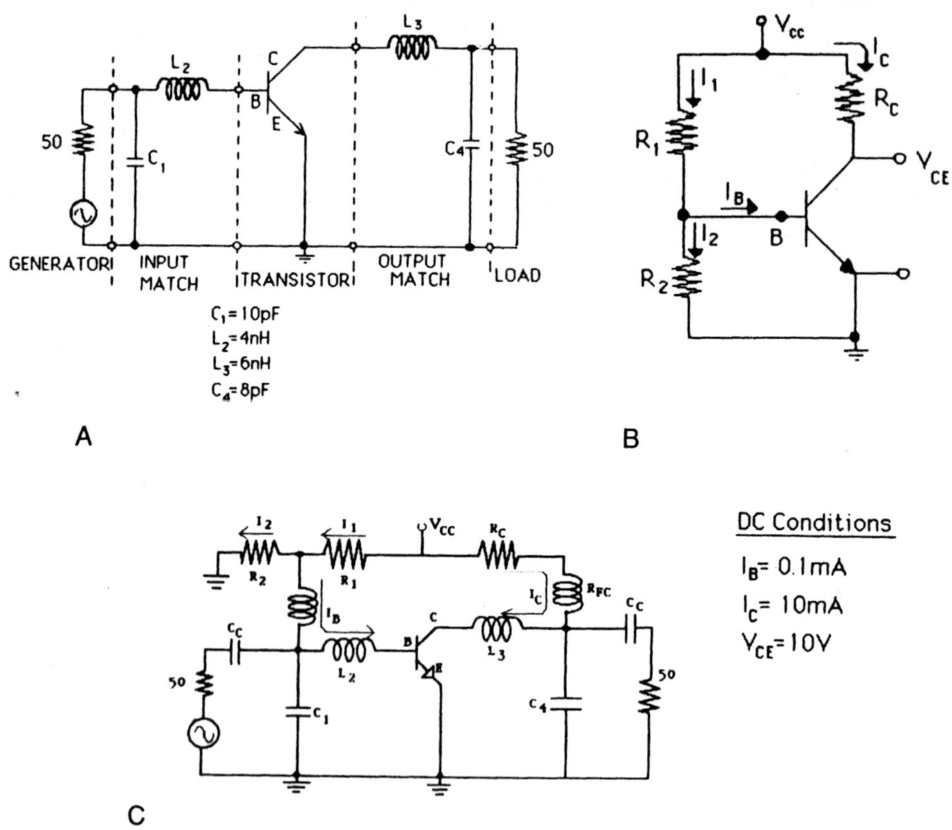

Figure 11.25 (A) A typical RF bipolar transistor circuit with input and output matched. (B) The dc conditions of the bipolar transistor circuit. (C) The integrated bipolar transistor circuit.

biasing circuit of R_1 and R_2 must be connected to the base circuit through an RF choke. Current I_1 flows through R_1 and current I_2 flows through R_2. Base current I_B flows through the RF choke which must go to the base of the transistor. The RF choke prevents any RF signals leaking into the base circuit. Base current does not flow through C_1 but can flow through L_2, which has no effect on the dc circuit. To prevent any base current from leaking into the 50 ohm generator, a coupling capacitor must be placed between the input matching circuit and the generator.

The collector resistor R_C must be connected to the matching circuit through an RF choke to prevent any signals leaking into the bias circuit. Collector current I_C does not flow through C_4 but can flow through L_3 into the collector. To prevent any collector current from leaking into the load resistor, a coupling capacitor must be placed between the load resistor and the matching circuit. Checking the dc conditions it is possible to have 0.1mA of current in the base, 10mA of current in the collector, and 10V from collector to emitter.

Integrating both the RF circuits and dc circuits must be done with care. Through the use of RF chokes and coupling capacitors, we can allow each circuit to perform its function without affecting the other.

Now that we have shown how a complete bipolar transistor amplifier operates in the microwave frequency range, some practical consideration will be discussed. With regard to matching, in practice we cannot achieve a perfect match in a production environment. Also, practical amplifiers generally operate over some bandwidth of frequencies, e.g., 1.0 GHz to 1.56 GHz. In general, a matching circuit provides a match so as to reduce the input and output reflected signals. When this is done, the gain will increase by G_1 and G_2, respectively. It is not necessary to match all frequencies in the bandwidth of interest to the center of the chart. If all the matched frequencies lie within the 1.3 *VSWR* circle on the *Z-Y* chart, then almost all of the G_1 and G_2 gain increase will be realized. Although this is a good rule to follow, it is by no means the final word. Each application must meet its own design objectives.

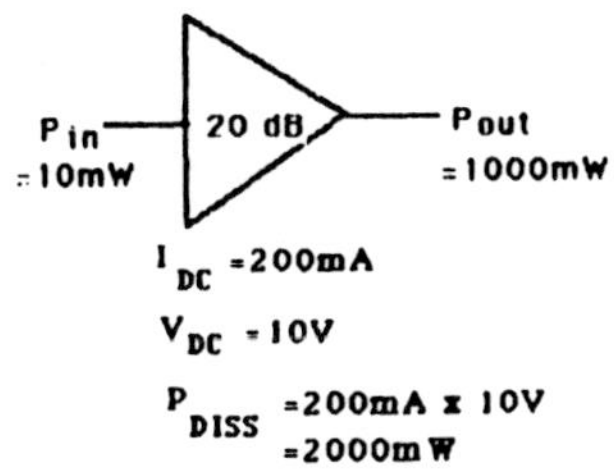

Figure 11.26 Conditions given to calculate an amplifier's efficiency.

11.2.9 Efficiency

The efficiency of a class A amplifier is simply the ability of an amplifier to produce an output signal power given an input signal power and a dc input power. Class A amplifiers are those which have their collector dc current on 100 percent of the time. Therefore, efficiency expressed as a percentage is

$$\text{Efficiency} = \frac{\text{output signal power}}{\text{input signal power} + \text{dc power dissipated}} \times 100\% \quad (11.2.12)$$

where the dc power dissipated by the amplifier is the total current from the power supply times the total voltage from the power supply. Note that the powers in Eq. (11.2.12) must be in numerical form, not dBm.

Example (11.2.9.1)

Given an amplifier with 20 dB gain and input signal power of 10mW, as shown in Fig. 11.26, calculate the output signal power. If the total dc current is 200mA, and the total dc voltage is 10V entering the amplifier, calculate the efficiency.

Solution: If the gain is 20 dB, then the output power is 100 times the input power. Therefore,

$$P_{out} = 100 \times 10\text{mW} = 1000\text{mW}$$

Note that the dissipated power, P_{DISS} is I_{DC} times V_{DC} which is 200mA × 10V = 2000mW. Then, the efficiency is

$$\text{Efficiency} = \frac{\text{output signal power}}{\text{input signal power} + P_{DISS}} \times 100\%$$

$$= \frac{1000\text{mW}}{10\text{mW} + 2000\text{mW}} \times 100\%$$

$$= \frac{1000\text{mW}}{2010\text{mW}} \times 100\% = 49.75\%$$

Remember that all the units must be the same and that efficiency is a unitless quantity. An efficiency of 49.75% means that for each 1mW or so of output signal power, it takes about 2mW of total input power, where the total input power is the sum of the input signal power and dc dissipated power.

11.2.10 Transfer Curves

A transfer curve is a curve on a graph which describes the output signal power *versus* the input signal power. Every microwave device has a transfer curve. It is called a transfer curve because we can determine the output signal power given as input signal power, and *vice versa*. Because we have been discussing bipolar amplifiers, lets us use one to define a transfer curve.

Figure 11.27 shows a typical transfer curve for a bipolar transistor amplifier. The output signal power is plotted on the vertical scale and the input signal power is plotted on the horizontal scale. There is a linear, i.e., straight line, region of the curve and a nonlinear portion. The transfer curve is divided into two important regions. This division is determined graphically. First, draw a straight line projection of the linear portion of the curve as in line *A*. Next determine where this straight line projection deviates from the actual output signal power by 1 dB (point *B*). The output power of +25 dBm, line *B*, deviates by 1 dB from the straight line projection *A*. Line *B* is the output power boundary at 1 dB gain compression. Point *C* is called the one-dB gain compression point.* Line *D* is

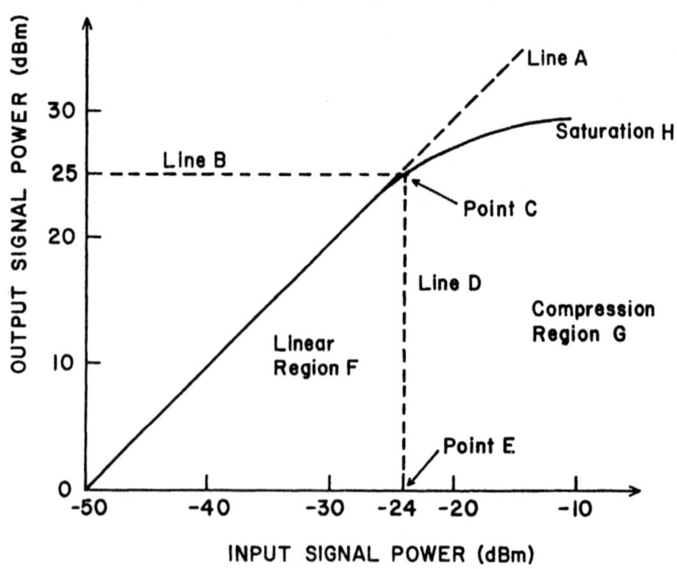

Figure 11.27 The transfer power curve of a bipolar transistor. Note that point C is the one-dB compression point.

*One-dB compression point was introduced in Ch. 5.

then drawn vertically from point C to the horizontal input power scale. Line D crosses the input power scale at -24 dBm, which is called the input power at one-dB gain compression.

Line D is the boundary between the linear region E and compression region F. The amplifier is considered to operate in the linear region for all signal powers limited to $+25$ dBm at the output or -24 dBm at the input. From all input signal powers from -24 dBm to -10 dBm or output signal levels from $+25$ dBm to $+30$ dBm, the amplifier is operating in the compression region. This region is called the compression region because the output signal power is compressed from where it should be if the amplifier were linear. Point G is called saturation because it is where the output neither increases nor decreases as the input power changes.

The uses of a transfer curve can best be illustrated by way of some examples. Refer to Fig. 11.27 for the following examples.

Example (11.2.10.1)

What is the gain if the input power to the amplifier is -40 dBm?

Solution: Using the transfer curve we can determine the output power given the input power. If $P_{in} = -40$ dBm, then $P_{out} = +10$ dBm. The gain is

$$\begin{aligned} \text{Gain (dB)} &= P_{out}\ \text{(dBm)} - P_{in}\ \text{(dBm)} \\ &= +10\ \text{dBm} - (-40\ \text{dBm}) \\ \text{Gain} &= 50\ \text{dB} \end{aligned}$$

We refer to this gain as the linear gain because it is in the linear region.

Example (11.2.10.2)

What is the gain if the input power to the amplifier is -10 dBm?

Solution: Using the transfer curve we can determine the output power given the input power. If $P_{in} = -10$ dBm, then $P_{out} = +30$ dBm. The gain is

$$\begin{aligned} \text{Gain (dB)} &= P_{out}\ \text{(dBm)} - P_{in}\ \text{(dBm)} \\ &= +30\ \text{dBm} - (-10\ \text{dBm}) \\ \text{Gain} &= 40\ \text{dB} \end{aligned}$$

Note that this gain is at saturation and is less than the linear gain of 50 dB. We call this condition gain compression, meaning that the gain is compressed from the linear gain.

Example (11.2.10.3)

What is the gain of the amplifier at the one-dB gain compression point?

Solution: We must first know the input and output powers. Remember that the output power at 1 dB gain compression is $+25$ dBm and the input power at 1 dB gain compression is -24 dBm. The gain is

$$\begin{aligned} \text{Gain (dB)} &= P_{out}\ \text{(dBm)} - P_{in}\ \text{(dBm)} \\ &= +25\ \text{dBm} - (-24\ \text{dBm}) \\ \text{Gain} &= 49\ \text{dB} \end{aligned}$$

One could have determined this by realizing that the gain at the one-dB compression point must be 1 dB less than the linear gain. Therefore,

$$\begin{aligned} \text{Gain at one-dB compression point} &= \text{linear gain} - 1\ \text{dB} \\ &= 50\ \text{dB} - 1\ \text{dB} \\ &= 49\ \text{dB} \end{aligned}$$

Example (11.2.10.4)

If the dc current and voltage entering an amplifier are 250mA and 10V, calculate the efficiency at the one-dB compression point.

$$
\begin{aligned}
P_{in} &= -24\text{ dBm} = 3.98 \times 10^{-3}\text{mW} \\
P_{out} &= +25\text{ dBm} = 316.2\text{mW} \\
P_{DISS} &= I_{DC} \times V_{DC} \\
&= 250\text{mA} \times 10\text{V} \\
P_{DISS} &= 2500\text{mW}
\end{aligned}
$$

Therefore,

$$\text{Efficiency} = \frac{P_{out}}{P_{in} + P_{DISS}} \times 100\%$$

$$= \frac{316.2\text{mW}}{3.98 \times 10^{-3}\text{mW} + 2500\text{mW}} \times 100\%$$

$$\text{Efficiency} = 12.65\%$$

Note that in the efficiency calculation, the powers must be converted to numerical form from dBm.

Example (11.2.10.5)

If the dc current and voltage entering an amplifier are 250 mA and 10V, calculate the efficiency at saturation.

$$
\begin{aligned}
P_{in} &= -10\text{ dBm} = 0.1\text{mW} \\
P_{out} &= +30\text{ dBm} = 1000\text{mW} \\
P_{DISS} &= 2500\text{mW}
\end{aligned}
$$

$$\text{Efficiency} = \frac{P_{out}}{P_{in} + P_{DISS}} \times 100\%$$

$$= \frac{1000\text{mW}}{0.1\text{mW} + 2500\text{mW}} \times 100\%$$

$$\text{Efficiency} = 40\% \text{ (approximately)}$$

Note that the efficiency is higher at saturation than at the one-dB compression point. Because the dc power entering the amplifier is a constant, the efficiency changes with the output power. As the output power rises, so does the efficiency. Therefore, the efficiency of an amplifier operating in the compression region will be greater than operating in the linear region, but the signal distortion will also be greater.

All amplifiers produce some signal distortion. Signal distortion can most easily be determined for a linear amplifier by measuring the amount of signal in the fundamental frequency *versus* its harmonics. Harmonics are integer multiples of the fundamental frequency. If the fundamental frequency f_0 is 2.0 GHz, then the second harmonic f_2 is $2 \times f_0$ or 4 GHz, and the third harmonic f_3 is $3 \times f_0$ or 6 GHz, et cetera. An amplifier operating in the linear region will have most of its output power at f_0 and very little in f_2, f_3, et cetera. As the amplifier is operated closer to the compression region, some of the signal power from f_0 will shift to the harmonics, f_2, f_3, *et cetera*. This results in increased signal distortion, which is generally undesirable for an amplifier. There are some special applications where signal distortion is highly desirable and these will be discussed later in the chapter.

PART III: FIELD-EFFECT TRANSISTORS

Salvatore Algeri

11.3 INTRODUCTION

11.3.1 MESFET Amplifiers

FET amplifiers are used in the middle portion of the microwave frequency range. Practical amplifiers are currently used from 1 GHz to 30 GHz, but advances in semiconductor techniques, like those of bipolar transistors, continue to challenge the upper frequency limits.

The particular type of FET discussed here is a MESFET, which can be described as a metallic Schottky-barrier field-effect transistor. A metal, usually aluminum, is used to form a Schottky diode at the gate which acts like a junction field-effect transistor (JFET). The source and drain are ohmic contacts.

This section will concentrate on RF matching techniques using microstrip transmission line matching elements. A brief review of dc biasing will be discussed but the readers are assumed to have a general understanding of FET biasing techniques. The FET RF circuit will be modelled as a two-port network so that the four scattering parameters can be used to define this network. As with bipolar transistors, a particular scattering parameter can be plotted on a *Z*–*Y* chart and discrete inductor and capacitor matching circuits are developed. Once these discrete components are developed, they will be transformed into microstrip transmission line elements. Finally, a complete microstrip FET amplifier will be analyzed.

11.3.2 Biasing Review

There are two common types of FET biasing circuits, negative gate bias and self-bias. Figure 11.28 shows a negative gate bias circuit. In this configuration, the gate to source voltage, V_{GS}, is minus V_1 battery voltage. There, the battery voltage V_1 controls the gate to source current as the *I*/*V* curve predicts. As V_{GS} becomes more negative (increasing V_1), the drain to source current, I_{DS}, will decrease. The drain to source voltage, V_{DS} is controlled by the battery voltage in the drain V_D. Because the source is grounded, the drain to source voltage V_{DS} is simply the battery voltage V_D. As V_D increases, so does V_{DS}. The dc operating point of the FET is set by I_{DS} and V_{DS}.

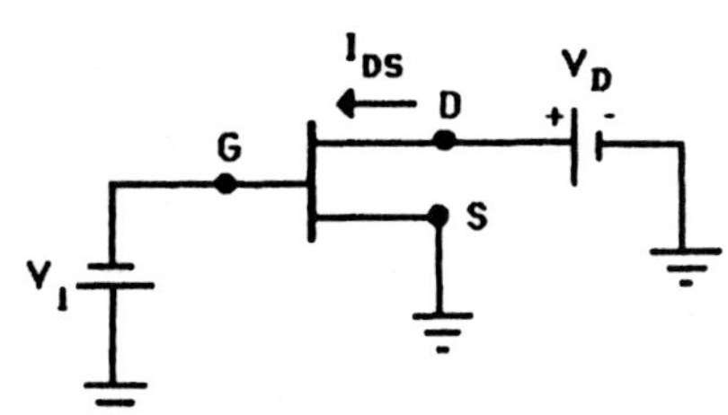

Figure 11.28 A negatively biased FET circuit with the source grounded.

The self-bias mode is shown in Fig. 11.29. For this configuration, the voltage at the gate V_G is grounded, i.e., zero volts. The voltage at the source V_S is the voltage drop defined by current I_{DS} flowing through resistor R_S. Therefore, the gate to source voltage is

$$V_{GS} = V_G - V_S \quad (11.3.1)$$

$$V_{GS} = 0 - I_{DS}R_S = -I_{DS}R_S \quad (11.3.2)$$

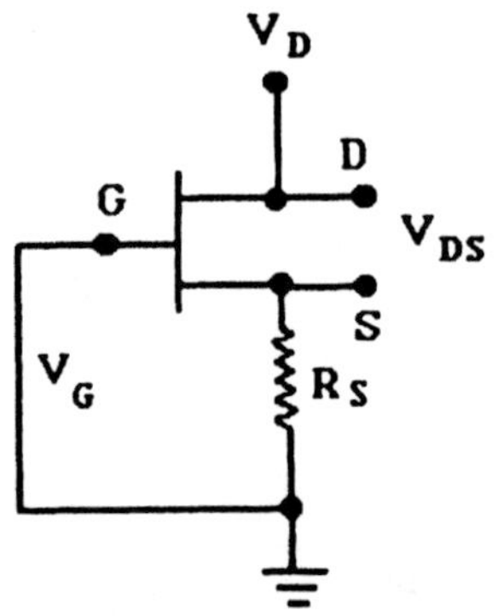

Figure 11.29 A self-biased FET circuit with the gate

The gate to source voltage V_{GS} sets the drain to source current I_{DS} according to the I/V curve of the device. The drain to source voltage V_{DS} is the voltage at the drain V_D minus the voltage at the source V_S. Once the current I_{DS} has been set by the source resistor R_S, the battery voltage at the drain V_D is used to set V_{DS}.

$$V_{DS} = V_D - V_S \quad (11.3.3)$$

$$V_{DS} = V_D - I_{DS}R_S \quad (11.3.4)$$

Again, the dc operating point of the FET is set by I_{DS} and V_{DS}. The configuration is called the self-bias mode because the drain to source current produces voltage drop V_S, which then defines the gate to source voltage V_{GS}. This gate to source voltage then sets the drain to source current I_{DS} according to the I/V curve. The drain to source current I_{DS} is used to self-bias the FET.

11.3.3 Low Frequency Equivalent Circuit

Before discussing the RF circuit, a brief review of low frequency gain will be beneficial. Figure 11.30 shows the low frequency equivalent circuit of a FET. Resistor R_i is the input resistance. Signal voltage V_{in} is across the gate to source and is the same as V_{GS}. The output signal voltage V_{out} is across the load resistor R_L. Signal current I_{DS} is the same as the controlled current source, g_mV_{GS}. A FET is a voltage-gain type of device.

$$\text{Voltage Gain} = V_{out}/V_{in} = -I_{DS}R_L/V_{GS} \quad (11.3.5)$$

Since

$$I_{DS} = g_mV_{GS} \quad (11.3.6)$$

Substituting g_mV_{GS} for I_{DS}, we have

$$\text{Voltage Gain} = -g_mV_{GS}R_L/V_{GS} \quad (11.3.7)$$

$$= -g_mR_L \quad (11.3.8)$$

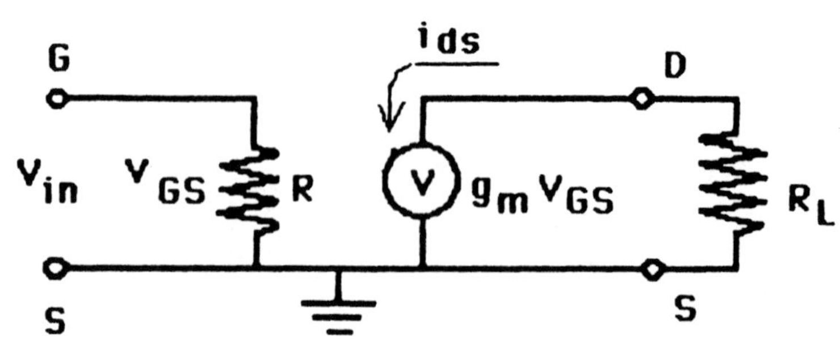

Figure 11.30 The low frequency equivalent circuit of a FET.

The voltage gain is the transconductance g_m of the FET times the load resistor R_L. The significance of the negative sign is that the output signal voltage is 180° out of phase with the input signal voltage.

The power gain is

$$\text{Power gain} = P_{out}/P_{in} = V_{out}^2 R_{in}/V_{in}^2 R_L \tag{11.3.9}$$

Substituting $V_{out} = I_{DS}R_L = g_m V_{GS} R_L$ and $V_{in} = V_{GS}$ in Eq. (11.3.9), We have

$$\text{Power gain} = g_m^2 R_L R_{in} \tag{11.3.10}$$

The power gain in a FET is dependent upon the transconductance, the load, and the input resistance.

Example (11.3.3.1)
Let $g_m = 20\text{mmho}$ (milli-mho), $R_i = 1\text{k}\Omega, R_L = 500\Omega$.

Find the voltage gain and power gain.

Solution:

$$\begin{aligned}\text{Voltage gain} &= -g_m R_L \\ &= -20 \times 10^{-3}\text{mho} \times 1 \times 10^3\text{ohms} \\ &= -20\end{aligned}$$

$$\begin{aligned}\text{Power gain} &= g_m^2 R_L R_i \\ &= (20 \times 10^{-3}\text{mho})^2 \times 0.5 \times 10^3\text{ohms} \times 1 \times 10^3\text{ohms} \\ &= 200\end{aligned}$$

$$\text{Power gain (dB)} = 10 \log(200) = 23 \text{ dB}$$

11.3.4 Microwave Circuit

The equivalent circuit of a FET operating in the microwave frequency range is shown in Fig. 11.31. It is considerably more complex than the low frequency equivalent circuit. The only true similarity between the two is the controlled current source $g_m V_{GS}$. For the low frequency equivalent circuit of Fig. 11.30, the input impedance is simply the resistor R_i; but in the microwave equivalent, the input impedance is a parallel RC. This makes the input impedance change with frequency. As the frequency increases, the input impedance decreases. Also, note that the output (drain) is connected to the input (gate) through a parallel

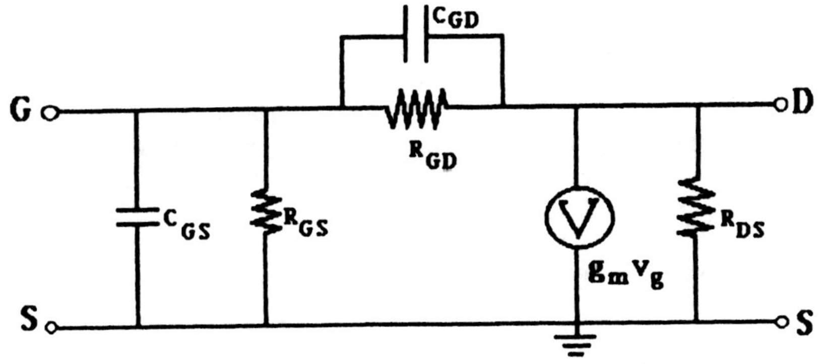

Figure 11.31 The high frequency equivalent circuit of a FET.

RC. Resistor R_{GD} provides negative feedback. Capacitor C_{GD} is in parallel with resistor R_{GD}, which also provides negative feedback. The negative feedback is frequency dependent. As the frequency increases, the reactance of C_{GD} decreases, thereby increasing the negative feedback, and thus reducing the gain. This capacitor is the single most important element limiting the upper frequency capability of a FET to produce gain. The typical gain of a FET, therefore, decreases with increasing frequency as shown by the solid curve of Fig. 11.32.

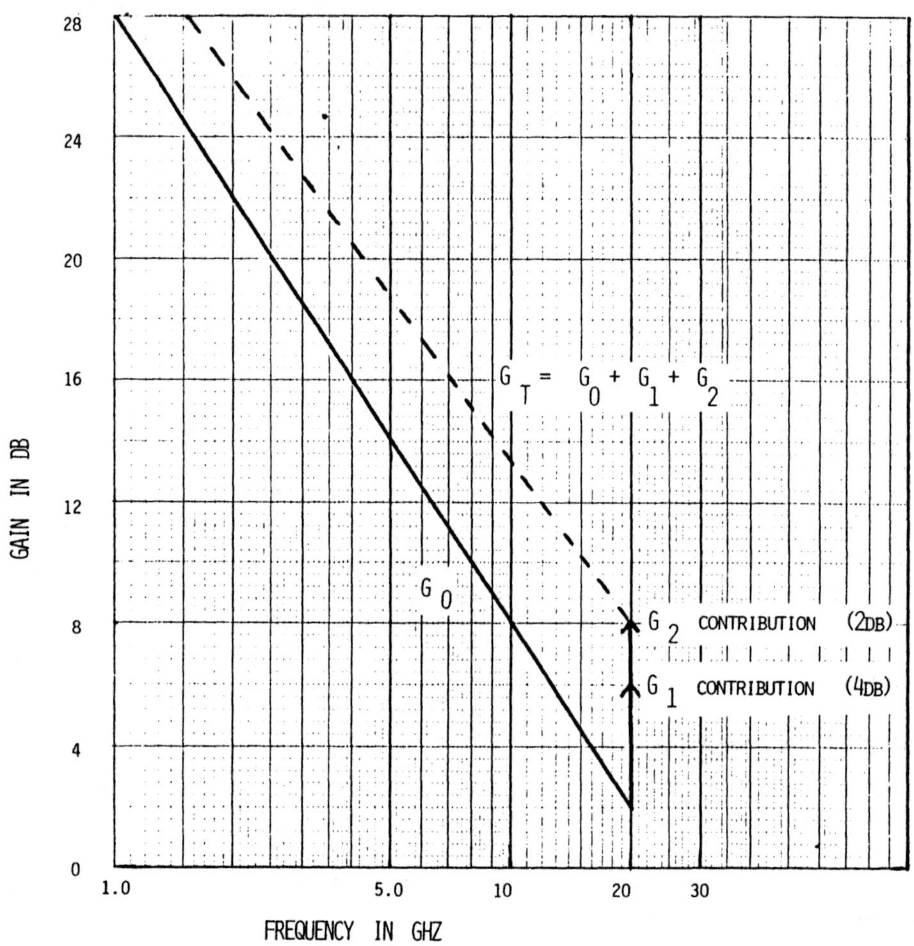

Figure 11.32 The gain response of a FET circuit versus frequency. The solid line is the gain of the FET itself. Matching the input and the output can increase the circuit gain substantially.

11.3.5 Scattering Parameters of a FET

The extreme complexity of the complete microwave equivalent circuit makes it very difficult to analyze problems using conventional ac circuit theory. Therefore, another approach is used to help simplify the analysis and the design of microwave FET amplifiers. As discussed in detail in Sec. 11.2.5, a two-port network can be completely characterized by four scattering parameters. These scattering parameters can be easily measured using microwave test equipment and then plotted onto a Smith chart or a *Z-Y* chart to develop input and output matching circuits.

11.3.6 Total Gain for a FET

Suppose that the S-parameters have been measured for a FET circuit and that both S_{11} and S_{22} are not equal to zero. This means that the FET has some input and reflected signal. The total gain defined in Sec. 11.2.6 can be applied here:

$$G_{total}(\text{dB}) = G_o(\text{dB}) + G_1(\text{dB}) + G_2(\text{dB}) \tag{11.3.11}$$

where all the gain values must be in dB. Remember that G_o is called the 50 ohm gain, and is simply the gain of the device when measured in a 50 ohm system. This translates to the S_{21} or forward transmission coefficient. The contribution of G_1 to the total gain is the additional gain measured for the FET by matching the input to the 50 ohm generator using a lossless network. Also, the contribution of G_2 to the total gain is the additional gain measured by the FET by matching the output to the 50 ohm load using a lossless network.

A FET may only have a 50 ohm gain of 2 dB at 20 GHz, which is the G_o contribution (see Fig. 11.32). The gain may increase by 4 dB by matching the input, which is the G_1 contribution. Also, the gain may increase by an additional 2 dB by matching the output, which is the G_2 contribution. As a result of matching both the input and output, the gain could be increased from a poor 2 dB at 20 GHz to a more reasonable 8 dB as shown in Fig. 11.32.

11.3.7 Matching Circuits

By matching both the input and output impedance of a FET Amplifier, the designer can optimize gain. This is accomplished by using lossless matching elements like capacitors and inductors. Resistors could be used, but they are lossy. As discussed in detail in Sec. 11.2.7, a *Z-Y* chart is used to simplify this procedure.

The four governing rules of using a *Z-Y* chart for a bipolar transistor also hold true for a FET. By following these four simple rules, we can develop matching circuit topologies at any one frequency more easily than using a standard Smith chart. To illustrate this, let us try a couple of examples which are similar to Sec. 11.3.7.1.

Example (11.3.7.1)

Assume that the S_{11} of a FET at 10 GHz is (normalized) $0.5 - j1.0$. Develop a matching circuit topology using three elements so that when placed between the input of the FET and a 50 ohm generator the circuit would produce no reflection.

Solution: First we must plot $S_{11} = 0.5 - j1.0$ using the black chart as shown in Fig. 11.33. One possible solution is to follow the (black) 0.5 resistance circle clockwise and stop at the red 0.7 circle. Element 1 maps out a series inductor L_1. Next follow the red 0.7 conductance circle counterclockwise and stop at the unity black circle. Element 2 maps out a shunt inductor L_2. Then follow the red 1.0 resistance circle counterclockwise to the center of the chart. Element 3 maps out a series capacitor C_3.

Therefore, the possible schematic is a series capacitor followed by a shunt inductor and then a series inductor as shown in Fig. 11.34. If the values of L_1, L_2, and C_3 are properly selected as described in Sec. 11.3.7.2, then the impedance Z_{in} will be $1.0 + j0$ and there will be zero reflection.

Example (11.3.7.2)

Assume that the S_{22} of a FET at 10 GHz is (normalized) $0.3 - j0.2$. Develop a matching circuit topology using three elements so that when placed between the output of the FET and a 50 ohm load would produce no reflection.

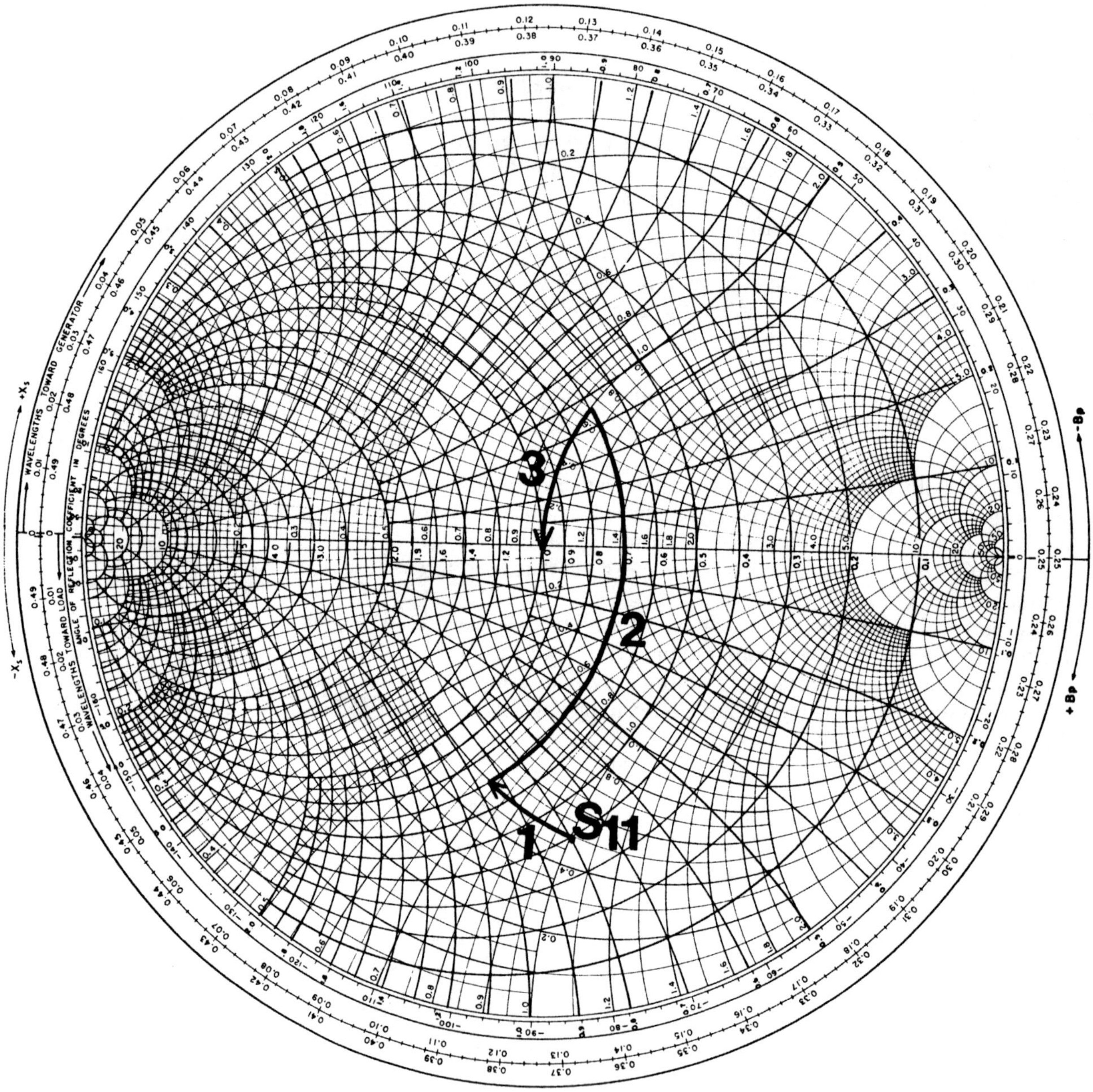

Figure 11.33 Using the Z-Y chart to match the FET input given S_{11} to be 0.5 − j1.0 at 10 GHz.

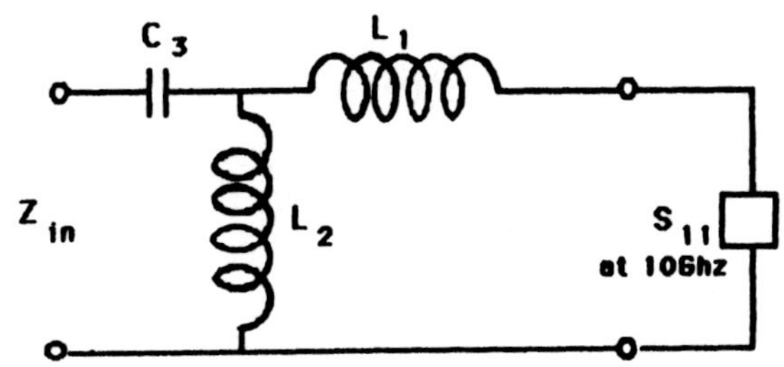

Figure 11.34 The input matching circuit for S_{11} = 0.5 − j1.0 at 10 GHz.

Solution: First we must plot $S_{22}=0.3-j0.2$ on the *Z-Y* chart as shown in Fig. 11.35. Follow the black 0.3 resistance circle clockwise and stop at the red 0.5 circle. Element 1 maps out a series inductor L_1. Next follow the red 0.5 conductance circle counterclockwise and stop at the unity black circle. Element 2 maps out a shunt capacitor C_2. Then follow the unity black circle counterclockwise to the center of the chart. Element 3 maps out a series capacitor C_3. Therefore, the schematic is a series inductor followed by a shunt capacitor and then a series capacitor, as shown in Fig. 11.36.

If the values of L_1, C_2, and C_3 are selected properly, as described in Sec. 11.2.7.2, the output impedance Z_{out} will be $1.0+j0$, i.e., no reflection.

11.3.8 Integrating the DC and the RF FET Circuit

Both the dc and the RF FET circuit have been discussed separately. The considerations in integrating these circuits are similar to those in integrating the dc and the RF bipolar transistor circuits as discussed in Sec. 11.2.8.

Figure 11.37A shows a typical RF circuit. The entire RF circuit consists of a 50 ohm generator, the input match, a FET with grounded source, the output match, and a 50 ohm load resistor. Values for the matching circuit may be as shown. Figure 11.37B shows a typical dc circuit where the dc conditions are $V_{GS}=-2V, V_{DS}=5V$, and $I_{DS}=20mA$.

Figure 11.37C shows the integration of both the RF and the dc circuits. As described in Sec. 11.2.8, the bypass capacitor passes RF signals but blocks the dc, i.e., it establishes an ac ground to the RF signal only.

Let us begin by analyzing the gate circuit. Inductors L_1 and L_2 represent the input match. The gate must be at zero volts for the dc circuit. Inductors L_1 and L_2 are simply two coils of wire connected to ground which conveniently satisfy the dc grounding condition.

In the drain terminal, inductor L_3 is connected to a bypass capacitor which makes point *A* RF ground. Therefore, inductor L_3 is a shunt inductor as required by the RF circuit. Capacitor C_4 is in series between L_3 and the load resistor as in the RF circuit. Current I_{DS} will not flow through the bypass capacitor at point *A*, but goes through L_3 to the drain. Capacitor C_4 serves a dual purpose as well. It is a matching element and also prevents any drain current from leaking into the 50 ohm load.

The RF circuit requires the source to be grounded, but we must also have 2Vdc there. This is achieved by the use of the bypass capacitors, which make the source RF ground for the signal, but do not upset the drain to source current flowing through R_S. The reason for using two bypass capacitors will be discussed in a later section. This schematic has successfully integrated the RF circuit and the dc circuit with the aid of bypass capacitors.

CH. 11/PART III

FIELD-EFFECT TRANSISTORS

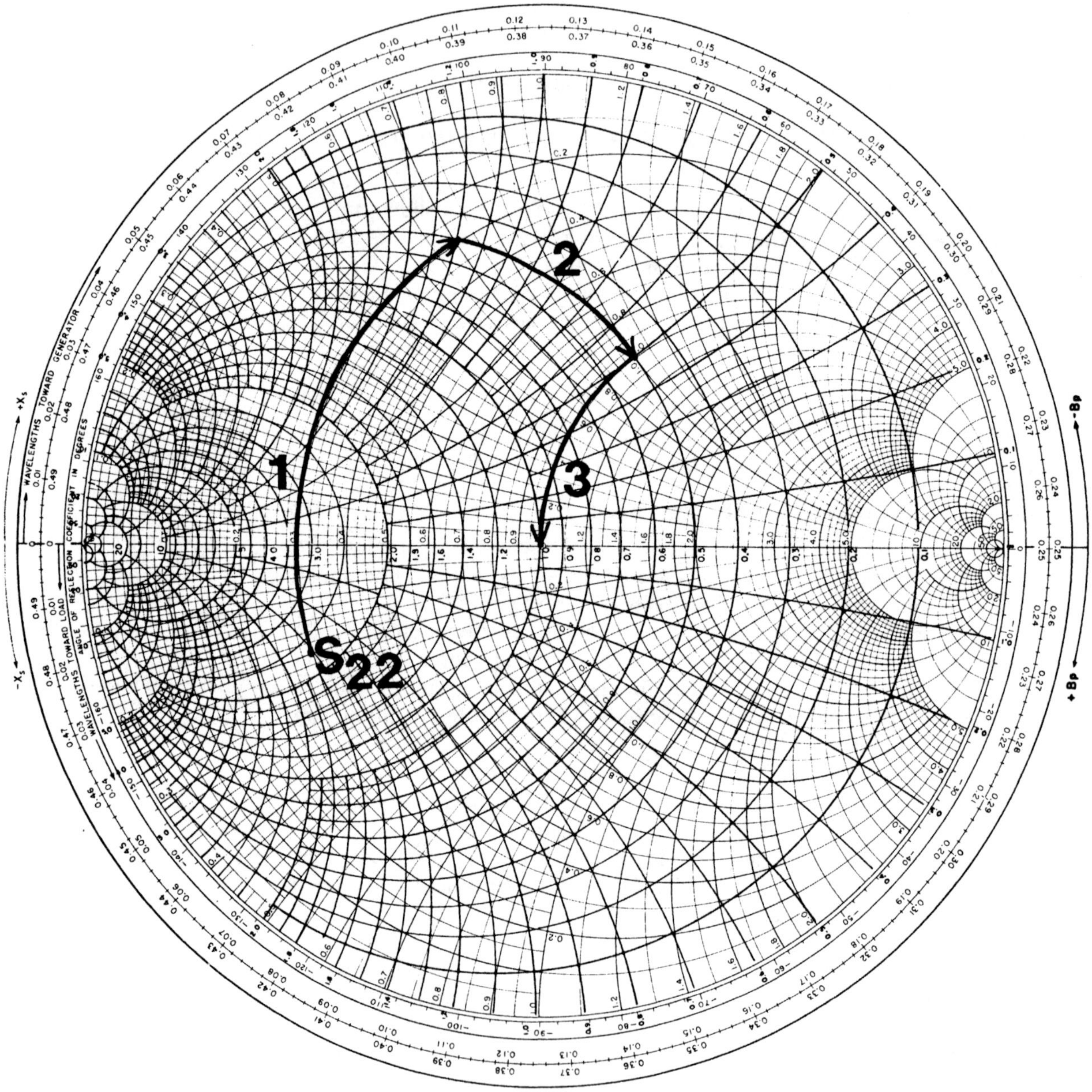

Figure 11.35 Using the Z-Y chart to match the FET output given S_{22} to be $0.3 - j0.2$ at 10 GHz.

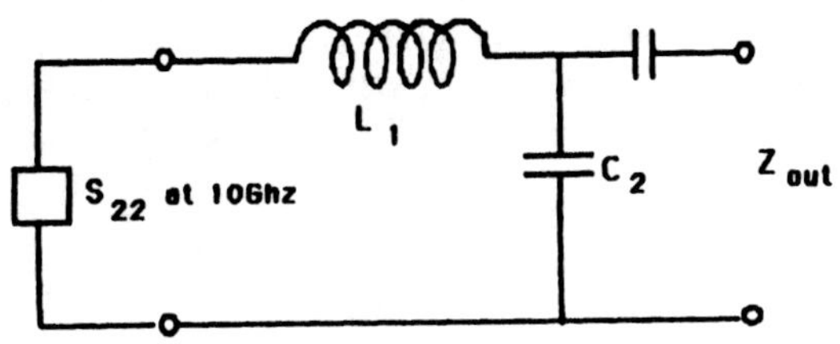

Figure 11.36 The ouput matching circuit for S_{22} = 0.3 − j0.2 at 10 GHz.

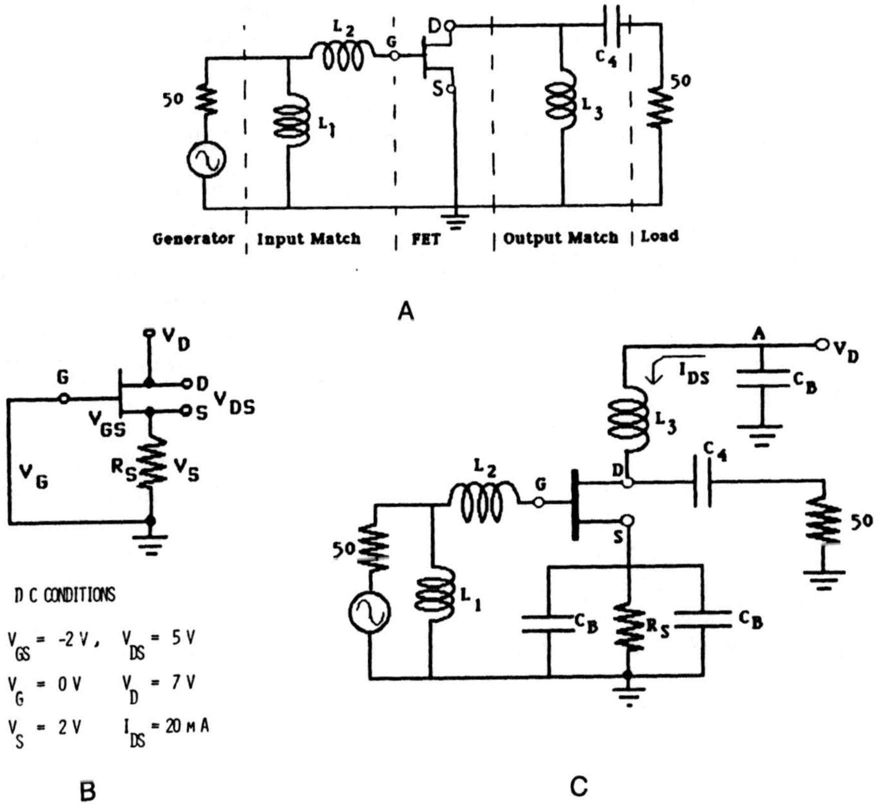

Figure 11.37 (A) A typical RF FET circuit with input and output matched. (B) The dc conditions of the FET biasing circuit. (C) The integrated FET circuit.

11.3.9 Microstrip Matching Elements

Several different types of transmission line media have been introduced, such as waveguide, coaxial cable, stripline, *et cetera*. Coaxial cable, stripline, and microstrip are part of the transverse electromagnetic (TEM) mode of propagation. When a signal propagates down a TEM transmission line, it means that neither the electric field nor the magnetic field has a component in the direction of signal propagation. The most common type of TEM transmission line is the microstrip.

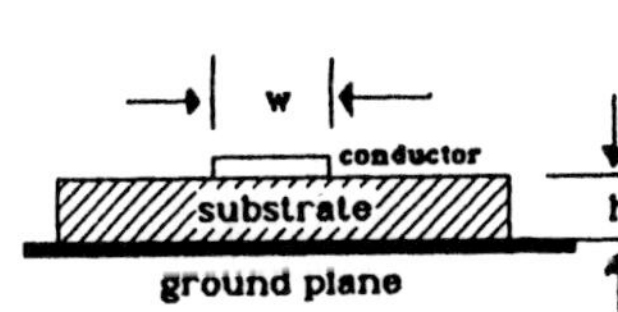

Figure 11.38 Cross section of a microstrip transmission line.

Miniature circuits including microstrips will be covered in more detail in Chapter 12. A cross section of a microstrip is shown in Fig. 11.38 where a substrate of dielectric constant ϵ_r is on a ground plane. The thickness of the substrate is h and a conductor of width w is on top of the substate. Note that

CH. 11/PART III

FIELD-EFFECT TRANSISTORS

the direction of propagation for a signal is coming out the paper. An important quantity of the microstrip is its characteristic impedance Z_0. The characteristic impedance of the microstrip will not be derived but it is important to know that it is proportional to the thickness to width ratio, i.e.

$$Z_0 \propto h/w \tag{11.3.12}$$

The characteristic impedance can be determined or varied by selecting the right value of the substrate thickness h and the width of the conductor w.

Correspondence of Discrete and Microstrip Matching Elements

In many high frequency matching circuits such as FET amplifiers, microstrip matching elements are preferred over discrete elements. This section presents a correspondence between discrete matching elements (L and C) and their more convenient microstrip counterparts. For our purposes of understanding microstrip circuits, taking a microstrip circuit and sketching its equivalent circuit with discrete elements often simplifies the circuit analysis.

Figure 11.39A shows a microstrip transmission line T_1, which is between two other transmission lines of width w and thickness h. The width of T_1 is w_1 and is smaller than w, the width of the two lines. The ratio h/w is chosen to yield a line impedance of 50Ω. Because the width w_1 is less than the width w, then the characteristic impedance Z_1 of T_1 is greater than 50Ω. This microstrip transmission line circuit is equivalent to a series "high" impedance transmission lines, and it tends to act like a series inductor.

The microstrip version of a perfect capacitor does not exist. One possibility is to use a simple parallel plate capacitor placed on top of the microstrip conductor to make a series capacitor. These simple capacitors are made from a piece of dielectric material whose opposite sides are metallized to form the plates of a capacitor (see Fig. 11.39B).

Figure 11.39C shows a transmission line, T_3, in shunt with a 50 ohm transmissioin line. Because the width w_3 of transmission T_3 is less than w, our reference, then the characteristic impedance Z_3 must be greater than 50Ω. The transmission line circuit for this microstrip circuit is a shunt "high" impedance transmission line, which tends to act like a discrete shunt inductor.

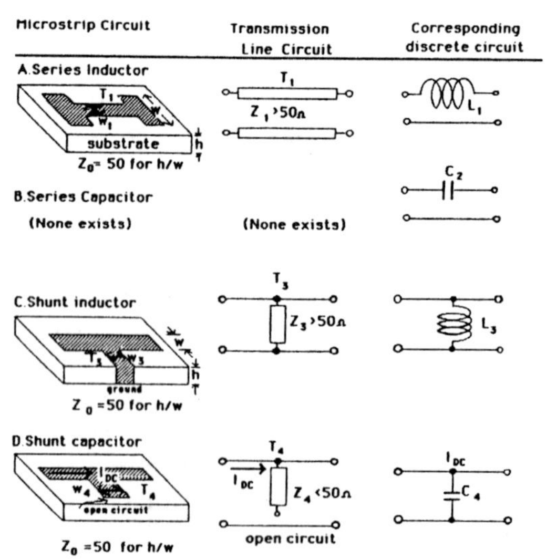

Figure 11.39 Four common microstrip circuits and their corresponding transmission line circuits and discrete circuits.

A transmission line acting like a shunt capacitor is shown in Fig. 11.39D. Transmission line T_4 is in shunt with a 50 ohm transmission line. Line T_4 is open-circuited and has a width w_4 greater than w, our reference. Therefore, the characteristic impedance of T_4 is lower than 50Ω. The transmission line circuit for this microstrip circuit is a shunt "low" impedance open-circuited transmission line which tends to act like a shunt capacitor. This correspondence is sometimes difficult to visualize, but a simple dc test may help. If there is a dc current flowing in our discrete circuit, it could not pass through capacitor C_4, which acts like a dc block. Likewise, transmission line T_4 must be open-circuited to ground to prevent dc current from flowing to ground.

It should be emphasized that microstrip circuits only tend to act similarly to their corresponding discrete circuits with respect to a Smith chart for matching purposes. This is illustrated by the following example.

Example (11.4.9.1)

For the microstrip circuit shown in Fig. 11.40A, sketch the corresponding discrete circuit which represents the designer's intention. Character impedance for h/w is 50Ω.

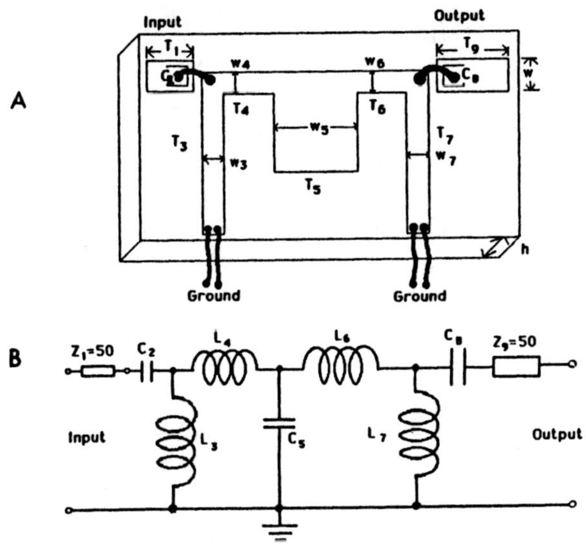

Figure 11.40 (A) A microstrip circuit and (B) the corresponding discrete equivalent circuit.

Solution: Refer to Fig. 11.40B, transmission line T_1 has width w so that its impedance Z_1 is 50Ω. Because T_1 does not correspond to one of the discrete elements of Fig. 11.39, it can be represented as a 50Ω transmission line. Element C_2 is a parallel plate capacitor mounted on T_1 and in series with the signal flow. Therefore, C_2 is a discrete series capacitor. A shunt high impedance transmission line can be represented as the shunt inductor L_3. Transmission line T_4 has a narrower width w_4 and is a high impedance line. It appears that T_4 is in series with the signal flow. A high impedance series line can be represented as the series inductor L_4. Transmission line T_5 has a width w_5, which is wider than w, and therefore is a low impedance line. Transmission line T_5 is in shunt with the signal flow and is also open-circuited. A shunt low impedance open circuited line can be represented as the shunt capacitor C_5.

Transmission line T_3 has width w_3, which is narrower than w, and therefore is a high impedance transmission line. T_3 is in shunt with the signal flow because it goes to ground.

11.3.10 The Balanced Amplifier Configuration

FET amplifiers are very difficult to match satisfactorily over wide bandwidths. If the input match or the output match had a *VSWR* larger than 2.0, then the user could not cascade two amplifiers, say 10 dB each, and expect an overall gain of 20 dB. For FET amplifiers covering bandwidths greater than 1.5 to 1, the balanced configuration is usually used. The major purpose of the balanced configuration is to achieve reasonable *VSWR*s at the input and the output so that amplifiers can be easily cascaded. A special circuit called a quadrature, 3-dB, 90° coupler, also known as the Lange coupler, is used to make the balanced amplifier configuration.

The Quadrature, 3-dB, 90-Degree Coupler

This coupler is a four-port circuit. An input signal is equally split between two output ports so that the signal at one of the output ports is 90-degree out of phase with the other output. Figure 11.41A is the microstrip version of the quadrature, 3-dB, 90-degree coupler. Microwave energy is coupled from one port to another due to the close proximity of the parallel lines. A suitable spacing between the lines will couple half of the input energy to each of the two output ports, hence the term 3-dB. Figure 11.41B is a schematic representation of the physical microstrip circuit.

Transmission line T_6 is a series high impedance line like T_4 and can be repesented as a series inductor. Transmission line T_7 is a shunt high impedance like like T_3 and can be represented as the inductor L_7. Parallel plate capacitor C_8 is in series with T_7 and T_9, and is simply a series capacitor. Line T_9 is 50Ω and has no corresponding discrete element, it remains as a 50 ohm line.

The discrete circuit of Fig. 11.40B is a representation of the designer's intent of the microstrip circuit, and can usually give an understanding of the circuit function.

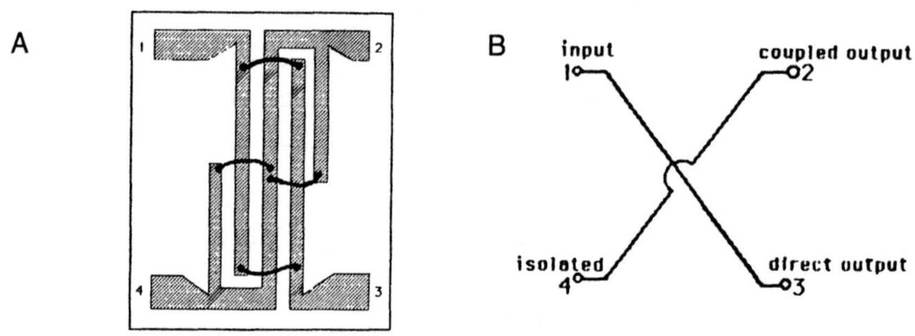

Figure 11.41 (A) The microstrip construction of a quadrature, 3 dB, 90° coupler and (B) its schematic representation.

The coupler can be used as a power combiner as shown in Fig. 11.43. An amount of power, which we shall call $P_{IN}/2$, at 90° enters port 1, and another power $P_{IN}/2$ at 0° enters port 4. The signal at port 1 will suffer an additional phase shift of 90°, making it a total of 180° relative to the signal at port 4. Upon combination, the amount of output power at port 2 is nil, and that at port 3 is P_{IN} at 0° as shown in Fig. 11.43.

To understand how the coupler operates, let us place a signal into port 1 of P_{IN} at 0° as illustrated in Fig. 11.42. Half of the input power, i.e., $P_{IN}/2$ at 0° will appear at port 3, which is the direct port. The remaining half of the input power will appear at port 2, the coupled port, phase shifted by 90°. Nothing will appear at port 4, which is the isolated port. Therefore, the coupler acts as a power splitter.

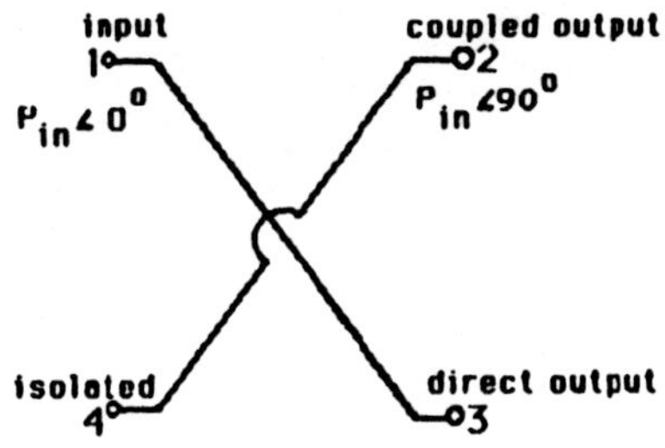

Figure 11.42 The distribution of power and phase at the output ports if an input is applied to port 1 of a quadrature coupler.

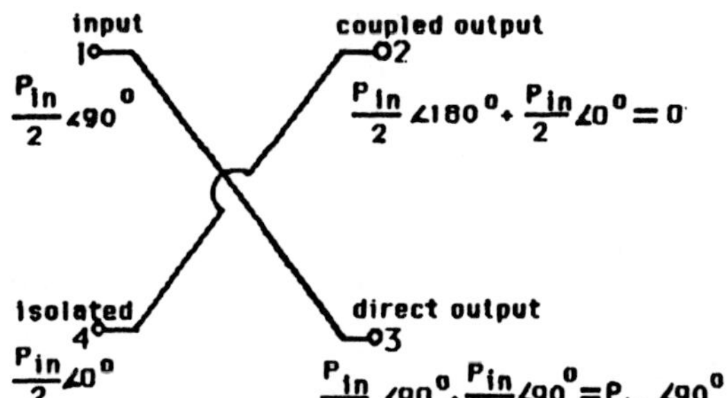

Figure 11.43 The quadrature coupler can be used as a combiner. Inputs at port 1 and port 4 will be combined as shown.

The primary purpose of the quadrature, 3-dB, 90-degree coupler in a FET amplifier is to improve the *VSWR* when the amplifiers are cascaded. This is illustrated in Fig. 11.44. At port 1, the power of the incident signal is $P_{IN}/2$ at 0°. This signal is split evenly so that the output power at port 2 is $P_{IN}/2$ at 90°, and that at port 3 is $P_{IN}/2$ at 0°. Assume that ports 2 and 3 are connected to identical FET amplifiers each of which, due to slight mismatch, reflects an amount of power P_R. The signal reflected back into port 2 is P_R at 90°, and that into port 3 is P_R at 0°. The coupler is now forced to act as a combiner for the reflected power. Due to the relative phase relationship of the two reflected powers, they sum at port 4 and are then absorbed by a 50 ohm termination. Similarly, no net reflected power appears at port 1, the input port.

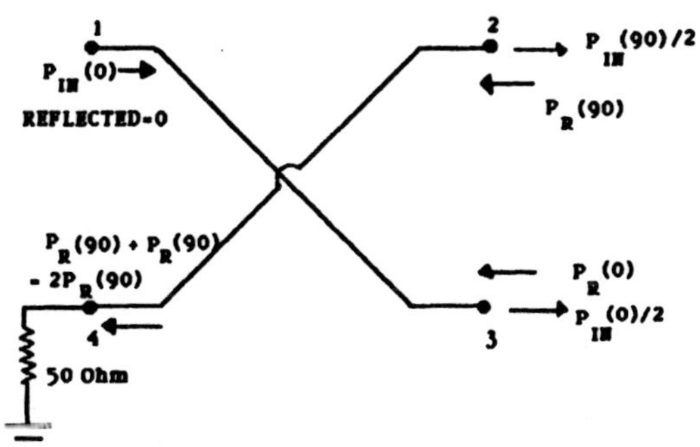

Figure 11.44 Any reflected power occurring at ports 2 and 3 due to an input applied at port 1 will end up at port 4.

The advantage of the above circuit is that although the *VSWR* of the individual FET amplifiers may be poor, the input to the balanced configuration (port 1) will not be affected because the reflected powers appear at port 4 and are then absorbed. Figure 11.45 shows a complete balanced amplifier configuration.

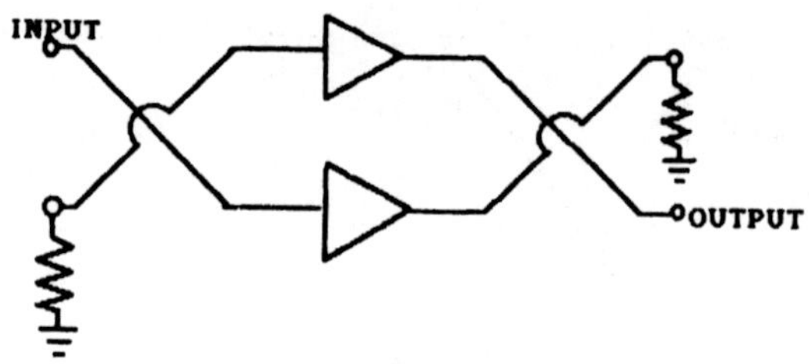

Figure 11.45 A balanced amplifier configuration using a quadrature.

11.3.11 The Hybrid Microstrip FET Amplifier

All the separate parts of a FET amplifier discussed so far will be integrated in this section. We shall analyze a typical hybrid microstrip FET amplifier circuit as an example. The analysis will take the form of a translation of the physical circuit into a schematic representation which can be easily understood. After the translation is made, circuit alignment techniques will be discussed.

The hybrid microstrip FET amplifier to be analyzed is shown in Fig. 11.46A. Let us first discuss the construction. Two substrates with transmission lines are mounted to a metal carrier, which is usually made of copper or Kovar, an iron nickel alloy. A rib, usually made of copper, is mounted to the carrier between the input and the output substrates. Components such as capacitors, resistors, and FETs are mounted onto the rib. The rib also provides a convenient ground location. Because this is a balanced amplifier configuration, the circuit is symmetrical about the center horizontal line going through capacitor C_7.

Before the analysis begins, a reference 50 ohm transmission line must be identified. Usually, the input or output to the circuit is a 50 ohm line. Therefore, the reference h/w at the input is a 50 ohm line. Now, we can determine whether other transmission lines are higher or lower impedance than the 50 ohm reference.

Let us proceed by sketching in the schematic for a quadrature, 3-dB, 90-degree coupler at the input and the output, as shown in Fig. 11.46B. We shall only analyze the top half of the circuit because the lower half is identical. Resistors R_3 and R_4 are connected from the isolated ports of the couplers to ground, they are 50 ohm thin-film terminating resistors for the couplers.

We must think about signal flow. Starting with transmission line T_1, we must ask if T_1 is in series or shunt with the signal flow. Line T_1 is in shunt with the signal flow since the signal is entering the gate of the FET on the rib. Line T_1 appears to be slightly wider than our reference width, w. Therefore, this shunt, open-circuited, low impedance transmission line acts like a shunt capacitor, and is represented as C_{T1}. Now the signal travels through C_1, which is a parallel plate capacitor mounted on T_1. Capacitor C_1 is in series with the signal flow and is represented by itself, C_1.

Transmission line T_2 is in shunt with the signal and is connected to ground. It also appears to be narrower than the reference width, w. Therefore, line T_2 is a shunt, high impedance line going to ground, and can be represented by the shunt inductor L_{T2}. Transmission line T_3 is a series of wires connected together. They are in series with the signal and are high impedances since their width is narrower than w. A series, high impedance line can be represented by the series inductor L_{T3}. The components L_{T3}, L_{T2}, C_1, and C_{T1} make up the input matching network to the FET. Again, this is only a schematic representation of the actual physical circuit.

At the drain to the FET, there is a group of wires connected to T_9. This group of wires is a series, high impedance transmission line, which is represented as the series inductor L_{T9}. Line T_{10} is in shunt with the signal and appears to be high impedance. It is connected in series with capacitor C_7 which is connected to ground. Line T_{10} is a shunt, high impedance line, and can be represented as the shunt inductor L_{T10}. T_{10} is in series with capacitor C_7. Capacitor C_7 is large and acts as a bypass capacitor to provide a convenient place to bring in the drain voltage V_D. Capacitor C_7 acts as a short circuit to the RF signal making L_{T10} effectively a shunt inductor for the signal only. Capacitor C_8 is a parallel plate capacitor and is represented by itself. The circuit of L_{T9}, L_{T10}, and C_8 make up the output matching network.

At the source, C_3 and C_4 are connected to the source of the FET, and they are both parallel plate capacitors mounted on the rib (ground). Therefore, C_3 and C_4 are the source bypass capacitors. Resistor R_1 is connected in parallel with C_4 to ground and is used as the source resistor to set the dc operating point.

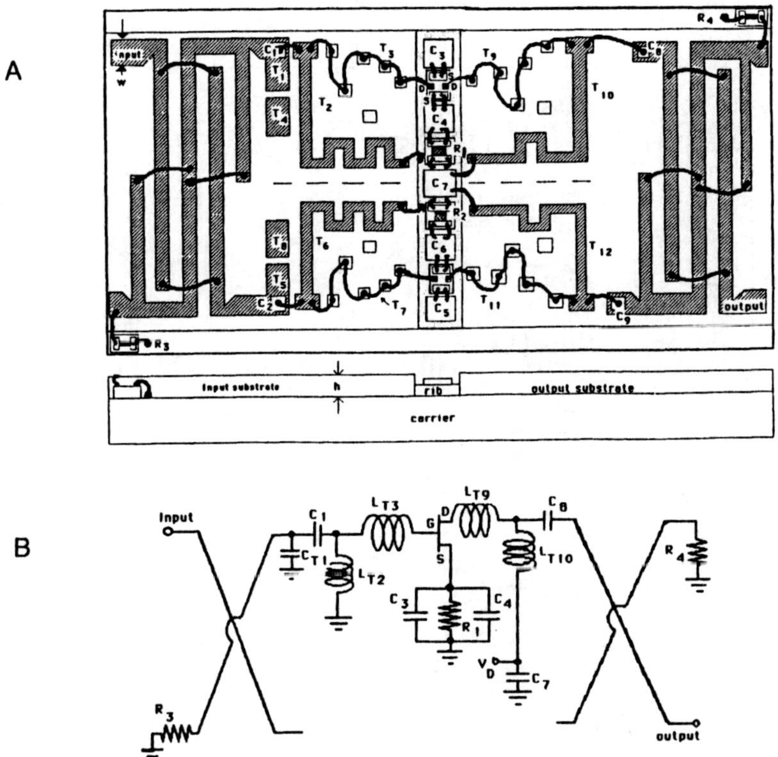

Figure 11.46 (A) A hybrid microstrip FET circuit. The circuit is symmetrical about a horizontal line through C_7. (B) The discrete elements of one-half of the balanced circuit.

Let us identify the basic parts of the circuit. Due to the two couplers, it is a balanced FET amplifier configuration. The FET is set up for the self-bias mode by noting that the gate is dc grounded through L_{T3} and L_{T2}. The input and output matching circuits have been identified. The supply voltage V_D enters the circuit at the bypass capacitor C_7. Remember that the lower half of the circuit is identical to the upper half. By sketching a schematic representation of the actual physical circuit, we can greatly simplify our understanding of the circuit.

11.3.12 Alignment Techniques

Hybrid microstrip FET amplifiers can only be manufactured to come close to meeting the desired specifications. Variations in the S-parameters from FET to FET play a major role in manufacturing tolerances. Thin-film microstrip circuits also have a limited tolerance range that can be controlled. Additionally, there is an assembly tolerance due to human error. Component location and bonding wire placement also affect the performance of the circuit in the microwave frequency range. As a result of these practical considerations, the circuits must be designed so that they can be aligned during electrical test.

To illustrate some common alignment techniques, we shall refer to the microstrip circuit in Fig. 11.47A, which is essentially the same as Fig. 11.46A. When transmission line T_1 is connected to T_4, the overall length of the low impedance open-circuited transmission line is increased. This will increase the shunt capacitance C_{T1} to C_{T1+T2} as shown in Fig. 11.47B. Therefore, the shunt capacitor is variable. If some bonding wires are placed on T_2 as shown, it will have the effect of making the line electrically shorter. When a shunt, high impedance short-circuited transmission line is shortened, it will reduce the shunt inductance of L_{T2}. If line T_3 is bonded in a different configuration to increase its length, then the series inductance L_{T3} will increase. Hence, three of the four elements making up the input matching circuit are easily adjustable.

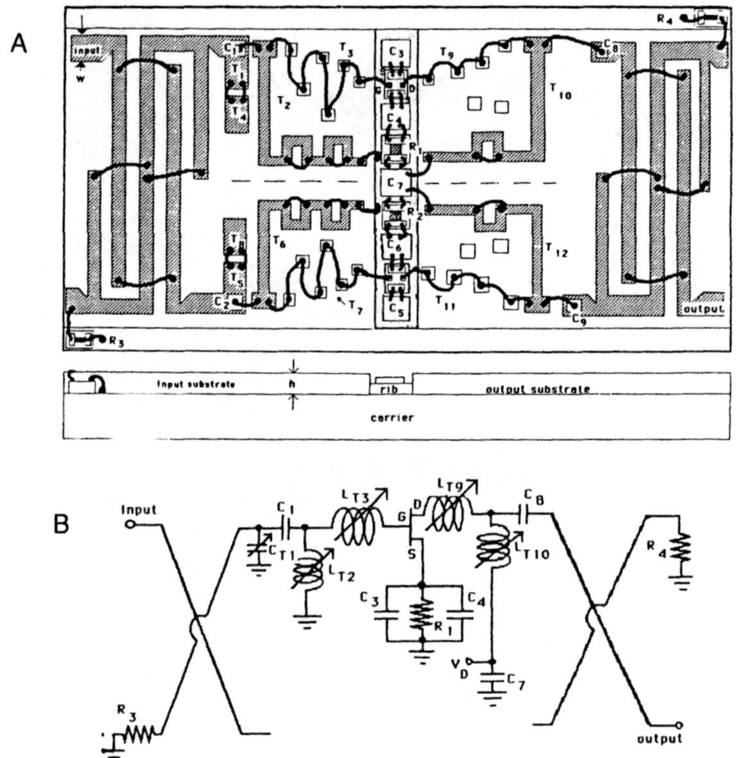

Figure 11.47 (A) The hybrid microstrip circuit in Fig. 11.46 with fine alignments. (B) The discrete elements of one-half of the balanced circuit.

If line T_9 is bonded so that it is shortened, the effective series inductance L_{T9} will be reduced. By placing a bond wire on line T_{10}, its electrical length can be reduced resulting in reduced effective shunt inductance L_{T10}. Two of the three elements making up the output matching circuit are easily adjustable.

The dc conditions on the FET can also be changed. Remember that for the self-bias mode, the gate to source voltage is the negative of the source voltage. By changing the source voltage, the gate to source voltage will change, resulting in a changed drain to source current. By replacing R_1 with another value, we can change the source voltage and adjust the drain to source current. As R_1 is increased, the gate to source voltage will become more negative, thereby reducing the drain to source current.

Alignment of microwave FET amplifiers is one of the most difficult tasks for a technician. To perform this job effectively, the technician must have a thorough understanding of microwave principles, dc basing, microstrip circuits, and matching. It is hoped that this section will give the reader the insight required in the design and manufacture of microstrip FET amplifiers.

PART IV: OSCILLATORS AND TWO-TERMINAL DEVICES

Salvatore Algeri

11.4 INTRODUCTION

11.4.1 Two-Terminal Devices

This section will introduce two-terminal devices such as a varactor diode, a transferred electron device, and an IMPATT diode. These solid-state devices have only two terminals, unlike transistors which have three terminals. Two-terminal devices tend to be more difficult to use than three-terminal devices, but transistors do not operate well above 30 GHz or 40 GHz. These devices can be made to operate between 10 GHz and 300 GHz.

Much of the section will be devoted to microwave oscillators. Oscillators are components which generate signals. Bipolar and FET oscillators are capable of operating up to 30 GHz, while two-terminal devices will be used to make oscillators which can operate up to 300 GHz. Oscillators will be divided into three classes: feedback, negative resistance, and natural pulsation. The end of the section will summarize the various types of oscillators discussed. Also, a section will be devoted to reflective amplifiers using two-terminal devices. Reflection amplifiers are used to amplify signals in the 10 GHz to 200 GHz frequency range.

11.4.2 Types of Resonators

Before we discuss oscillators in detail, an understanding of some common microwave resonators will be helpful. Every oscillator must have a resonator to set the frequency of the oscillator and an active device to produce the instability. Once an unstable condition is produced by the active device, then the signal current and voltage will have a frequency which is set by the resonant frequency of the resonator.

Lumped Element Resonator

Two of the most common resonators are lumped-element resonators, shown in Fig. 11.48. These circuits are commonly referred to as *RLC* circuits. Resonance occurs when capacitive reactance, X_C, equals the inductive reactance, X_L. At resonance,

$$X_C = X_L \tag{11.4.1}$$

Since

$$2\pi f_R L = \frac{1}{2\pi f_R C} \tag{11.4.2}$$

we have

$$f_R = \frac{1}{2\pi\sqrt{LC}} \qquad (11.4.3)$$

When a lumped element resonator is excited by an active device, the resultant signal voltage or current will have a resonant frequency as in Eq. (11.4.3).

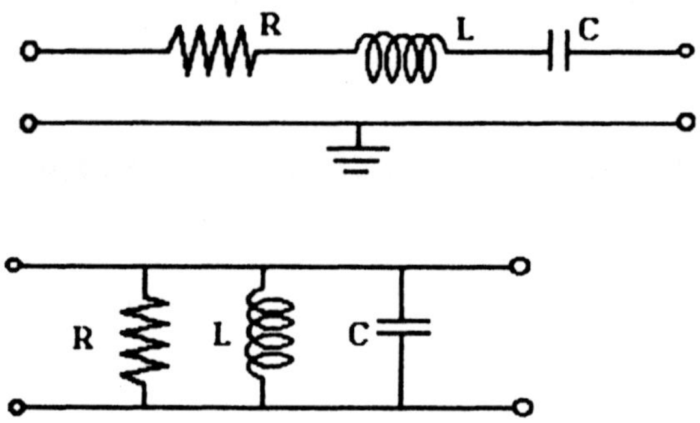

Figure 11.48 Two simple RLC resonant circuits—series and parallel.

Cavity Resonator

Cavity resonators are the general classification of resonators which can support a wave or signal of a certain frequency and tend to attenuate waves or signals of all other frequencies. The frequency at which the wave is supported is called the resonant frequency. Cavity resonators are usually air-filled metal structures. The geometry of the metal structure determines the boundary condition of the wave to be supported.

A cylindrical cavity resonator is shown in Fig. 11.49A. The geometry of the resonator is determined by its radius *a* and length *l*. Figure 11.49B shows the cross section where the electric field is maximum in the center and zero at points 1 and 2. The electric field along the length (Fig. 11.49C) must be zero at both sides and maximum at the center. Therefore, the resonant frequency for this cylindrical cavity operating in the TE_{101}* mode is given by [1]:

$$f_{res} = \frac{c\sqrt{(1 + [2L/3.41\ a]^2)}}{2L} \qquad (11.4.4)$$

where *c* is the speed of light.

A waveguide resonator is similar to a cylindrical cavity resonator in that a wave can be supported at a certain frequency only. The boundary conditions on the wave are determined by the geometry of the metal waveguide structure. A rectangular waveguide resonator is shown Fig. 11.49D. As with a cylindrical cavity, the electric field must fulfill the boundary conditions on opposite sides of the metal rectangle. There can be no electric field (voltage) along the short circuit. Therefore, on sides 1, 3 and 2, 4 the electric field must be zero. Dimensions *a* and *d* then set up the guide half-wavelength conditions for the resonant wave. There can be only one frequency for the lowest order mode (TE_{101}) that can fulfill this condition, which is given by [2]:

*Editor's note: TE_{101} is the simplest *TE* mode allowed for a circular waveguide.

$$f_{res} = \frac{c\sqrt{(a^2 ; d^2)}}{2ad} \qquad (11.4.5)$$

where c is the speed of light.

In Fig. 11.49E, the electric field must be zero at sides 2 and 4 so that the length a must be one-half wavelength in the x-direction. Also, the electric field must be zero at sides 1 and 3 (Fig. 11.49F) so that the length d must be one-half wavelength fd in the z-direction.

Tuning a cavity resonator can only be accomplished by changing the boundary conditions of the confined wave. This is usually done by changing one of the dimensions, i.e., the length of the cylindrical cavity. Because physical dimension must be changed in order to change the resonant frequency of the cavity, the tuning is generally slow (seconds or minutes).

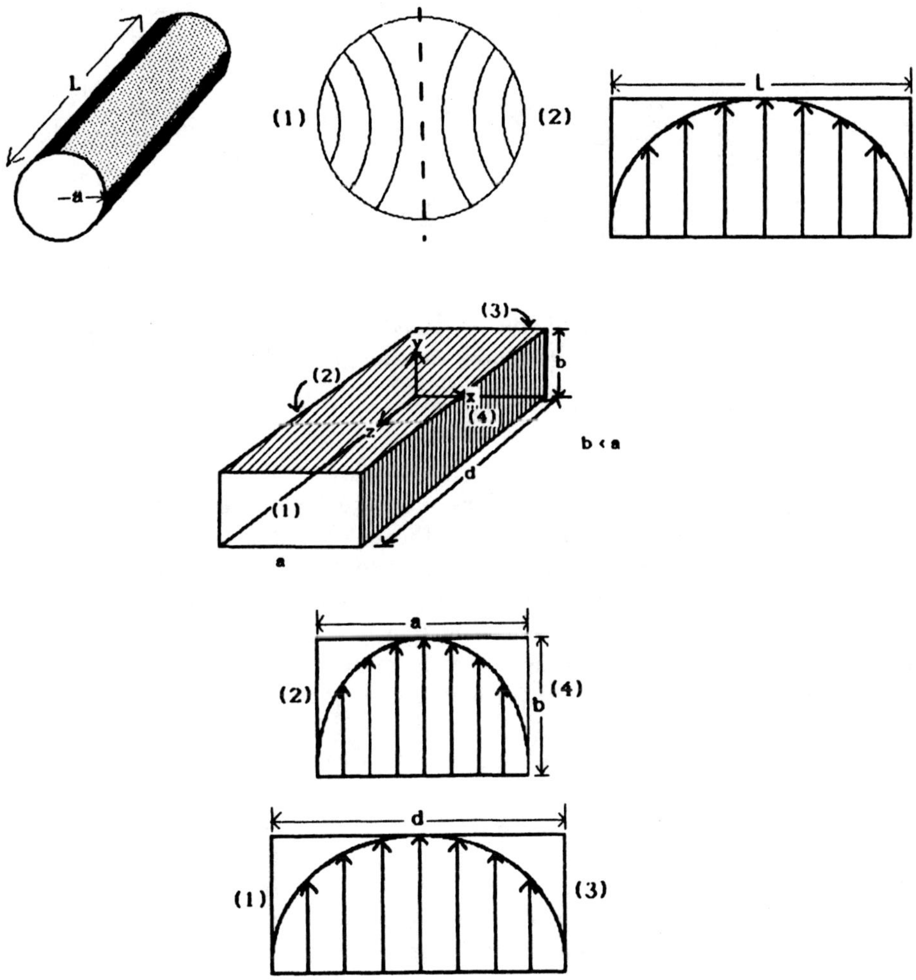

Figure 11.49 The distribution of electric field strength inside a circular and rectangular waveguide.

Dielectric Resonator (DR)

In 1939, R. D. Richtmyer [3] showed that unmetallized dielectric objects can function similarly to metallic cavities, which he called dielectric resonators. Practical applications of dielectric resonators to microwave oscillators did not appear until the early 1970s. Recent developments in reducing the losses and improving temperature stability of ceramic materials have led to the wide usage of dielectric resonator oscillators (DRO).

The dielectric material is the resonant element in the DRO. It operates similarly to a cavity resonator, except that all of its fields are not contained within the ceramic resonator. The ceramic material used is called barium tetratitanate, which has a dielectric constant of about 37. The resonator is found by locating this ceramic material of a certain geometry, usually a cylinder, between two grounded plates as shown in Fig. 11.50A.

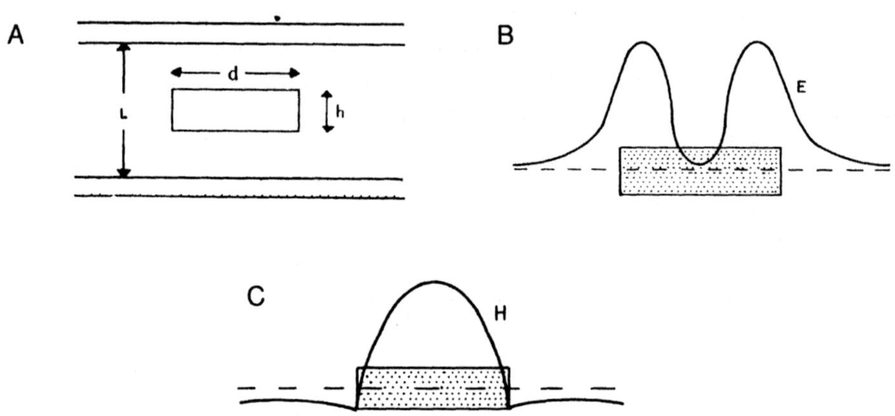

Figure 11.50 (A) The construction of a dielectric resonator. (B) and (C): The distribution of electric and magnetic fields inside the dielectric resonator.

The electric and magnetic field distributions are shown in Fig. 11.50B and 11.50C. Unlike a cavity resonator where all the fields are contained inside its conducting walls, about 20 percent of the fields extend beyond the ceramic material. These external fields are the key to tuning the dielectric resonator frequency. Tuning can be done by changing the spacing *L* of the grounded plates. As the plates move closer together, decreasing *L,* the resonant frequency decreases. Energy from the DR is generally coupled magnetically to the circuit containing the active circuit to form either a feedback or negative resistance oscillator.

Calculation of the resonant frequency is extremely complex. It is dependent upon the geometry of the ceramic material, i.e., its diameter and thickness, and the amount of magnetic coupling to the active circuit [4].

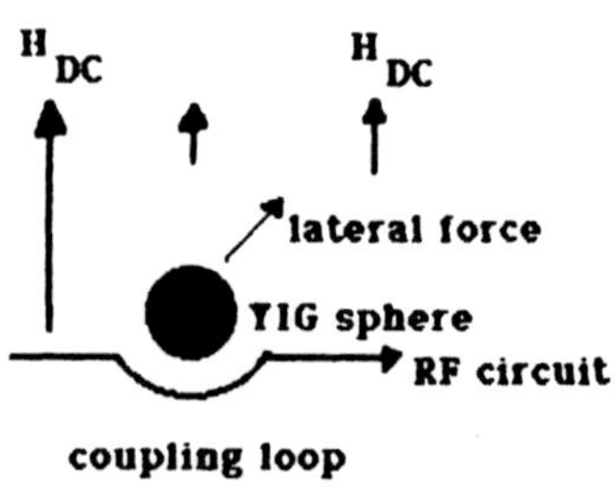

Figure 11.51 The conceptual construction of a YIG resonator.

Magnetic Resonator (YIG)

One type of magnetic resonator makes use of a ferrite material called yttrium iron garnet, or YIG, and is part of the magnetic circuit. YIG spheres are manufactured from a crystal growth similar to silicon and then polished into very small spheres.

A magnetic resonator consists of three parts, a YIG sphere, a coupling loop, and a static magnetic field as shown in Fig. 11.51. The static magnetic field can be simply two poles of a permanent magnet, but it is usually two coils of wire forming an electromagnet. The YIG sphere is magnetized and acts like the needle of a compass. If permitted, it would rotate and align its magnetic dipoles in the direction of the static magnetic field. In this case, the YIG sphere is attached at the end of a rod and oriented in the static magnetic field so that there is some lateral force on the sphere. The rod is then fixed so that it cannot rotate. This is similar to holding the needle of a compass and preventing it from rotating. The needle would exert a lateral force trying to align itself to the earth's magnetic field.

A small gauge wire is placed very close to the YIG sphere to form a coupling loop. When a signal passes through the coupling loop wire, it will magnetically couple some of its energy to the YIG sphere. If the frequency of the signal in the wire is the same as the natural frequency of the sphere, a strong interaction will occur. There is no interaction between the magnetic field of the signal in the wire and the YIG sphere at all other frequencies. This type of interaction is called resonance.

The resonant frequency of the YIG resonator can be changed by changing the natural frequency of the YIG sphere. This is done by changing the force on its magnetic dipoles. If the current in the electromagnet is changed, the static magnetic field will also change. Therefore, by changing the current in the electromagnet, the interaction of the signal's magnetic field and the YIG sphere can be varied. This is how the resonant frequency of a YIG resonator can be electronically tuned.

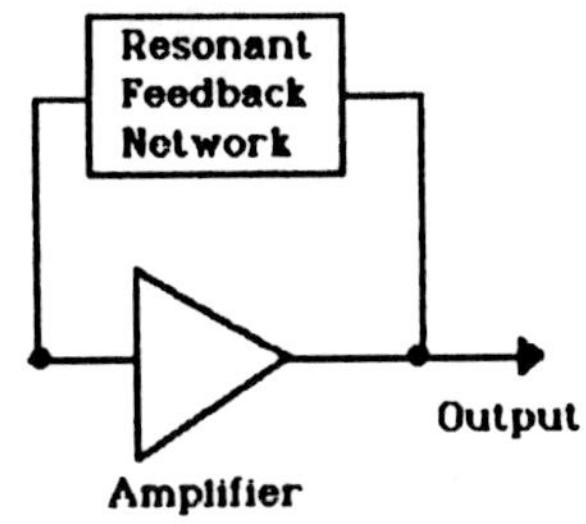

Figure 11.52 The general principle of a feedback oscillator.

11.4.3 Feedback Oscillators

Feedback oscillators are a general classification of oscillators where a resonant network is connected from the output to the input of an amplifier, as shown in Fig. 11.52. The amplifiers can be either bipolar or FET transistors, as previously discussed in the chapter. Lumped elements generally compose the resonant network. Feedback oscillators make use of positive feedback to produce an unstable condition.

1) Positive Feedback

Positive feedback occurs when the output of an amplifier is connected to the input with a 180° phase shift. Negative feedback occurs when the output of an amplifier is connected to the input with no additional phase shift. The difference may be subtle, but the result is very different. Figure 11.53A shows a block diagram of negative feedback and Fig. 11.53B shows a block diagram for positive feedback.

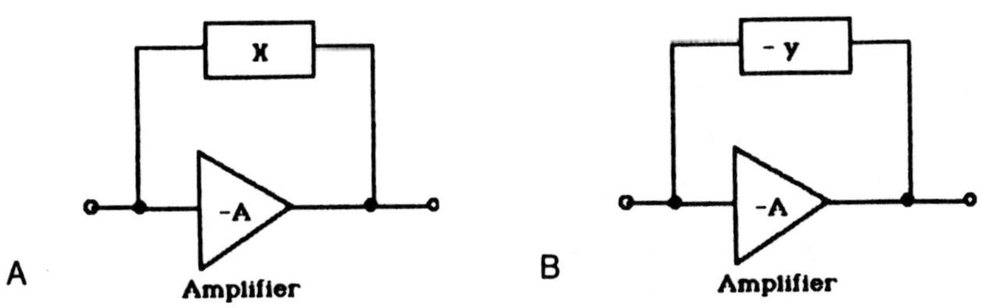

Figure 11.53 (A) Negative feedback is when the output of an amplifier is fed back to the input with no phase change. (B) There is a 180° phase shift between the output and the fedback input in a positive feedback system.

A series of timing diagrams will be discussed to explain positive feedback. Four simple signal voltage equations will be used to describe the signal flow in Fig. 11.54. The input signal is V_{in} and the output signal is V_{out}. The signal voltage V_f is the feedback voltage. Therefore,

$$V_f = -y\, V_{out} \tag{11.4.6}$$

and the sum voltage is

$$V_S = V_{in} + V_f \qquad (11.4.7)$$

and the output voltage is

$$V_{out} = -AV_S \qquad (11.4.8)$$

To illustrate what happens with positive feedback, let us select the following values. Let V_{in} vary from $+1V$ to $-1V$ with time, and $A=2$ and $y=0.5$. Then

$$V_f = -0.5\ V_{out}$$

and

$$V_{out} = -2\ V_S$$

where

$$V_S = V_{in} + V_f$$

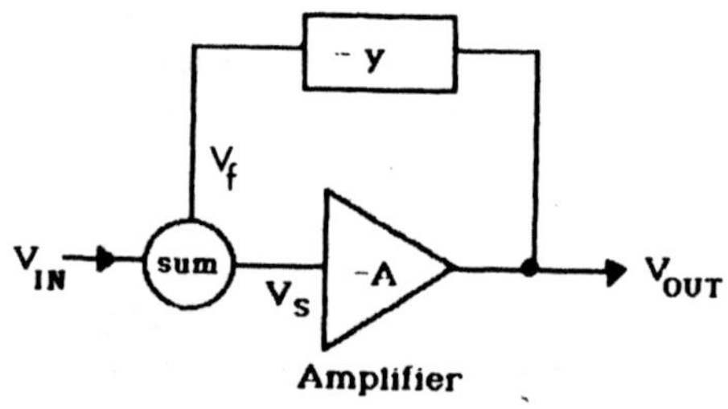

Figure 11.54 In an amplifier with positive feedback, the output of the amplified input is fed back so that the output amplitude is even larger than the previous cycle.

Figure 11.55A is a timing diagram for three cycles of V_{in}. Assume that there is zero output voltage before the first cycle. If V_{out} is zero, then V_f must also be zero, as shown in Fig. 11.55B. The sum voltage V_S is simply V_{in} as shown in Fig. 11.55C. The output voltage is -2 times the sum voltage as shown in Fig. 11.55D. The significance of the negative sign is a $-180°$ phase shift of the signal.

For the second cycle, V_{in} is the same, but V_f is -0.5 times from the first cycle as shown in Fig. 11.55B. Note the phase shift of 180°. Both V_{in} and V_f are in phase, and sum as shown in Fig. 11.55C. This results in the V_{out} for the second cycle as shown. Note that the output voltage for the second cycle is larger than the first cycle.

For the third cycle, V_{in} is the same again and V_f is -0.5 times V_{out} from the second cycle as shown in Fig. 11.55B. Because V_{in} and V_f are in phase, they will sum as shown in Fig. 11.55C. This results in an even larger output voltage than the second cycle, as shown in Fig. 11.55.D.

The timing diagram illustrates how positive feedback works. The feedback voltage is a portion of the output voltage with 180° phase shift. This feedback voltage is in phase with the input voltage so that the sum voltage entering the amplifier continues to increase. If the input to the amplifier increases, then the output will increase. Positive feedback produces an unstable condition by allowing the amplifer output to continue to increase.

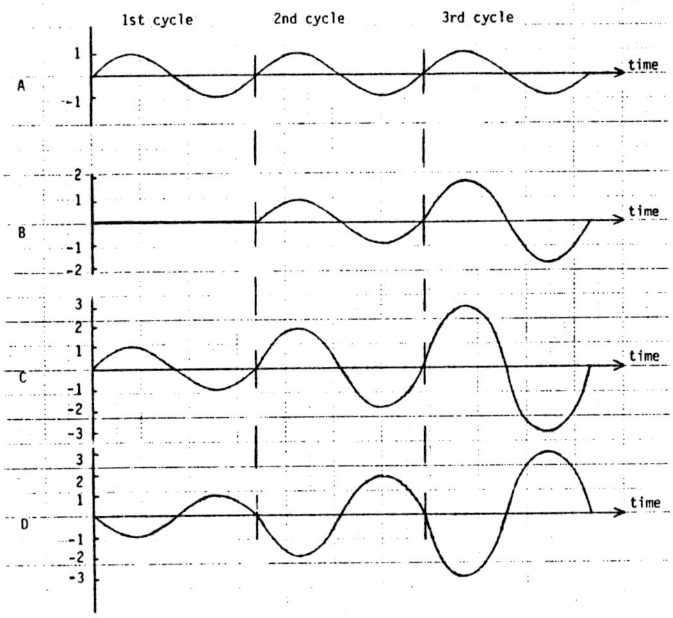

Figure 11.55 Timing diagrams showing that the amplitude of the output grows.

There are some limits which prevent this unstable circuit from increasing without bound. Generally, the power supply voltage will limit the peak signal voltages. The self-saturation of the transistor will also limit the output voltage. Figure 11.56 shows how an amplifier with positive feedback is unstable and when instability becomes controlled.

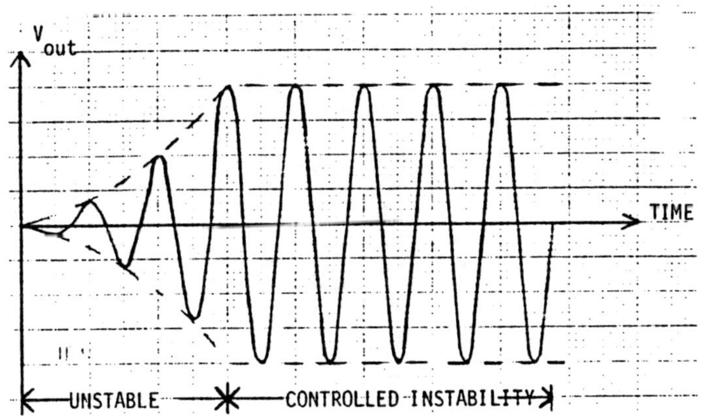

Figure 11.56 Growth of a signal due to positive feedback is usually unstable. Circuit limitations eventually control the instability.

2) Colpitts Oscillator

A Colpitts oscillator is shown in Fig. 11.57. Notice the feedback path from the collector to the base of the transistor through the resonant circuit. The resonant circuit is formed by C_1, C_2, and L. All the other circuit elements are needed for transistor biasing. The resonant frequency of this feedback oscillator is

$$f_{res} = \frac{1}{2\pi}\sqrt{\frac{(C_1 + C_2)}{LC_1C_2}} \tag{11.4.9}$$

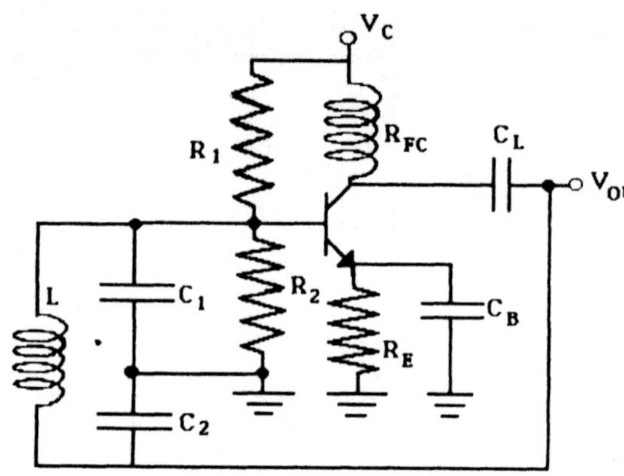

Figure 11.57 Circuit connection of a Colpitts oscillator using an npn transistor.

3) Hartley Oscillator

A Hartley oscillator is shown in Fig. 11.58. The feedback path from the collector to the base of the transistor is through the resonant network. Circuit element C_T, L_1, and L_2 compose the resonant network. All other circuit elements are needed for transistor biasing. The resonant frequency for this feedback oscillator is

$$f_{res} = \frac{1}{2\pi\sqrt{C_T (L_1 + L_2)}} \tag{11.4.10}$$

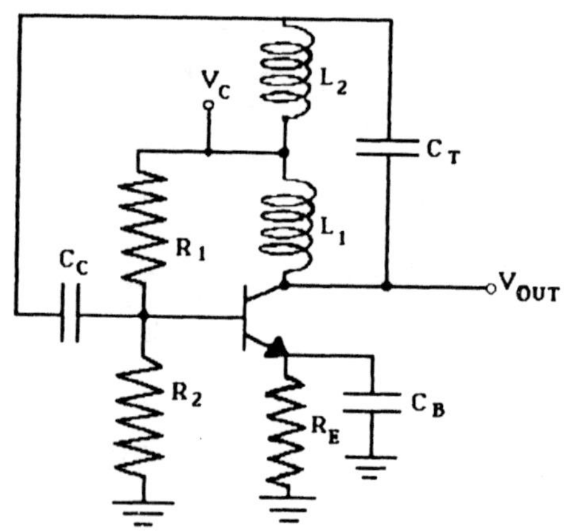

Figure 11.58 Circuit connection of a Hartley oscillator using an npn transistor.

11.4.4 Negative Resistance Oscillators

Before oscillations can be produced, we must first have an unstable condition. With feedback oscillators, the positive feedback forces the active device to produce higher and higher outputs. This unstable condition is the heart of a feedback oscillator. Negative resistance oscillators also must have some unstable condition before oscillations are produced.

Figure 11.59 shows a block diagram of a negative resistance oscillator. It is composed of a resonant circuit, an active device, and an output circuit. All resonators have some residual resistance at resonance which contributes to its loss. The purpose of the active device is to produce a negative resistance which is greater than the positive resistance of the resonator. Therefore, the first condition for oscillation is

$$R_{device} > R_{res} \tag{11.4.11}$$

This condition produces the instability necessary for oscillation.

Now that we have produced the instability, we must somehow control it. This is usually done by using a combination of the self-saturation characteristics of the active device, bias voltage limitations and by matching. Controlling the unstable condition is as important as producing the instability.

The purpose of the output circuit is to couple RF energy to the load. This is done in a variety of ways. If a cavity resonator is used, the output circuit would either be an inductive loop or a capacitive probe. Two common types of negative resistance oscillators will be illustrated: YIG and varactor negative-resistance oscillators.

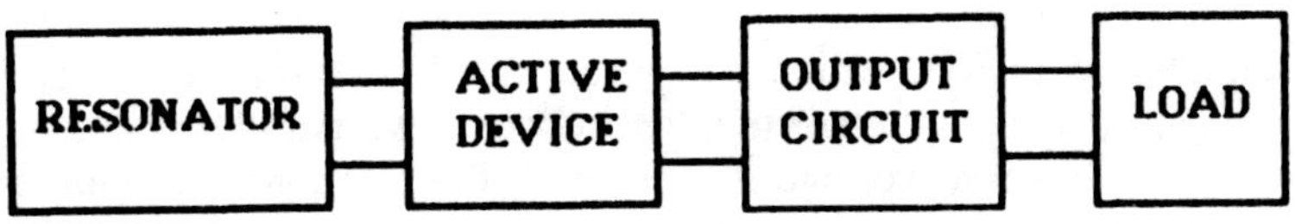

Figure 11.59 A block diagram of a negative resistance oscillator.

1) YIG Negative-Resistance Oscillator (YTO)

A YIG negative-resistance oscillator is sometimes called a *YIG-tuned oscillator* (YTO). In a negative-resistance oscillator, an active device produces the instability by generating a negative resistance. The resonator sets up the frequency of the oscillations, and the output circuit couples RF energy to the load. Figure 11.60 is a schematic representation of a YTO. The active device is a FET, which produces a negative RF resistance through the use of a small inductor from the gate to ground. The mathematics of how this occurs will not be covered here. An output matching circuit couples RF energy from the output of the FET to the load. For oscillations to occur the magnitude of the negative resistance from the FET r_s must be greater than the positive resistance of the YIG resonator R_o.

A YIG resonator can be easily tuned by changing the static magnetic field on the YIG sphere. This is usually done by changing the dc current in an electromagnet. YIG-tuned negative-resistance oscillators can be tuned over very wide bandwidths, e.g., 2-8 GHz and 6-18 GHz.

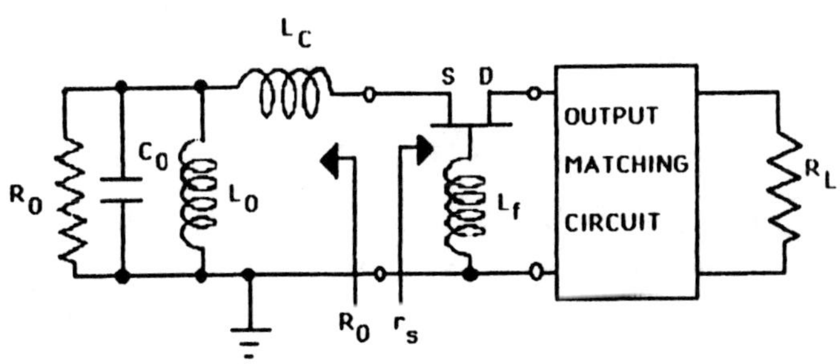

Figure 11.60 A schematic representation of a YIG-tuned oscillator.

2) Varactor-Tuned Negative-Resistance Oscillator (VCO or VTO)

Varactor-tuned negative-rsistance oscillaors are called *voltage-tuned oscillators* (VTO) or *voltage-controlled oscillators* (VCO) in the industry because the

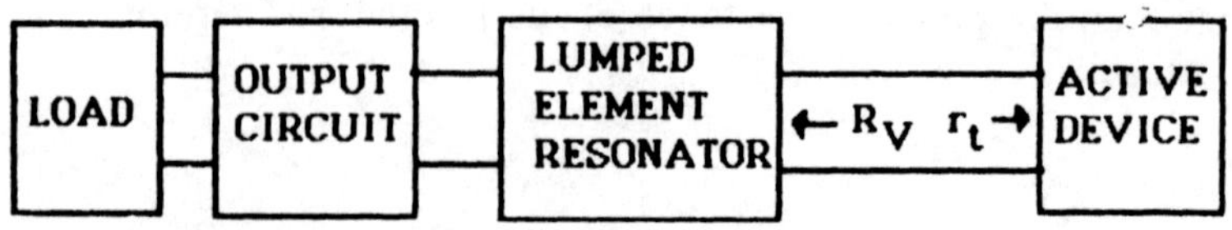

Figure 11.61 A block diagram of a voltage-tuned oscillator. The oscillating frequency is controlled by a voltage.

resonant frequency of the resonator can be changed by changing the voltage across a varactor diode. A block diagram of a VCO is shown in Fig. 11.61.

Before discussing the complete oscillator, let us review how a varactor diode operates. A varactor diode is a *pn*-junction, as shown in Fig. 11.62A, where the *p*-type material has excess holes (+) and the *n*-type material has excess electrons (−). With application of a reverse voltage across the diode as shown in Fig. 11.62B, the electrons in the *n*-type are attracted to the positive battery terminal. Also, the holes in the *p*-type are attracted to the negative battery terminal. This creates a depletion region, which is a direct result of the reverse voltage. This depletion region acts as a dielectric layer between two parallel plates of a simple capacitor. As the reverse voltage is increased, the two parallel plates appear to be pulled apart, thus reducing the effective capacitance of the junction. A typical junctioin capacitance curve *versus* voltage is shown in Fig. 11.63.

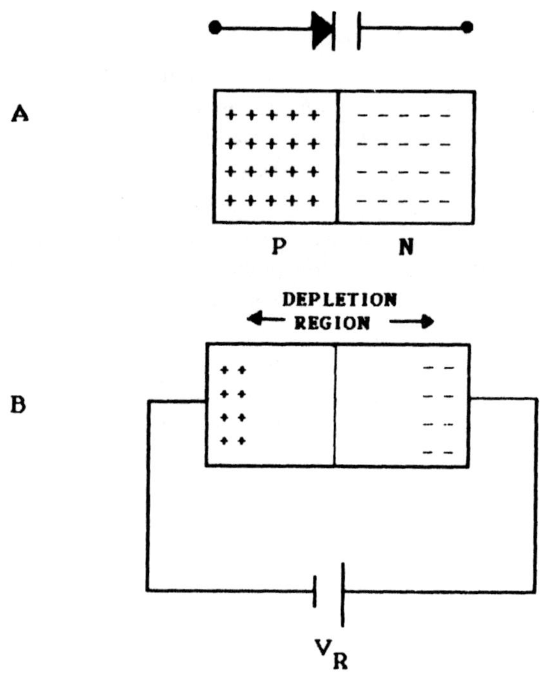

Figure 11.62 (A) A varactor diode is a heavily doped pn diode. (B) the capacitance of the diode is controlled by the reverse biased voltage.

A complete circuit schematic of a VCO is shown in Fig. 11.64. The varactor diode D_1 is reverse biased by applying a positive voltage V_T through RF chokes. V_T is sometimes called the tuning voltage, which will be explained shortly. The bipolar transistor is the active device, which provides the negative resistance necessary for oscillations to occur. Coil L_p is part of the series resonant lumped element circuit and coil L_s is the secondary of a transformer to couple energy out of the resonator.

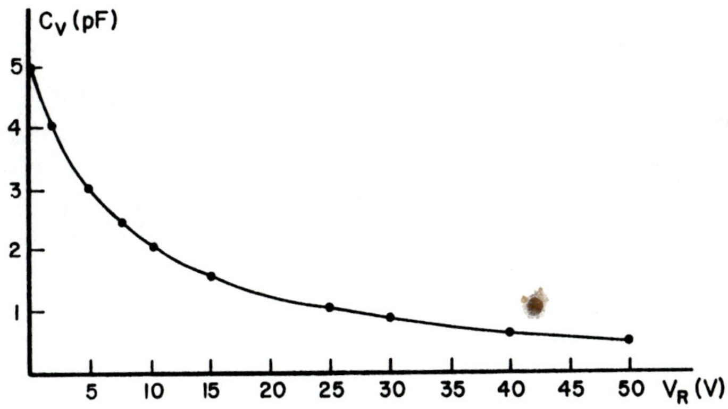

Figure 11.63 Capacitance of the varactor diode versus applied voltage.

A closer examination of the resonator can be seen using Fig. 11.65, which is a schematic of the RF equivalent circuit. The varactor diode has two RF components, a voltage variable junction capacitance and a fixed RF resistance. The equivalent circuit for the bipolar transistor is a series capacitor, C_t, and a negative resistor, $-R_t$. If the magnitude of R_t is greater than the magnitude of R_v, then an unstable condition exists and oscillations will occur at the resonant frequency of the resonator.

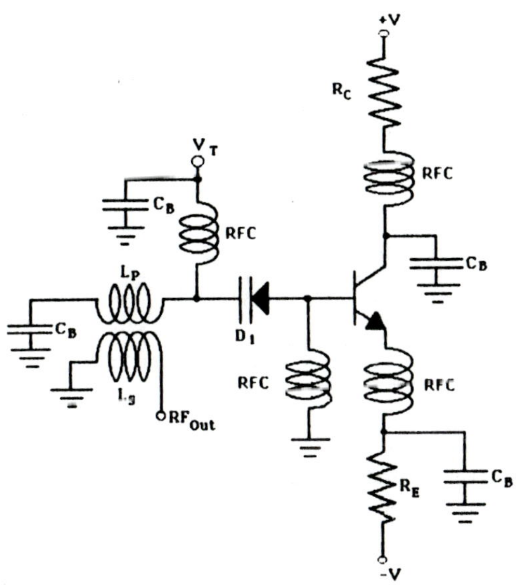

Figure 11.64 The circuit diagram of a voltage-tuned oscillator using a varactor diode.

The resonator is a lumped element type and is shown in Fig. 11.66. The resonant frequency of a series lumped element resonator is given by

$$f_{res} = \frac{1}{2\pi\sqrt{L_p C_{eq}}} \tag{11.4.12}$$

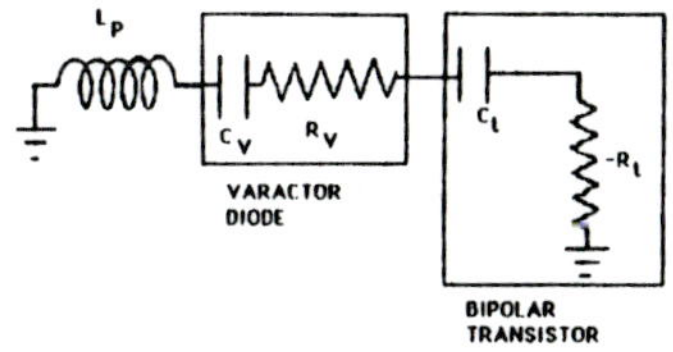

Figure 11.65 The RF equivalent circuit of the voltage-tuned oscillator.

The equivalent capacitance C_{eq} is the series combination of C_v and C_t, which is

$$C_{eq} = \frac{C_v C_t}{C_v + C_t} \tag{11.4.13}$$

How does a VCO change its frequency with tuning voltage? Remember from Fig. 11.63 that the junction capacitance C_v changes with reverse or tuning voltage. When C_v changes, C_{eq} changes correspondingly according to Eq. (11.4.13). The resonant frequency f_{res} in Eq. (11.4.13) will change when C_{eq} changes. Therefore, we can change the voltage across the varactor diode, and therefore change the frequency of oscillation, hence the name voltage-controlled oscillator (VCO). This is illustrated by an example.

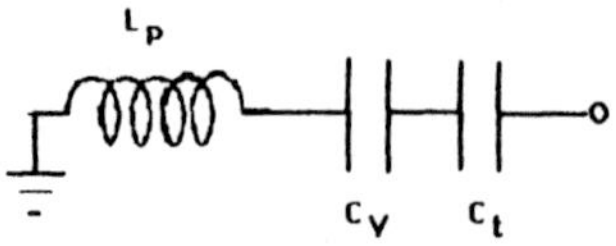

Figure 11.66 The series connection of L_p, C_V, and C_t results in a resonant circuit.

Example (11.4.4.1)

For a VCO, assume that $V_T = 10V$, $C_t = 2pF$ and $L_p = 1nH$. Use the C_v *versus* V_R (V_T) curve in Fig. 11.63 to find the resonant frequency and assume the oscillator is unstable.

Solution: If the oscillator is unstable, we need not worry about the bipolar transistor provided that there is enough negative resistance.

The quantity C_v must be determined before the resonant frequency can be calculated from Eq. (11.4.12). Using Fig. 11.63 and assuming that the tuning voltage or reverse voltage is 10V, we can read C_v from the chart to be 2 pF.

According to Eq. (11.4.13), the equivalent capacitance is

$$C_{eq} = \frac{2pF \times 2pF}{2pF + 2pF} = 1pF$$

The resonant frequency can be calculated from Eq. (11.4.12),

$$f_{res} = \frac{1}{2\pi\sqrt{1nH \times 1pF}}$$

$$= 5.03 \text{ GHz}$$

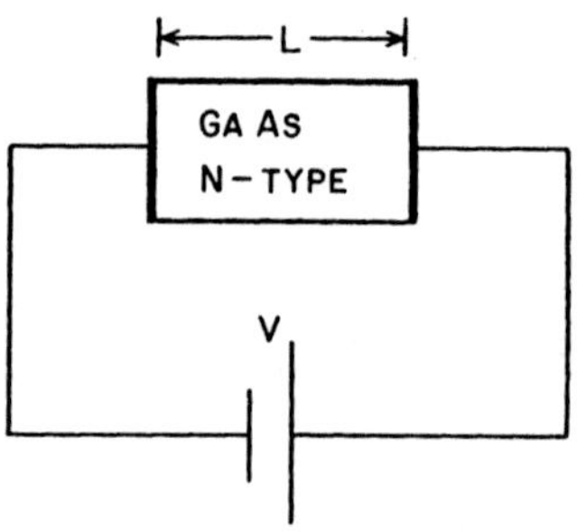

Figure 11.67 A piece of bulk GaAs as a TED.

11.4.5 Transferred Electron Devices

Transferred electron devices are used to amplify and generate signals in the frequency range from 10 GHz to 100 GHz. Transferred electron devices (TED) are sometimes referred to as Gunn devices, named after the person who discovered the transferred electron effect. A TED is a piece of bulk semiconductor usually fabricated from gallium arsenide (GaAs). Figure 11.67 shows a piece of bulk gallium arsenide semiconductor. Note that there are no junctions and that it is forward biased.

The unusual feature about transferred electron or Gunn devices is the electron velocity through the gallium arsenide semiconductor. Figure 11.68 shows the electron velocity or current *versus* electric field or voltage across the semiconductor. Silicon has the characteristic that the electron velocity (current) increases as the electric field (voltage) increases across the device until saturation is reached. Gallium arsenide has the characteristic that the electron velocity increases with increasing electric field until a peak is reached. After the peak is reached, increasing electric field (voltage) has the effect of reducing the electron velocity (current) until saturation is reached. The reason for this phenomena is extremely complex. One simple explanation is that the electron energy momentum states of gallium arsenide molecules are transferred into higher order states so that their mobility decreases with increasing electric field.

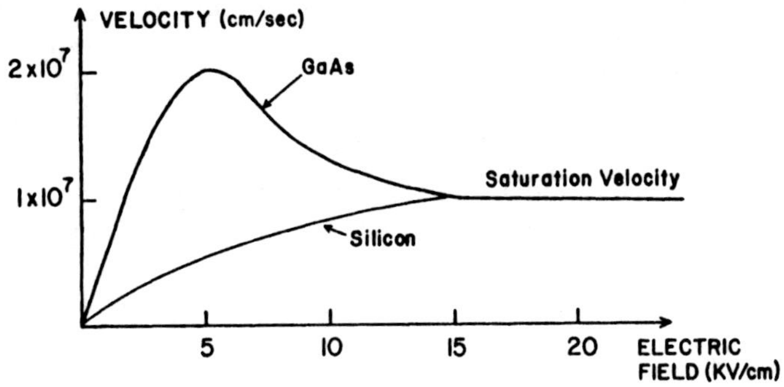

Figure 11.68 Comparison of electron velocity versus applied electric field for Si and GaAs.

1) Negative Dynamic Resistance

The negative resistance mode is one of the two modes of a TED. The concept of a negative dynamic resistance will be reinforced through the use of an example. Consider the diagram in Fig. 11.69, which shows the electron velocity or current *versus* electric field or voltage. The term *dynamic resistance* is used to differentiate between static resistance as in a normal resistor. A dynamic resistance is only apparent to an RF signal, and cannot be measured in a dc sense using Ohm's law.

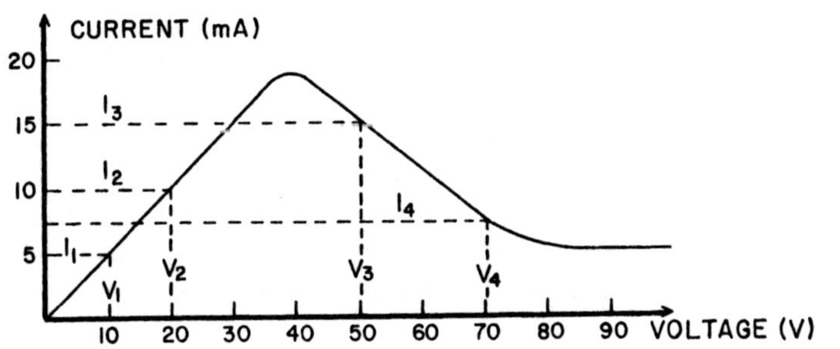

Figure 11.69 A special I/V curve illustrating a negative dynamic resistance region.

Dynamic resistance R_a is similar to the slope of a line. R_a can be calculated as the change in the voltage divided by the corresponding change in current. Therefore,

$$\begin{aligned} R_a &= \Delta V/\Delta I = (V_2 - V_1)/(I_2 - I_1) \\ &= (20\text{V} - 10\text{V})/(10\text{mA} - 5\text{mA}) \\ &= 2\text{k}\Omega \end{aligned}$$

The dynamic resistance for this part of the curve is 2 kΩ.

Dynamic resistance R_b can be calculated as the change in voltage divided by the corresponding change in current. Therefore,

$$\begin{aligned} R_b &= \Delta V/\Delta I = (V_4 - V_3)/(I_4 - I_3) \\ &= (70\text{V} - 50\text{V})/(7.5\text{mA} - 15\text{mA}) \\ &= -2.67\ \text{k}\Omega \end{aligned}$$

R_b is a negative dynamic resistance which simply means that the signal curent decreases as the voltage increases. Stated another way, the signal current and voltage are 180° out of phase.

2) Negative-Resistance Reflection Amplifier

Now that it has been shown how a TED can produce a negative dynamic resistance, we can make a reflection amplifier. This type of amplifier is primarily used at frequencies above 20 GHz where FETs do not operate well. Figure 11.70 shows a block diagram for a circulator coupled TED negative-resistance reflection amplifier.

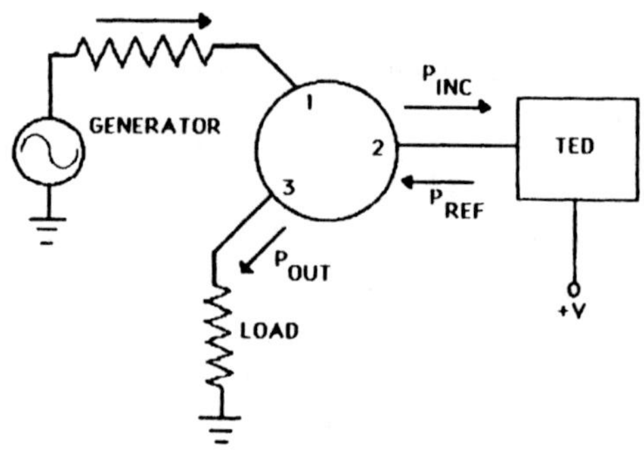

Figure 11.70 A circulator coupled TED negative resistance reflection amplifier.

A circulator is a three-port device that directs a signal to the adjacent port in a circular direction. For the purposes of this illustration, assume that the insertion loss of the circulator is 0 dB. The TED is biased in the negative resistance region of its *I*/*V* curve. Power from the generator goes through the circulator and is incident at the TED. Two things happen to the signal, some will be absorbed and some reflected. The reflected signal, P_{ref}, goes through the circulator to the load. For this circuit to be an amplifier, the output power P_{out} must be greater than the incident power P_{inc}; but P_{out} is the same as the reflected power P_{ref}. Therefore,

$$P_{ref} > P_{inc}$$

for the circuit to be an amplifier. This appears to be inconsistent with the general concept of reflected signals (see Ch. 5). Let us examine this more closely by using two illustrations.

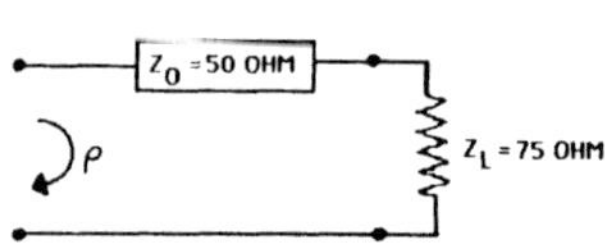

Figure 11.71 A transmission line terminated by a real resistance results in a reflection coefficient less than 1.0.

Consider the circuit in Fig. 11.71 where a 50 ohm transmission line is terminated by a 75 ohm load. Reflection occurs since the load's impedance is different from the characteristic impedance of the line. The magnitude of the reflection coefficient $|\rho|$ can be calculated to be

$$\begin{aligned} |\rho| &= (Z_L - Z_o)/(Z_L + Z_o) \\ &= (75\ \Omega - 50\ \Omega)/(75\ \Omega + 50\ \Omega) \\ &= 0.2 \end{aligned}$$

When $|\rho|$ is less than 1.0, it means that the reflected signal voltage is less than that of the incident signal. Therefore, for any load impedance with a positive (dynamnic) resistance, the reflected power will be less than the incident power.

Now consider the circuit in Fig. 11.72, where a 50 ohm transmission line is terminated by a −75 ohm load. Let us calculate the magnitude of the reflection coefficient, $|\rho|$,

$$|\rho| = (Z_L - Z_o)/(Z_L + Z_o)$$
$$= (-75\ \Omega - 50\ \Omega)/(-75\ \Omega + 50\ \Omega)$$
$$= 5.0$$

When $|\rho|$ is 5.0, this means that the reflected voltage is five times the incident voltage. Therefore, for any load impedance with a negative (dynamic) resistance, the reflected power will be greater than the incident power.

When the TED is biased to produce a negative dynamic resistance, it can act as a reflection amplifier. The purpose of the circulator is to direct the signals from the generator to the TED and from the TED to the load. Amplification depends upon reflections from the TED which are greater than the incident signal. From an input-output viewpoint, it appears as a simple amplifier.

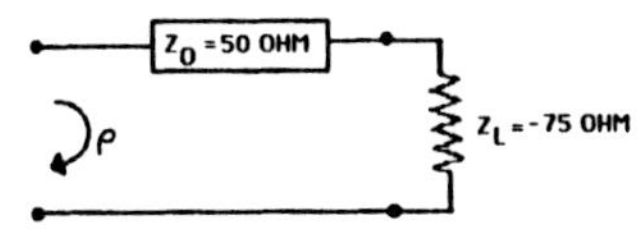

Figure 11.72 A transmission line terminated by a negative resistance results in a reflection coefficient more than 1.0, i.e., amplification.

3) Domain Mode in TED

The domain mode in a TED is used to produce CW oscillators. This mode provides a natural pulsating effect which, when coupled to a resonator, will make an oscillator. As with the Gunn effect, which slows the electron velocity after some peak is reached, the domain mode is extremely complex to conceptualize. The following analogy will be used to help simplify the explanation.

Suppose you have just purchased a new sports car. You drive out of the dealer's lot slowly at 10% pressure with a speed of 10 mph. It is rush hour, so you can only move at 30 mph which is 30% pressure. You enter the ramp to the highway and accelerate to 50 mph, or 50% pressure. Traffic is heavy and you are going with the flow. You look at the rear view mirror and see an old pickup truck behind you. You do the natural thing and press the accelerator to 60% so that your speed will increase to 60 mph. The truck is still behind you and has turned its lights on. Now is the time to see what your newly acquired sports car can do. You step on the accelerator to 70% and a strange thing happens. Your car slows down to 50 mph! The pickup truck is now flashing the lights and blowing the horn. You begin to panic and press the accelerator to 100% and your car slows to 40 mph. Because you are now traveling slower than the traffic flow, the cars in front of you have moved far ahead while the cars behind you are all bunched up.

A piece of gallium arsenide can be thought of as the highway and the cars as the electrons flowing through the material. As the electron velocity increases beyond the peak, the electrons begin to slow down and bunch. They travel at the saturation velocity (the maximum allowed for the applied voltage) in bunches, just like the cars on the highway. When the electron bunch exits the GaAs material, a pulse of current occurs. This electron bunch is called a domain. This domain is produced repeatedly as a current pulse with a certain frequency. This frequency is dependent upon the length of the GaAs material, as illustrated by the following example.

Example (11.4.5.1)

A piece of GaAs material of length 10μm is operating in the domain mode. Calculate the frequency of the domains exiting the material.

Solution: From Fig. 11.68, the saturation velocity is 10^7 cm/s. The frequency of the domain traveling at this velocity is

$$f = v_{sat}/L$$

$$= \frac{10^7 \text{ cm/s}}{10^{-3} \text{ cm}}$$

$$= 10^{10}\text{Hz} = 10\text{ GHz}$$

4) YIG-Tuned TE-Oscillator

A YIG-tuned transferred electron oscillator consists of a YIG resonator with a natural frequency set at the domain frequency of the TED. Figure 11.73 shows a cross section of the oscillator. Note that the YIG resonator consists of the YIG sphere, an inductive loop, and a static magnetic field. The active device is the TED operating in the domain mode. The output coupling loop inductively couples energy out of the YIG resonator. If the current through the electromagnet producing the static magnetic field is changed, the resonant frequency of the resonator will change. Therefore, we can electronically change the resonant frequency by simply changing the current through the electromagnet.

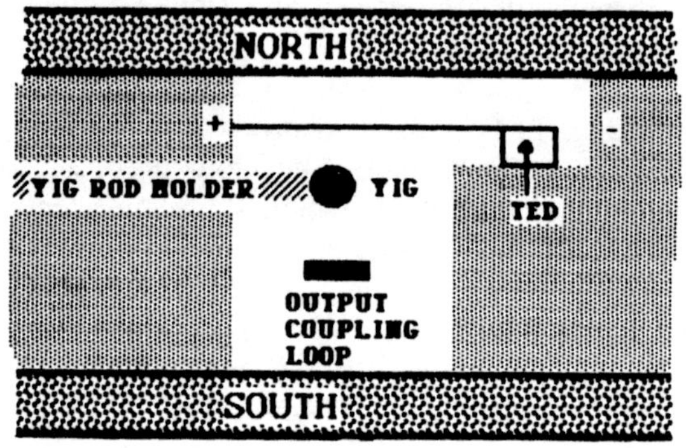

Figure 11.73 The cross section of a YIG-tuned transferred electron oscillator.

11.4.6 IMPATT Diode

An IMPATT diode stands for *impact avalanche* and *transit time.* Figure 11.74 shows the cross section of an IMPATT diode. It consists of a *pn*-junction, an intrinsic region (pure semiconductor) and an n^+ region to form a contact. The *pn*-junction acts as an avalanche generator, and the length of the intrinsic, or *I*-, region has a transit time to delay the electron current flow through the semiconductor. IMPATT diodes are complex devices, but are one of the very few devices capable of generating or amplifying signals in the 20 GHz to 300 GHz frequency range.

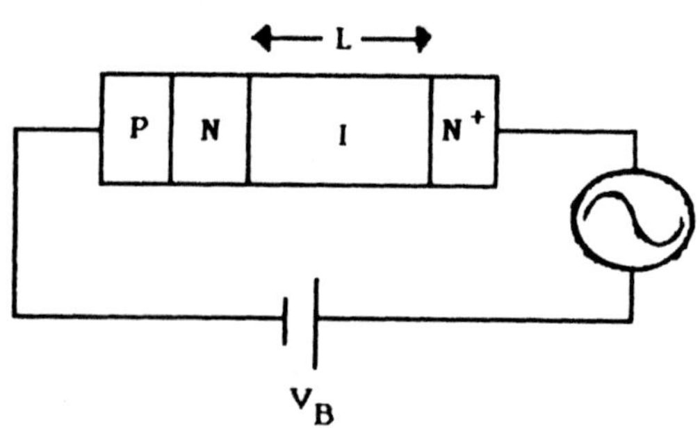

Figure 11.74 The cross section of an IMPATT diode.

IMPATT diodes are biased near the avalanche region, as shown in Fig. 11.75. With the addition of an RF signal voltage applied to the device with the bias voltage, the *pn*-junction is driven into avalanche for a short part of the RF cycle. When this occurs, a large number of current carriers are generated which must travel through the *I*-region toward the n^+ contact.

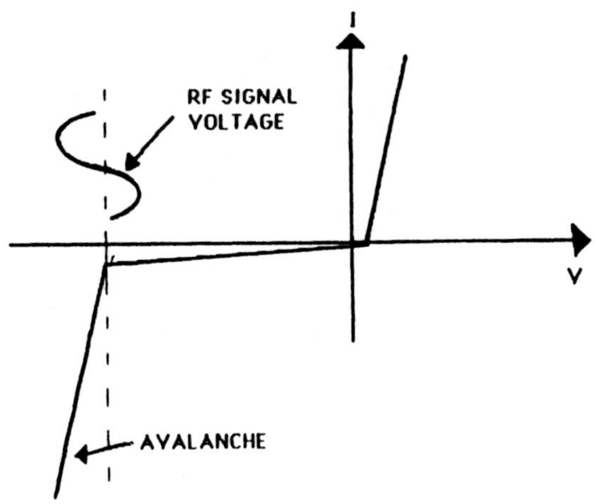

Figure 11.75 An IMPATT diode is biased near the avalanche region and the signal is applied so that part of it will result in avalanche.

A series of diagrams will be used to help illustrate the flow of the electron current through the semiconductor. Figure 11.76A–D will be used in conjunction with Fig. 11.77 as a reference. Figure 11.76A corresponds to point *A* of the timing diagram in Fig. 11.77. At point *A*, the RF signal adds nothing to the bias voltage and the IMPATT is in the negative bias region (inactive). When the RF signal voltage adds to the bias voltage the *pn*-junction is driven into avalanche. Point *B* of Fig. 11.77 shows a large number of electrons generated. These electrons must then travel through the entire length of the *I*-region. Point *C* of Fig. 11.77 shows when the electrons are halfway through the *I*-region. Finally, when the RF signal voltage subtracts from the bias voltage, point *D* of Fig. 11.77, the electrons exit the *I*-region and appear at the battery terminals as shown in Fig. 11.76D. Because the electrons have just appeared, the RF signal current will be a pulse, as shown in Fig. 11.77.

Note that the RF signal current and signal voltage are 180° out of phase. This appears as a negative dynamic resistance which can be used to produce reflection amplifiers and negative resistance oscillators. How do we set the length of the *I*-region? The length *L* is set so that the electron transit time required to travel through the *I*-region is one half of a cycle of the RF signal voltage. This is the key parameter to making the signal current 180° out of phase with the signal voltage to produce a negative dynamic resistance.

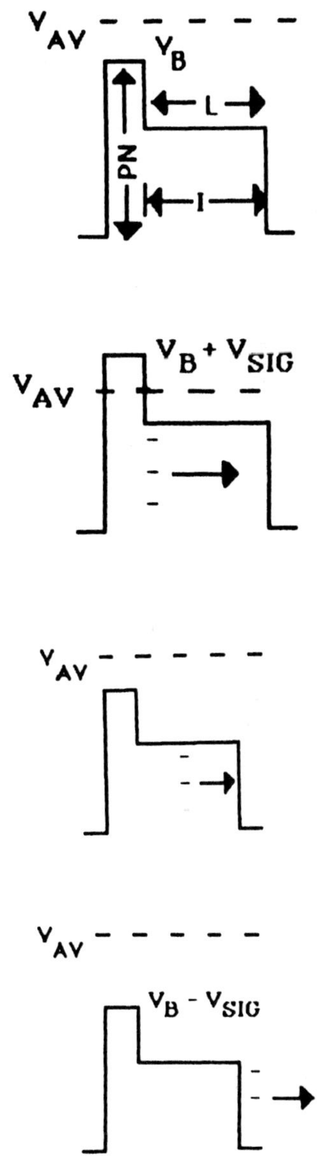

Figure 11.76 The change in electron current flow an RF signal is applied to the IMPATT diode biased near the avalanche.

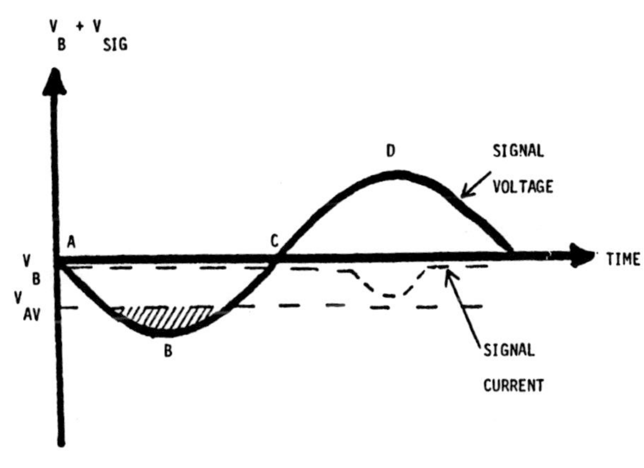

Figure 11.77 The bias voltage V_B and the signal voltage V_{SIG} applied to an IMPATT diode. Use this diagram in conjunction with Fig. 11.76.

1) IMPATT Oscillator

An IMPATT oscillator uses a resonator and an active device operating as a negative dynamic resistance, as discussed in Sec. 11.4.4. Figure 11.78 shows an IMPATT oscillator with a waveguide cavity resonator. The bias voltage enters through the center conductor as shown, while the coaxial short circuit determines one boundary of the waveguide cavity. The tuner can mechanically change the dimensions of the cavity to adjust the resonant frequency. The signal exits the cavity through a tapered waveguide to the output. The tapered waveguide is nothing other than an impedance-matching network.

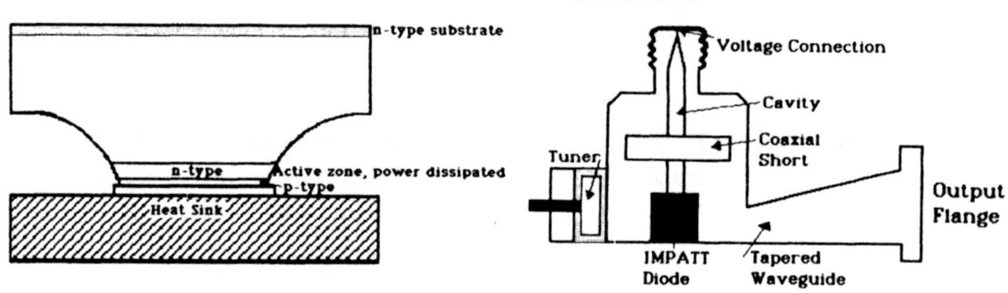

Figure 11.78 An IMPATT oscillator in a waveguide cavity.

2) IMPATT Negative-Resistance Reflection Amplifier

IMPATT diodes can be used to make reflection amplifier in exactly the same way as discussed in Sec. 11.4.5. These amplifiers can be cascaded into chains. Figure 11.79 shows a three stage cascade reflection amplifier chain. The first stage is a TED followed by two IMPATT stages. The TED is usually the first stage because it has lower noise contribution than IMPATT diodes. Because IMPATT diodes can produce higher powers than TEDs, they are usually used at the output of the chain. This chain has 24 dB gain.

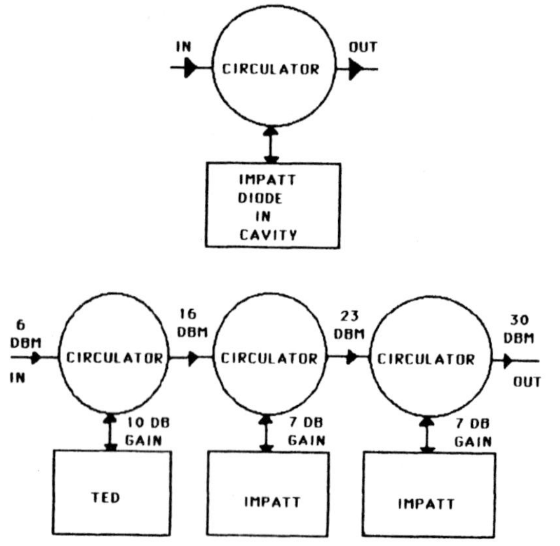

Figure 11.79 Three amplifiers in cascade. The first one is a TED and the next two are IMPATT reflection amplifiers.

11.4.7 Conclusion

This section has introduced microwave solid-state oscillators, which use both three-terminal and two-terminal devices. Three-terminal devices are limited to approximately 30 GHz. Three different techniques have been discussed to produce oscillators: positive feedback, negative resistance, and natural pulsation (domain modes). Some common oscillator performance terms are defined at the end of this chapter. These terms are used to specify a microwave oscillator.

A summary of oscillator performance trade-offs is shown in Table 11.1. Four common types of oscillators are summarized. The YIG resonator has the widest tuning bandwidth. For a TED and an IMPATT no resonator is indicated because a proper resonator must be selected for the appropriate frequency range.

This section also showed how signals can be amplified using a negative dynamic resisance. A negative resistance reflection amplifier produces a reflected signal greater than an incident signal due to the negative dynamic resistance of the active device.

Table 11.1
Oscillator Performance Trade-Offs

Oscillator Type	Frequency Range (GHz)	Bandwidth	RF Power (W)	Phase Noise
VCO	0.01 to 20	2:1 or less	~10mW	3
YIG	0.5 to 26.5	4:1	~10mW	2
TED(Gunn)	10 to 80	1.5:1	~10mW	1
IMPATT	10 to 250	1.2:1	1W	4

Phase noise: 1(low), 4(high)

Oscillator Performance Terms

1) *Post-Tuning Drift:* The change of frequency with time after setting resonant frequency.
2) *Bandwidth:* The frequency range the oscillator operates.
3) *Frequency Drift Over Temperature:* The change of frequency due to temperature.
4) *Frequency Pulling:* The change of frequency due to mismatch loads at RF output.
5) *Harmonic Signals:* Signals whose frequencies are n times the fundamental frequency where n is an integer greater than 1.
6) *Spurious Signals:* Signals not integer related to the fundamental and caused by the active device.
7) *Setting Time* (of an electronically tuned oscillator): The time required for the oscillator to change fequency referenced to the control signal.
8) *FMing:* The frequency modulation (FM) generated by the modulation on the power supply.
9) *Phase Noise:* Noise generated by random processes and white noise (see Fig. 11.82 for phase noise *versus* frequency).

Acknowledgement

The author S. Algeri wishes to dedicate this writing to his wife, Ruth, and his children, Catherine and Matthew.

References

1. Ramo, S., J. R. Whinnery, T. Van Duzer, *Fields and Waves in Communication,* New York, John Wiley and Sons, 1965.

2. *Ibid.*
3. Richtmeyer, R. D., "Dielectric Resonators," *Journal of Applied Physics*, vol. 10, June 1939, pp. 391–398.
4. Guillon, P., "Calculation of Resonant Frequency for Dielectric Resonators," *MTT Microwave Symposium Digest*, 1976, pp. 197–199.

Suggested Reading

1. Bodway, G. E., "Two-Port Power Flow Analysis Using Generalized Scattering Parameters," *Microwave Journal*, vol. 10, No. 6, May 1967, pp. 6–1 to 6–9.
2. Pengelly, R. S., *Microwave Field-Effect-Transistors—Theory, Design and Applications*, New York, Research Studies Press, 1982.
3. Gonzalez, G., *Microwave Transistor Amplifiers—Analysis and Design*, Englewood Cliffs, NJ, Prentice-Hall, 1984.
4. Krauss, H., C. W. Bostian, and F. H. Raab, *Solid State Radio Engineering*, New York, John Wiley and Sons, 1980.

CHAPTER 12

MINIATURE TRANSMISSION LINES

W. Stephen Cheung

12.1 INTRODUCTION

Advances in integrated circuitry for low frequency electronics have found applications in modern microwave electronics. A category of microwave transmission lines is that comprised of the various miniature transmission lines to be discussed in this chapter. Conducting paths are formed on printed circuit boards, and components and devices are then either mounted or fabricated during the film processing. This is known as a *hybrid* connection. Monolithic microwave integrated circuits (MMICs) have all the necessary devices, components, and conducting paths embedded in different layers, thus they require very little tuning.

The high frequency nature of microwave hybrid and integrated circuits makes circuit design more difficult than its low frequency counterpart. With the advances in CAD/CAM (computer-aided design and computer-aided-manufacturing), the circuit designer can usually accurately predict the circuit performance. Microcircuits are mathematically complicated, so it is not the intention of this book to cover the topic extensively.

12.2 MICROSTRIPS

Insights can be gained by seeing how a special miniature line is constructed from a coaxial cable. Figure 12.1A shows the cross section of a common coaxial cable. If the cylindrical geometry of the coaxial conductors are now compressed to become rectangular, an embedded microstrip is formed (Fig. 12.1B). Other versions that can be derived from Fig. 12.1B are shown in Figures 12.1C, D.

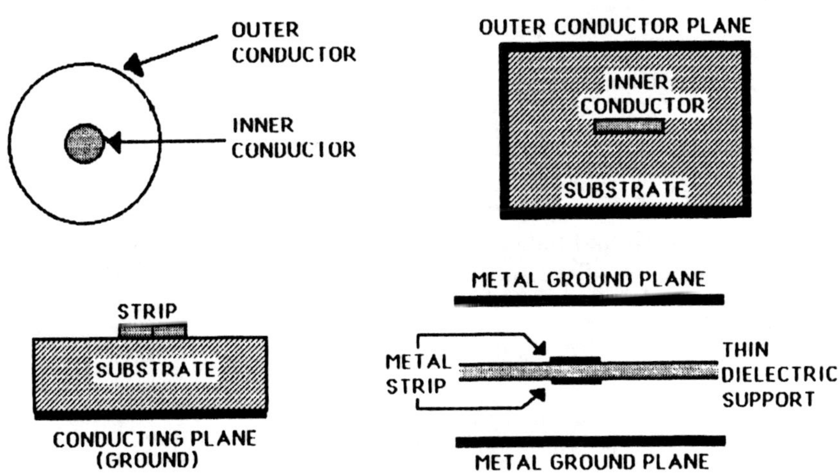

Figure 12.1 Cross sections of (A) a coaxial cable, (B) a suspended (or embedded) stripline, (C) a microstrip, and (D) a triplate stripline.

SEC. 12.2
MICROSTRIPS

Recall from Ch. 4 that low frequency signals propagate in the transverse electromagnetic (TEM) mode. For signals whose wavelengths are significantly larger than the dimension of the microstrip, the TEM mode is the main propagation mode. In fact, quantities such as characteristic impedance, capacitance, and inductance can be calculated using the conventional low frequency approach. At the low end of the microwave spectrum, bulk components such as capacitors and inductors are simply inserted into the hybrid circuit. Such bulk components, however, must be made of high quality dielectrics with very small dielectric losses.

In most of the centimeter wavelength range where the signal wavelength is still larger than the circuit dimension, the propagating characteristics of the TEM mode can be satisfactorily predicted by using low frequency techniques with some corrections for high frequency effects. This is called the *quasi-TEM* method. In the millimeter wavelength range, a thorough analysis using Maxwell's equations will be necessary; this is the *full-wave* method.

The electric field patterns of the coaxial cable and the miniature lines in Fig. 12.1 are shown in Fig. 12.2. Note that in the open-air microstrip, each electric field line suffers a kink at the boundary between air and the substrate due to the discontinuity of the material at the boundary.

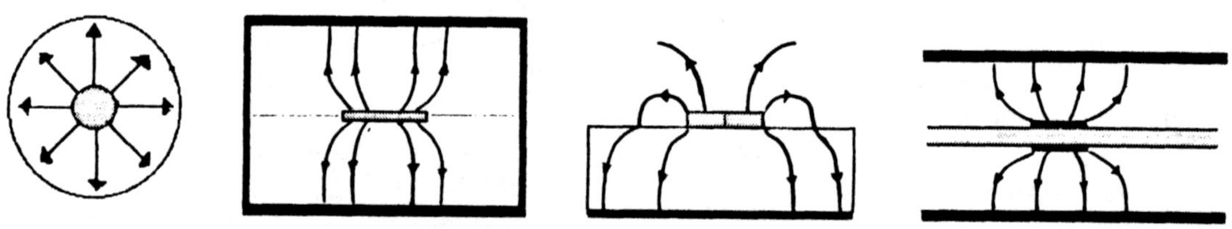

Figure 12.2 The electric field patterns of the coaxial cable, suspended stripline, microstrip, and triplate stripline.

The calculations for the relevant quantities for miniature lines range from simple to extremely difficult depending on the geometry, the degree of accuracy, and the method used (quasi-TEM or full-wave). We shall show some simple calculations as demonstrations.

As shown in Ch. 4, the guide wavelengths of some miniature lines have the same format as that of the coaxial cable. For example,

coaxial cable $\lambda_g = \lambda_0/\sqrt{k}$ (12.2.1)
embedded stripline $\lambda_g = \lambda_0/\sqrt{k}$ (12.2.2)

where λ_0 and k are the free-space wavelength and the dielectric constant of the substrate, respectively.

For an open-space microstrip, the dielectric constant is the average value beween that of air and the dielectric. Therefore, taking the dielectric constant of air as 1.0, and that of the substrate as k_s, the guide wavelength is

$$\lambda_g = \lambda_0/\sqrt{k_{eff}} \quad (12.2.3)$$

where

$$k_{eff} = (1.0 + k_s)/2 \quad (12.2.4)$$

SEC. 12.3
ATTENUATION IN STRIPLINES

The characteristic impedance calculations for various striplines are usually complicated and are now done by computers. The width of the conductor, the separation between the conductor and the ground plane, and the dielectric constant of the substrate involved (Fig. 12.3) are all pertinent quantities. One simple case will be presented here, and the interested reader should consult the references cited at the end of this chapter for more extensive discussion.

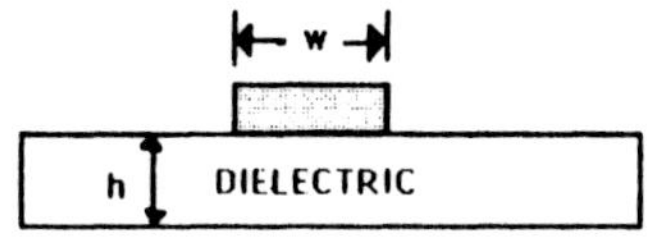

Figure 12.3 The pertinent parameters of a microstrip.

Recall that the characteristic impedance of a pair of parallel lines is given by

$$\text{(parallel line) } Z_0 = \sqrt{L/C} \tag{12.2.5}$$

where L and C are inductance and capacitance per unit length for the parallel lines. The electric field distribution of the parallel line is shown in Fig. 12.4. Imagine now that a conductive plane is inserted between the two parallel lines and the lower line is removed as shown in Fig. 12.5. This configuration resembles a microstrip. The conducting plane is known as the image plane in electrostatics because the plane functions just like the electrostatic image of the upper conducting line but at a different distance as shown in Fig. 12.6. In other words, a conducting line together with a conducting plane below are electrically equivalent to two parallel lines. The equivalence is valid only if the width of the upper conducting line w is small compared to the gap distance between the line and the plane h. Most microcircuit formulism involves the ratio w/h. If w/h is much smaller than 1.0, i.e., the line width is negligible, the circuit analysis will be relatively simple. If w/h is almost equal to or larger than 1.0, the circuit analysis will become quite complicated.

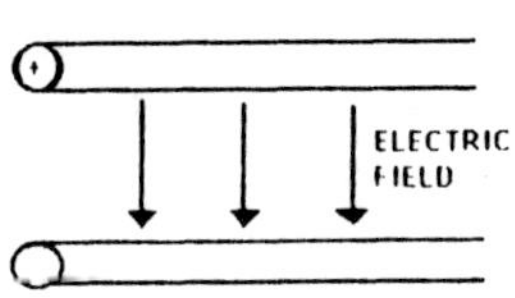

Figure 12.4 The electric field pattern of a pair of parallel lines.

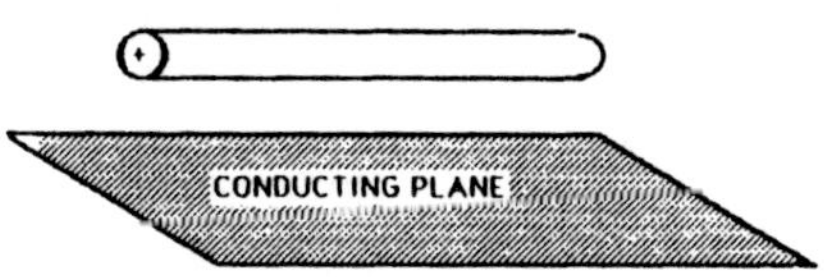

Figure 12.5 A conducting plane replaces the lower line of the parallel lines.

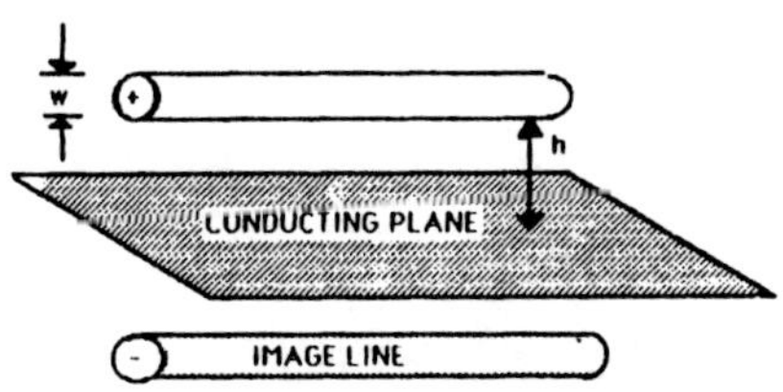

Figure 12.6 A single line above a conducting plane is equivalent to two parallel

In the simple case discussed above, the characteristic impedance of the conducting line and the conducting plane is

$$Z_0 = 0.5\sqrt{L/C} \tag{12.2.6}$$

where L and C are the inductance and capacitance per unit length for a pair of parallel lines separated by a distance of $2h$.

12.3 ATTENUATION IN STRIPLINES

Attenuation is loss of signal power. Attenuation can be categorized as ohmic, dielectric, and radiative. Ohmic loss is simply the motion of charges in metallic strips of finite resistivity; heat is usually the end product. Dielectric loss is power loss to the repeating rearrangement of charges in the dielectric medium (air or other dielectrics) separating the conducting lines. Radiative loss is power loss to the environment by the signal conductor which is effectively acting as an antenna.

Attenuation in striplines, especially the microstrip, is much higher than that in a waveguide or a coaxial cable. The ohmic loss is frequency dependent because the overall resistance of a conducting strip is related to the signal's skin depth which is frequency dependent. The dielectric loss can be reduced by using low-loss substrates such as sapphire, alumina, and some new dielectrics. The radiative loss is inevitable because a great portion of the conducting strips is un-shielded.

12.4 COMMON STRIPLINE PATTERNS

The concept of characteristic impedance of an infinitely long transmission line was discussed in Ch. 4. Any sudden change in physical dimension and straightness along the line is equivalent to a mismatch. Such discontinuities introduce reactive impedances (inductive or capacitive) at the points of the discontinuities. This is because a discontinuity causes a sudden change in electric field due to area change and fringe effects, and any possible change in magnetic field due to change in current density.

Several patterns* of discontinuities are shown in Fig. 12.7, and they are briefly discussed as follows:

a) *Step in width*: While a sudden change in width can cause a mismatch, it can also be employed to match an existing mismatch. Sometimes, the length of the new section can be chosen to be one-fourth of a wavelength to form an impedance transformer. The edge effects are usually small enough to be negligible in most cases.

b) *Gap*: A gap is equivalent to a capacitor connected in series. The capacitance value is related to the gap separation.

c) *Open end*: An open end is *not* a perfect open circuit, as that in low frequencies, but a small radiator of microwave from the end of the line. Different patterns can be designed to make different antennas.

d) *Stub*: A matching stub can be easily fabricated on a stripline circuit as a matching component (capacitor or inductor).

e) *Corner*: An abrupt right-angle corner is strategically unwise because the acute fringe effect of the electric field is equivalent to a shunt capacitance. The shunting effect increases with frequency, so the corner is effectively the same as a low-pass filter. To minimize the shunting effect, a turn must be made smooth with a radius that should be a few times larger than the width of the conductor.

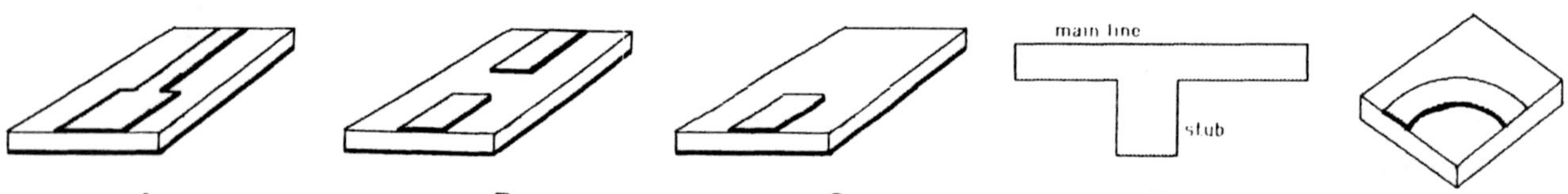

A B C D E

Figure 12.7 Microstrip discontinuity patterns: (A) step change; (B) gap; (C) open end; (D) stub; and (E) corner.

Several practical hybrid circuits using microstrips and discrete components have been used as design examples for FET amplifiers in Ch. 11. The readers are, therefore, referred to that chapter for more detailed applications.

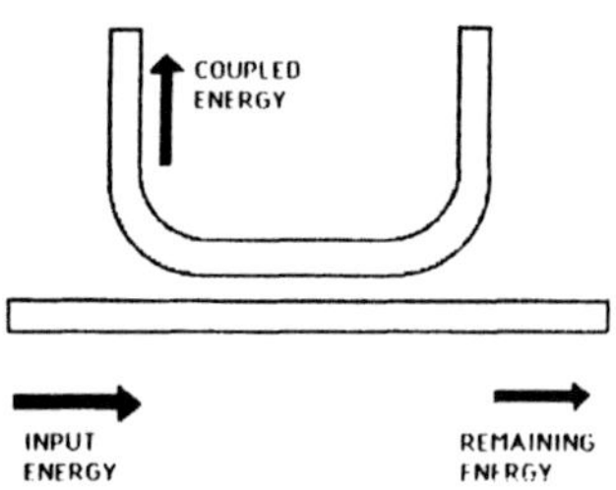

Figure 12.8 A microstrip coupler.

12.5 OTHER MICROSTRIP DEVICES*

a) *Coupler*: Figure 12.8 shows the microstrip version of a coupler. When a signal propagates in line 1, its electric and magnetic field see a nearby conductor, coupling occurs and some of the carried energy is coupled to the nearby conductor. The length of the nearby conductor can be chosen to form couplers of different coupling factors, for example, 3 dB, 10 dB, 20 dB, 30 dB, *et cetera*.

b) *Hybrid ring*: The hybrid ring shown in Fig. 12.9 is 1.5 wavelengths long and the separations between the four ports as shown. Signals entering at two of the four ports will come from other ports, either out of phase or in phase. The combinations are given in Table 12.1.

c) *Y-Junction circulator*: A microstrip circulator using a mounted ferrite and an external magnet is shown in Fig. 12.10. The gyromagnetic property of ferrite causes signals entering port 1 to exit from port 2, but not from port 3.

*See reference 2.

Table 12.1
Power Distribution of a Hybrid Ring

Enter port	Signals out	No signal out
1	2 and 4: each half-power	3
4	1 and 3: each half-power	2
1 and 3	2: sum; 4: difference	

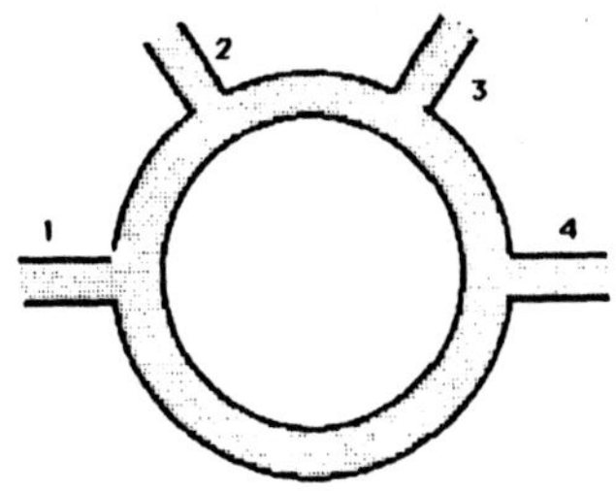

Figure 12.9 A microstrip hybrid ring.

d) *YIG oscillator*: Yittrium iron garnet (YIG) is a special ferrite material which exhibits a high-Q resonant property when made into spheres of millimeter size. Both the size of the sphere and the strength of the applied magnetic field determine the sphere's resonant frequency. The oscillating frequency of a YIG oscillator can be tuned by controlling the applied magnetic field. A YIG oscillator is shown in Fig. 12.11.

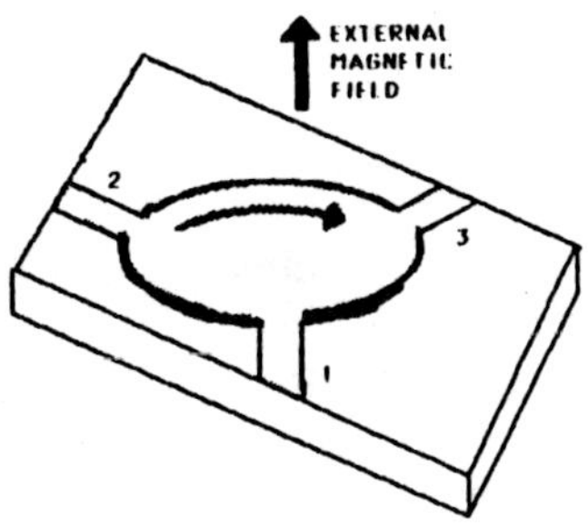

Figure 12.10 A microstrip Y-junction circulator.

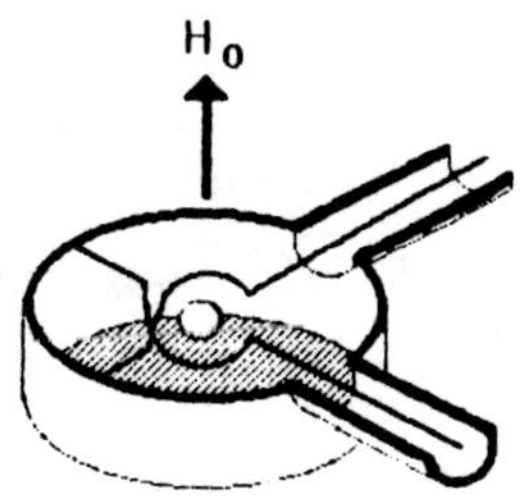

Figure 12.11 A yttrium iron garnet (YIG) oscillator.

12.6 LUMPED CIRCUIT COMPONENTS

Components such as resistors, capacitors, and inductors employed in the integrated circuit industry can be fabricated as conducting strips of different sizes, or as deposits of other materials on the substrates. For the low end of the microwave frequency spectrum, these components can also be fabricated as discrete components using high quality dielectrics and then inserted into the hybrid circuit board.

Resistors can be fabricated by depositing resistive material on the substrate in a controlled manner. A 50 ohm resistor, for example, can be deposited at the end of a microstrip transmission line of 50 ohm characteristic impedance as a termination.

A capacitor of large capacitance value is formed by sandwiching a dielectric (the substrate, for example) between two conductors. A sub-picofarad capacitor can be fabricated either by forming a gap between two conductors (see Sec. 12.4), or by using an extended configuration of the gap capacitor, known as the interdigital gaps, as shown in Fig. 12.12.

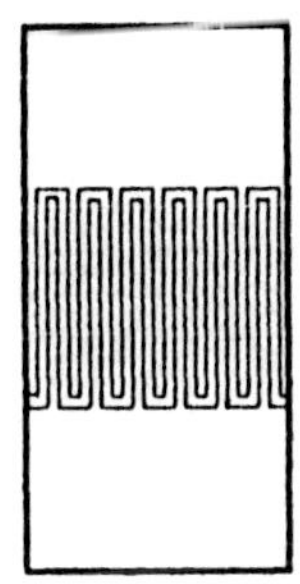

Figure 12.12 An inter-digital capacitor.

Finally, sub-nanohenry inductors can be formed by fabricating loops, or multiple layers of loops, on the substrate as shown in Fig. 12.13.

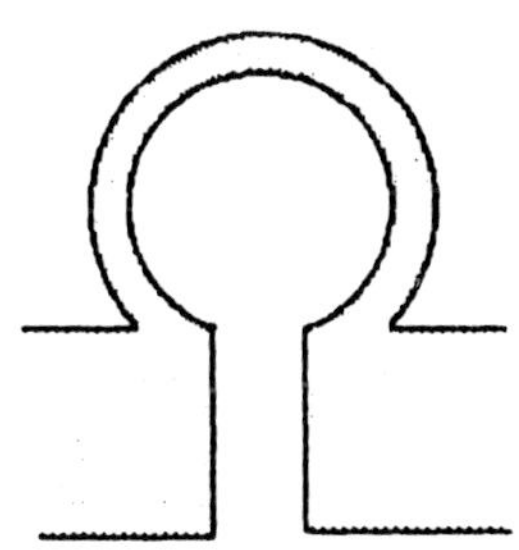

Figure 12.13 A stripline inductor.

12.7 MATERIAL PROPERTIES

Ordinary printed circuit board material such as phenolic laminates are not suitable for high frequency work due to their high dielectric loss at high frequencies. High grade material such as quartz, sapphire, alumina, and glass are typically used as substrates for hybrid circuitry. Monolithic microwave integrated circuits (MMICs) use gallium arsenide (GaAs), and sometimes silicon, as the substrate material.

Extensive discussions of the properties of microwave stripline material are available in the references given at the end of this chapter. A good comparison

SEC. 12.7
MATERIAL PROPERTIES

of different striplines and waveguides is summarized by Caulton and is shown in Table 12.2. The numerical rating is such that 1 is the best and 5 is the worst. Note the ratings on coaxial and waveguides in their cost, size and weight, and circuit loss. Waveguides are irreplaceable in terms of power handling and low loss.

Table 12.2
Comparison of Transmission Lines

Type	Microstrip MIC	Slotted Line Coplanar MIC	Lumped Element	Conventional Slotted Line	Coaxial and Waveguide
Size and weight	2	3	1	4	5
Cost	2	2	1	2–3	4
Reproducibility	1	1	2	2–3	4
Reliability	1	1	1	2	3
Circuit losses	2–3	3	2–3 (for C-band) 4 (higher frequency)	2	1

Properties of different substrate material are also summarized by Caulton in Table 12.3. The dielectric loss of a dielectric is typically described by the loss tangent of the material. The smaller the loss tangent, the less lossy is the material. For example, the loss tangent of alumina is 2×10^{-4}, according to Table 12.3, as compared to the higher loss tangent of GaAs, which is 1.6×10^{-3}. Alumina and sapphire are common substrate material in hybrid circuitry but their hardness causes them to break easily and they are difficult to machine.

Thermal conductivity is an important consideration when power is involved. Beryllia has good thermal conductivity, but berrylium dust is poisonous and requires special handling facilities. Beryllia is therefore used in conjunction with other substrate material, i.e., in compound substrates.

Table 12.3
Properties of Substrate Dielectrics

Material	Loss tangent*	Dielectric constant	Conductivity (W/cm) *C*	Applications
Alumina	2×10^{-4}	10.0	0.30	Microstrip, suspended-substrate
Sapphire	$<10^{-4}$	9.3–11.7	0.40	Microstrip, lumped-element
Glass	$>2 \times 10^{-3}$	5.0	0.01	Lumped-element, quasi-monolithic MIC
Berrylia	10^{-4}	6.0	2.50	Compound substrates
Rutile	4×10^{-4}	100	0.02	Microstrip
Ferrite garnet	2×10^{-4}	13–16	0.03	Microstrip, slotline, coplanar, compound substrates; non-reciprocal components
GaAs	1.6×10^{-3}	13	0.30	High frequency microstrip monolithic MIC

*at 10 GHz

12.8 MONOLITHIC MICROWAVE INTEGRATED CIRCUITS (MMICs)

Today, low frequency microchips contain hundreds and thousands of transistors and components on multilayer substrates about the area of a dime. Similar efforts for microwave have been under intense development in the last fifteen years. The concept of MMICs is to fabricate most, if not all, of the components and devices on one chip. Another clear advantage of MMIC is high speed due to a reduction in signal processing time. The compactness and performance consistency of MMICs have found wide prospective applications in space and military communications and weaponry.

Recall that the TEM mode dominates and circuit analysis is relatively simple when the dimensions of the transmission line are much smaller than the wavelength of the signal (except for waveguides where no TEM mode is permitted). Because the width and length of a conductor on a microchip can be made very small, the TEM approach is satisfactory, even for microwaves of tens of gigahertz.

Applications of MMICs should be extremely wide. For example, identical antenna ICs can be fabricated for a phased-array radar. The wrist watch type transmitters will no longer be a science-fiction device.

The advantages of MMICs include low cost (if demand is high), small size, reliability (for a functioning chip), and minimum tuning. The disadvantages, which may disappear in a few years, are low manufacturing yield and relative inflexibility for design.

The substrate material for MMICs is typically gallium arsenide (GaAs). Transistors capable of handling several to tens of gigahertz must be fabricated on the substrates. The special kind of field-effect transistor (FET) used for tens of gigahertz is the MESFET (metal-semiconductor FET), which is fabricated on the substrate in a planar configuration. The stray capacitance of the MESFET is also very low, meaning that high speed signal processing is feasible. Wideband amplifiers, amplifiers in cascade, and mixers are typical MMIC items.

REFERENCES

1. Edwards, T. C., *Foundations For Microstrip Circuit Design*, New York, John Wiley and Sons, 1981.
2. Fuller, A. J. B., *Microwaves: An Introduction To Microwave Theory and Techniques*, New York, Pergamon Press, 1979.
3. Caulton, M., "Microwave Integrated Circuit Technology—A Survey," *IEEE Journal of Solid State Circuits*, Vol. SC-5, No. 6, 1970, pp. 292–327.
4. Pengelly, R. S., "Hybrid vs. Monolithic Microwave Circuits—A Matter of Cost," *Microwave Systems News*, January 1983, pp. 77–114.
5. Abbot, D. A. *et al.*, "Monolithic Gallium Arsenide Circuits Show Great Promise," *Microwave Systems News*, August 1979, pp. 73–92.
6. Liu, L. C. T. *et al.*, "An 8–18 GHz Monolithic Two-Stage Low Noise Amplifier," *IEEE 1984 Microwave and Millimeter-Wave Monolithic Circuits Symposium*, May 1984, pp. 49–51.
7. Scott, B. N. *et al.*, "A Family of Four Monolithic VCO MIC's Covering 2–18 GHz," *IEEE 1984 Microwave and Millimeter-Wave Monolithic Circuits Symposium*, May 1984, pp. 58–61.
8. Watanabe, S. *et al.*, "GaAs Monolithic MIC Mixer-IF Amplifiers for Direct Broadcast Satellite Receivers," *IEEE 1984 Microwave and Millimeter-Wave Monolithic Circuits Symposium*, May 1984, pp. 19–23.
9. Osbrink, N. K., "Key Concepts in EW Microelectronics Technology," *Journal of Electronic Defense*, Vol. 7, No. 3, 1984.
10. Osbrink, N. K. *et al.*, "Surface-Mounted MICs and MMICs Play Vital Role in Military Communications," *Microwave Systems News*, January 1985, pp. 72–85.

CHAPTER 13

ANTENNAS

W. Stephen Cheung

13.1 INTRODUCTION

An antenna, according to *Webster's Dictionary*, is "an arrangement of wires, rods, etc. used in sending and receiving electromagnetic waves." A more general definition of an antenna is basically any device that can transmit and/or receive electromagnetic waves.

This chapter covers several major antennas for low (MHz range) and high (microwave) frequencies. Dipole antennas will be discussed first because the concept of plane waves can be easily revealed from the radiation pattern of a dipole antenna. The isotropic (or point) antenna is an important, though fictitious, antenna for a simple mathematical development of the power density concept from which antenna gain can be easily understood. Other low frequency antennas such as loop, helix, and array antennas will be discussed.

From a layman's standpoint, the principle of operation for a microwave antenna can be easily visualized by analogy to optics. This simplified viewpoint will be adopted here to help the reader gain some insight about microwave antennas. It is not the intention of this book to go into the complex behavior of working antennas. Key items such as beamwidth, main lobe and sidelobes, antenna gain, and a receiver's capture cross section will be discussed.

13.2 ANTENNA SIZE

When we take a look at the "tall" transmitting antenna for an AM radio station, we get the impression that it is "big." Such an impression is due to human beings' tendency to compare the size of an object to our physical size.

In expressing the size of an antenna, an appropriate approach is to compare the antenna's dimension with that of the signal's wavelength. As it turns out, in solving the necessary equations for antennas, the quantity D/λ, where D is the typical dimension of the antenna such as length (for a rod) or diameter (for a dish) and λ is the wavelength of the signal, is often found in the final solution.

An antenna is large if its physical size is about 100 times or more larger than the wavelength of the radio waves it is propagating. An antenna is small if its physical size is about the same, or even smaller than, one wavelength of the radio waves.

According to the above definition, a 500ft (40-story high) AM antenna is actually small because the wavelength of a typical AM signal (1 MHz) is 300m, i.e., 984 ft. A 5m (15ft) dish antenna is large if it is used to transmit or receive, let us say, a 10 GHz signal whose wavelength is 3cm, i.e., 0.03m.

13.3 DIPOLE ANTENNA

A dipole antenna is composed of two metallic rods pointing away from each other. Its operational principle can be visualized by first assuming that the rods are parallel to each other, like a capacitor. As shown in Fig. 13.1A, an ac generator

SEC. 13.3
DIPOLE

is connected to the parallel rods. Due to the periodic change in voltage, the rearrangement of positive and negative charges results in an alternating current (ac). At the capacitor, the electric field *E* is varying (both in magnitude and direction) at the same frequency as the generator output. The alternating current produces an alternating magnetic field *H*, which follows circular traces around the conductors.

Imagine the parallel rods are slowly opened with one side as the pivot point (see Fig. 13.1B, C). The final electric and magnetic field patterns are shown in Fig. 13.1C. Note that the *E* field is always perpendicular to the *H* field.

The *E-H* pattern of Fig. 13.1C is incomplete. When vigorously analyzed with Maxwell's equations, it turns out that the periodic variation of the *E* and *H* fields results in a "denial of termination" of the electric field at both rods. Consequently, the *E-H* energy is "kicked out" by the antenna in an energy packet. We can visualize this process from the analogy of blowing bubbles. When the lips of the mouth move up and then down, a bubble is forced out. This is due to a variation of stress applied to the bubble's surface. Electric and magnetic fields are some form of stress. The correct *E*-field pattern is shown in Fig. 13.2. The magnetic field pattern is omitted in this figure to avoid confusion of *E* with *H*.

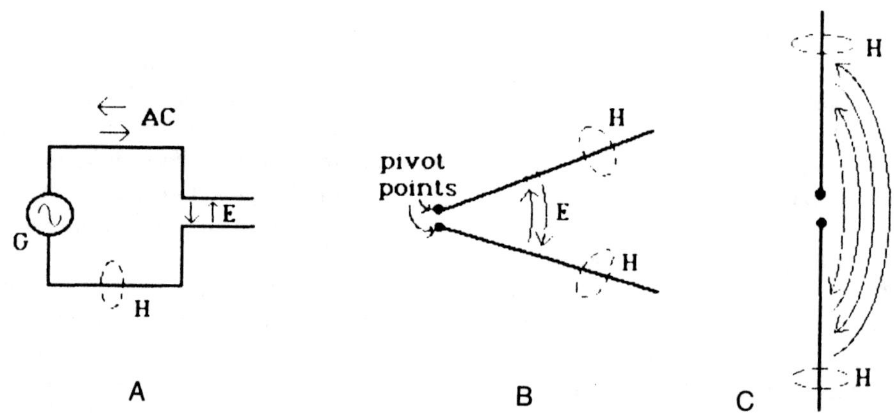

Figure 13.1 A dipole antenna can be constructed by (A) connecting a parallel rod capacitor across the generator, and slowly opening up the two rods (B) and (C). The electric and magnetic field patterns are labelled by E and H.

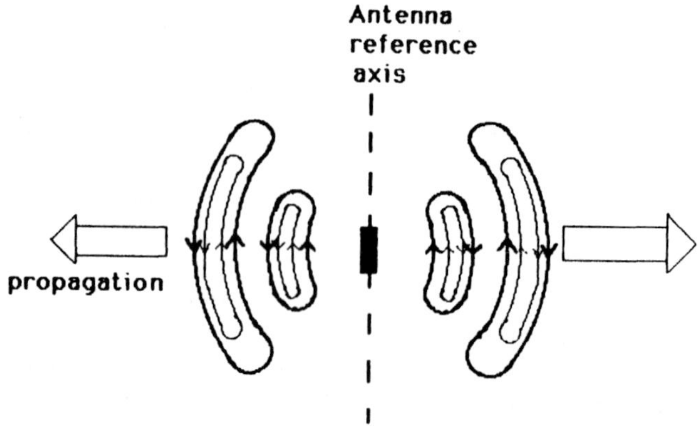

Figure 13.2 The electric field coming out of a dipole antenna is in packets of energy "ejected" by the antenna.

13.4 FAR FIELD AND NEAR FIELD

Upon close examination of the electric and magnetic field pattern in Fig. 13.2, we find that the electric field is a curve from one rod of the antenna to the other while the magnetic field traces a circle surrounding the antenna. Such curvatures for E and H will diminish as the distance from the antenna increases.

Loosely speaking, the area where the abovementioned curvatures are still significant is called the near field, or Fresnel zone. Beyond the near field, the curvatures are negligible and the area is known as the far field, or Fraunhofer zone. As a rule of thumb, distances beyond 10 times the wavelength of the signal define the far field. Hence, the far field for a 1 MHz AM signal is $10 \times 300\text{m} = 3\text{km}$, and that for a 1 GHz signal is $10 \times 30\text{cm} = 300\text{cm}$.

In the far field, the curvatures for the electric and magnetic field are small as far as a receiving equipment is concerned. For a local detector in the far field (Fig. 13.3), the electric field appears to be vertical, the magnetic field appears to be horizontal, and the electric field, the magnetic field, and the direction of propagation are mutually perpendicular to one another. Such an approximated waveform is known as a *plane wave*. A trace of the electric and magnetic field pattern during propagation is given in Fig. 13.4. In most antenna applications, the transmitter and the receiver are separated by a long enough distance that far field is implicitly assumed.

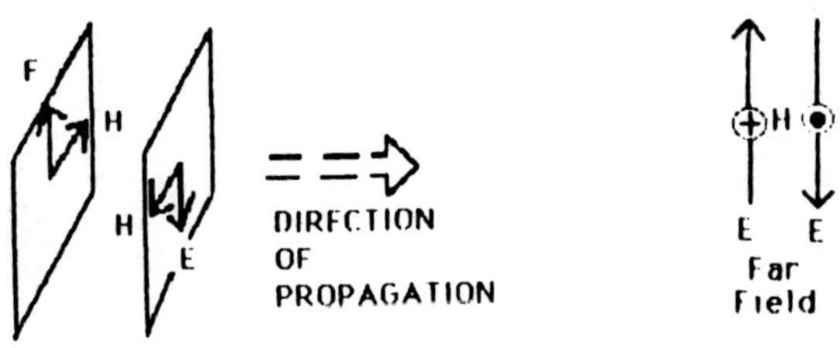

Figure 13.3 Near field is the region where the electric and magnetic fields of the radiated wave are curved while those in the far field are straight.

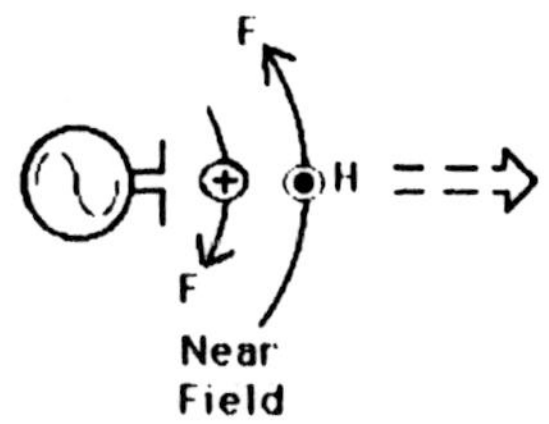

Figure 13.4 The waveform in the far field is that of the plane waves.

13.5 RADIATION PATTERN OF A DIPOLE ANTENNA

Consider a dipole antenna radiating electromagnetic signals outward as shown in Fig. 13.5. The power level of the signal at different distances and angular locations with respect to the center of the antenna can be measured by a detector. The power level *versus* distance will be discussed in a later section. For convenience, let us choose a nominal distance of, let us say, 1 mile from the center of the antenna. In Fig. 13.5, Points A, B, D, are all at a distance of 1 mile from the center of the antenna, C. The power levels at different locations on the vertical plane, all the same distance away from C, constitute the radiation pattern of the antenna.

For a dipole antenna, the power levels detected at A will be larger than that detected at B. This can be seen from a simple geometrical argument. Pick two small segments, P and Q, symmetrically located at the upper and lower ends of the dipole antenna. The distance PA is the same as QA. Hence, any signal emitted by segment P and detected at A is in phase with signal emitted by Q and detected at A. Therefore, the total power level at A is additive. For location B, the distance PB is less than QB. It takes the signal emitted by P less time to reach B than that by Q. This is equivalent to a phase difference. When all the contributions from the antenna are considered, it is easy to see that location A has the maximum detected power and progressively less as the detector is moved to other angular locations such as B and D.

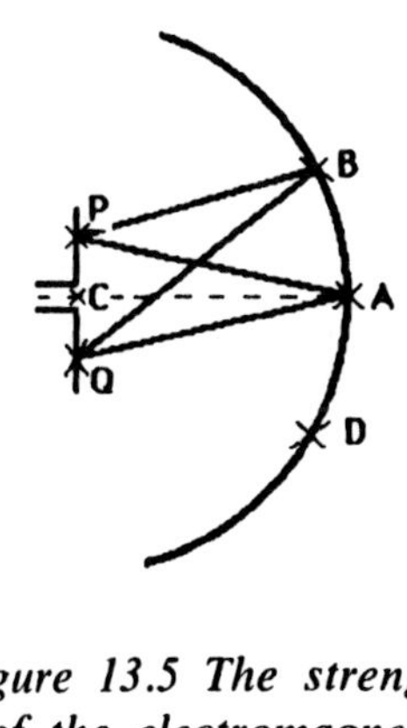

Figure 13.5 The strength of the electromagnetic wave at different locations relative to the antenna can be easily analyzed by geometry.

SEC. 13.6
POLARIZATION

The radiation pattern of the dipole pattern is shown in Fig. 13.6. In a vertical plane, the pattern is that of the "figure eight." Because the dipole antenna exhibits azimuthal symmetry, meaning the antenna looks the same as one travels around it, the three-dimensional configuration of the radiation pattern is like a doughnut. The readers must be careful in interpreting the radiation pattern. Recall that the pattern is for a nominal distance. If another nominal distance is chosen, let us say, 2 miles from the center, the doughnut shape still holds and the power detected at locations similar to *A* (Fig. 13.5) is still the largest compared to locations similar to *B* and *D*. Note that when the detector moves beyond a certain boundary, e.g., near the top or the bottom of the erected antenna, no power will be detected.

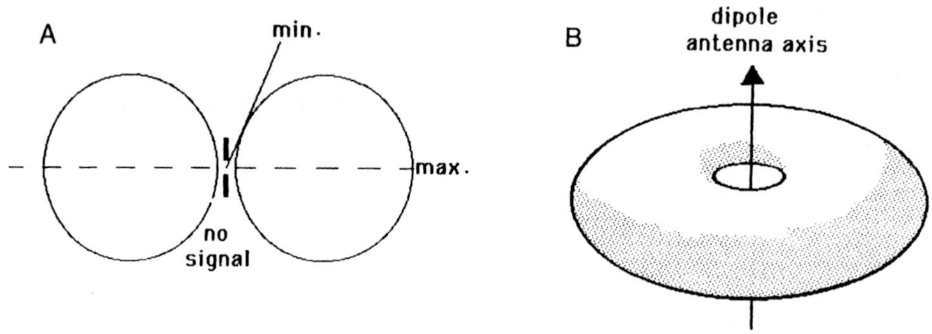

Figure 13.6 The radiation of a dipole antenna is (A) the "figure eight" two-dimensionally and (B) the shape of a doughnut three-dimensionally.

The doughnut-shaped radiation pattern is valid for a dipole antenna only when it is operating optimally. A dipole antenna functions best when its entire length is half the wavelength of the signal to be transmitted. This optimum length is related to the phase relationship between the voltage and the current as well as reflection problems (when the length is too long). Assuming the separation between the two near ends of the antenna rods to be negligible, the optimum length requires that the length of each rod to be one-quarter of the signal wavelength. Hence, for an AM signal of 1 MHz (wavelength = 300m), using a dipole antenna will require two rods, each 75m (246ft) long, with a total erected height of 150m (492ft)!

13.6 POLARIZATION

The polarization of an electromagnetic wave refers to the orientation of its electric field, E. For a dipole antenna erected vertically, the E-field is also vertical so the emitted signal is vertically polarized. If the dipole antenna is mounted horizontally, the emitted signal will then be horizontally polarized.

From the receiving standpoint, the signal's electric field excites the electrons, or charge carriers, in the receiving antenna which is assumed to be rod shaped. If the E vector is parallel to the length of the rod, then the electrons will undergo maximum motion resulting in a large received signal. Hence, a receiving rod antenna must be mounted vertically in order to maximally detect a vertically polarized signal. A receiving antenna mounted horizontally cannot detect any vertically polarized radio wave. This is illustrated in Fig. 13.7.

In general, radio waves from AM stations are vertically polarized, so the receiving antenna must be erected vertically. Radio waves from television stations are horizontally polarized, this is why roof-top television antennas are mounted horizontally. In the case of FM, the electric field is what is known as circularly polarized. For a dramatized analogy, we can visualize that a dipole antenna is mounted on a turning windmill. The orientation of the electric field will rotate as time progresses (Fig. 13.8). The polarization vector rotates at the same frequency as the signal. Circular polarizations have two types: clockwise and counterclockwise. The direction of rotation is viewed from the transmitter.

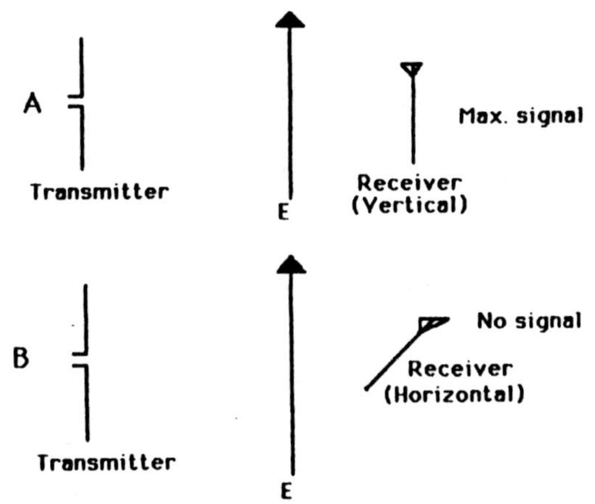

Figure 13.7 Polarization is the orientation of the electric field. The polarization of a vertical transmitter is vertical. A receiver mounted vertically will receive maximum signal, while mounted horizontally will receive minimum signal.

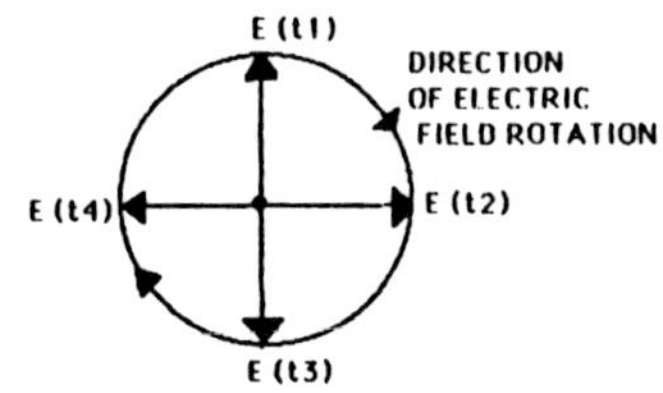

Figure 13.8 The electric field of a circularly polarized wave appears to rotate as time progresses.

13.7 QUARTER-WAVE (MARCONI) ANTENNA

A quarter-wave antenna employs only one of the two rods of the dipole antenna; the other rod is grounded. A quarter-wave antenna is shown in Fig. 13.9. The ground replaces the lower antenna rod.

To understand the role of ground in the antenna action, we should recognize that earth is a conductor. The earth is, of course, not a perfect conductor-like metal, but it is not an insulator either; its conductivity varies with the soil and water content. The reader may recognize the fact that the household "ground" is ohmically traced to a thick (3/4") copper rod driven 8ft into the earth. The dampness of the soil at that depth enhances the conductivity.

Figure 13.10A shows a point charge (positive) Q placed above a conductor which is the ground. The positive charge attracts negative charges from the conductor. From the theory of electrostatics, the geometrical distribution of the negative charges is equivalent to a negative charge Q^- of the same magnitude as the positive charge Q placed at a distance below the conductor's surface. The equivalent charge Q^- is known as the image charge. Imagine now that the charge Q is moving toward and away from the conductor (Fig. 13.10B). Such a movement will cause the image charge Q^- to move correspondingly. It can now be seen that the moving charge Q is equivalent to a current. Together with the movement of the image charge Q^- a dipole antenna is formed (Fig. 13.10C).

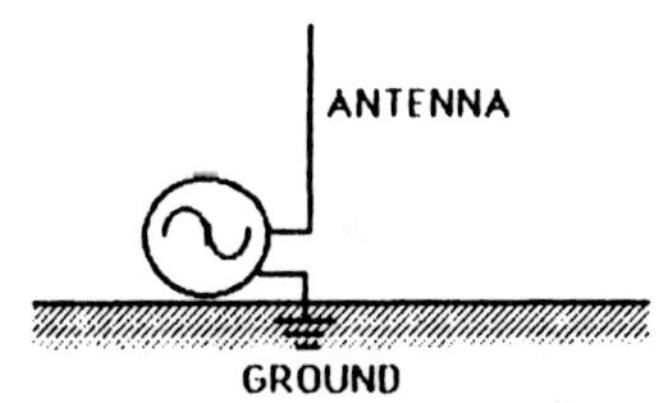

Figure 13.9 A quarterwave antenna. The lower rod of a dipole antenna is replaced by earth ground.

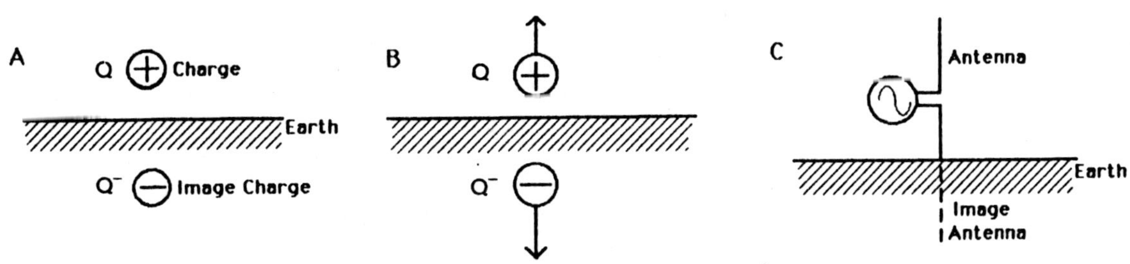

Figure 13.10 The charge distribution of a quarterwave antenna can be seen by assuming the earth to be a perfect conductor.

The radiation pattern of the quarter-wave antenna is shown in Fig. 13.11, and is basically the top half of a dipole antenna, as expected. The quarter-wave antenna is more practical in several ways. Using a quarter-wave antenna, the height of an AM antenna at 1 MHz needs only to be 75m (246ft). The generator

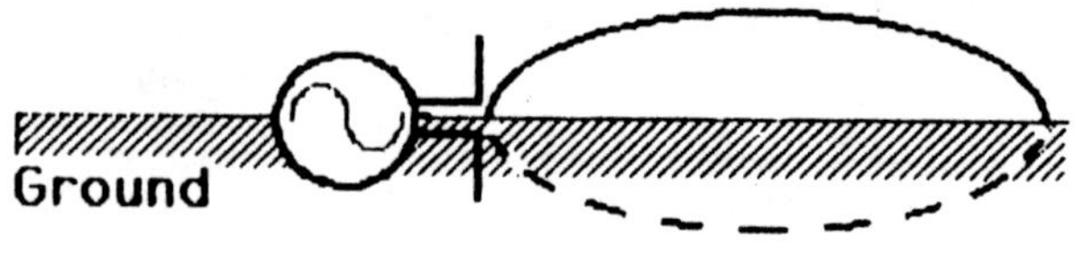

Figure 13.11 The radiation pattern of a quarterwave antenna.

or the antenna driver can be placed right at the ground level. Most importantly, the radiation pattern of the quarter-wave antenna shows that maximum power level takes place at the ground level and is, therefore, adequate for most receiving households.

13.8 LOOP ANTENNA

A loop antenna and its reference axes are shown in Fig. 13.12A. If the *z*-axis of the loop antenna is vertical, as shown in Fig. 13.12B, its radiation pattern is identical to that of a dipole antenna. A qualitative approach to the loop antenna is to imagine that the loop antenna is actually a polygonal arrangement of wires (Fig. 13.12C shows an octagonal arrangement). It can be seen that when a current flows in a particular segment, there is always a current segment on the opposite

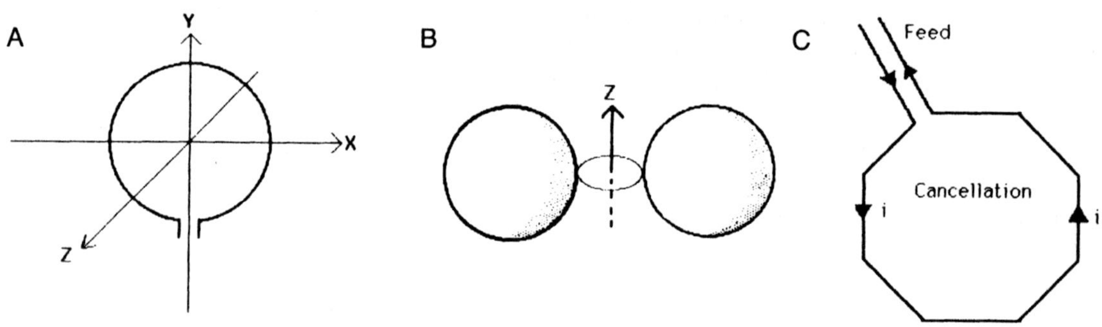

Figure 13.12 (A) The xyz orientation of a loop antenna. (B) The radiation pattern of a loop antenna. (C) The circular loop antenna can be thought of as a polygon; the electromagnetic effect inside the loop sums up to be nil.

side flowing in the opposite direction. Hence, there is a total cancellation of electric and magnetic field inside the loop. The radiation pattern outside the segment will follow half of the "figure eight" pattern. Hence, the three-dimensional pattern is again that of a doughnut.

The polarization of a loop antenna is parallel to its *z*-axis. Remember that a receiving antenna has a maximum pick-up when its polarization is parallel to that of the transmitted signal. Hence, a loop antenna as a receiver registers a maximum pick-up *not* with the loop surface facing the signal as we intuitively think, but with the loop surface facing perpendicular to the signal. When the loop surface directly faces the signal, no power is detected (Fig. 13.13A). Such a null reading is employed in direction-finding. Extra measure is necessary to eliminate the left-right uncertainty, as illustrated in Fig. 13.13B.

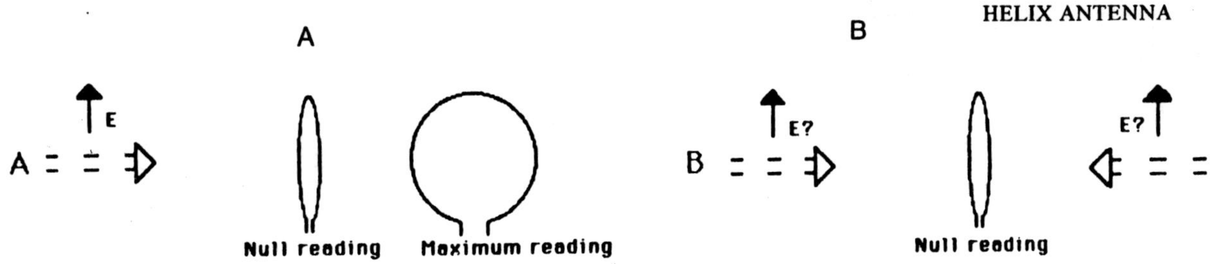

Figure 13.13 The incoming signal is vertically polarized. (A) The receiving antenna registers no power when facing the incoming signal and maximum power when facing away perpendicularly. (B) Even a null reading still means uncertainty concerning which direction the signal is coming from.

13.9 HELIX ANTENNA

A helix antenna consists of a conductor wound in the form of a spring. As illustrated in Fig. 13.14, important parameters are the length and the diameter of the helix, the spacing between each turn, the number of turns, and the pitch angle of the turn. A helix antenna is usually fed by a coaxial cable as shown.

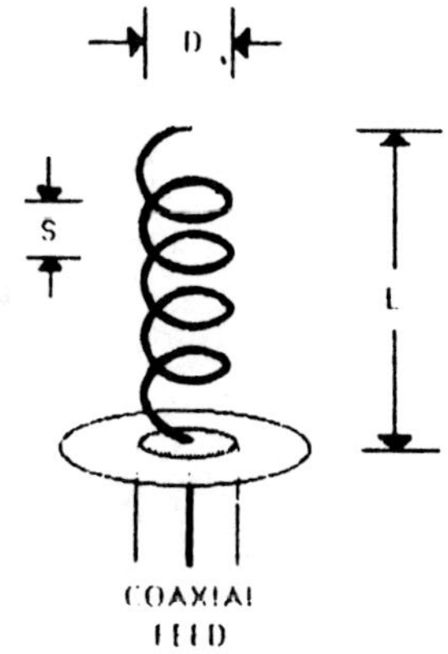

Figure 13.14 The para meters involved in a

The helix antenna can operate in two modes: normal and axial. The normal, or broadside, mode is when the length of the helix is much smaller than that of the transmitted signal's wavelength. Under this condition, the helical nature of the antenna is virtually invisible and it behaves just like a dipole antenna, as illustrated by its radiation pattern shown in Fig. 13.15A.

The axial mode is when the length of helix is comparable to the signal's wavelength. Radio waves emitted by the antenna in this mode are circularly polarized. The corresponding radiation pattern, as shown in Fig. 13.15B, indicates a strong directional behavior. This type of radiation pattern will be characterized in the next section.

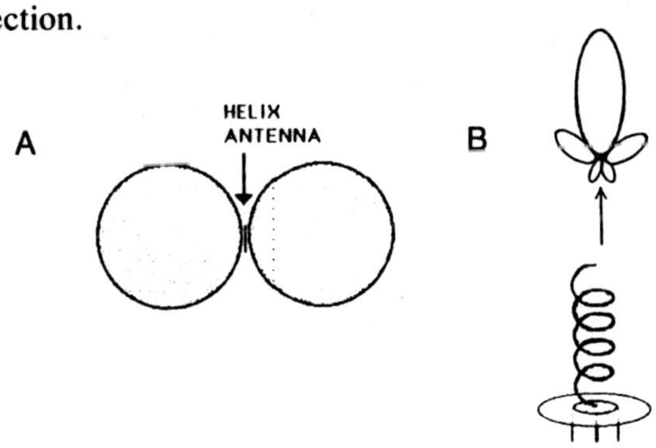

Figure 13.15 Two radiation patterns of a helix antenna are normal and axial. The pattern of (A) the normal mode is a doughnut, and (B) the axial mode is circularly polarized.

13.10 LOBES AND BEAMWIDTHS

The radiation pattern of an axial-mode helix antenna is a typical one for many microwave antennas. Therefore, the pattern will be used in this section to illustrate a few important terms.

When a transmitting antenna has a strong directional preference, a major lobe usually exists accompanied by sidelobes and back lobes. These lobes are illustrated in Fig. 13.16. Some radiation patterns have more complicated sidelobes.

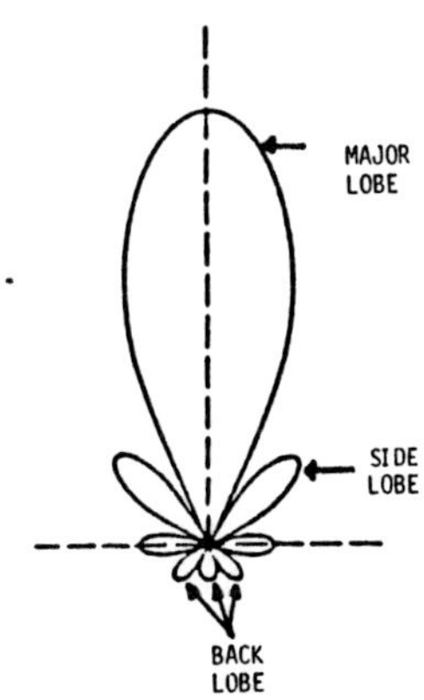

Figure 13.16 The pattern of an antenna with directional preference consists of major lobes, sidelobes, and back lobes.

Note that no major lobe can be defined in the doughnut-shaped pattern of a dipole antenna, and neither can sidelobes and backlobes. Sidelobes and backlobes can be loosely thought of as the results of "spilled over"" power in different directions.

Two types of beamwidths exist: half-power and first-null. The half-power, or 3-dB, beamwidth is the angular width between two directions where the power levels detected are half of, or 3 dB less than, the peak level. The first-null beamwidth is the angular width between two directions where the power levels detected just become null. These two types of beamwidth are illustrated in Fig. 13.17A.

It should be clear that the first-null beamwidth is somewhat larger than the half-power beamwidth. The half-power beamwidth can be employed to define a territory within which a receiver will receive at least half of the maximum power level.

The radiation pattern shown in Fig. 13.17A is given in what is known as a polar plot. The reference line for 0° is conveniently chosen to be the direction of maximum power. An alternative plot is known as the rectangular plot, as shown in Fig. 13.17B. The rectangular plot can be visualized as applying a pair of scissors and cutting along the 180° line of the polar plot.

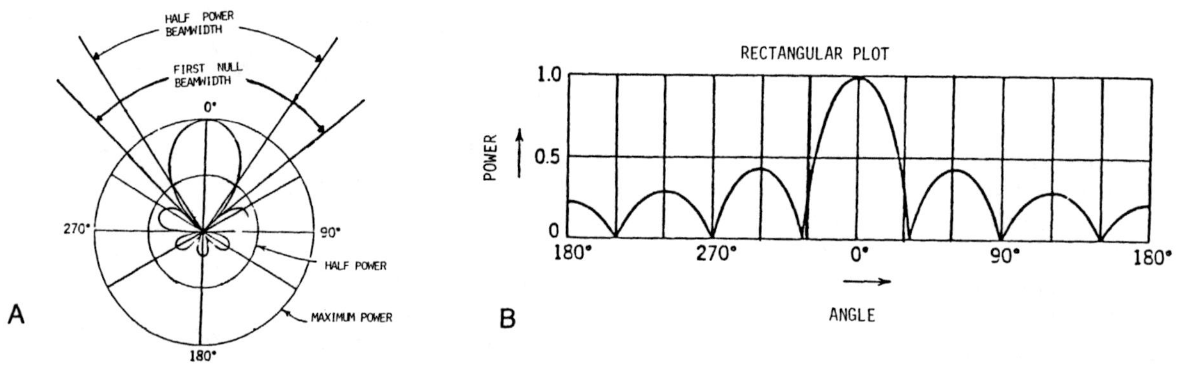

Figure 13.17 (A) The half-power beamwidth is the angular width between two half-of-maximum power locations, while the first null beamwidth is that between two null locations. (B) The power pattern is the polar plot can be equally described by the rectangular plot.

13.11 ISOTROPIC ANTENNA

An isotropic antenna, sometimes known as a point antenna, is a fictitious transmitter that does not exist in practice. It is a point source occupying a negligible amount of space. More importantly, it exhibits spherical symmetry, i.e., there is absolutely no directional preference.

Signals emitted by an isotropic antenna are spherical waves, much like that of an expanding balloon. At a distance far enough away from the antenna, the curvature of a tiny section of the spherical wave is so small that the wave can be approximated as a plane wave.

The power carried by a spherical wave is spread evenly over the entire spherical surface. As the wave propagates outward, i.e., as the radius of the expanding balloon increases, the same amount of power is now spread over an increasing area. The power density of a spherical wave, i.e., that of an isotropic antenna, at a radius R from the center (the antenna) is defined as

$$\begin{array}{c} PD \text{ (Power Density)} \\ \text{for isotropic antenna} \end{array} = \frac{P_t}{4\pi R^2} \tag{13.11.1}$$

where P_t is the total power transmitted by the antenna. Note that $4\pi R^2$ is simply the surface area of a sphere of radius R. Hence, the power density for an isotropic antenna is simply the total power spread over the surface area. Note also that the power density defined in Eq. (13.11.1) is consistent with that given in Ch. 3, i.e., $PD = E \times H$ *(electric field multiplied by magnetic field).*

Example (13.11.1)

Given the total power carried by the wavefront of an isotropic antenna to be 1W, find the power density at a distance 100 meters away from the antenna.

Solution: Using Eq. (13.11.1),

$$\begin{aligned} PD &= \frac{1}{4\,\pi\,100^2} \quad \text{W/m}^2 \\ &= 8 \times 10^{-6} \quad \text{W/m}^2 \end{aligned}$$

The power densities at two locations can be compared. Let us say locations A and B are at distances R_A and R_B from the antenna and assume that R_B is larger than R_A. Then, the ratio of the power density at A to the power density at B is

$$\frac{PD\,(A)}{PD\,(B)} = \frac{P_t/4\,\pi\,R_A^2}{P_t/4\,\pi\,R_B^2} = \left(\frac{R_B}{R_A}\right)^2 \qquad (13.11.2)$$

Because Eq. (13.11.2) compares two power values, the dB notation can be applied as illustrated by the following example.

Example (13.11.2)

Compare the power densities at two locations A and B which are 200 meters and 500 meters away from the point antenna, respectively. Express the ratio in number as well as in dB.

Solution: Using Eq. (13.11.2),

$$\frac{PD\,(A)}{PD\,(B)} = \left(\frac{500}{200}\right)^2 = 2.5^2 = 6.25 \text{ (8 dB)}$$

According to Eq. (13.11.1), the power density falls off inversely as the square of distance. This is known as the inverse-square law and is commonly observed in many science and engineering disciplines. It should be noted that while the power density decreases as the square of distance, the total power integrated over the entire balloon is always the same, assuming no loss of power due to absorption takes place.

13.12 ANTENNA GAIN

An isotropic antenna emits radio waves evenly in all directions like an expanding balloon. A practical antenna usually has one or more preferred directions. Along the preferred direction, the powers detected are usually more than powers detected elsewhere.

The conventional method of comparison is to compare the power density of a given antenna detected at a designated location with the power density of a would-be isotropic antenna at the same location, provided that both the given antenna and the isotropic antenna have the same transmitted power. The provision that both antennas must have the same transmitted power is crucial, otherwise no sensible comparison can be made.

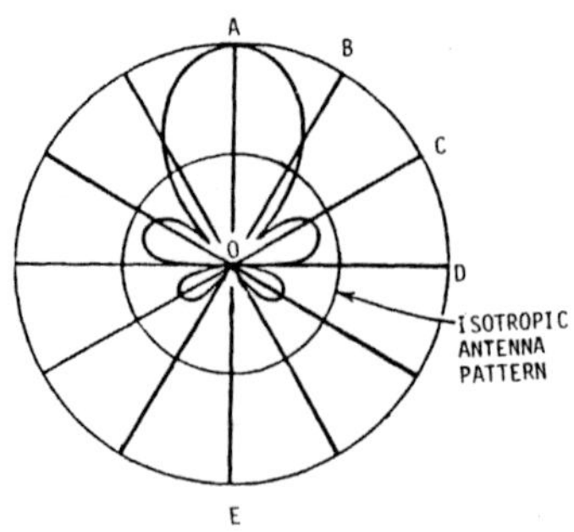

Figure 13.18 Power pattern comparison between a helix and an isotropic antenna.

Figure 13.18 compares the radiation pattern of an isotropic antenna and that of a helix antenna. Both antennas transmit the same amount of power. Along the direction of OA and within the reasonable beamwidth of the main lobe the helix antenna has more power density, but less elsewhere (directions OC, OD, OE, *et cetera*).

The factor by which a given antenna's power density is larger than that of an isotropic antenna in a selected direction is defined as the antenna *gain* along that direction. The antenna's gain can also be expressed in dB because two powers are being compared. Hence, a helix antenna having a gain of 10 at the tip of the main lobe means along that direction the power density of the helix antenna is 10 times larger than that of an isotropic antenna.

It is clear from the definition of an antenna's gain that it can be larger as well as smaller than 1.0. This is a typical example of a "gain some, lose some" situation. The maximum gain of an antenna, e.g., along the tip of the main lobe, is known as the *directive gain* of the antenna.

The reader may wonder if comparing a realistic antenna with a fictitious antenna is a practical choice. Even though an isotropic antenna does not exist, its mathematical expression is straightforward. Recall that the power density of the isotropic antenna is

$$PD \text{ (isotropic antenna)} = \frac{P_t}{4\pi R^2} \tag{13.12.1}$$

at a distance R away from the isotropic antenna. If the power density of an antenna under test is measured at the same distance R, the ratio of the measured value to the calculated value of the power density of an isotropic antenna using Eq. (13.12.1) is the gain of the antenna under test.

Example (13.12.1)

An antenna under test is to transmit 1W of signal. The power density at a particular location of distance $R = 1000$m away from the antenna is measured to be 1.6×10^{-6} W/m^2. What is the antenna's gain in that direction?

Solution: Using Eq. (13.12.1), the power density of an isotropic antenna measured at the same location is

$$PD \text{ (isotropic antenna)} = \frac{1}{4\pi 1000^2} \text{W/m}^2$$

$$= 8.0 \times 10^{-8} \text{ W/m}^2$$

The power density of the antenna under test is 1.6×10^{-6} W/m^2, which is 20 times larger than that of the isotropic antenna. The antenna gain along the selected direction is 20, or 13 dB.

It has been misunderstood by some that an antenna with a certain gain means the antenna can amplify the input power to the antenna. This is not true. An antenna is a passive device (only its mechanical motions may require power) whose geometry greatly determines the preferred direction. The preferred direction does better than an isotropic antenna, and other directions do worse. The performance of a certain antenna is sometimes compared with another common antenna, such as a dipole.

13.13 ARRAYS

An array of antennas is a collection of antennas, usually dipoles, placed at equal distances along a common line of reference known as the array axis. We can imagine that the radiation patterns of individual dipoles will combine under certain phase relationships. The overall radiation pattern is different from the "figure eight" shape in the plane of the antenna.

Figure 13.19 shows a broadside array and the corresponding radiation pattern. Note the phase inversion from one dipole connection to the following one, and the separation between the dipoles. The contributions from each dipole add constructively at the broadsides as shown.

Figure 13.20 shows an end fire array and its radiation pattern. The separation between each dipole and their phase relationships are now different. Figure 13.21 shows a turnstile array which is simply two dipole antennas perpendicular to each other. The resulting radiation is also shown in Fig. 13.21.

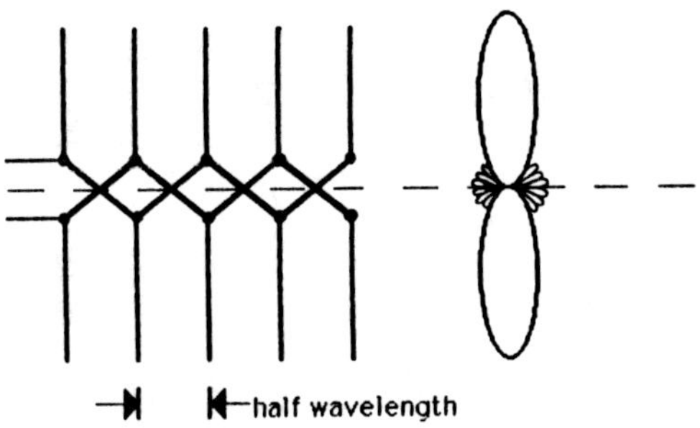

Figure 13.19 The construction and radiation pattern of a broadside array.

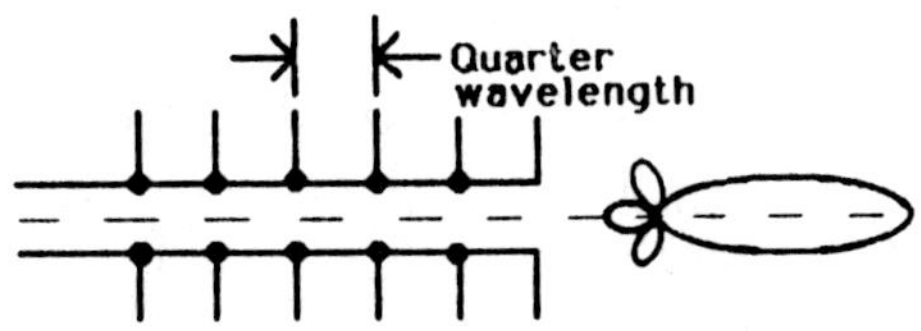

Figure 13.20 The construction and radiation of an end-fire array.

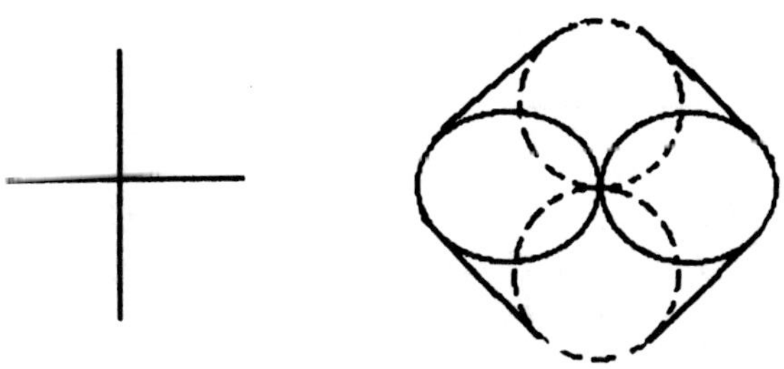

Figure 13.21 The construction and radiation pattern of a turnstile array.

13.14 OPTICAL PRINCIPLES APPLIED TO MICROWAVE ANTENNAS

The wavelength of visible light is about 0.5 microns (μm), i.e., 5×10^{-7} m. Most optical equipment such as a lens and a prism is much bigger than the wavelength. Therefore, in describing the propagating characteristics of (visible) light, it is usually graphically drawn as a beam of rays (pencil beams) following straight lines. Little attention is paid to the "wavy" nature of light.

Geometrical optics is very effective in beam tracing in cases such as the light beam coming out of a flash light, the focusing of parallel sunbeams to a small dot by a magnifying lens, the formation of images, *et cetera*. Common properties of a light beam in the context of geometrical optics are:

1. It follows a straight path;
2. Upon encountering a flat surface, a portion of it will reflect and a portion of it will refract (into the new medium);
3. Upon encountering a curved surface such as parabola, it will follow the law of reflection and trace a path dictated by the geometry of the curve.

In the case of a parabola, the third property causes a parallel beam to be focused at a point known as a focal point. The same property is responsible for using a lens to focus a sunbeam to a dot which is the lens' focal point.

The wave nature of light accounts for additional properties not predicted by geometrical optics. Such properties are interference and diffraction. Interference in optics results in bright and dark fringes. Diffraction is responsible for a light beam turning a sharp corner or edge. These wave properties have significant effects in microwaves. The standing wave is a result of interference. The diffraction property permits reception of radio waves on the other side of a mountain.

Optical principles can be applied to the construction of microwave antennas. The wavelength of a microwave signal is typically in the centimeter range, the degree of approximation that the signal can be treated as a beam is not as close as that of visible light. Hence, the radiation pattern of a microwave antenna can be obtained to a first approximation by assuming the microwave signal as straight lines. Secondary effects due to the wave nature of microwaves in addition to the finite nature of the wavelength to equipment dimension ratio are responsible for the deviation of the radiation pattern from the first approximation. Sidelobes and backlobes can be thought of as being formed by the secondary effects.

13.15 PARABOLIC ANTENNA

A parabola is a mathematical curve such that its reflection property causes a parallel beam of incoming rays to be focused to one point (Fig. 13.22A). Conversely, a point signal placed at the focal point will be reflected by the surface to form a parallel outgoing beam (Fig. 13.22B). Hence, a parabolic antenna can be employed as a transmitter-receiver.

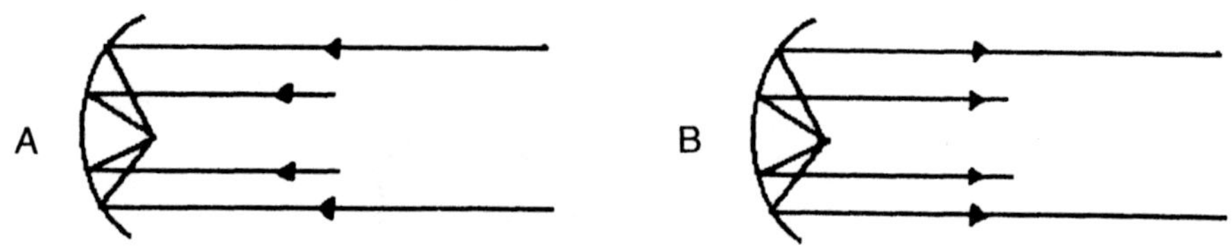

Figure 13.22 An ideal parabola can (A) focus all parallel incoming signals to one point and (B) radiate signals emitted from the focus out in a parallel manner.

There are several three-dimensional versions of the parabolic antenna: paraboloid, truncated paraboloid, cylindrical paraboloid, and orange-peel paraboloid (Fig. 13.23). A paraboloid is formed by rotating a parabola about its axis. It therefore exhibits rotational symmetry. The ideal and realistic radiation patterns

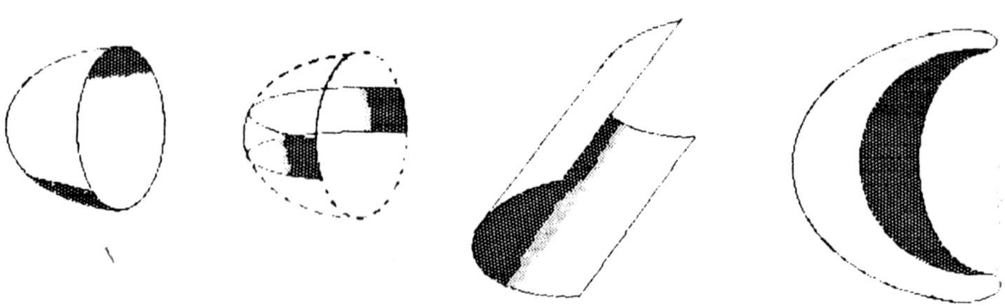

Figure 13.23 Different three-dimensional constructions of parabolic antennas.

of a parabolic antenna are shown in Fig. 13.24. Ideally, a parabola gives a perfectly parallel beam. The spill-over effect due to the wave nature of the signal results in the realistic radiation pattern as shown. The directive gain of a parabolic is high (typically 30 dB) and its beamwidths are relatively narrow. The realistic radiation pattern shown can be rotated to form the three-dimensional radiation pattern for a paraboloid. However, such rotation cannot be done with, for example, a cylindrical paraboloid.

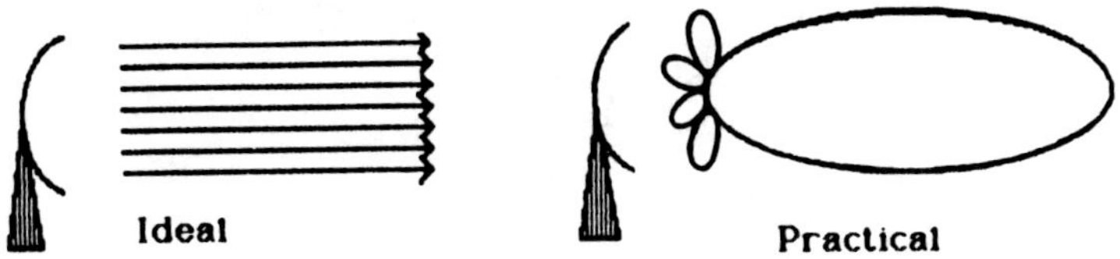

Figure 13.24 The ideal and realistic radiation pattern of a parabolic antenna.

Two types of parabolic mounts exist. The first type is to put the transmitter-receiver at the focal point as shown in Fig. 13.25A. The second type, which is mechanically more stable, is to mount the transmitter-receiver at the parabola apex and place a reflecting surface at the focal point (Fig. 13.25B). The second type is more practical for large-diameter antennas.

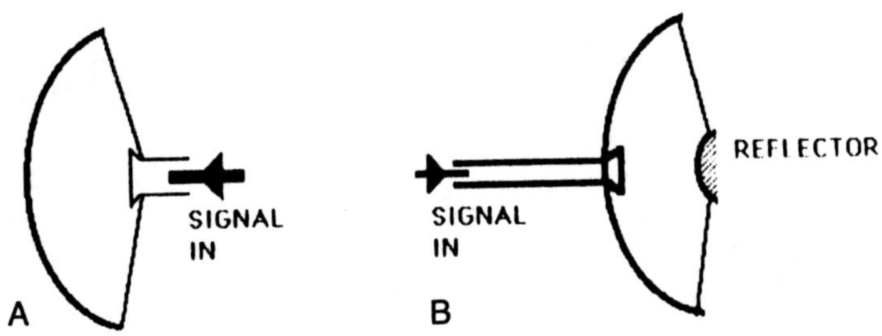

Figure 13.25 The feed to a parabolic antenna can be mounted in two ways.

Parabolic antennas have very high gains and narrow beamwidths. Their antenna gain is given by

$$G\text{ (number)} = 6\,(D/\lambda)^2. \qquad (13.15.1)$$

The half-power and first-null beamwidths in degrees are

$$HPBW\text{(degree)} = 58\ \lambda/D \qquad (13.15.2)$$

$$FNBW\text{(degree)} = 70\ \lambda/D \qquad (13.15.3)$$

It can be seen from the three equations that as the ratio λ/D decreases, i.e., the wavelength of the signal is negligible compared with the antenna's physical dimension, the beamwidths approach zero and the antenna gain becomes infinitely large. In other words, the optical principle applies and the parabolic antenna radiates straight ahead with virtually no spill-over or power spread.

Example (13.15.1)

The diameter of a parabolic antenna is 3m and the wavelength of the signal is 2cm (15 GHz). Find the *FNBW*, *HPBW*, and antenna gain.

Solution: According to Eq. (13.15.3), the first-null beamwidth is

$$FNBW = 70 \times \frac{0.02}{3} = 0.47°$$

The half-power beamwidth, according to Eq. (13.15.2), is

$$HPBW = 58 \times \frac{0.02}{3} = 0.38°$$

The antenna gain, according to Eq. (13.15.1), is

$$G\,(\text{number}) = 6 \times (3/0.02)^2 = 1.35 \times 10^5$$

$$G\,(\text{dB}) = 51.3 \text{ dB} \cdot$$

13.16 HORN ANTENNA

A horn antenna can be formed by flaring one end of a waveguide, rectangular or circular. Without the flare, a microwave signal making a sudden transition from the waveguide's impedance to free-space impedance would suffer a mismatch. The flare can be designed in various geometrical ways to smooth the transition. Some of the various geometries of horn antennas available are shown in Fig. 13.26.

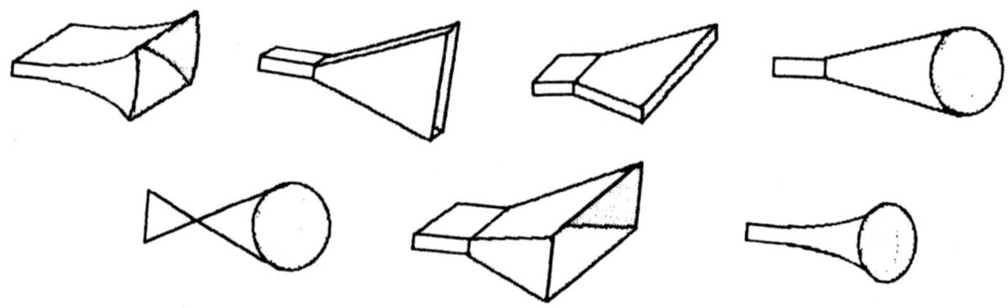

Figure 13.26 The various geometries of a horn antenna.

Horn antennas have lower gain than the parabolic antennas, and the beamwidths are comparable. Horn antennas are, however, easier to manufacture and to tune for matching. The expressions will not be given here because the gain and beamwidth are highly dependent upon the geometry of the horn.

13.17 RADIATION RESISTANCES

Radiation resistance is a useful quantity for measuring the effectiveness of an antenna. This concept will be approached by using a dipole antenna.

The total amount of power radiated into space by a dipole antenna is related to the value of the alternating current in the antenna. The exact expression is

$$P_{total} = \frac{790\, I^2 L^2}{\lambda^2} \text{ Watts} \tag{13.17.1}$$

where

I = *rms* current value in amperes,
L = total length of the dipole antenna in meters,
λ = wavelength of the signal in meters.

Rearrangement of Eq. (13.17.1) yields

$$P_{total} = I^2 R_r \tag{13.17.2}$$

where R_r is known as the radiation resistance of the antenna. For the dipole,

$$R_r = \frac{790\ L^2}{\lambda^2} \qquad (13.17.3)$$

The purpose of rewriting Eq. (13.17.1) in the form of Eq. (13.17.2) is that the second equation reminds us of the power dissipation in an ordinary resistor due to current flow I. The important distinction here is that in an ordinary resistor the power dissipation usually results in heat while the dissipation of power in a radiation resistance is into space in the form of electromagnetic wave. The larger an antenna's radiation resistance, the more effective it is in radiating signal power into space.

Example (13.17.1)

Use different lengths to evaluate the radiation resistance of a dipole antenna operating at 1 MHz (AM).

Solution: The wavelength of the signal is 300m. According to Eq. (13.17.2),

$$R_r = \frac{790\ L^2}{\lambda^2} = \frac{790}{300^2} \times L^2 = 8.8 \times 10^{-3}\ L^2 \text{ ohms}$$

For L = 1 meter, $R_r = 8.8 \times 10^{-3}$ ohms;
For L = 100 meter, $R_r = 88$ ohms;
For L = 150 meter, $R_r = 198$ ohms (optimum length);
For L = 300 meter, $R_r = 790$ ohms.

From the above example, it appears that the longer the antenna length, the larger the radiation resistance, and therefore power can be radiated more effectively. However, the formation of a standing wave becomes a problem due to reflection when the antenna length is too long.

Finally, the reader should note that the radiation resistance of an antenna is not the ohmic resistance of the metal, it cannot be measured by an ohm-meter. It is a useful quantity to bear in mind when evaluating the effectiveness of radiation for an antenna under test.

13.18 CAPTURE AREA

The capture area of a receiving antenna is an important quantity in determining the amount of power received. In the case of a dish or horn antenna, the capture area is the effective area facing the incoming signal. For example, if a dish antenna is facing north but the signal is coming from the west direction, little signal will be detected regardless of the physical size of the dish antenna. See Fig. 13.27 for illustrations.

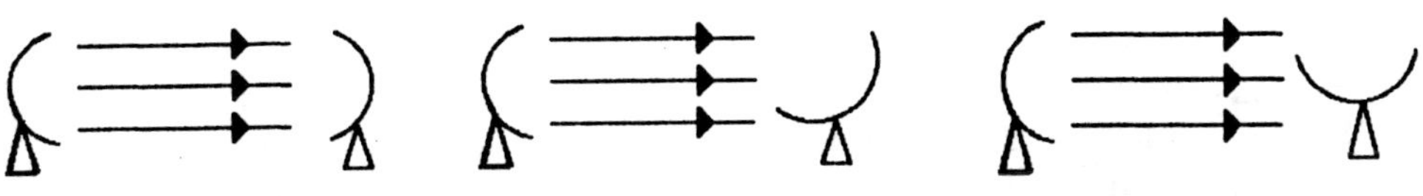

Figure 13.27 A receiving antenna can detect a maximum. intermediate, or minimum amount of incoming signals depending on the orientation it is facing relative to the signal.

In general, if A_p is the physical area of an antenna and it is facing the incoming angle by angle θ with respect to the axis, the capture area A_c is

$$A_c = A_p \cos\theta \qquad (13.18.1)$$

SEC. 13.19
SLOT RADIATORS

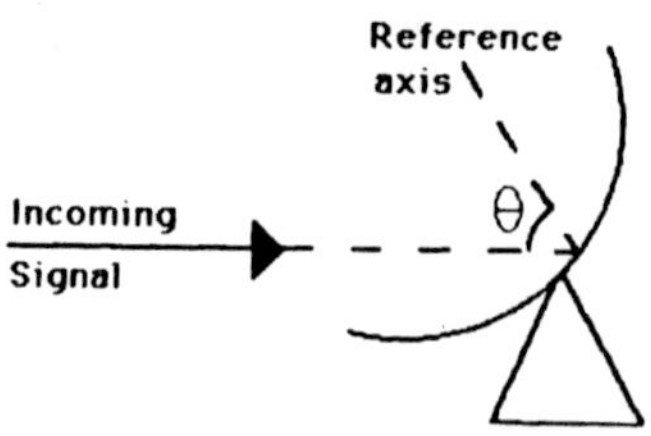

Figure 13.28 The capture area of a receiving antenna is the physical area multiplied by the cosine of the angle between the incoming signal and the reference axis of the antenna.

The angle θ is illustrated in Fig. 13.28.

Example (13.18.1)

A dish area has a physical radius of 5m. It is facing an incoming signal by 10°C. Calculate the capture area.

Solution: The physical area of the dish antenna is

$$A_p = \pi\ r^2 = \pi \times 5^2 = 78.54\text{m}^2$$

According to Eq. (13.19.1), the capture area is

$$A_c = 78.54 \cos 10° = 77.35\text{m}^2$$

Knowing the capture area and the power density of the signal at the receiver location, the received power can now be calculated:

$$P_{received}\ (\text{Received Power}) = PD \times A_c \qquad (13.18.2)$$

Example (13.18.2)

An isotropic antenna transmits 10W of signal. At a distance 1000 meters away, a dish receiver of physical area 10m^2 is installed, but it faces the signal at an angle of 45°. What is the received power?

Solution: The power density of the isotropic antenna at $R = 1000$ meters can be calculated by using Eq. (13.11.1),

$$PD = \frac{10}{4\pi 1000^2}\ W/m^2$$

$$= 7.96 \times 10^{-7}\ W/m^2$$

According to Eq. (13.18.1), the capture area is

$$A_c = 10 \times \cos 45°\ \ \text{m}^2$$
$$= 7.07\ \ \text{m}^2$$

The received power can now be calculated by using Eq. (13.19.2):

$$P_{received} = 7.96 \times 10^{-7} \times 7.07\text{W}$$
$$= 5.63 \times 10^{-6}\ \text{W}$$

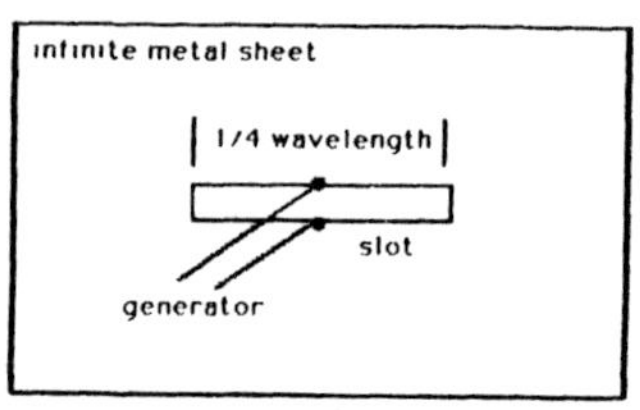

Figure 13.29 A slot cut into a large sheet of metal is equivalent to a dipole antenna. The E and H planes are, however, inverted.

13.19 SLOT RADIATORS

Figure 13.29 shows a slot cut into a large flat sheet of metal and RF power is connected to it. It turns out that the slot can now function as a dipole antenna and RF is radiated. The radiation pattern is similar to that of a dipole except that the directions of electric and magnetic field vectors are interchanged. The length of the slot is the length of the dipole antenna. The slot behaves exactly like a dipole antenna when the metal sheet is infinitely large.

Qualitatively, an open slot in a sheet of metal is the inverse of a metal rod in open space. Radiation resulting from a slot is predicted by Babinet's principle.

We can think of the slot linking the main line to the coupling arm of a directional coupler as the slot radiator. Microwave power in the main line is coupled, or radiated, into the coupler arm. The polarization of the radiation pattern depends on the orientation of the slot, while the power radiated depends on the location of the slot and whether RF current flows in that location. This is illustrated in

SEC. 13.19
SLOT RADIATORS

Fig. 13.30. Several slots can be cut to form slot arrays, as shown in Fig. 13.31, in any mathematical pattern. The slot arrays, as an antenna, is structurally and aerodynamically advantageous in aircraft.

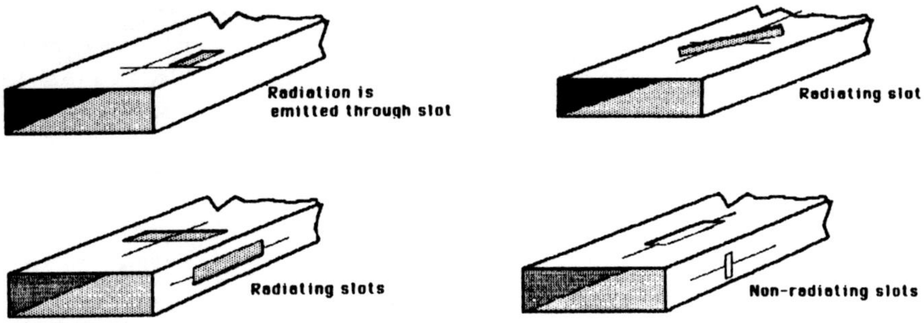

Figure 13.30 Slots can be cut along a waveguide to be radiating and non-radiating.

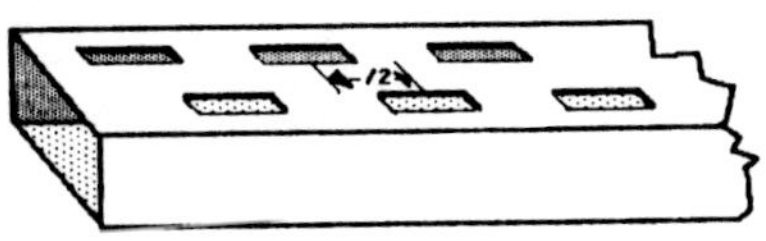

Figure 13.31 A slotted array is a series of slots cut along a waveguide.

CHAPTER 14

RADAR SYSTEMS

Frederic H. Levien

14.1 HISTORICAL BACKGROUND

14.1.1 Early Developmental Work

The word *radar* is an acronym from the description "*radio detection and ranging*." The earliest recorded mention of the concept was put forth by a German engineer, Hulsmeyer, in a 1904 patent after noting the presence of reflected waves from a ship a mile away. He proposed a demonstration for the German Navy, which declined with an official "No interest."

The first officially sanctioned experiment of "radio echos" occurred in the US in 1922 at the David Taylor Naval Research Laboratory in Washington, D.C. Other experimenters at the Naval Research Laboratory and Bell Telephone Laboratories continued their efforts in 1930 and recorded their ability to detect aircraft; however, no other target information or position was obtainable.

The US Navy, recognizing the value of this new science, continued to push the development, and in the spring of 1937, with a system mounted on the *U.S.S. Leary*, was able to detect a target at a distance of 12 miles. By January 1938, with a system mounted on the battleship *U.S.S. New York*, the range had been increased to 50 miles.

The US Army was also pushing development of a pulsed radar system. By December 1936, they had developed the SCR-268 which had the electronics coupled to a searchlight, which provided the angular tracking capability for anti-aircraft fire control. Work continued, resulting in the development of the SCR-270 in 1939. This long-range radar developed for early warning was indeed able to detect the presence of the Japanese air armada as they approached Pearl Habor in 1941. However, the refusal of the army to believe the signals this new technology provided, showing the incoming Japanese air armada, is now history.

14.1.2 Basic Early Principles

The basic theory behind radar was developed in 1831 by Michael Faraday with his key experiments in electromagnetism. Later in 1864, James C. Maxwell synthesized the properties of electricity and magnetism into a set of equations now popularly known as Maxwell's equations. These early scientists, whose work is valid to this day, laid the theoretical foundations for our understanding of radar.

The fundamental observations upon which all radar systems are based, are highlighted and presented here:

- Time varying electric currents

  . . . produce
- Time varying electric and magnetic fields

  . . . in
- Free space

  . . . and

SEC. 14.1
HISTORICAL BACKGROUND

- These fields would induce time varying electrical currents
  . . . in
- Materials that they encountered
  . . . and that
- These currents would *in turn* generate electric and magnetic fields of their own,
  . . . then
- These fields would propagate in free space at the speed of light.

14.1.3 Functions and Applications

A radar system is used to determine the velocity, range, azimuth, elevation, and geometry of a target that is not necessarily visually observable, and to do so in the presence of large interfering reflections, multiple targets, and potential jamming by an enemy system.

Radar frequencies are generally located in the electromagnetic spectrum beginning from the UHF to the EHF bands. A pictorial indication of this relationship is presented in Fig. 14.1. Standard radar frequency letter band nomenclature is shown in Table 14.1.

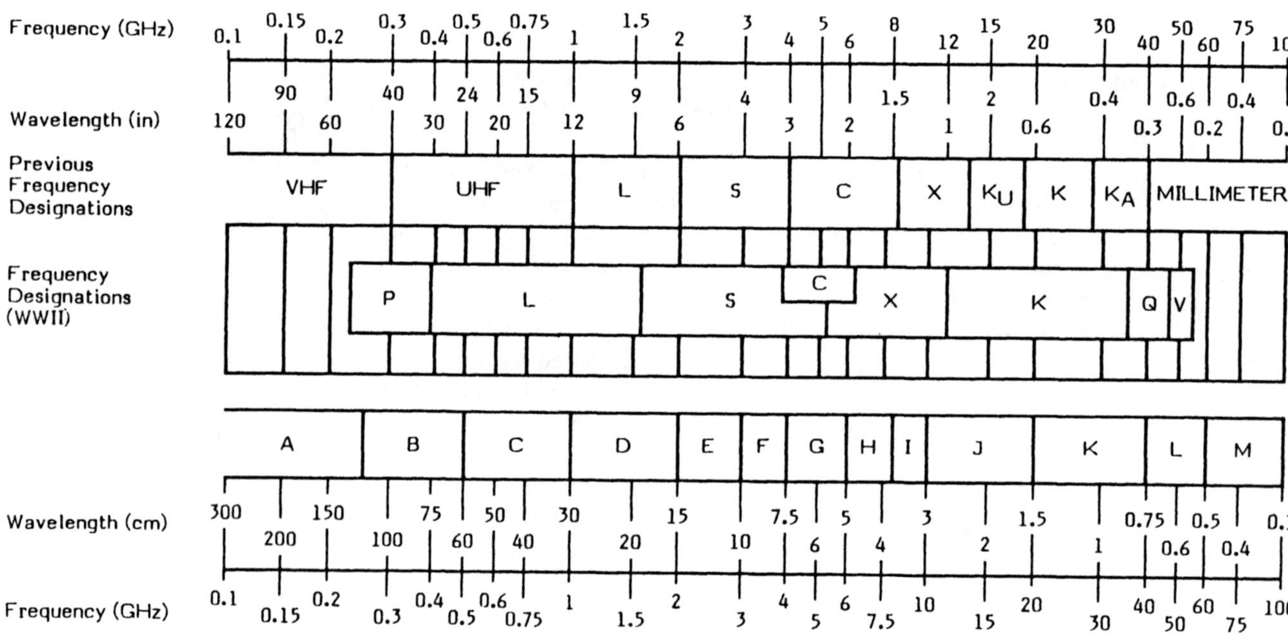

Figure 14.1 Standard frequency designations.

Table 14.1
Standard Radar-Frequency Letter-Band Nomenclature

Band Designation	Nominal Frequency Range	Specific radar bands based on ITU Assignments
L	1000–2000 MHz	1215–1400 MHz
S	2000–4000 MHz	2300–2500 MHz
C	4000–8000 MHz	5250–5925 MHz
X	8000–12,000 MHz	8500–10,680 MHz
K_u	12.0–18 GHz	13.4–14.0 GHz
		15.7–17.7 GHz
K	18–27 GHz	24.05–24.25 GHz
K_a	27–40 GHz	33.4–36.0 GHz

These frequencies are generally in the range called the microwave region. The reason for having radar systems operate in this range is that the short wavelengths of the microwave region allow for highly directive antennas of reasonable size such that they can be located on moving platforms like missiles, aircraft, and ships. It is also possible to generate high levels of power in this frequency range.

14.1.4 Basic System Block Diagram

A block diagram for a typical radar system is presented in Fig. 14.2. Each of the elements of this system functions as follows:

Duplexer: A circulator that allows microwave energy to pass from the transmitter to the antenna without entering the low-noise radio frequency amplifier.

Transmitter: Generates and amplifies the microwave signal.

Pulse Modulator: Switches the transmitter on and off causing pulses of microwave power to emit from the antenna.

Low-Noise RF Amplifier: Amplifies the weak received echo signal.

Mixer, Local Oscillator: Converts the microwave signal to a lower more convenient intermediate frequency.

IF Amplifier: Amplifies the intermediate frequency signal.

Second Detector: Eliminates the intermediate frequency, leaving only baseband information.

Video Amplifier: Amplifies the baseband signal.

Display: Presents the received radar echo in some form (usually visual) useable to an operator.

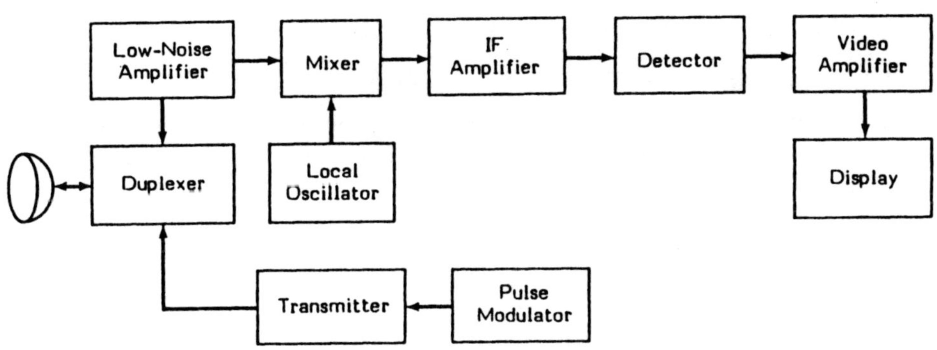

Figure 14.2 Radar system block diagram.

14.1.5 Radar Range Measurement

Radio energy travels at the speed of light in space.

$$\text{Velocity of Signal} = 3 \times 10^8 \text{ meters/second} = c \text{ (velocity of light)} \tag{14.1.1}$$

$$\begin{aligned}\text{Distance Traveled} &= \text{velocity} \times \text{travel time} \\ &= 3 \times 10^8 \text{ (meters/second)} \times t \text{ (seconds)}\end{aligned} \tag{14.1.2}$$

With a radar, however, the microwave signal must travel out to the target and then back (as an echo) to the radar transmitter in order to be detected as a target.

The distance to the target (the range) is then one-half the distance the signal must travel.

$$\text{Range (meters)} = \tfrac{1}{2} \text{ distance signal travels} \tag{14.1.3}$$

$$\text{Range} = (\tfrac{1}{2})(3 \times 10^8) \text{ meters/second} \times t \text{ (seconds)}$$

This equation can be modified to let us insert more useful radar terms: range in kilometers and travel time in microseconds.

$$\text{Range to Target (kilometers)} = .15 \times t \text{ (microseconds)} \tag{14.1.4}$$

Or, using miles and microseconds, we obtain

$$\text{Range to Target (nautical miles)} = .081 \times t \text{ (microseconds)} \tag{14.1.5}$$

Note that a nautical mile is slightly different than the statute mile we are more familiar with, i.e.,

$$1 \text{ nautical mile} = 1.15 \text{ statute miles} \tag{14.1.6}$$

If we were to convert these times into round trip miles of the signal, the values are as indicated below.

Round Trip Times

1 Nautical Mile:	12.34 microseconds
1 Statute mile:	10.73 microseconds
492 feet:	1 microsecond
150 meters:	1 microsecond
5 feet:	10 nanoseconds

14.2 DERIVATION OF THE RADAR EQUATION

14.2.1 Manipulation of the Basic Radar Equation

A derivation of the basic radar equation begins with a study of the power density from an isotropic (equal radiation in every direction) antenna (see Ch. 13). The equation is given by

$$PD = \frac{P_t}{4 \pi R^2} \tag{14.2.1}$$

where

P_t = transmitted power,
R = distance from transmitter to target.

If we make the antenna directive, the power density at the target is given by

$$PD = \frac{P_t}{4 \pi R^2} \times G \tag{14.2.2}$$

where

G = antenna gain along the direction of the target.

The transmitted signal is collected by the target and then reflected by the target and returns to the transmitter. The power density of this reflected signal (echo) back at the radar is given by

$$PD_r = \left(\frac{P_t\,G}{4\,\pi\,R^2} \times \sigma\right)\frac{1}{4\,\pi\,R^2} \tag{14.2.3}$$

where

σ = radar cross section of the target.

Note that the quantity inside the parenthesis is the power striking the target and reflected (assumed 100%) while the term, $1/4\pi r^2$ is the power density at the radar transmitter due to the reflected signal (assumed isotropic).

The echo is received by the radar's antenna. The power that is received by the radar is then given by

$$P_r = \frac{P_t\,G\,\sigma}{(4\,\pi)^2R^4} \times A_e \tag{14.2.4}$$

where

A_e = effective receiving antenna area.

The simple form of the radar equation is then found by taking

$$P_r = \frac{P_t\,G\,\sigma\,A_e}{(4\,\pi)^2R^4} \tag{14.2.5}$$

and solving for R

$$R^4 = \frac{P_t\,G\,\sigma\,A_e}{(4\,\pi)^2P_r} \tag{14.2.6}$$

then R equals

$$R - \left[\frac{P_tG\,A_e\,\sigma}{(4\,\pi)^2P_r}\right]^{1/4} \tag{14.2.7}$$

The maximum range of the radar will be determined by the minimum received power that can be detected:

$$R_{max} = \left[\frac{P_tG\,A_e\,\sigma}{(4\,\pi)^2S_{min}}\right]^{1/4} \tag{14.2.8}$$

where

S_{min} = minimum detectable received signal.

The above equation [Eq. (14.2.8)] is called the simplest form of the radar equation.

In summary, then, the radar equation is stated as follows:

$$R_{max} = \left(\frac{P_tG\,A_e\sigma}{(4\,\pi)^2S_{min}}\right)^{1/4}$$

where

R_{max} = The maximum range in meters;
P_t = The transmitted power in watts;
G = The transmitting antenna gain factor;
σ = The radar cross section of the target in m^2;
A_e = The effective area of the receiving antenna in m^2;
S_{min} = The minimum detectable signal in watts.

Manipulating the radar equation and solving for various other key characteristics of this system, we have the following set of governing equations:

$$R_{max} = \left[\frac{P_t G \sigma A_e}{(4 \pi)^2 S_{min}}\right]^{1/4}$$

$$R_{max}^4 = \frac{P_t G \sigma A_e}{(4 \pi)^2 S_{min}} \tag{14.2.9}$$

Solving for P_t:

$$P_t = \frac{R_{max}^4 (4 \pi)^2 S_{min}}{G \sigma A_e} \tag{14.2.10}$$

Solving for G:

$$G = \frac{R_{max}^4 (4 \pi)^2 S_{min}}{P_t \sigma A_e} \tag{14.2.11}$$

Solving for S_{min}:

$$S_{min} = \frac{P_t G \sigma A_e}{(4 \pi)^2 R_{max}^4} \tag{14.2.12}$$

Solving for σ:

$$\sigma = \frac{R_{max}^4 (4 \pi)^2 S_{min}}{P_t G A_e} \tag{14.2.13}$$

We can view these relationships and determine the trade-offs between desired system performance and the cost associated with the performance requirements of the components upon which the system is designed and built. Thus, antenna size, transmitter power, and the size of the target detectable at a given range can be varied. The resulting analysis is better able to provide the optimum cost performance trade-off.

14.2.2 Detection Level and False Alarms

In determining the minimum detectable signal for a radar display, we can examine Fig. 14.3, which is a display of radar information as seen on an "A" scope. (That is, a presentation of amplitude as a function of time.) The display shows signals, *A*, *B*, and *C* imbedded in noise. Some of these noise spikes (*D* and *E*) look like weak target returns. In order to avoid confusion, we set a voltage level in the receiver that is called the "threshold." Then, if a signal crosses the threshold, it is called a target echo, and everything below the threshold is called noise and is eliminated from the display. The level of the threshold determines the minimum detectable signal for the radar system. If the threshold is set too low, some noise spikes will cross the threshold and be detected as targets (Fig. 14.4). This type of an occurrence is called a "false alarm." If the threshold is set too high, some of the weaker target echoes will not be able to cross the threshold and will not be detected. This type of occurrence is called a "missed detection."

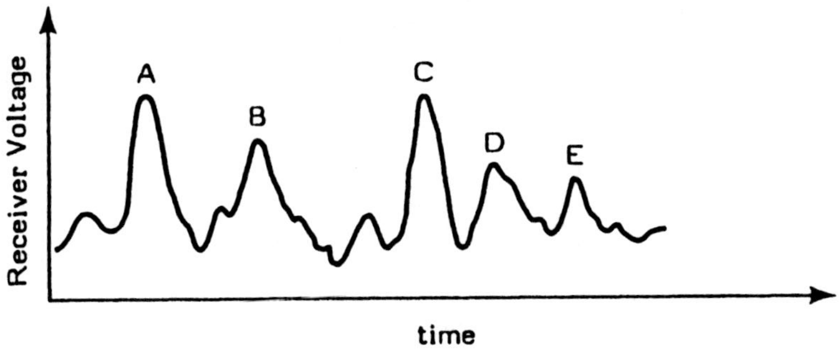

Figure 14.3 Noise on "A" scope.

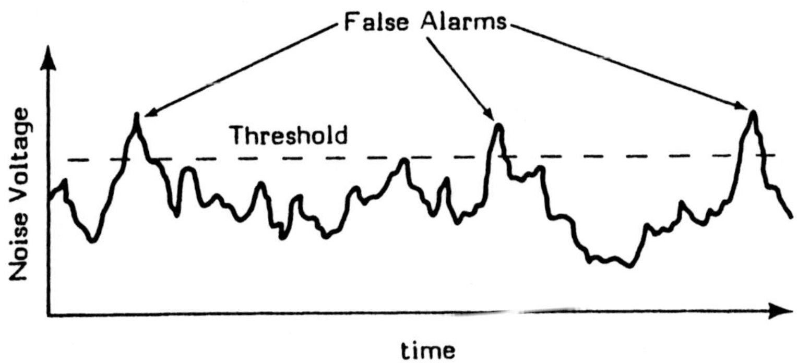

Figure 14.4 Threshold setting on "A" scope.

The sensitivity of a radar is determined by its ability to maximize the signal-to-noise ratio of the received echo. The signal must be amplified without adding a lot of noise in the process. In other words, noise must be minimized. The noise that appears in a radar receiver will not always have the same level (amplitude) with time. The probability that a noise spike will reach a certain level at the output terminals of a receiver is given by a probability curve called a Rayleigh distribution. The probability function of a signal imbedded in noise (signal-plus-noise) is given by a distribution called a Gaussian distribution.

If we superimpose both the Rayleigh and Gaussian distributions on a graph of probability *versus* signal amplitude, the situation exists as shown in Fig. 14.5. Here, by moving the threshold (that is, the lowest detectable signal level) toward a smaller signal level, we intersect more of the signal-plus-noise curve. We would detect more targets, but this would also intersect more of the noise curve and thus encounter more false alarms. By moving the threshold toward a larger signal level, we will avoid the noise, but we will miss more of the targets.

We can see that the setting of the threshold value represents a compromise between the number of false alarms that we are willing to live with as opposed to the number of targets that the system is going to miss. We can plot the trade-off between the probability of detection from a target echo as opposed to noise as function of the signal-to-noise ratio of our system. This graph, indicating the rate at which we will be receiving false alarms, is displayed in Fig. 14.6.

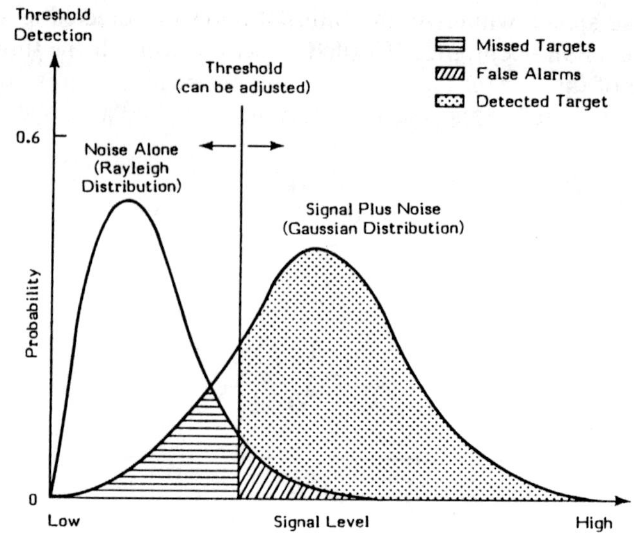

Figure 14.5 Effect of threshold variation.

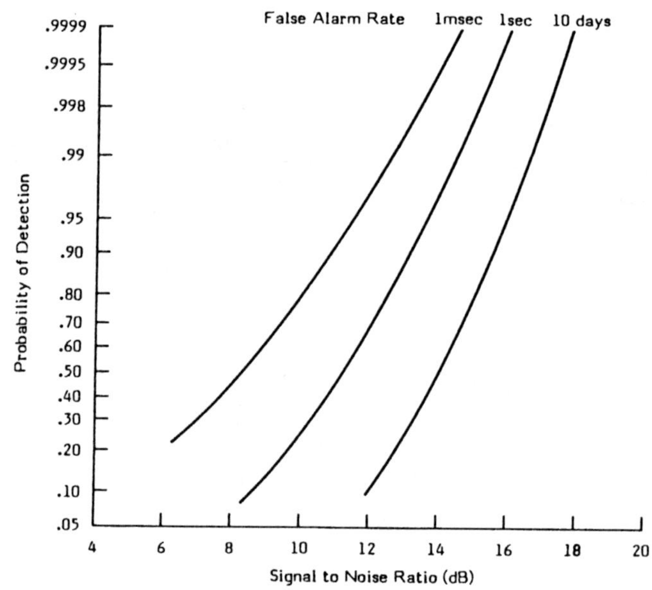

Figure 14.6 False alarm rate.

14.2.3 Radar Receiver Noise

The electrical noise caused by the thermal movement of electrons was given in Ch. 9 by

$$\text{Thermal Noise Power} = k\ T_o\ B_n \tag{14.2.14}$$

where

k = Boltzmann's constant $= 1.38 \times 10^{-23}$ J/K,
T_o = Temperature in degrees Kelvin (K)
B_n = The receiver bandwidth in Hertz (Hz).

This noise is present in any receiver. The noise that is present at the output terminals of a receiver will be greater than the thermal noise alone because the receiver contains components that also add noise. The ratio of the noise present at the output of the receiver to the noise due to thermal effects alone is called the receiver's noise figure.

$$F_n = \frac{\text{NOISE OUT}}{\text{THERMAL NOISE}} = \frac{N_o}{k\, T_o\, B_n\, G_a} \tag{14.2.15}$$

Note that the thermal noise has been increased by the gain of the receiver.

In a radar system, the minimum signal is determined by the requirements of a certain probability of detection and a certain maximum false alarm rate. These two requirements directly determine the minimum signal-to-noise ratio [(S_o/N_o) *min*] out of the receiver.

The minimum signal-to-noise ratio determines the minimum signal that the system can detect.

$$S_{min} = F_n\, k\, T_o\, B_n \left(\frac{S_o}{N_o}\right)_{min} \tag{14.2.16}$$

With this relation, we can modify the simple form of the radar equation to include the effects of receiver noise:

$$R_{max} = \left(\frac{P_t G\, A_e\, \sigma}{(4\,\pi)^2 k\, T_o B_n F_n \left(\frac{S_o}{N_o}\right)_{min}}\right)^{1/4} \tag{14.2.17}$$

Notice that the receiver temperature, bandwidth, and especially noise figure, are just as important as output power and antenna parameters in determining the range a radar will have. Because of this, a great deal of development effort has been put into the radar receiver.

14.2.4 Integration of Radar Pulse

The antenna pattern of a radar has a finite beamwidth. As the antenna scans back and forth across a target, the target is illuminated by more than one pulse of energy.

$$N_b = \frac{\theta_b \times Prf}{\theta_s} \tag{14.2.18}$$

or

$$N_b = \frac{\theta_b Prf}{6\, \omega_m} \tag{14.2.19}$$

where

N_b = Number of pulses hitting target,
θ_b = Antenna beamwidth in degrees (half-power, 3 dB down),
Prf = Pulse repetition frequency in Hz,
θ_s = Antenna rotation rate in degrees per second,
ω_m = Antenna rotation rate in rpm.

A typical set of search radar parameters would be

Prf = 300 Hz
θ_b = 1.5 degrees
ω_m = 5 rpm

SEC. 14.2
DERIVATION OF THE RADAR EQUATION

Then the number of hits on a target during one revolution or antenna scan would be

$$N_b = \frac{(1.5°) \times (300 \text{ Hz})}{(6) \times (5 \text{ rpm})}$$

$$= 15 \text{ Hits/Scan}$$

Integration is the summing of these hits on the target. By summing the individual hits on the target, the strength of the received echo pulse can be made effectively greater. In the example, perfect integration of the 15 pulses that hit the target would make the effective echo strength 15 times stronger, or, in terms of dB, 11.7 dB larger. Integration usually provides a significant increase in the effective received signal strength, and, as a consequence, it is employed in many radars.

A perfect integrator increases the signal-to-noise ratio by adding signal pulses without adding more noise to the output.

$$SNR \text{ (due to integration)} = n\ (S/N)_{\text{single pulse}}$$

where n = number of pulses integrated.

The integration improvement factor can be determined from Fig. 14.7. Therefore, looking at the maximum range in the radar equation [Eq. (14.2.17)], the new formula that we come up with is

$$R_{max} = \left(\frac{P_t\, G\, A_e\, \sigma\, nE_i(n)}{(4\,\pi)^2\, kT_oB_nF_n(S/N)_{min}}\right)^{1/4}$$

where

$nE_i(n)$	=	Integration improvement factor,
P_t	=	Transmitted power,
G	=	Antenna gain,
A_e	=	Effective antenna aperture area,
$kT_oB_nF_n$	=	System noise level,
$(S/N)_{min}$	=	Minimum signal-to-noise ratio.

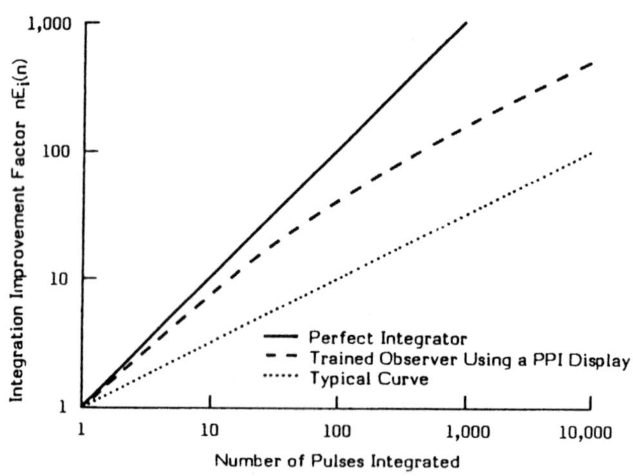

Figure 14.7 Integration improvement factor.

Notice again that the integration of pulses adds to the maximum range of a radar with the same degree of importance as increasing the transmitted power, improving the antenna, or decreasing the noise level in the receiver.

Integration of pulses is a very cost-effective means of improving radar range performance.

14.2.5 System Losses

Defined below are the most important types of system loss:

Plumbing: One of the most predictable losses of power in a system is that due to the absorption of signal by the waveguide itself, the so-called "plumbing loss." These losses caused by the microwave signal travelling through coaxial or waveguide transmission lines can be easily calculated by examining Fig. 14.8.

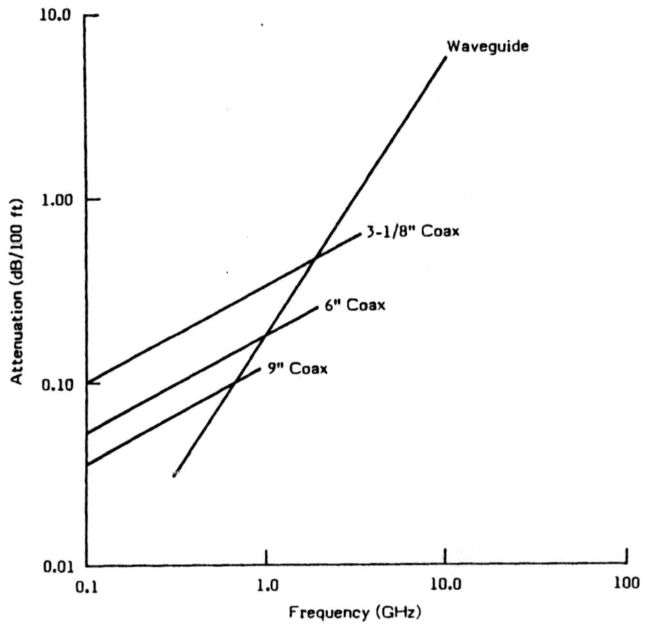

Figure 14.8 Plumbing losses.

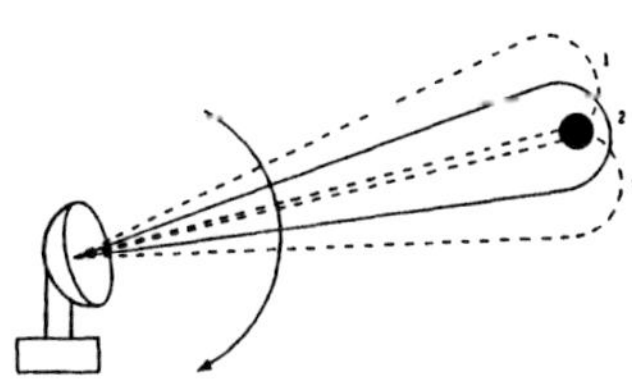

Figure 14.9 Beam shaping loss.

Beam Shape: The target is not always illuminated by the maximum strength of the antenna beam. This affects the actual received signal strength if integration of pulses is being used. Figure 14-9 shows how this beam shape loss comes about.

Limiting: There is also a limiting loss which occurs when the amplitude of the received echo pulse is clipped (the top portion is lost) or limited in the signal detection circuitry, causing the processed signal amplitude to be smaller than that actually received.

Collapsing: A collapsing loss occurs when pure noise pulses are added to the desired signal pulse noise pulses in some types of radar signal processors.

Real World: Finally, there is a "real world" loss. This occurs because real equipment and radar environments are usually not as good as computed or simulated situations. This can be due to non-ideal equipment, field usage degradation, and often by non-ideal (tired) human operators.

Combining the radar equation with system losses included, we have

$$R_{max} = \left(\frac{P_t\, G\, A_e\, \sigma}{(4\,\pi)^2 S_{min} L_s}\right)^{1/4} \tag{14.2.20}$$

where

L_s = Plumbing loss + Beam shape loss + Limiting loss + Collapsing loss + Real world loss.

14.2.6 Atmospheric Attenuation

There are atmospheric effects on a radar signal due mainly to attenuation, caused by water vapor and atmospheric gases, mostly oxygen.

The radar equation with atmospheric attenuation effects included is seen to be

$$R_{max} = \left(\frac{P_t\,G\,A_e\,\sigma}{(4\,\pi)^2\,\alpha^2 S_{min}}\right)^{1/4} \qquad (14.2.21)$$

where

α = Loss factor, noting that α appears in the equation as α^2 because the loss is two-way, out and back.

Note that the loss factor α is in numbers, not dB.

14.2.7 Radar Cross Section

Radar cross section (σ) is the area a target would have to occupy to produce the amount of reflected power (echo) that is detected back at the radar.

The radar cross section of a target is dependent upon several factors, including viewing direction, radar frequency, physical size, geometry of the object, and the composition of the object.

Although target cross section is known to be dependent upon target size, it has been determined that for a simple sphere (like a raindrop) when the circumference is greater than 10 times the wavelength of the illuminating radar (almost always the case), the cross section is essentially constant. The cross sections of other objects, however, depend not only on their actual size, but upon the direction from which they are viewed by the radar.

For comparison, some typical radar cross sections are presented here:

Object	*Radar Cross Section (m^2)*
Missile	0.5
Single Engine Aircraft	1.0
Fighter Aircraft	2.0
Medium Size Jet Aircraft (737)	20
Large Size Jet Aircraft (707)	40
Jumbo Jet (747)	100
Small Open Boat	0.02
Small Pleasure Boat	2.0
Large Ship	5000
Pickup Truck	200
Car	100
Bicycle	2
Man	1
Bird	0.01
Insect	0.00001

There is also a variation of radar cross section with viewing direction, as shown in Fig. 14.10. By carefully studying the effects of aircraft geometry on the radar cross section, and then modifying the geometry to reduce the peaks, the radar cross section of an aircraft can be reduced by several orders of magnitude.

This work, coupled with the application of materials to the aircraft surfaces with reduced radar reflectivity, has resulted in low radar visibility or "stealth" design technique for military aircraft.

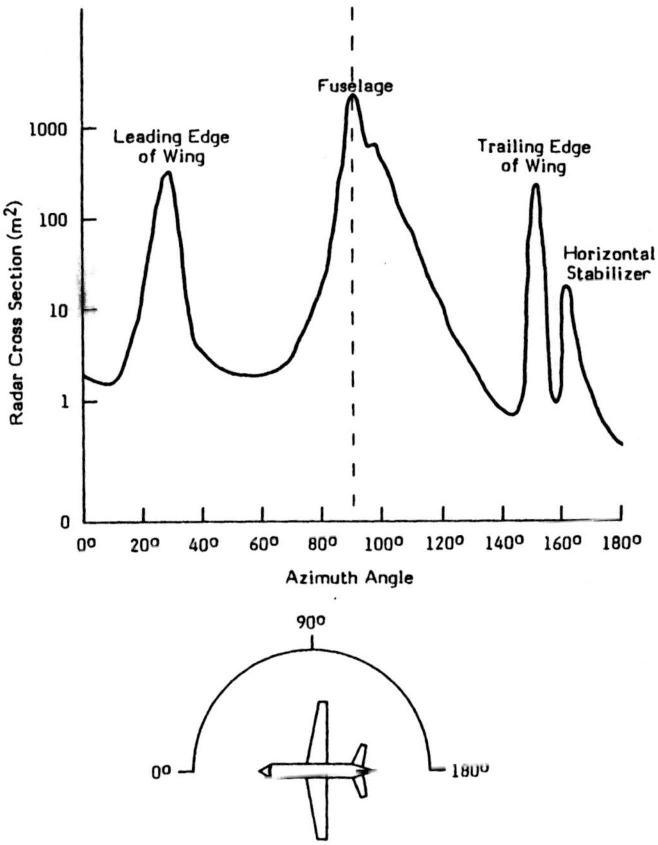

Figure 14.10 Radar cross section variation of an aircraft.

Because the radar cross section of a target fluctuates, the value of the minimum signal-to-noise ratio (SNR_{min}) must be increased to maintain the desired probability of detection and false alarm rate. This increase is roughly given by the relationship shown in Fig. 14.11.

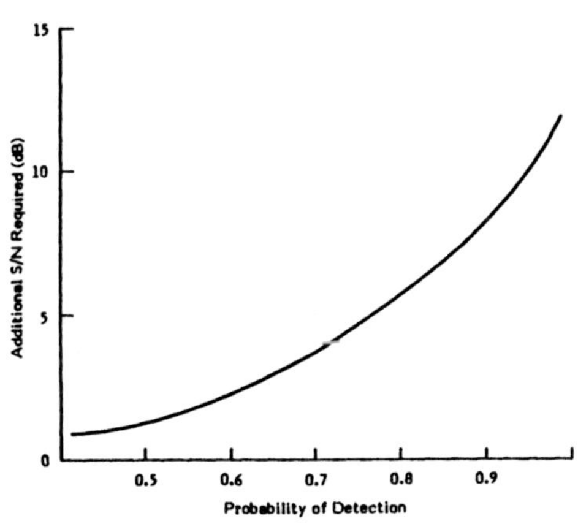

Figure 14.11 Increased signal-to-noise ratio required for desired probability of detection due to cross section fluctuation.

SEC. 14.3
CW RADAR

14.3 CONTINUOUS WAVE (CW) RADAR

14.3.1 The Doppler Effect

The Doppler effect is a familiar phenomenon to anyone who has ever sat at a railroad crossing waiting for a high-speed train to pass by. As the train approaches, blowing its whistle, the pitch (frequency) of the whistle seems to increase. Suddenly, as the engine passes, with all the noise and dust and the whistle still blaring, a strange change occurs in the whistles' pitch. It suddenly starts to drop in frequency. The drop in pitch is an acoustic example of the Doppler effect, or Doppler shift.

The Doppler effect also occurs with electronic signals. If a radar transmits a signal at a frequency f_0 the signal is reflected back to the radar by a target moving toward the radar, with a new frequency $f_0 + f_d$.

The signal returning to the radar is shifted up in frequency due to the target's velocity toward the radar. This shift is called the *Doppler shift*. The amount of frequency shift is given by

$$f_d = \frac{1.03\ V_r}{\lambda} \qquad (14.3.1)$$

where

f_d = Doppler frequency (Hz),
V_r = Velocity toward or away from the source (Kts, nautical miles per hour),
λ = Source signal wavelength (meters).

In the audio frequency range, this sounds like a change in pitch.

If the source is operating in the microwave region at a frequency f_0 (GHz), the equation can be rewritten as

$$f_d(\text{Hz}) = 3.9\ V_r(\text{mph})\ f_0(\text{GHz}) \qquad (14.3.2)$$

or

$$f_d(\text{Hz}) = 3.43\ V_r(\text{Kts})\ f_0(\text{GHz}) \qquad (14.3.3)$$

Example (14.3.1.1)

A 5 GHz radar receives an echo that is shifted up in frequency by 10 KHz. Calculating the target's velocity in miles per hour.

According to Eq. (14.3.2), we can solve for V_r:

$$V_r = \frac{f_d}{3.9\, f_0}$$

$$= \frac{(1 \times 10^4\ \text{Hz})}{3.9\ (5\ \text{GHz})}$$

$$= 512.8\ \text{mph}$$

Doppler frequency is a measure of the relative velocity between the target and the radar. It will only be a measure of the absolute velocity if the target is moving directly toward or directly away from the radar, and if either the transmitter or the target is stationary.

Note that a target moving about the radar in a circle gives $f_d = 0$, and a target moving toward or away from the radar gives a maximum f_d. The higher the operating frequency of the radar, the greater the frequency of the Doppler shift for a given velocity of the target. This relationship is plotted in Fig. 14.12.

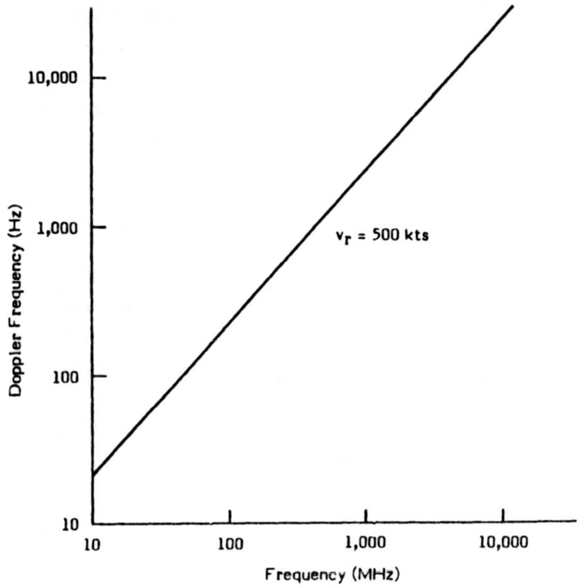

Figure 14.12 Doppler shift versus radar frequency.

A simple CW radar using the Doppler effect is shown in Fig. 14.13. In this block diagram, we show that f_0 is the transmitted signal, and $f_0 \pm f_d$ is signal returned from the target. The Doppler shifted signal ($f_0 \pm f_d$) is mixed with f_0 to give a beat frequency of f_d. Note that the algebraic sign is lost upon beating.

The doppler frequency f_d is amplified by an amplifier specially designed to have a lower cut-off to eliminate noise or clutter near zero frequency, and an upper cut-off that is above the highest expected Doppler frequency.

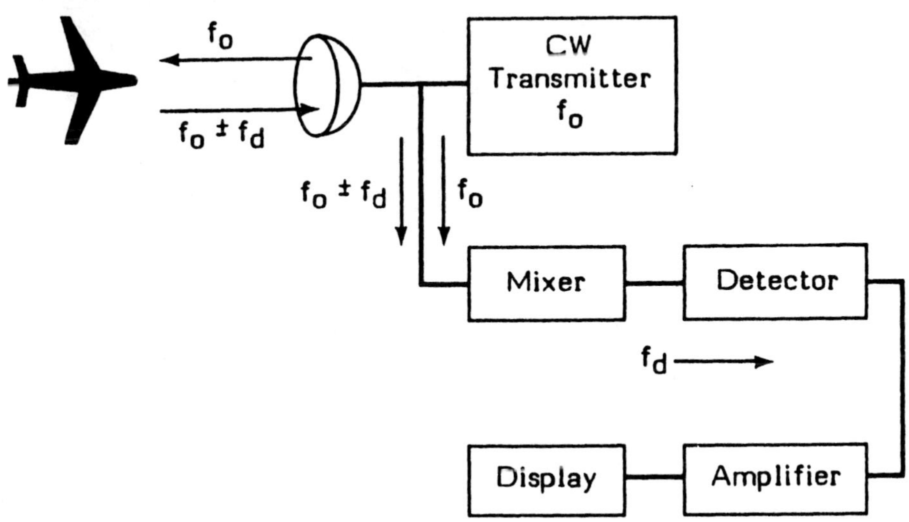

Figure 14.13 CW Doppler-effect radar.

14.3.2 Application

The major drawback of straight CW radar is its inability to measure range. It can only measure relative velocity to the transmitting signal station. However, there are many applications where this information by itself is quite useful. One of the most common commercial applications is use by law enforcement agencies to record vehicle speed in excess of maximum limits. Some of us have had first-hand experience with this effective system. Also in law enforcement, there is a growing market for intrusion alarm systems to detect movements of intruders in restricted areas. In another application, many large ships have several radar points on the side of the vessel that can detect how rapidly the ship is closing in on a pier because control of the ship's speed during the last few feet of travel is critical when docking operations are in progress, and visibility from the bridge of the ship is often severely restricted because of the size of the vessel. This allows the captain of the ship to change his speed as necessary.

Some applications still in the laboratory stage include collision avoidance radar in both automobiles and aircraft environments. Here, coupled with range measurements, rates of closing speed are detected and when the combination of both exceed pre-agreed safety limits, alarms are sounded.

14.3.3 Frequency Modulated, Continuous Wave Radar

Theory (FM-CW Radar)

Range cannot be measured with a simple CW radar because there is no way to measure how long it takes the signal to travel out to the target and back. There are no timing marks as with pulse radar. If the signal is frequency modulated, however, the travel time can be determined. Referring to Fig. 14.14, we see a signal frequency transmitted at T_A, returns as an echo at T_B. This allows the travel time to be determined, which in turn gives the target range. If the two signals, both transmitted and received, are mixed with each other, it becomes easy to establish the travel time.

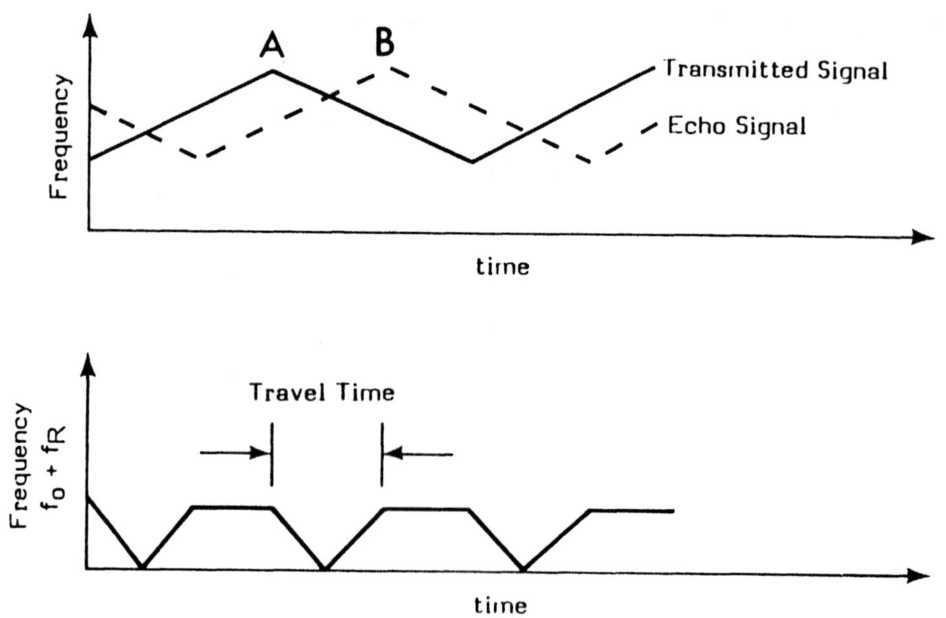

Figure 14.14 The time difference between the transmitted signal and the echo signal determines the range.

By examining the difference signal shown in Fig. 14.15, we can observe the instant in time when it starts to change. This time interval, between T_A and T_B, which is a function of the time it takes the original frequency to return as an echo, can be converted to distance since we have a known frequency, thereby determining the range.

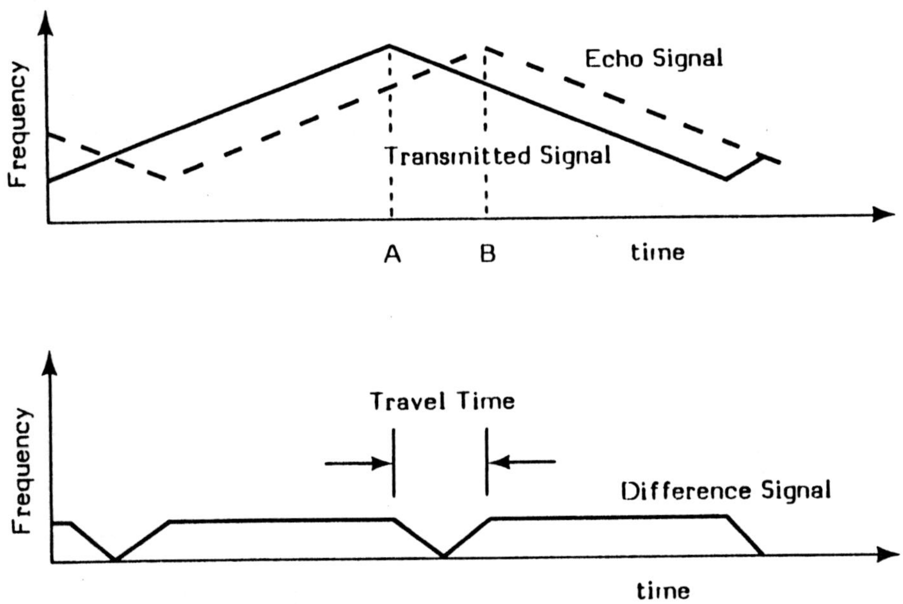

Figure 14.15 Using FM to determine range.

As a result of all the above complications, FM-CW radar is generally used in applications where there is at most one moving target and where the target is moving slowly with respect to the radar. A typical application meeting these constraints is a radar altimeter requirement where there is only one target (the earth), and the target is usually not moving rapidly toward or away from the radar.

If the target is also moving with respect to the radar, there will be a Doppler shift added to the returning signal. This makes the situation much more complex. If, in addition, there is more than one moving target, the situation becomes extremely complicated. Additional problems arise if the Doppler frequency is not significantly lower than the transmitted frequency.

14.4 PULSE DOPPLER RADAR AND MOVING TARGET INDICATOR

The Doppler effect can be utilized in a pulsed radar system. By combining special delay line techniques, the system can be made to determine target velocities, and distinguish moving targets from stationary targets. This improved system is called a *moving target indicator* or MTI radar system. A simple CW radar system is shown in Fig. 14.16.

A simple pulse radar is diagrammed in Fig. 14.17. Notice that both echo signals return with some Doppler shift of the original frequency.

The problem can be visualized more clearly by examining Fig. 14.18. Here we see that an aircraft target cannot be seen by the range display such as an "A" scope, which presents amplitude as a function of time (called "time domain" display). The targets echo is masked by the very large echo return from the mountain. However, if we look at the returning echos in the frequency domain (see Fig. 14.19), amplitude plotted as a function of frequency, a different picture is obtained. The aircraft target is easily seen in the frequency domain display due to its Doppler shifted echo frequency.

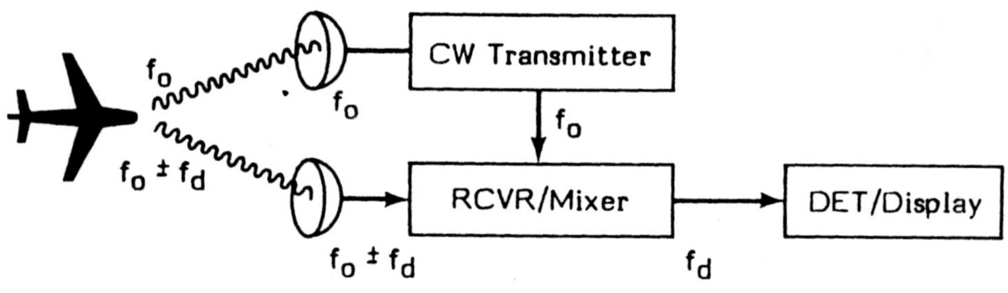

Figure 14.16 CW radar system.

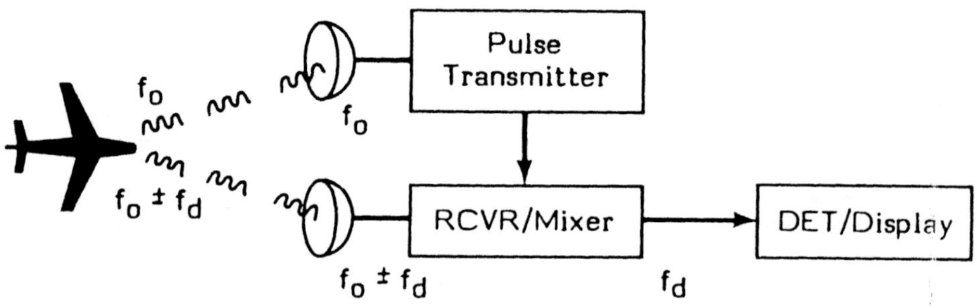

Figure 14.17 Pulsed radar system.

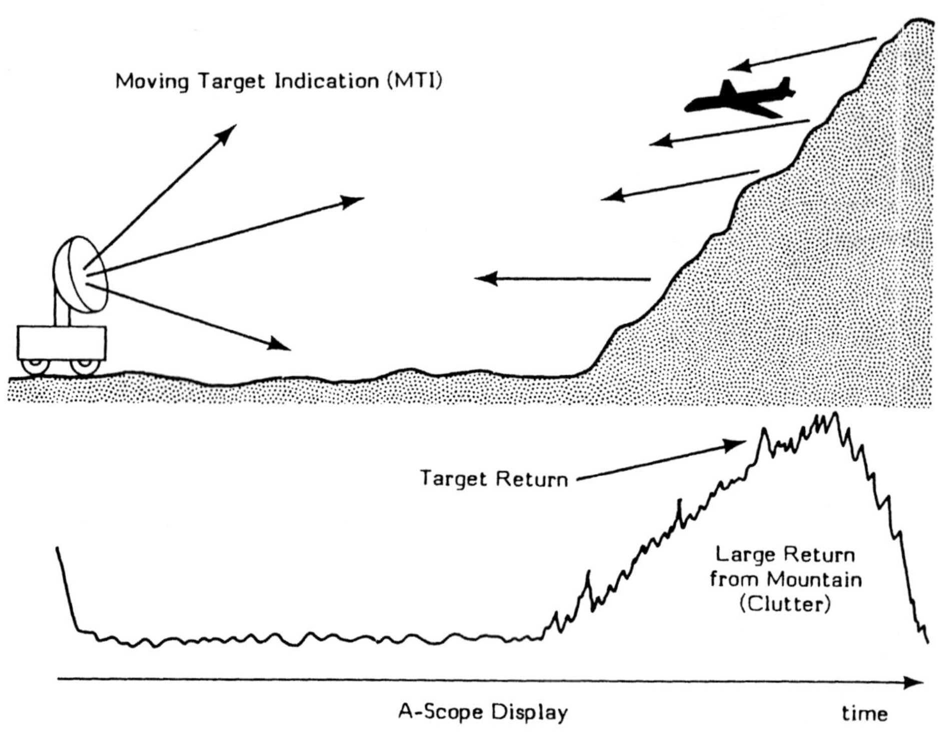

Figure 14.18 Masking of target echo.

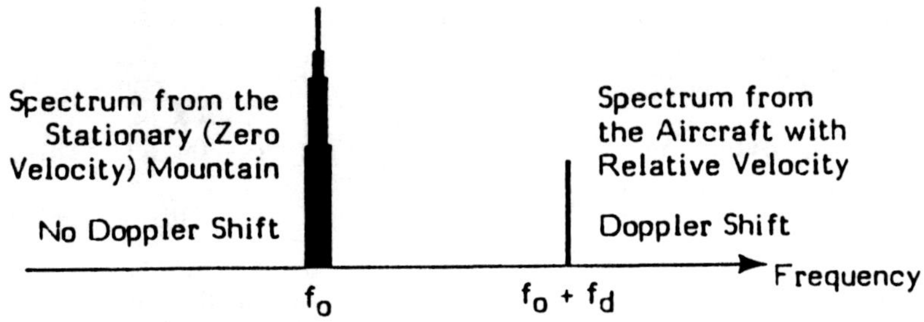

Figure 14.19 Target echos in frequency domain.

The phase and amplitude of the echos of stationary targets do not change from pulse to pulse. However, the phase and amplitude of moving target echos will change. Now, if we develop a method where the echo signal from one pulse is subtracted from the echo return of the previous pulse, everything will cancel out *except* the moving targets. Performing this subtraction-cancellation process (shown in Fig. 14.20) and displaying only the remainder, the moving target part, constitutes the basis of MTI radar. An MTI radar detector block diagram is given in Fig. 14.21.

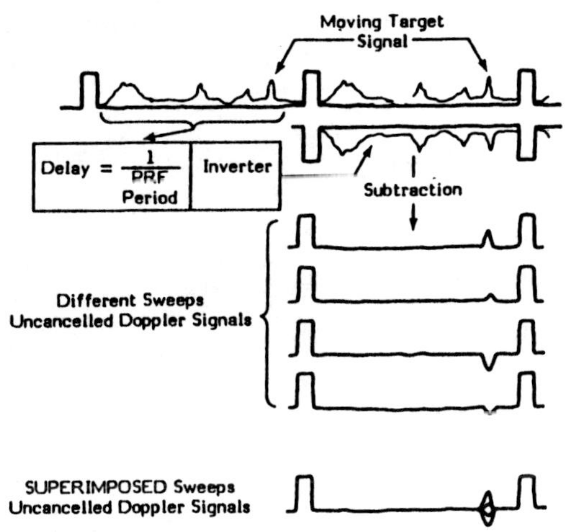

Figure 14.20 MTI radar techniques.

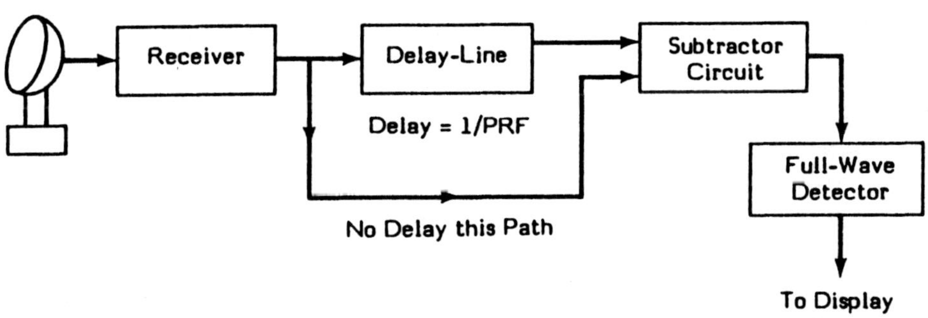

Figure 14.21 MTI radar block diagram.

PROBLEMS

PROBLEMS

Problem 1:

A target is moving toward a radar at a velocity of 100 Kts. The radar is operating at 10 GHz. What is the Doppler shift?

Problem 2:

A 5 GHz radar receives an echo that is shifted up in frequency by 10 kHz. Calculate the target's velocity in mph.

Problem 3:

What Doppler shift will be measured by a 3 GHz radar tracking a 1000 Kts target? Kts = nautical miles per hour.

Problem 4:

If a 30 GHz radar measures a +200 kHz Doppler shift, what is the target's velocity in Kts?

Problem 5:

If a 50 kHz Doppler frequency is measured and a target's velocity is known to be 1282 Kts, what must the transmitter frequency of the radar be?

Problem 6:

A radar has a maximum detection range of 400 nautical miles. How long after the pulse is transmitted will an echo be detected at this range?

Problem 7:

Find R_{max}, given the following:

P_t = 100,000W
G = 30 dB or ratio = 1000
A_e = $10m^2$
S_{min} = $1 \times 10^{-10}W$
σ = $1m^2$

Problem 8:

Find the maximum range of a radar having the following parameters:

P_t = 1×10^6W
G = 35 dB
A_e = $10m^2$
S_{min} = $1 \times 10^{-10}W$
σ = $3m^2$

Problem 9:

Find the maximum range of a radar having the following parameters:
Transmitted power = 500kW
Antenna gain = 37 dB
Effective antenna area = $7m^2$
Target cross section = $2m^2$
Minimum detectable signal = $1 \times 10^{-10}W$

Problem 10:

If the minimum detectable signal for a radar is $1 \times 10^{-10}W$, with the antenna gain equal to 35 dB, and the effective antenna aperture equal to $10m^2$, what must

PROBLEMS

the transmitted signal power be to detect a target whose radar cross section = $5m^2$ at a range of 20 kilometers?

Problem 11:

If only 10 kilowatts of transmitted power is available, what antenna gain is needed to fulfill the radar requirements of the preceding problem?

Problem 12;

What signal-to-noise ratio (*SNR*) is needed for the following cases?

a) Probability of detection = 0.99
False alarm rate = 1 per second — *SNR* = ______ dB

b) Probability of detection = 0.50
False alarm rate = 1 per 10 days — *SNR* = ______ dB

c) Probability of detection = 0.9995
False alarm rate = 1 per 10 days — *SNR* = ______ dB

Problem 13:

For a signal-to-noise ratio of 10 dB, and a false alarm rate of 1 per second, what is the probability of detection? Use Fig. 14.6.

Problem 14:

An echo is detected microseconds after a radar pulse is transmitted. What is the target range in kilometers and nautical miles?

CHAPTER 15

SATELLITE COMMUNICATIONS

Frederic H. Levien

15.1 HISTORY

The earliest hint of satellites for potential use in communications appeared in *Wireless World* in February 1945, in an article by Arthur C. Clarke.

The first published information exploring in-depth, technical considerations for the possible usefulness of satellites for communications was presented in 1955 by John R. Pierce of the Bell Telephone Laboratories. His paper discussing the relationship between power, bandwidth, antenna gain, and orbit parameters, concluded with the statement that it was possible to use satellites as relay stations in the heavens. Then, Sputnik's launch by the Soviets in 1957 proved the feasibility of man-made satellites. The pieces of the puzzle began fitting together.

Advances in rocketry, coming as it were from the research done during world War II, had a great deal to do with the science of this new concept. However, of even more significance was the knowledge gained during the war years and immediately thereafter of the entire field of microwave technology. Understanding the basic mechanisms of how to generate a microwave signal, the effects of earth's atmosphere on the propagation characteristics of the signal and, finally, efficient methods of receiving and processing the incredibly faint pulses of energy that contained the useful data needed, together formed the key to unlocking the world of satellites communications.

The first actual space communication using a satellite however, was done with the natural satellite of Earth, the Moon. On November 23, 1959, the first live voice transmission was bounced off the moon between Bell Telephone Laboratories (BTL) in New Jersey and the Jet Propulsion Laboratory (JPL) in California. It was not until 1960 that the first man-made satellite was used in an actual successful broadcast between two locations on earth using a satellite as a reflector. This was Project Echo. Echo I, a 100ft diameter aluminum-coated balloon, was launched on August 12, 1960, and at 11:40 *a.m.* GMT, the first demonstration of transmission via a man-made satellite was completed with President Dwight D. Eisenhower's recorded message sent from California to New Jersey. The first two-way audio transmission was sent the next day between the JPL and BTL using messages from President Eisenhower and Senator Lyndon B. Johnson. Also, on that day, the first two-way live voice communication was held. The reality of John Pierce's dream was unfolding. So far, however, the satellite had been used only as a passive reflector. The missing link of technology was finally set in place when for the first-time transmission from earth to an active satellite was achieved. A broadcast was received, converted to another frequency, and retransmitted early in the morning of July 10, 1962, when the Bell Telephone System TELSTAR satellite demonstrated both speech and television transmission between Andover, Maine and Washington, D.C. That particular broadcast was televised for the national networks, and the entire world suddenly became aware of the potential for this new technology, microwave relay stations in the sky.

Because of the constraints due to the availability of launch vehicles and rockets, early satellites were launched into an orbit fairly close to earth (2,000 miles). Consequently, they circled the globe at a fairly rapid rate of speed. Therefore, they were usable only during that period of time when they were visible above one horizon and being tracked until they dropped beneath the opposite horizon. When the satellite disappeared over the curvature of the earth, it was useless as a relay station until it reappeared once again over the opposite horizon. The period of visible time was approximately one hour. This caused severe disruption of its ability to function as a communication device. It also required huge antennas that were capable of picking up the satellite's track and then moving rapidly to follow it across the sky.

More sophisticated rocket technology has now given us the ability to place satellites in a "geostationary" orbit. This orbit (also called geosynchronous or fixed orbit) is where the centrifugal acceleration of the hurtling satellite exactly balances the earth's gravitational pull, thus enabling the satellite to orbit the earth at the same relative rate of speed as the earth's rotation, i.e., one revolution per 24 hours. The distance at which a satellite will become geostationary is 22,279 statute miles above the equator. Satellites in such an orbit appear to remain fixed in relation to a specific point on earth. All commercial communication satellites currently in use are in geostationary orbits. That is what enables them to relay signals without interruption from one point on earth to another. The ability to maintain a satellite in one fixed position in relation to a point on earth greatly simplified the ability to locate and lock onto the satellite.

The first really effective communication satellite was launched in 1963. It was manufactured by the Hughes Aircraft Corporation and was named SYNCOM. The key to its effectiveness was the fact that it was launched into a geostationary orbit.

SYNCOM was an experiment. It provided the capacity to relay either a single TV channel or 50 separate telephone voice channels. From its orbit above the equator between Africa and South America, it interconnected North America and Europe with their first "real-time" live television transmissions. By 1965 the geostationary satellite concept had been proved as a practical approach.

15.2 APPLICATIONS

The use of satellites as a means of relaying information via microwave transmissions has matured rapidly in just a few decades. Applications have expanded to include: communications (telephone, radio, television, data, and facsimile); navigation; weather forecasting; earth resource management; and, probably one of the most significant areas, defense (reconaissance, aircraft and missile detection, guidance, and control).

15.2.1 Communications

Communications is currently the sole commercial satellite space activity of any magnitude. This includes domestic, regional, and global satellite system networks that carry telephone, radio, television, and data transmissions, having estimated global annual revenues in excess of $2 billion (US), with demand expected to grow at 30% per year at least into 1987.

Between 1970 and 1975, there were only 17 civilian telecommunications satellites launched throughout the world. Progress was not much more rapid between 1975 and 1980, with only 20 launches reported. However, between 1980 and 1990, it is expected that the world will launch over 200 communication satellites. A staggering five-fold increase of satellites in orbit. At present, there are a little over 150 geostationary satellites in orbit, about half of them still working.

There is a developing political problem as countries vie for allocations of space for geostationary orbits. There is only one arc for true geostationary positioning, and satellites separated by less than about half a degree of arc in space can interfere with each other if they broadcast using the same frequencies. Therefore, the number of slots available is limited. The geostationary arc above North America is already nearing saturation because each region wants to select its choice of satellites for different time zones and different uses.

15.2.2 Navigation

Starting in 1967, the US Navy launched a series of five satellites in their TRANSIT system, designed to provide navigational fixes for their fleet ballistic missile submarines.

The government began allowing civilians to use these navigation signals, and in recent years the worldwide non-military market for SAT-NAV receivers has ballooned. Receiving equipment that cost $70,000 (US) ten years ago, can now be obtained for less than $5,000 (US). Even the cheapest sets provide updated latitude and longitude readings accurate to within 200 yards. The users of this equipment cover the span from ocean yachts wanting to know where they are in a race, to tuna boat fishermen wanting to know how to return to the exact spot at sea where the fishing is best, to offshore oil explorers who want to find their pipelines at the bottom of the sea.

The successor to this now outdated system is the new NAVSTAR Global Positioning System. It will include 24 satellites, eight in each of three orbits. When fully deployed, this truly artificial galaxy will provide position information accurate to 33ft. Some enhanced procedures will improve this accuracy to within 3m, or about 10ft. This information will include not only latitude and longitude, but also altitude above sea level. The size of the receiver can be reduced to a box no larger than a portable radio, making it available for individual carrying.

15.2.3 Weather Forecasting

"People always complain about the weather, but never do anything about it." This old complaint has so far resulted in few successful achievements to change the weather. However, thanks to satellite technology, a great deal has been learned about forecasting the weather. Nature on the rampage often dwarfs man-made activities when it comes to destruction of life and property. The ability to predict violent storms, often only dreamed of in earlier times, is now coming closer to reality thanks to advancement in satellite and microwave technology.

The National Aeronautics and Space Administration (NASA) placed the world's first meteorological satellite, Tiros I, into orbit in April 1960, about 450 miles above the earth. This increased by several orders of magnitude man's information about the weather. Circling the globe about once every hour and thirty minutes, this spin-stabilized satellite carried two small television cameras. Panoramic views from space of the cloud patterns of earth were breath-taking to those who watched. Early Tiros revealed spiral types of hurricane and typhoons in progress, and it became abundantly apparent that visual imagery from satellites could be of significant importance in a warning service to advise people of the birth and possible paths of the world's most dangerous storms.

The most dramatic demonstration of the success of this service was in the tracking of tropical storm Camille in the summer of 1969. This was the fiercest tropical storm ever recorded in the history of the US. Its winds reached up to 200 miles per hour and it brought with it a 25ft tidal wave. As the storm began to gather strength in the Caribbcan on August 11, 1969 it was discovered and picked up for tracking by a weather satellite. This knowledge of the storm's path allowed sufficient time for about 75,000 persons to be evacuated from the storm's path before it struck the US mainland near Biloxi, Mississippi, saving possibly 50,000 lives.

Since the advent of weather satellites in the late 1960s, millions of satellite cloud pictures have been received and analyzed, giving meteorologists a vastly improved three-dimensional concept of the skies. Of great help, too, are the weather data obtained by special instruments, such as the high resolution radiometer and the spectrometer carried by some satellites. In a single day, the infrared spectrometer on Nimbus III, provides as much information as 10,000 radiosondes, or sounding rockets, could have yielded.

Nimbus IV, one of the largest, most complex and versatile of NASA's early experimental weather satellites, weighed 1,366 pounds. It was launched on April 8, 1970 into a polar sun-synchronous orbit. This "high noon" orbit kept the sun directly behind the space craft providing good light for pictures of all sunlit parts of the earth that its TV cameras takes at local noon every day. Nimbus IV is also equipped to take pictures at local midnight daily, using an advanced infared radiometer in the process of measuring the earth's absorption of solar energy and its thermal radiation. In addition to this atmospheric probing, Nimbus IV relays data collected hundreds of miles below and transmitted to it by remote robot ground stations, oceanogaphic buoys and balloons.

15.2.4 Remote Sensing

Remote sensing is the science, technology, and art of finding out about the earth from space. It is also developing as the second space technology to become commercially viable after satellite communications. Remote sensing began initially from the desire of the US to improve both its weather forecasting capability and to obtain better intelligence, particularly about economic data in the USSR; especially about the Soviet grain harvest. From this early desire, remote sensing has turned into a series of techniques and technology which can provide low cost detailed information about almost all parts of world.

The most sophisticated remote sensing satellites are still used by the US and Soviet armed forces, and are powerful enough to provide photographs in which soldiers regiments can be read from their uniform decorations.

The remote sensing technologies which are now available for civilian use are almost entirely based upon optical light or infrared radiation, which is slightly longer in wavelength than light perceived by the human eye. Remote sensing data is only available in the world market from the US national remote sensing effort. No information is currently available in the Soviet Union.

Until now, civilian remote sensing has grown up around the Landsat system run by the public sector in the US. Landsat satellites are developed and launched by NASA and controlled once in orbit by the National Oceanic and Atmospheric Administration (NOAA). One of the major Landsat objectives is still the determination of the Soviet grain harvest. However, the public availability of data from the earliest days of Landsat onward has meant that Landsat images have also been used by geologists, urban and rural planners, civil engineers, agricultural and forestry authorities as well as many other people. In some of the Third World countries, particuarly in South America and Africa, for the first time in history, rivers, lakes and villages have been reliably placed on maps.

A Landsat photograph is taken with a four-color camera, a combination of four images taken with cameras operating in different colors of light. They depict an area of the earth's surface 185 × 185 kilometers square. One critical aspect of Landsat is that they have all been in sun synchronous orbits. This means that the satellite is always photographing terrain where the time is 9:30 *a.m.* Early enough for long shadows and good definition, but usually late enough for full daylight. There is an enormous amount of information recorded in a Landsat image. The resolution is officially 80m, but smaller objects can be distinguished.

Landsats I, II, and III were launched in 1972, 1975, and 1978, respectively, and produced a total of over 700,000 such images of every part of the earth's surface outside of the polar regions. This flood of data has given rise to a significant industry. A Landsat data center at Sioux Falls, South Dakota, sells tens of thousands of images every year to US customers and there are 13 ground stations worldwide to receive Landsat data and distribute it to those agencies and people who desire its use.

In addition, there are consultants who specialize in finding and interpreting Landsat data for people, such as geologists and city planners.

The useful potential for Landsat data (already considerable before) will be even greater since the launch of Landsat IV in 1982. This is partly because Landsat IV has a lower orbit, 700km as opposed to 900km above the earth's surface for the three earlier Landsats. This lower orbit will bring resolution down to perhaps 30m for the four-color camera. More importantly, Landsat IV was the first "thematic mapper" yet flown. This is a seven wave band camera originally devised to distinguish between wheat and barley on the steppes of Asia. It will also provide fine detail on snow and ice cover, clay and other mineral composition, urban and rural development, and state of rivers, lakes and seas to a depth of 10m.

The world of remote sensing is becoming more diverse with more sources and with a wider variety of data. The same is happening on the ground with more bodies being set up to process remote sensing data. The task is heavily computerized, but a lot of subjective skill is used to interpret the photographs obtained from space. The result is that the less developed countries of the world, in particular, already have a powerful, inexpensive, new information resource available to them, which is growing all the time and is also available, for example, to community and environmental groups. Facilities to interpret remote sensing data are cheap to establish. Pollution monitoring, assessment of desertification and forest loss, fish stock and water supply measurements, and many other tasks are eased by remote sensing along with the difficult job of obtaining first-hand knowledge of geology and mineral potential.

15.2.5 Defense

A major use of military satellites is for reconnaissance of enemy activity and visually reporting this activity for interpretation on the ground by the host country. Very often, satellites play leading roles in events and activities that sometimes appear in media stories and at other times never get reported. One such event concerned the Middle East confrontation of 1973 when combined Arab national army units pressed Israel back toward her borders. Fearing invasion, Israel's Jericho surface-to-surface missiles were armed with nuclear warheads in a frantic three days of activity centered in the Negev Desert. These 20 kiloton warheads could have been launched against targets in Cairo or Damascus. The political decision had already been made that, if Israel's borders were violated, the country would unleash a quick flurry of nuclear missiles to warn Arab forces that continuation of the war would result in their own destruction. These activities were observed and confirmed by US reconnaisance satellites. The US realized that, had this nuclear capability been exploited by Israel, it could have brought Soviet forces to the aid of Arab states, which in turn may have brought the US into the war with the Soviets. A nuclear holocaust might have ensued. This potentially explosive situation was resolved by Soviet and US use of the international hot line. This peaceful resolution may not have happened if satellites had not initially detected this activity.

In the last half of the 1970s, there have been many satellite launchings by both the US and the USSR for the purposes of defense and communications. In the area of satellite systems with maritime intelligence functions alone, the numbers are impressive:

Satellite Applications	*Launches 1976–1981*	
	US	*USSR*
(1) Photo-Reconnaissance	15	209
(2) Electronic Surveillance	4	40
(3) Early Warning	9	18
(4) Ocean Surveillance	13	20
(5) Navigation	9	41
(6) Communications	29	189
(7) Meteorological	15	14

Military satellites have up until now been used to gather information by various sensors and then pass it on, either physically or electronically to a land-base center for interpretation, analysis, and dissemination. The need for up-to-the-minute information at sea is now becoming so urgent, however, that systems are being developed whereby satellites will pass on their information in real time by digital downlinks directly to ships at sea. Satellite communications brings together some of the most sophisticated technologies in existence. Advanced computer systems, rocketry science, space engineering, and solar energy technology are all part of today's satellites and, of course, absolutely key to all of this are continued advances in microwave electronics.

15.3 SYSTEM CONSIDERATIONS

15.3.1 Satellite Advantages

The communication system engineer must consider many variables when deciding whether to use a satellite or some other means of communication when planning to solve his information transmission needs. A few of the key parameters, however, that will always be reviewed, include the following:

Reliability: The ability to place long-life dependable equipment into space has been steadily improving. Useful lifetimes for many modern satellites are guaranteed for 10 years. Further, satellite transmissions are virtually unaffected by changes in the weather, time of day, or sun spot activity.

Capacity: It is axiomatic in electronics that the higher the frequency, the greater the information capacity. Thus, with microwave frequencies in the billions of cycles per second (hundreds of gigahertz), and the ever-increasing ability to build larger and more sophisticated satellites, traffic-carrying capacity has grown dramatically.

Consider this account of where the world was in 1858:

> . . . The first transatlantic cable laid by the H.M.S. Agamemnon in 1858 had a spectacular but short life. After more than a year of heart-breaking failures on the high seas, it was eventually laid successfully. The signals that trickled through it were so minute that only the most sensitive suspended mirror galvanometer could detect them.
>
> The first message of ninety words from Queen Victoria took 16½ hours to transmit. Press headlines were sensational beyond precedent. However, three days after the Queen's message, the insulation of the cable failed and it never worked again . . .

Compare this to satellites planned today. AT&T's Telstar 3, for example, when optionally loaded, will be rated at 93,600 voice circuits; all transmitted, of course, at the speed of light.

Cost: The expense of putting a satellite system in place is very high. For example, the approximate cost for putting the TDRSS system (six satellites plus earth stations) into operation was approximately $1 billion (US). However, despite their high launching costs, satellites are by far the lowest cost means of medium to long distance communications, when compared to earth-bound microwave relay systems, or undersea cables.

15.3.2 Antennas

Satellite Antennas

We will examine the most common satellite-to-earth receiving system in the world today, the *television receive-only* (TVRO) home entertainment system, to gain some insight into how any receive-transmit antenna pair interacts.

There are three basic elements that enter into the calculations for determining how much signal reaches a receiver on earth from a satellite:

1. The power level at the satellite transmitting antenna;
2. The overall gain and efficiency of the earth-station receiving antenna;
3. The quality of the earth-station receiver.

On the satellite itself, the receiving antenna is a fairly wide-beam, sculptured antenna that covers the necessary uplink frequency range with reasonable efficiency. It is a directional antenna which receives signals with a 5.9 to 6.4 GHz broadband front end.

If a satellite is equipped with a broadbeam transmitting antenna (a "global beam"), slightly more than one-third of the earth's surface will be covered by the signal from the satellite. Typically, however, satellite transmitting antennas are highly directional, high gain, narrowbeam antennas, which are focused on a narrow area like the lower half of the US, for example. This avoids the problem of wasting the satellite's valuable solar power by radiating unnecessary signals into space. Another benefit of the narrow beam is that it provides enough gain to reduce the satellite transmitting power requirements to just a few watts per channel.

Because both the receiving and transmitting satellite's antennas are directional, they both have a pattern. The center of this pattern, where maximum gain occurs, is called the "boresight." The pattern of the transmitting antenna becomes particularly important when attempting to determine the strength of a satellite signal reaching the earth. As the transmissions leave the satellite, they form a beam that covers a specific area of the earth. The energy levels of this beam are called *effective isotropic radiated power* (*EIRP*). Upon reaching earth, this energy is distributed in a pattern, with the signal strong in the center and then weaker at the edges. This pattern, referred to as a "footprint," is generally shown on a map with contour lines which connect equal levels of *EIRP* together. The levels of *EIRP* are expressed in "decibels above one watt" (dBW), and they tend to fall away from the center of the footprint pattern in descending values.

The gain of each satellite's on-board antenna is expressed in dB. Dish gain measurements are made in reference to a standard isotropic antenna, which is an imaginary antenna conceived by engineers. The isotropic antenna theoretically transmits signals equally in all directions at once. It has a gain of 0 dB. Because the real-life satellite antenna is shaped to concentrate the transmitted signal within a narrow corridor out in front of it, it can provide a much higher signal level in the desired direction than if the signal were allowed to be dispersed in all directions. For example, in a typical satellite, Westar 5, the gain provided by its antenna at boresight is 31 dBi (decibels referenced to an isotropic antenna). This provides a signal level that is 1,260 times greater than what the theoretical isotropic antenna could produce.

To determine satellite transmitted power (the *EIRP*) level of Westar 5, the 31 dBi gain of the on-board dish antenna must be added to the power level of the travelling wave tube amplifier (TWTA) of a single transponder. In Westar 5, there are 12 of these transponders, each delivering 8W at a given frequency (or 9 dBW).

Therefore, for Westar 5, the *EIRP* at beam center is:

$$\text{antenna gain} + \text{amplifier power}_{OUT} = EIRP$$
$$= 31 \text{ dB} + 9 \text{ dBW} = 40 \text{ dBW}$$

Only the vertically polarized (we will examine this concept later on in the chapter) transponder set on Westar 5 is referenced to the 40 dBW beam-center *EIRP*. The 12 horizontally polarized transponders have a portion of their signals split off and fed to a separate spot beam center on Hawaii. This separate beam reduces by about 2 dBW the North American footprint contour levels for the horizontally polarized (odd numbered) transponders.

A footprint map of Westar 5 is presented in Fig. 15.1. It shows a boresight strength of 38 dBW with concentric lines indicating 38, 37, 36 dBW, and so on towards the outer fringes. Note that these values represent power at the satellite and do not take into account the path losses incurred between the satellite and the receiving antennas. These losses run as high as 200 dB.

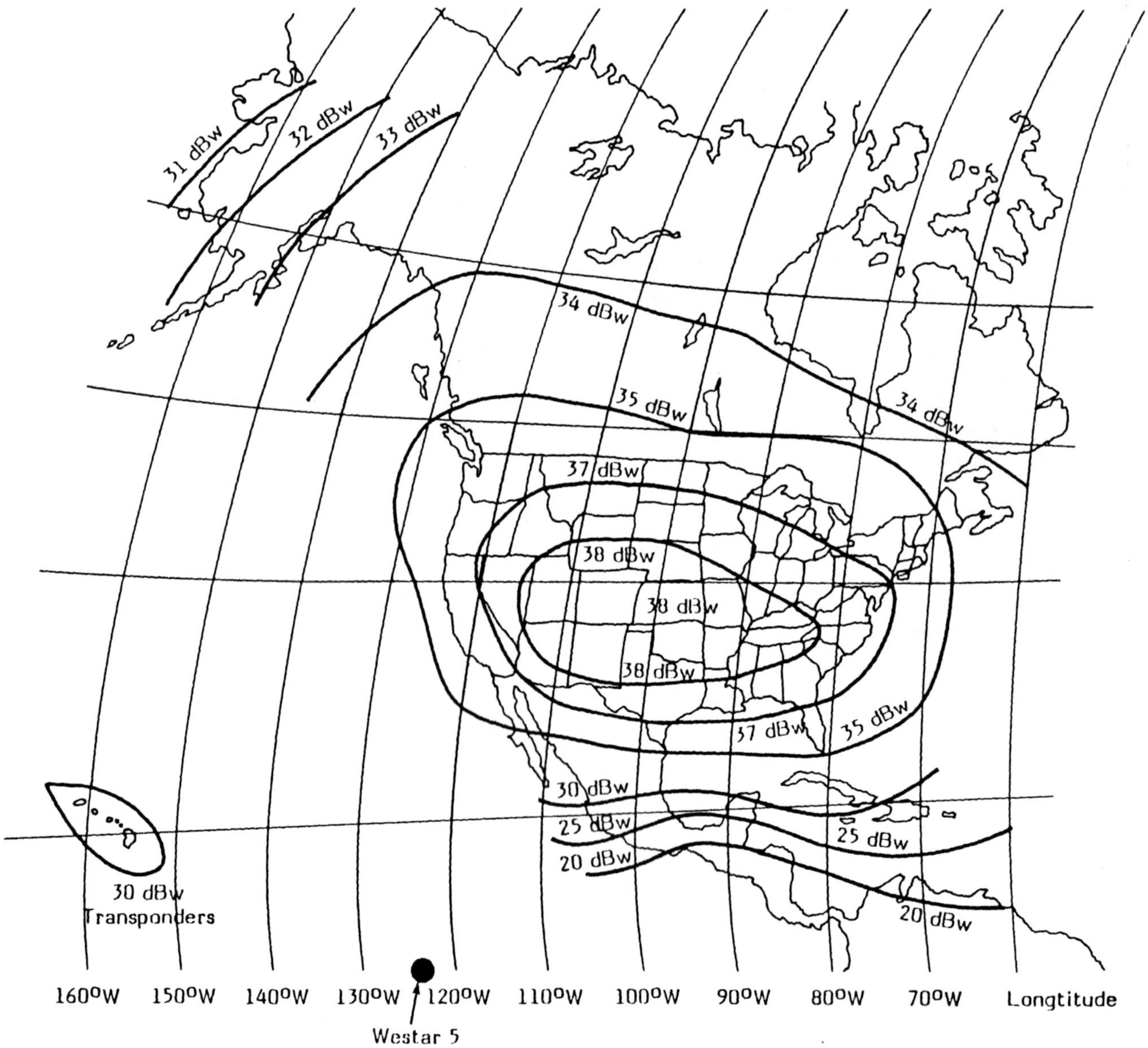

Figure 15.1 Westar 5 footprint.

SEC. 15.3
SYSTEM CONSIDERATIONS

Earth Station Antennas

The receiving antenna, collecting as much energy from the satellite signal as its size will allow, acts as the first gain stage of the earth-station receiver (Fig. 15.2 is a typical installation). Although the dish is the most obvious part of this equipment, the overall efficiency (expressed in a ratio *G/T* to be elaborated on later) is really a combination of all of the elements in the antenna, including feed horn, cabling, *et cetera*. For the great majority of all systems currently in use in the TVRO industry, the number runs about 55% at around 4 GHz.

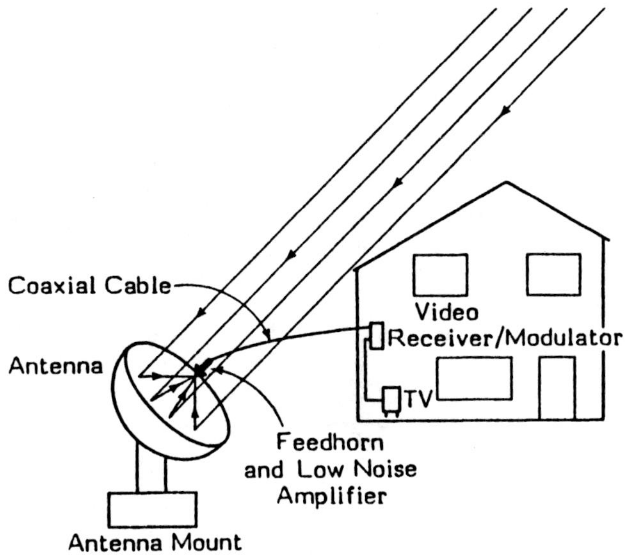

Figure 15.2 Typical TVRO installation.

The first active element to amplify the signal, the *low noise amplifier* (LNA), also generates unwanted signals (noise) and this noise contribution, expressed in equivalent *noise temperature* (degrees Kelvin) or *noise figure* (*NF*), all contribute to System Noise (see Ch. 9 on noise). Last, the elevation angle of the antenna system above the horizon can contribute to system noise as well, picking up more earth noise the lower it looks.

All these items, adding up to become the *G/T* (antenna gain minus antenna system temperature), provide a measure of merit of an antenna system. Table 15.1 shows how the gain and size of an antenna can effect *G/T* with a fixed LNA of 100K.

Table 15.1
Antenna Size vs. Gain

Antenna Diameter (ft)	Gain (dBi)	G/T (°K)	C/N (dB)
6.0	35.0	12.0	4.7
7.5	37.0	14.2	6.9
8.0	37.5	15.1	7.8
9.0	38.5	16.4	9.1
10.0	39.4	17.5	10.2
11.0	40.2	18.4	11.1
12.0	41.0	19.4	12.1
14.0	42.3	20.8	13.5

Note 1: Antenna gain computation is based on 3950 MHz and 55% efficiency.
Note 2: G/T calculations assume a 100K LNA and a look angle of 30 degrees.

SEC. 15.3
SYSTEM CONSIDERATIONS

If we were to hold antenna gain constant and lower the noise by installing a more expensive but quieter LNA, or, conversely, keeping the LNA fixed but using a larger antenna, Table 15.2 gives us guidance for deciding what level of *EIRP* is required to get an acceptable quality video signal with a typically mid-priced receiver system with an 8 dB carrier-to-noise ratio (*CNR* or *C/N*) threshold. Table 15.3 shows the relationship between antenna size (reflector diameter) and LNA temperature.

Table 15.2
EIRP to Dish Size Conversion Chart
(4 GHz Satellite *EIRP* Contours)

8 dB *CNR* Threshold Receiver

	LNAs		
EIRP (in dBW)	120K	100K	80K
40	6ft	5ft	5ft
39	6ft	6ft	5ft
38	7ft	7ft	6ft
37	8ft	7ft	7ft
36	9ft	8ft	8ft
35	10ft	9ft	9ft
34	11ft	10ft	9ft
33	12ft	11ft	11ft
32	14ft	13ft	12ft
31	15ft	14ft	13ft
30	17ft	16ft	15ft
29	20ft	18ft	16ft
28	22ft	20ft	18ft
27	24ft	22ft	21ft
26	27ft	25ft	24ft

*At 55% dish efficiency, 30 MHz receiver bandwidth, and average free-space loss values. Reception of those satellites which require a dish elevation setting below 20 degrees will need slightly larger antenna sizes than those listed above.

Table 15.3
G/T Antenna (Reflector) Size vs. LNA Temperature

	Reflector Diameter* (ft)				
LNA Temp.	8.0	9.0	10.0	11.0	12.0
120K	16.0	16.6	17.7	18.7	19.4
100K	16.5	17.2	18.4	19.3	20.1
80K	17.2	17.8	19.1	20.1	20.8
70K	17.5	18.3	19.5	20.5	21.2

*Reflector gain is rated at a conservative 55% efficiency, and reflector noise at 45° elevation.

15.3.3 Frequency Selection

Satellites operate in the microwave frequency range. This allows them sufficient bandwidth necessary to handle several television channels and thousands of voice and data transmissions simultaneously. Most commercial communications satellites operate in the C-band, 6/4 GHz frequency range (some of the newer satellites are using K_u band, 14/12 GHz range). The uplink signals are beamed to the satellite from the ground station at between 5.9 and 6.4 GHz while the satellite converts these signals to between 3.7 to 4.2 GHz and sends them downlink to earth. The satellite uses a series of repeaters or transponders which convert the signals down from 6 to 4 GHz allowing a two-way, or duplex, communication where the incoming and outgoing signals do not interfere with each other.

Both the uplink and downlink frequency bands are 500 MHz wide. This permits 12 TV channels of 36 MHz each with a 4 MHz "guard bands" between them. The rest of the band is used for ground-to-satellite command signals, satellite-to-ground acknowledgement signals, and a couple of beacons to help ground control measure the exact position of the satellite at any given moment. Generally, a satellite can be expected to relay one TV channel per transponder for a total of 12 transponders within its assigned frequency range. A new technique has recently been developed called "opposite sense polarization." Transmitted carrier signals at the same frequency are split into two senses of oscillation, one in the vertical plane, the other in the horizontal. Each is then coupled with the information stream it is to carry, and transmitted to the receiving antenna. Here, by filtering, each of the two polarizations are separated and, once again, processed independently. This procedure is schematically shown in Fig. 15.3. This technique uses the same 36 MHz wide frequency for two channels by processing one with a horizontal polarization and the other with a vertical polarization, thus allowing the satellite to double the number of TV channels it can handle. This frequency re-use scheme has proved to be highly effective.

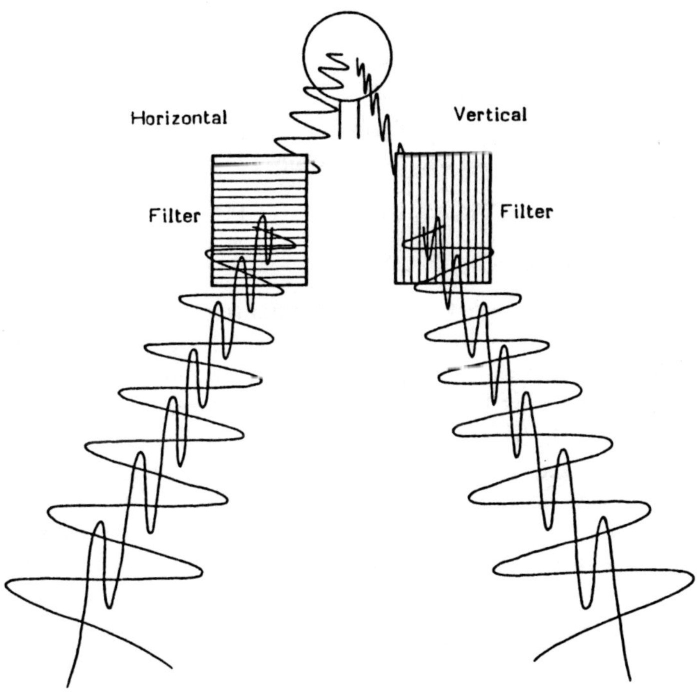

Figure 15.3 Filtering for opposite sense polarization.

15.4 SYSTEM CONFIGURATIONS

Examining the various combinations that are available to a system designer concerning how to arrange the interactions of satellites and earth stations, and possible combinations reduce to four major choices.

The main types of synchronous communications satellite systems can be divided into these groups (Sec. 15.4.1 through 15.4.4) according to the way they are configured.

15.4.1 Point-to-Point Relay Satellite

These satellites transmit between large, fixed earth stations where high capacity groups of frequency-division multiplex (FDM) signals are relayed between these points, as over the common carrier major terrestrial channels. The satellites provide two-way links which can carry voice, television, data, or any other signals that can fit into the multiplex bandwidths on earth. These links are regarded as part of a common carrier routing facility.

15.4.2 Multipoint Relay Satellites

In this application, many earth stations are able to use a given satellite with calls being switched between different satellites as desired. In this system, the satellite will relay many different groups of signals. Each earth station transmits only one group or a small number of the groups. It may receive all the transmitter groups and then send only those addressed to it over the terrestrial trunks.

15.4.3 Distribution Satellite System

This is a satellite for distributing television, radio, or other one-way radio signals such as news services or stock market information. The signal will be picked up by relatively inexpensive antennas which relay them to local broadcasting stations or directly to businesses and homes. In this system, there are a small number of transmitting stations, for example, one or two nationally, and others for regional applications. The number of receiving stations here is much larger, and these stations are lower in cost than the transmitting stations.

15.4.4 Direct Broadcasting Satellite

In this system which depends upon more powerful satellites, broadcasts are made directly into a dish antenna which may be as small as only 5ft in diameter and mounted on a user's own roof top. In the long run, it may even be possible to use rabbit-ear antennas on the living room TV set.

15.5 SATELLITE-TECHNICAL CONSIDERATIONS

15.5.1 Communication

The electronics inside most of today's synchronous communication satellites is fairly similar in design until we reach the output stages and the transmitting antennas. All received signals are processed by a broadband 5.9 to 6.4 GHz front end. The signals are amplified and fed into a converter stage that translates incoming frequency directed down to the appropriate area in the 3.7 to 4.2 GHz range. Both the uplink and downlink frequencies are divided into 40 MHz wide channels, and because the up and down frequency bands are 500 MHz wide, there is room for 12 such channels, both up and down, with each channel to 40 MHz wide. In addition, there is some room for ground-to-satellite command signals, satellite command signals, satellite-to-ground acknowledgement signals, and a couple of beacons to help ground controllers measure exactly where in space the satellite is located at any given moment.

Figure 15.4 is a block diagram of a typical 12-transponder satellite. The example shown is the Canadian Anik series. It should be noted that the input side is redundant as a security measure in case something in the broadband circuit area should fail prematurely. Once the signals have been translated down to the 4 GHz range, they are fed into the appropriate output amplifier stages which are individual traveling wave tubes. The peak power at this point is 5W (+7 dBW), and from there the 5W are coupled into the appropriate downlink transmit antennas. At boresight on the transmitter antenna, the range of power generated aided by the antenna gain is equivalent to +34 dBW to +37 dBW. This varies only slightly from satellite to satellite.

SEC. 15.5

SATELLITE—TECHNICAL CONSIDERATIONS

All of the synchronous communication satellites lie along the earth's celestial equator at four-degree spacings in longitude that allows them to transmit to the North American continent. As viewed from outer space, they would appear as in Fig. 15.5. There are similar satellites serving the rest of the world as well. Table 15.4 indicates those international satellites operating at C-band. In addition, higher frequency satellites are also being used throughout the world, and those are shown in Table 15.5.

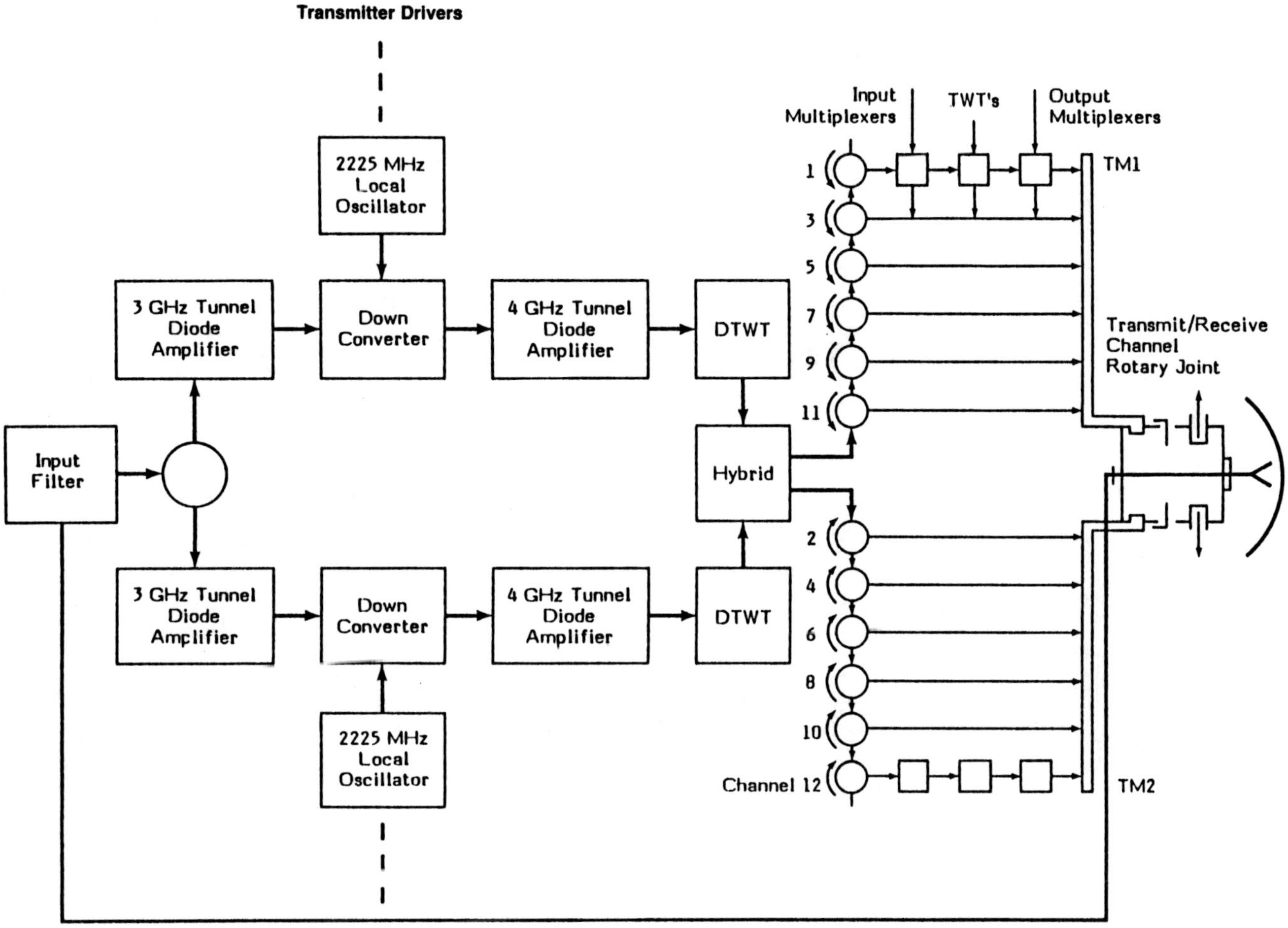

12 Channel Satellite designed by Hughes Aircraft

Figure 15.4 Typical electronics for communication satellite.

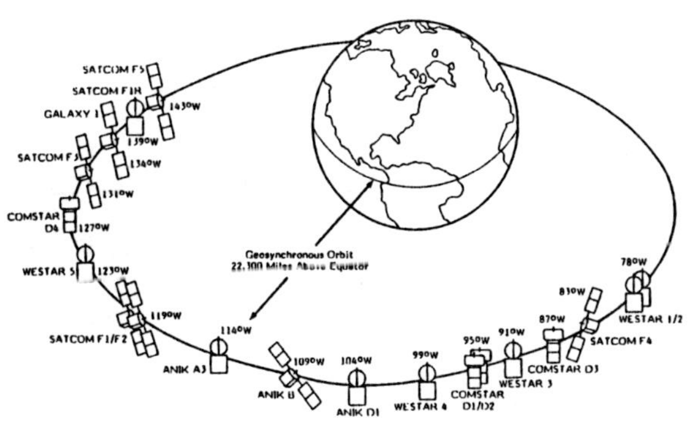

Figure 15.5 Equatorial satellite spacing over North America.

Table 15.4
International C-Band Satellites

Satellite Name	Operator	Launch Date(s)	Number of Transponders per Satellite	Uplink (GHz)	Downlink (GHz)	*EIRP* (dBW) at Edge.	Lifetime (Years)	Orbit Location (Longitude)
ARABSAT 1 & 2	ARAB SAT. COMM. CONSORTIUM	1984	24	5.495–6.425	3.7–4.2	31	7	19 & 26 E
CS SAKURA	NASDA JAPAN	1977	2	5.975–6.315	3.7–4.2	29	3	135 E
CS 2A–2B	NASD/MTT JAPAN	1983	2	5.925–6.425	3.7–4.2	29	5	130 & 135 E
INSAT 1B	INDIAN SPACE RES. ORG.	1983	12	5.35–6.475	3.7–4.2	32	7	94 E
INTELSAT IV	INTELSAT	1971–75	12	5.925–6.425	3.7–4.2	20–31	7	174 & 179 E 17 & 53 W
INTELSAT IV-A	INTELSAT	1975–79	20	5.925–6.425	3.7–4.2	20–29	7	60 & 63 E 21, 24, 34 W
INTELSAT V (9 SAT)	INTELSAT	1980–82	25	5.925–6.425	3.7–4.2	23–29	7	60, 63, 66E 18.5, 21, 22, 27, 34 W
INTELSAT V-A (6 SAT)	INTELSAT	1984	32	5.925–6.425	3.7–4.2	23–29	7	Unknown
INTELSAT VI (16 SAT)	INTELSAT	1986	36	5.925–6.425	3.7–4.2	23–29	10	Unknown
PALAPA A1–A2	INDONESIA	1976–77	12	5.975–6.425	3.7–4.2	32	7	77 & 83 E

Table 15.4
Internaitonal C-Band Satellites
(continued)

Satellite Name	Operator	Launch Date(s)	Number of Transponders per Satellite	Uplink (GHz)	Downlink (GHz)	*EIRP* (dBW) at Edge.	Lifetime (Years)	Orbit Location (Longitude)
PALAPA B1–B3	INDONESIA	1983	24	5.925–6.425	3.7–4.2	34	8	108, 117, 118 E
TELECOM 1A–1B	FRANCE	1983	2	5.925–6.425	3.7–4.2	28–35	7	7 & 10 W
RADUGA	USSR INTERSPUTNIK	1978–81	7	5.675–6.125	3.450–3.900	25–36	5	35, 53 & 90 E
GORIZONT	USSR	1978	6	5.875–6.175	3.650–3.950	35–45	5	51, 53 & 90 E
SATMEX	MEXICO	1985	24	5.925–6.425	3.7–4.2	36	7	85, 102 W
SATCOL	COLOMBIA	1984	24	5.925–6.425	3.7–4.2	36	7	70, 75 W
BRAZILSAT (SBTS)	BRAZIL	1985	24	5.925–6.425	3.7–4.2	35	7	60–70 W
SYMPHONIE 1 & 2	FRANCE GERMANY	1975	2	5.925–6.425	3.7–4.2	32	5	11.5, W
STW-2	CHINA	1986	1	6.0	4.0	50	3	70 E

Table 15.5
Other International Satellites

Satellite Name	Operator	Launch Date(s)	Number of Transponders per Satellite	Uplink (GHz)	Downlink (GHz)	*EIRP* (dBW) at Edge.	Lifetime (Years)	Orbit Location (Longitude)
CS SAKURA	NASDA JAPAN	1977	6	27.55–30.05	17.75–20.25	37	3	135E
CS 2A–sB	NASDA/MTT JAPAN	1983	6	27.55–30.05	17.75–20.25	37	5	139 & 135 E
EUTELSAT ECS 1–5	EURO CONF. OF TELECOM	1983	12	10.95–11.2 14.0–14.5	11.45–11.70	34	7	10 & 13 E
INSAT 1B	INDIAN SPACE ORG.	1983	2	Unknown	2.5–2.7	42	5	94E
INTELSAT V & V-A		1980–82	6	14.0–14.5	10.9–11.7	44	7	66, 63, 66E, 1, 18, 21, 22, 27 & 34 W
INTELSAT	INTELSAT	1986	10	14.0–14.5	10.9–11.7	44	10	Unknown
OT5-2	ESA	1978	6	14.2–14.5	11.5–11.8	37–45	5	5 E
TELECOM 1A–1B	FRANCE	1983	6	7.98–8.095 14.0–14.25	7.25–7.37 12.5–12.75	33–43	7	7 & 10W
AUSSAT 1–3	AUSTRALIA	1985–86	4	14.0–14.5	11.7–12.2	47	7	156, 160, 164E

Table 15.5
Other International Satellites
(continued)

Satellite Name	Opeator	Launch Date(s)	Number of Transponders per Satellite	Unlink (GHz)	Downlink (GHz)	*EIRP* (dBW) at Edge.	Lifetime (Years)	Orbit Location (Longitude)
EKRAN	USSR	1976–82	1	6.2	.714	50	2	99E
L-SAT	ESA	1986	2	17 & 30	12, 20	63	5	19W
TDF 1A–2B	FRANCE	1985–87	3	17.0	12.0	64	7	19W
TV-SAT A3	GERMANY	1985	3	17.0	12.0	66	7	19W
UNISAT 1–3	UNITED KINGDOM	1986–7	6	14.0–14.5	11.7–12.2	64	7	31W

SEC. 15.5
SATELLITE—TECHNICAL CONSIDERATIONS

15.5.2 Specialized Satellites

Representing the latest in state of the art satellite sophistication is TDRSS. To describe this "bird" built by TRW, and one of the first to be launched by the space shuttle, requires extensive use of superlatives. A diagram of TDRSS is shown in Fig. 15.6.

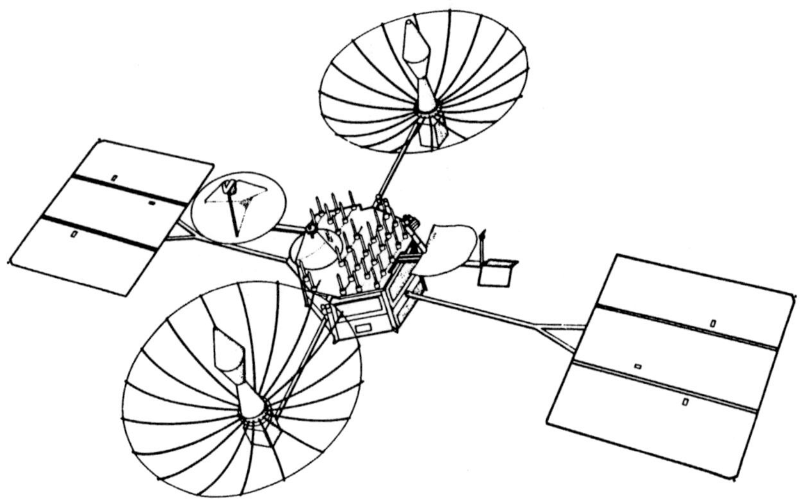

Figure 15.6 Tracking and data relay satellite system.

The size of the satellites is staggering. The solar cell array panels when deployed span the height of a six-story building. This antenna array is so large that special "sails" are employed to compensate for the pressure from sunlight that will be exerted on the antennas. Without these compensating solar sails, the flow of light from the sun (the so-called "solar wind") would exert enough pressure on the antennas to push the spacecraft out of orbit.

The weight of TDRSS is in excess of two tons (2250kg). The power generated by the solar cell array is close to 2kW. The cost of the system with 6 birds is close to $1 billion (US).

The TDRSS mission requires operation at a host of frequencies:

Frequency (GHz)	*Transponders*	*Transponder Spares*
2	1 at 26W	2
	1 at 28W	–
6/4	12 at 5.5W	–
14/12	4 at 30W	2
	2 at 1.5W	2
14.5–15.3	2 at 1.5W	2
13.4–14.05	2 at 25W	4

Antennas include two 16ft diameter, fully-steerable K- and S-band antennas for reception from other satellites in space for relay to earth. In addition, there is both a 32-element body-mounted phased array antenna as well as a small K-band steerable antenna for space-to-ground communications with control at White Sands, New Mexico. Finally, there is a small, 4ft diameter K-band antenna for commercial digital communications.

SEC. 15.6

GROUND STATIONS—TECHNICAL CONSIDERATIONS

15.6 GROUND STATIONS—TECHNICAL CONSIDERATIONS

The ground-to-satellite signal path in the 6 GHz range requires substantial transmitter power. For example, one to three kilowatts plus a large antenna gain (50 to 60 dB) is required to saturate the input of a satellite with high quality, noise-free signals. Like any relay station, the signal quality returning to earth is only as good as that initially transmitted to the satellite. On the uplink path, free-space loss approximates 198 dB.

Satellites operating today use microwave frequencies that are also used terrestrially. Problems can arise with interference between the satellite and terrestrial systems. For example, most systems use 6 GHz to transmit to the satellite and 4 GHz to transmit from the satellite to earth. These are frequencies used by many terrestrial microwave links and interference may occur, as shown in Fig. 15.7. The degree of interference depends upon the relative positioning of the interferring antenna. If they are far apart, and the antenna dishes do not point at one another, there will be no problem. However, careful positioning of the antenna and the design of directional antennas are necessary. At higher frequencies, especially those above 10 GHz, radio interference is not likely to be a problem because the antennas are more directional and therefore there is less spill-over effect. Later on in this decade, the millimeter wave frequencies will become much more common for satellite transmission.

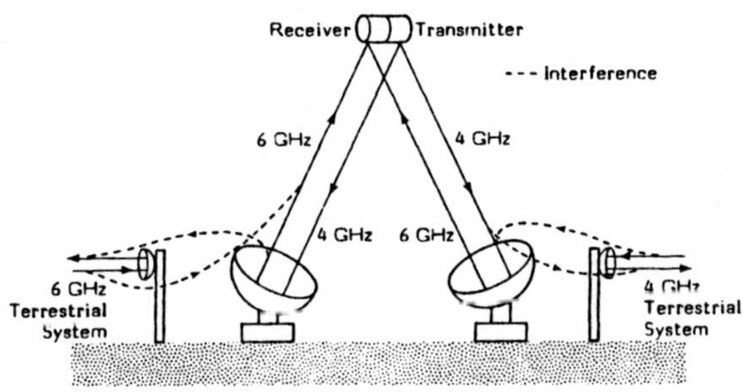

Figure 15.7 Possible interference between satellite and terrestrial links.

This capacity enhancement provided by millimeter wave systems will be necessary because the earth orbit is becoming crowded, not physically, of course, but electronically. To overcome this crowding, engineering improvements have concentrated on three ways to provide more electronic room in orbit. First, increase the directionality of the beams transmitted from the satellites, thereby reducing interference; second, increase the amount of useful information that can be transmitted within each frequency band; and, third, increase the number of frequency bands available by using higher frequencies than those in use today.

The first wave of capacity expansion consisted of using the brute force approach—simply placing more and more transponders in orbit. So, when electronic crowding began to loom as a potential problem, satellite manufacturers in the 1980s began to employ a wide variety of technological tricks to permit re-using the same frequency. First came the use of orthagonally polarized beams allowing each frequency to carry two entirely separate signals, one polarized horizontally and one vertically. Next came special separation or beam focusing, achieved by controlling the size, shape, and pointing angle of the antenna so that different beams at the same frequency can be transmitted to different ground stations. The INTELSAT V satellite now in orbit to serve the 108 nations of the International Telecommunications Satellite Organization (INTELSAT) was the first operational system to combine these frequency re-use mechanisms.

The largest increase in satellite communication capability by far will be achieved by using higher frequencies. Although most transmissions now utilize C-band, some current satellites use K_u band transponders (11 to 14 GHz). The next generation, however, will go to the much higher frequencies in the K_a band (20 to 30 GHz).

The big disadvantage of a millimeter wave frequency is that millimeter waves are heavily absorbed by clouds and rain, and to some extent by the atmosphere itself. Because of this absorption, higher satellite power will be needed. Nevertheless, the advantage of using millimeter wave frequencies instead of lower frequencies are great. For example:

Bandwidth: The bandwidths available are much higher. Therefore, satellites with very high channel capacity can be constructed.

Interference: Microwave interference increases as the radiated power from the satellite goes up, as it must do if we are to achieve low-cost earth stations. The microwave band will, in other words, most certainly become congested just as the VHF and UHF bands are today. This will require using those higher frequencies which are not being utilized today.

Frequency Strength: Because the millimeter wave beam from the satellite can be much more narrowly directed, this means that the receive signal will be stronger. It also means that the same frequency can be used at different earth stations without interference.

Satellite Spacing: The number of satellites in orbit can be much greater. With satellites using the same microwave frequencies as currently employed, it has been estimated that not more than a total of 75 could be in orbit without causing undue frequency interference. The spacing, however, can be closer using millimeter waves. It has been estimated that 95 satellites spaced 1° apart could serve the North American continent if millimeter waves are used. In addition, other satellites using frequency bands other than millimeter waves could also be operating in orbit without interference, and this would greatly enhance the total number of satellites available.

To overcome the disadvantage of millimeter waves caused by absorption of rain, clouds, and even water vapor and air, higher transmitter powers will have to be used. Even then, the beam will occasionally be attenuated badly or even blotted out by intense storms.

Gentle rain may spread over a wide area, but severe storms which cause the chief problem are almost always small in extent. There would be considerable benefit then, of being able to switch the transmission path between two earth stations some miles apart. Stations using two antennas that are several miles apart, linked to the same control center, are called "diversity" earth stations and may become common with millimeter wave systems. It is even possible that, for future networks, some transmissions will not be in real time like our telephone calls, but will include transmission and storage of data, movies, facsimile, mail, *et cetera*. If this is the case, the duplexing over earth stations to avoid storms will not appear too wasteful, and the drop in capacity caused by a storm will merely delay the non-real time transmission.

Even the enormous capacity of K_a band cannot satisfy the expected global demand for satellite communications services forever. Within a few decades, it will be necessary to introduce the multipurpose space platform, a new concept in satellite communications. These larger, more sophisticated, higher power spacecraft could deliver such advanced capabilities as electronic mail and news services, personal wrist-radio communications, and a personal navigation system contained in a wrist-watch sized earth station, able to pinpoint the location of an automobile, small boat, or an individual bicycle to within a few feet. Also planned is holographic teleconferencing in which a single life-like pseudo-three-dimensional meeting can be held using a number of widely dispersed conference rooms simultaneously, as well as universal individual two-way television.

CHAPTER 16

ELECTRONIC WARFARE

Frederic H. Levien

16.1 INADEQUATE EW MEANS DEFEAT

The term electronic warfare has begun to take on new meaning as more and more, the time frame to measure the length of a battle is more likely counted in seconds, rather than hours, days, or weeks as in times past.

The successful (and unsuccessful) consequences of the application of electronic warfare often grabs headlines. In the early summer of 1982, as a concerned world watched and waited, Israeli defense forces invaded Lebanon. Thus began a major conflict against military elements of the Palestinian Liberation Organization and the armed forces of Syria. By mid-June, Israeli forces reported the destruction of 86 Syrian aircraft, including Soviet-built Mikoyan MIG-23 fighters and five French-built Aero Spatiale Gazelle attack helicopters. The Israelis reported that they in turn had lost only two helicopters and one US-built A-4 fighter bomber. A stunning defeat and an incredible victory. What was the decisive factor? *Electronic warfare.*

On May 4, 1982, a British Type-42 destroyer, HMS Sheffield, was destroyed by a sea-skimming French-built Exocet Missile. The military world was shocked, for this type of ship was supposed to constitute a main fleet defense against air attack in cooperation with airborne early warning aircraft to detect low flying enemy aircraft. Unfortunately for Sheffield, there were no airborne early warning radars on board in operation on May 4th when the Exocet missiles were sighted close-in. The Sea-Dart missile system on board Sheffield, designed to engage aerial platforms at a distance, simply was unable to get on target in time because of the slower reaction time of Sea-Dart.

This costly lesson helped underscore the reality that electronic warfare is a new but utterly deadly battlefield, where victory or defeat may come in a matter of seconds, or even microseconds.

16.2 TYPES OF ELECTRONIC WARFARE

The term "electronic warfare" (EW) is now commonly accepted in the literature and in our day-to-day discussions. However, we must be careful to note that EW is not some means of combat using electrons as a weapon. Rather, it is a form of conflict which uses the entire spectrum of electromagnetic radiation as a battle field.

The standard definition used by NATO states that electronic warfare is: "That division of the military use of electronics involving actions taken to prevent or reduce an enemy's effective use of radiated electromagnetic energy, and actions taken to ensure our own effective use of radiated electromagnetic energy."

16.2.1 Electronic Countermeasures

Electronic countermeasures (ECM) is the general term used to describe the attempt by targets to deny information the radar is seeking. This denial of

SEC. 16.2
TYPES OF EW

information is achieved by using three basic types of technology (1) jamming; (2) deception; and (3) elimination.

16.2.2 Jamming

To help understand the basics underlying jamming technology, we will once again consider the basic radar equation. Also visualize the situation where you are the transmitting radar signal, and the target is transmitting a jamming signal that it hopes is of sufficient power to saturate your receiver with unintelligible noise. The fundamental question the radar system engineer asks is, "When will my signal be strong enough to see through the target's jamming signal?"

Given the conditions:
At the radar:

P_r = Power of the radar signal,
G_R = Antenna gain at the radar,
λ = wavelength of the signal.

At the target:

P_j = Power of the jamming,
G_j = Antenna gain of the jamming,
σ = target cross section.

Distance between radar and jammer is R.

It is possible to determine the power of the echo signal received by the radar from the target, P_{Rr}. The equation for this is given by

$$P_{Rr} = \frac{P_R G_R^2 \sigma \lambda^2}{(4\pi)^3 R^4} \qquad (16.2.1)$$

This is just the simple form of the radar equation for the case where the same antenna is used for both transmitting and receiving. We can also establish the power of the jamming signal received by the radar. It is given by the equation:

$$P_{jr} = \frac{P_j G_j G_R \lambda^2}{(4\pi)^2 R^2} \qquad (16.2.2)$$

If the jammer power must be radiated over a broader bandwidth than the bandwidth of the radar receiver (this would happen if the radar were frequency agile), the effective signal at the radar would be reduced by the ratio of the bandwidths:

$$P_{jr} = \frac{P_j G_j G_R \lambda^2 (B_R/B_j)}{(4\pi)^2 R^2} \qquad (16.2.3)$$

The ratio P_{jr} to P_{Rr} is called the jamming to signal ratio, or J/S,

$$J/S = \frac{4\pi P_j G_j R^2 (B_R/B_j)}{P_R G_R \sigma} \qquad (16.2.4)$$

When $J/S = 1$, the echo signal will be as strong as the jamming signal. This is called the "burn-through" point. We can calculate the range at which burn-through occurs:

$$J/S = 1 = \frac{4\pi P_j G_j R^2 (B_R/B_j)}{P_R G_R \sigma}$$

$$R = \left(\frac{P_R G_R \sigma (B_j/B_R)}{4\pi P_j G_j}\right)^{1/2} \tag{16.2.5}$$

Example 16.1.2.1

Using the above information in an example, given a radar transmits 100kW of power, its antenna gain is 30 dB, and the jammer bandwidth is 10 times the radar bandwidth. For an aircraft with a radar cross section of $1m^2$ and carrying a 100W jammer with a transmitting antenna gain of 10 dB, at what range will it be visible to the radar?

Solution:

$$R = \left(\frac{1 \times 10^5 W \times 1 \times 10^3 \times 1m^2 \times 10}{4\pi \times 1 \times 10^2 W \times 1 \times 10^1}\right)^{1/2}$$

$$= 282\text{m}$$

In a basic noise jammer system, the threat bearing frequency and priority are determined in the threat analysis system. If noise jamming is determined to be the appropriate countermeasure, noise is generated over the necessary band, amplified and transmitted out of the appropriate transmitting antennas. The ultimate purpose in jamming is to radiate power so that target echos are masked, and therefore not discernable by the radar. There are two types of jamming currently employed: *barrage jamming,* where the jamming energy is spread in frequency to deny the radar the use of multiple frequencies, and *spot jamming,* where the jammer's entire power output is concentrated in a very narrow bandwidth, ideally, identical to that of the victim radar.

The sensitivity of most microwave radar receivers is generally limited by noise. If one uses external means such as a jammer, to raise the noise level, the sensitivity of the radar is further degraded. When the main beams of a radar antenna are illuminating the jammer source, it becomes very difficult to keep the noise out. Although the direction to the jammer can be determined, its range and the ranges of any targets masked by the jammer noise cannot be determined.

In the broadband barrage jammer, the advantage is simplicity. It is useful when exact threat frequency is unknown and these jammers are generally small both in size and weight. The disadvantage is lower jammer power density. For example, 100W P_j over bandwidth of 1000 Mhz gives P_j density of only 0.1 W/MHz. The microwave devices used to generate this power are typically either high power crossed-field noise generators, or a noise source followed by a TWT or cross-field amplifier chain.

A jammer whose noise energy is concentrated within the radar receiver bandwidth is called a spot jammer. If allowed to concentrate large amounts of power entirely within the radar bandwidth, the spot jammer can be a serious threat to the radar. This threat can be overcome by forcing the jammer to spread its power over a much wider bandwidth. This can sometimes be accomplished by changing the radar frequency from pulse to pulse in a random fashion over the entire tuning band available to the radar. This type of rapidly changing frequency radar is called "frequency agile."

For the spot noise generator, the advantage is it can concentrate P_j on threat frequency (i.e., 100W over 20 MHz or 5 W/MHz). Disadvantages include the fact that it is more sophisticated and requires a receiver plus a larger and heavier system. The microwave devices used for this application generally are backward wave oscillators or a noise source followed by TWT or cross-field amplifier chain.

For an appreciation of how a jammed signal might look to an operator manning a PPI (plan position indicator) scope, we can refer to Fig. 16.1. A PPI scope is like a TV screen that is mounted horizontally where the operator is viewing it as a "picture" of what is happening on earth as though seen from outer space.

In Fig. 16.1A, we see the antenna pattern indicating the gain of the main (ML), side (SL), and rear (RL) lobes. The target is shown approaching from the east, with azimuth fixed as 090°. AZIMUTH 000° is at the top of the page.

Figure 16.1B is a plot of the antenna gain as a function of azimuth with 0° shown lying between the two side lobes facing the top of the page. A "fully jammed" signal, where $J/S = 1$, is shown saturating the main lobe. With this condition however, because the gain of the rear lobe is 20 dB down, the radar could still be effectively used on this azimuth. However, if the J/S is increased to $J/S = 100$, then the jamming signal would be great enough to render useless even sidelobe detection.

Figure 16.1C is how the scope face would look to an operator, peering at it with three different levels of jamming power ($J/S = 1$, 100, 1000) and the target approaching from the East, at an azimuth of 090°, or the right edge of the page.

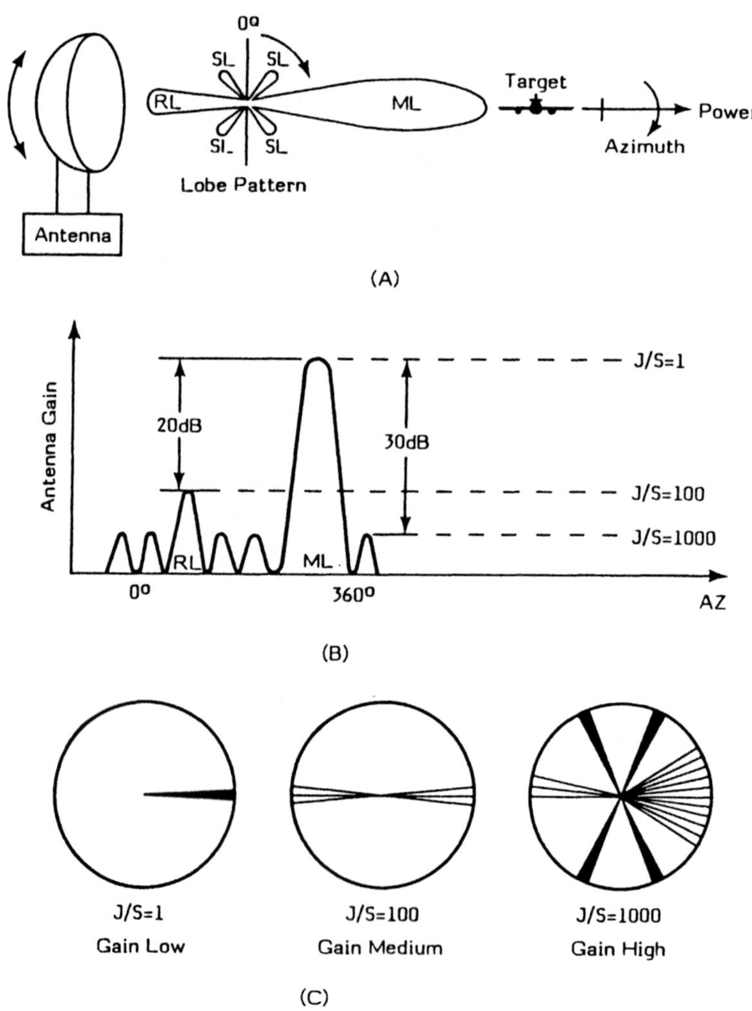

Figure 16.1 Analysis of radar PPI presentation while being noise-jammed: (A) antenna gain pattern; (B) antenna response versus azimuth; (C) PPI presentation with jamming.

16.2.3 Deception

Long before radar had become a mainstay of Naval operations, visual sighting was the only means of locating and aiming your guns at an enemy. One way to confuse enemy gunners was to "lay smoke!" That is to generate intentionally large clouds of billowing smoke, behind which friendly forces could then deploy, sight unseen, and thereby avoid enemy fire until ready once again to come forth and do battle.

There exists today the electronic equivalent to "smoke," it is called *chaff*. It consists of bundles of metallized strips, which are cut to a length so as to be resonant at a missiles radar frequency. These reflective pieces are carried in canisters aboard an aircraft, and are fired ahead of or behind when trying to elude enemy radars and missiles. By using an explosive charge, these canisters are ejected and expanded into a cloud over a large dispersion area, thus attracting the missile radar, which locks onto it. Behind this electronic smoke screen, the target aircraft then beats a hasty retreat.

In recent times, another means of deception that has become increasingly popular is the remotely piloted vehicle (RPV). Sometimes used as well are "drones." An RPV is an aircraft platform that is under remote but direct control, while a drone functions with a preset sequence and has no remote control.

These decoys are usually smaller than a typical aircraft target, but are made to appear larger electronically. The intention is to trigger "on" the enemy radar, thus forcing them to reveal their presence, location and operating characteristics. All of this information, of course, being vital to those forces trying to counter such a radar threat.

In late spring of 1982, during the Israeli invasion of Lebanon, it was announced that 19 Soviet-built Syrian SAM-6 surface-to-air missiles in the Bekaa Valley area had been destroyed by the Israeli airforce without losses. Here the Israelis had employed their Mastiff RPV (as well as drones) to ascertain the microwave radio frequencies used by the Syrian SAM-6s. Two Israeli Grumman E-2C Hawkeye aircraft obtained electronic bearings of the Syrian missile radar system, allowing them to plot their exact location. Israeli aircraft then destroyed the sites with rockets riding a microwave beam to the SAM-6 sites.

16.2.4 Elimination

Another approach to the problem of foiling enemy radars is to drastically reduce the radar cross-section of a target. A great deal of research has been done in this area, with the results conjured up in a single word "Stealth."

Techniques such as fabricating airframes from non-metallics, such as monofilament carbon-reinforced material, and new types of fiberglass have proved to be very effective. Ask any US Navy man who served in Korea or Vietnam as a Deck Officer, how often he nearly ran down a wooden junk or sampan, because they failed to show up on the ship's radar.

In addition, new aircraft are designed to minimize the amount of flat surface area they present which might act as a good radar signal reflector. Where possible, all surfaces are either cylindrical or conical, thus reducing possibility of reflections. Further, special electromagnetic absorbing coatings are being placed over the outside of planes. Thus, much like a man wearing a black suit at night, the greater portion of the energy (light rays for the man, radar waves for the aircraft) is absorbed instead of reflected, making it difficult to locate the object.

16.2.5 Active Deception

Two new electronic techniques have been developed whereby false signals are returned from the target to the victim radar. These signals have characteristics similar to those of the radar, thereby deceiving the radar into erroneous conclusions about range, azimuth, or size.

SEC. 16.2
TYPES OF EW

Range Deception

Range deception is used to foil missile guiding radar systems where the tracking radar guides the missile (or other defensive measure) to the target in range by "locking" a range gate onto the target. This range gate straddles the target echo and its position is relayed to the missile to be used for intercept information.

A deception jammer, called a "range gate stealer," attempts to break the tracking lock on itself by capturing the radar's range gate with a false echo and then moving it off to a false range (time) location. Initially, the jammer transmits a single pulse in synchronism with those pulses received from the tracking radar.

As shown in Fig. 16.2, the received pulse from the radar is repeatedly passed around a delay loop and sent back to the radar shifted in time and much larger in amplitude than the skin return (real echo). The radar range gate locks onto this false pulse. The next radar pulse is again received by the target's electronic equipment and passed around the delay loop. The delay time for this pulse is made to be longer than the last so that the echo is returned to the radar at an even more incorrect time. This process is continued until the radar range gate has been walked off to a safe distance. The jammer on the target is then turned off and the radar is left without a target.

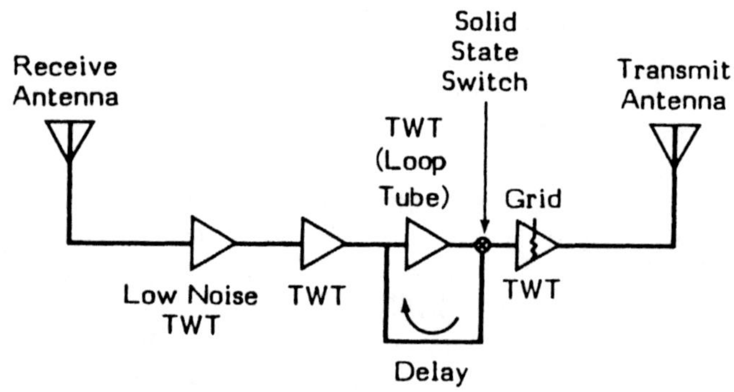

Figure 16.2 Range gate stealer system.

Velocity Range

In the deceptive velocity jammer operation, the CW illuminator signal is detected by the jammer and a false, stronger Doppler shifted signal is sent back to the radar. The radar locks onto the incorrect doppler signal and the jammer slowly sweeps the false signal's frequency away from the actual doppler frequency of the target. When the radar has been led far enough away in frequency, the jammer is turned off and the radar is once more left without a target.

Azimuth Deception

Conical scan tracking radars are very effective and are in widespread use for locating radar targets. The technique uses a pencil beam, rotating rapidly at a slight angle to its pointing axis. This is done by either rotating or nutating the feed horn or a subreflector, resulting in a cone like pattern being transmitted toward the target (see Fig. 16.3).

The echo returns from the target contain an amplitude modulation due to the nutation. The amplitude of the successive echos will be equal only when the target is exactly on the axis of the antenna. Figure 16.4A illustrates the amplitude of an echo pulse train with a target illuminated as shown on Figure 16.4B.

Figure 16.5 illustrates how an inverse gain repeater can be used to counter a conical scanning tracking radar, by transmitting false pulses. the method takes the conical scan rate detected by the jammer and responds with an inverse modulation, which is sent back to the radar with an amplitude much stronger than the skin return, thus the radar thinks the target is on boresight when it is really much off the boresight axis. The target is not where the radar thinks it is.

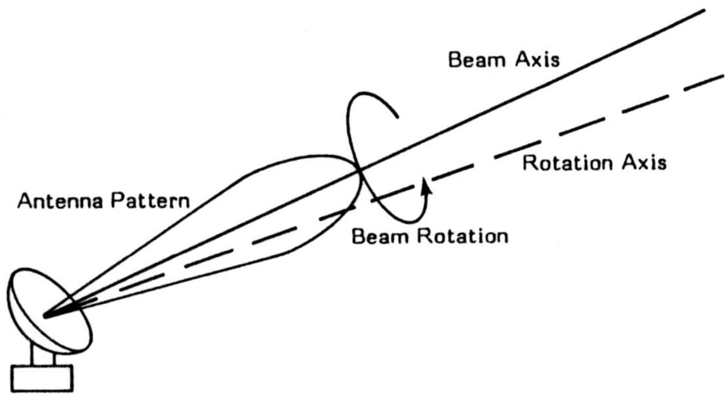

Figure 16.3 Conical scan tracking radar.

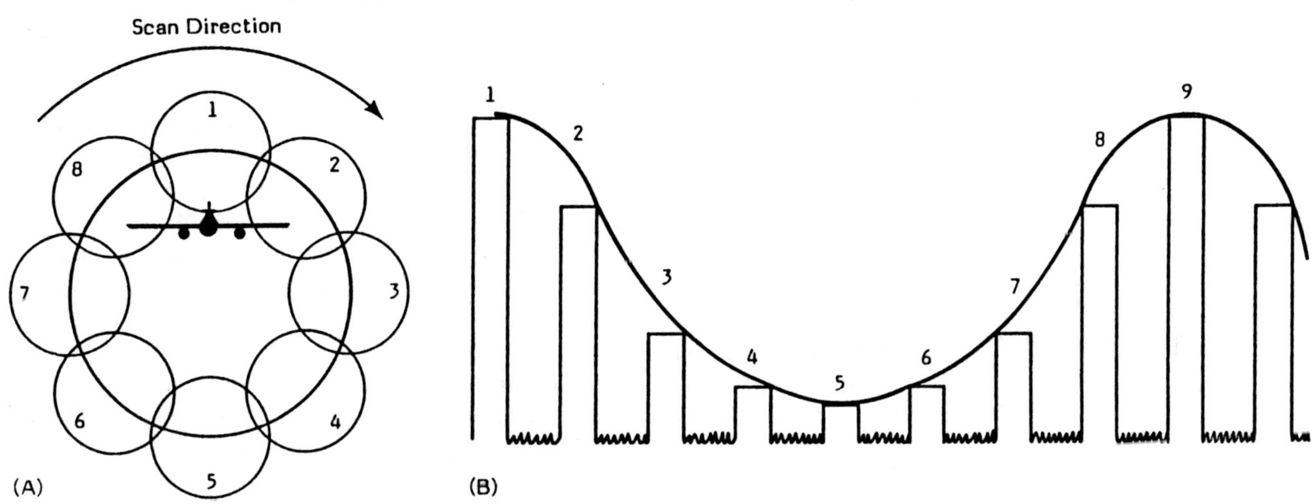

Figure 16.4 (A) Echo pulse train amplitude. (B) Conical scan target illumination.

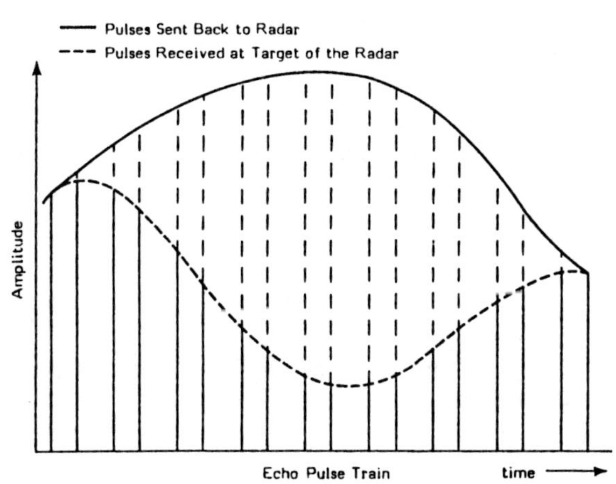

Figure 16.5 Inverse gain repeater.

SEC. 16.3
EWSM

16.3 ELECTRONIC WARFARE SUPPORT MEASURES

No chapter on EW would be complete without looking into the future. "Star Wars" is a term that holds as many interpretations as there are people learning the term. However, one aspect of it must be clearly stated. It is not a dream (or nightmare) of the future. It exists here and now, today.

Satellites of both the Soviet Union and the US continually circle the earth with their sole mission to search, intercept, locate, record and analyze radiated electromagnetic energy for the purpose of exploiting such radiations in support of military operations.

Photography was one of the earliest tasks assigned to these surveillance craft. In the late 1950s and early 1960s, film from satellite or reconaissance missions, was periodically ejected and aircraft specially equipped, snagged packages of film still in the air and slowed by parachutes. Now, however, the photography is becoming available through "real time" transmission. A Kodak imaging system working on the BIMAT process, feeds a reel of film through two optical systems with long and short focal length lenses, respectively. Held between the camera and a specially automated processor, the film when ready is wound on to a gelatin surface coated with a combined developer and fixer. The surface carrying a negative image is then withdrawn and moved to pads where moisture is soaked away prior to drying by a small heater. The negative is then wound on a take up reel and exposed to a CBS line scan tube. With this device a narrow beam of light is focused onto the film by a special lens which moved the spot across the film in a series of spiralling scans, moving down the film until the complete image has been scanned. The modulated beam is picked up on the opposite side of the film by a photo-multiplier tube which generates an electrical signal proportional to the intensity of the beam observed through the negative. Amplified in the communication system of the satellite, these signals are beamed by microwaves to receiving stations on the ground, where the live scan transfer of the photographic image is completed and they are turned back into pictures.

The photography becoming available today through real time transmission is expected to be at least 10 times more accurate than the satellite photography available during the 1970s and early 1980s.

In the initial series of SAMOS Satellites, launched in January 1961, the optimum resolution achieved was 8.2ft at no more than 124 miles distance. That is, a photo interpreter could, under the best of all conditions, discriminate two spots 8.2ft apart. By the early 1970s, this was reduced to 23.6in. By the early 1980s, technology and new satellites had once again improved this resolution to 5.9in at an altitude of 248 miles.

With this sort of resolution, a skilled military analyst could determine the type and caliber of a piece of field artillery or the size of engine nozzles of a fighter aircraft or the size of a missile on a warship.

16.4 KILLER SATELLITES

The control of outer space may in future conflicts tip the balance of battle in favor of one or the other side. The enemy, denied the electronic eyes and ears to continually monitor enemy force movements, the relay stations to communication, or the beacons by which to navigate his sophisticated weapons, could be rendered impotent. Clearly, the denial of the use of satellite information in the space segment may prove a fundamental blow to battle management and eventually determine the outcome of any conflict.

As with all weapons systems since the dawn of time, countermeasures have been developed to eliminate the potential for using satellites for electronic warfare in the event of conflict.

The USSR, beginning in the late 1960s has been heavily involved in its own program of anti-satellite development projects. The first successful demonstration of a killer satellite technique came in October 1968, when COSMOS 248 was put into approximately circular orbit about 323 miles above earth. A day later, COSMOS 249 followed into an elliptical 85 × 158 mile orbit. On command, the interceptor portion of the chase vehicle separated from the main rocket and moved to a highly elliptical orbit. This new orbit placed COSMOS 249 in a position to intercept COSMOS 248 from above with a high relative speed at close approach. Four hours after getting into this orbit, COSMOS 249 swept past its target and was detonated a safe distance away, proving that all the skills needed for a fully effective killer satellite were now in place. Since 1976, only one calendar year has passed without some form of additional Soviet killer satellite testing.

Toward the end of the decade, a flurry of accelerated spacecraft flight operations by the Soviets caused serious concerns in the US. Concerted efforts to agree on a limitation on the operational deployment of killer satellites led to nothing, the Soviets refusing to abandon a system they had already apparently perfected.

US technology developed for the anti-ballistic missile (ABM) role has been successfully applied to anti-satellite operations. By marrying the modified rocket motor of a short range attack missile with the fourth stage of a Scout launcher, the sequential thrust of the two motors accelerates a small impact head out of the earths atmosphere, and on course for a collision with some designated target in space. The anti-satellite (ASAT) device, approximately 18ft long and weighing 2645 pounds, is carried beneath the fuselage of a converted F-15 Eagle fighter. It is released close to the aircraft's operating ceiling (70,000ft). At this height, in the rarefied outer layers of the earth's atmosphere, the ASAT is coupled to a guidance system which uses an active infrared homing sensor in the nose of the impact head. Small rockets in the impact head are used to fine tune the trajectory and to fire collectively for a terminal accelleration, designed to provide ram velocity at the satellite. It takes only one direct hit from even the comparitively small ASAT head to disable a satellite.

The decade of the 1980s will present a major shift in use of satellites in space for military applications. From early conceptual ideas in the 1960s and 1970s, we have moved to a monitoring and support role and now to an active participation in space conflict. A final shift may come when killer satellites become permanent residents in space, both as guardians of the satellites they protect, and as potentially active participants in rapid destruction of the space capability for an opposing force.

Already, ground based deep-space surveillance cameras reveal unknown pieces of hardware that must be identified either as useless pieces of debris, or potentially active satellites waiting in silence for the day when they may be needed. Passive and active satellites, some of them killer satellites, may be hiding among the refuse of space flights past. It is known that steps have already been taken by the Soviets to guard against retrieval by systems like the shuttle. Explosive charges have been placed aboard certain defense satellites in the belief that, knowing this, Americans would never risk destroying a shuttle in a vain attempt to snatch a Soviet military satellite from orbit.

CHAPTER 17

MANUFACTURING ACTIVE MICROWAVE DEVICES

Ronald W. Lawler

17.1 INTRODUCTION

Fundamentally, there are three types of active microwave devices: electron vacuum tubes, hybrid integrated circuits and monolithic integrated circuits. The function of electron vacuum tubes is generally based upon modulating the velocity of an electron stream. They consist of two major types: linear beam and crossed-field devices. Klystrons, traveling-wave amplifiers, and backward wave oscillators are the major types of linear beam devices. Magnetrons and crossed-field amplifiers are the major types of crossed-field devices. Figure 17.1 illustrates a six-channel satellite communications uplink klystron (VA-936).

Figure 17.1 Six-channel satellite communications uplink klystron (VA-936).

Electron vacuum tubes are fabricated by joining precision fabricated parts. The vacuum envelopes are joined by a series of welding, brazing, and metalization operations. Figure 17.2 illustrates the cross section of a klystron.

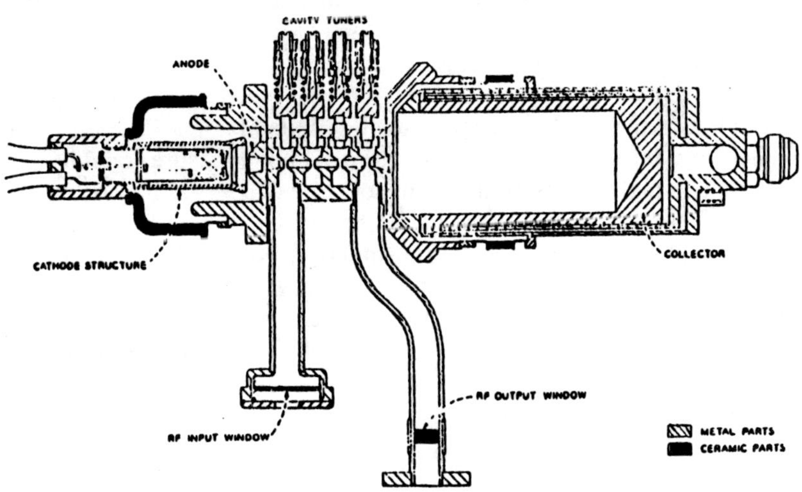

Figure 17.2 Cross-sectional Drawing of the VA-849. Each cavity is tuned by deforming a sidewall of the cavity. (Courtesy of Varian Associates)

Hybrid integrated circuits are manufactured on substrates with individual components attached. The components are interconnected by deposited metal transmission lines. These interconnections can be created by "thick-film" techniques such as silk screening or by "thin-film" techniques such as evaporation or sputtering. Monolithic integrated circuits have all active and passive components formed in and on the surface of a single semiconductor. Figure 17.3 illustrates a mixer preamplifier constructed by using hybrid integrated circuit technology.

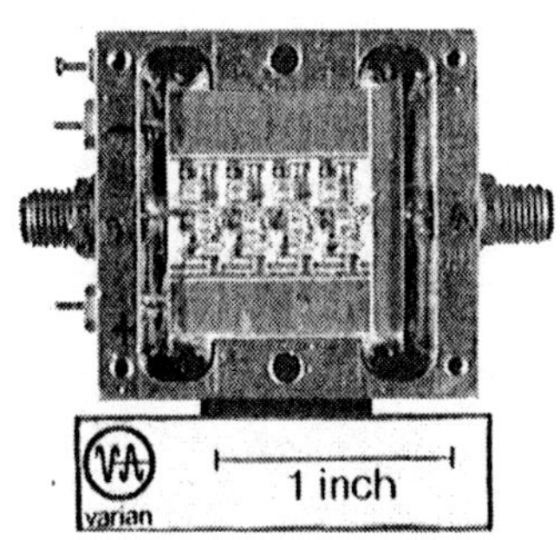

Figure 17.3 Hybrid mixer preamplifier.

SEC. 17.2

JOINING MICROWAVE DEVICES

These technologies utilize a vast number of materials and processes. These devices are used for a wide range of amplification and control functions. They provide a substantial range of power, frequency, bandwidth, efficiency, and noise figure options. Selection of the appropriate device is a matter of economics, space, operating requirements, and environment. Manufacturing these devices involves many different processes. Even within a given class of microwave devices, a substantial number of processes are both technically and economically practical. Despite this range of processes, several functions are common to manufacturing most active devices. This chapter will focus on three of these common functions: joining microwave components, yield management, and process control.

17.2 JOINING MICROWAVE DEVICES

All metals can be joined, some easier than others. Soldering is the process of joining two metal parts with a third one with a lower melting point. The temperature required for soldering is generally less than 450°C. Brazing is the process of joining two closely adjacent metal surfaces. The surfaces are joined when molten filler metal is drawn by capillary attraction into the space between the two adjacent surfaces. Brazing is performed at temperatures above 450°C. The brazing temperature should be between 50°C and 200°C lower than the melting point of the brazed parts. Welding is the process of creating a localized coalescence of metals or nonmetals produced either by heating the materials to suitable temperature, with or without the application of pressure, or by the application of pressure alone, and with or without the use of a filler material [1]. Welding temperature is a function of the melting point of the materials to be joined. Most welding is done at temperatures exceeding 1000°C.

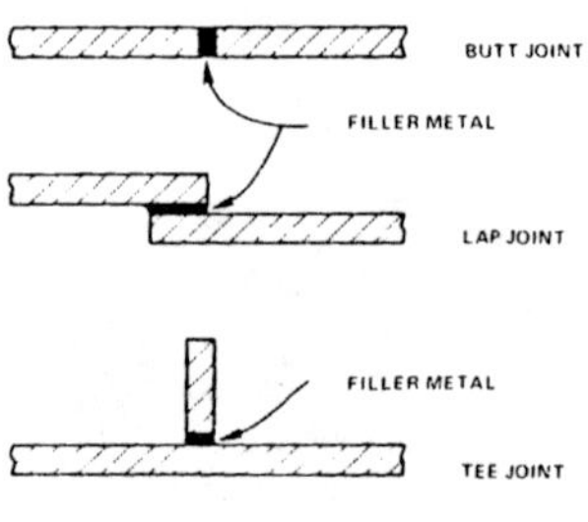

Figure 17.4 Joint types.

Careful joint design and proper fit are critical to successful joining. A successful joint provides the strength required, while limiting contamination to acceptable levels. Joints may generally be classified as butt, lap, or *tee* shaped. These joints are illustrated in Fig. 17.4.

Obviously the variations on these three types of joints are limitless. The joint design should reflect the characteristics of the base metals, the impact of the joining cycle, and the final joint strength required. Major base metal characteristics to be considered include melting point, vapor pressure, coefficient of expansion and thermal conductivity. The joint design should also consider the intended requirements of the operating environment such as mechanical performance, electrical conductivity, vacuum integrity, corrosion resistance, and operating temperature.

The base metals to be joined greatly influence the joint design. For example, if the metals are the same, then both may be the same thickness because they expand at the same rate. If the two metals are different, then the more ductile of the two should be made thinner. To illustrate this, an L-shaped joint that combines copper and 1020 steel is used. In this case it is proper to make the joint as illustrated in Fig. 17.5.

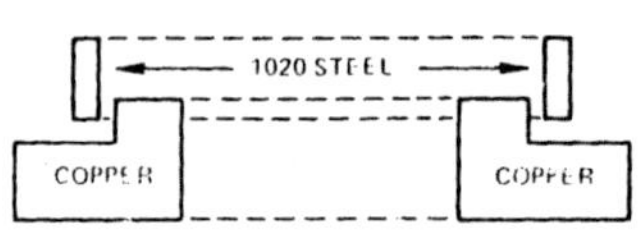

Figure 17.5 Relative placement and thickness of metals with dissimilar coefficients of expansion.

The reasoning behind this is that the joint is heated to join the metals, the copper will expand much more than the steel. At 1000°C copper will expand about 0.018″ per inch of length while 1020 steel will expand about 0.012″. The copper will also attempt to contract more than the steel as the joined metals cool. Being thinner, weaker, and more ductile than the steel, the copper will be overridden by the steel and will contract only as much as the steel. In this way the joint will be somewhat stressed, but the copper cannot exert much stress on the joint before it (the copper) starts to yield.

By reversing the position of the two metals and placing a section of copper on the outside of the steel (Fig. 17.6), different results would be expected.

Although the fit may be close at room temperature, the copper will expand much more per degree of temperature change than the steel, and the gap would be enlarged at brazing temperature. It is questionable whether capillary action would fill the enlarged gap with filler material. If the material were attracted to the gap, it would be spongy and would not achieve its maximum strength or vacuum integrity.

Figure 17.6 Relative placement and thickness of metals with dissimilar coefficients of expansion. (Compare with Fig. 17.5)

Filler metal usually flows toward the source of heat. Therefore, the filler metal is usually placed inside the assembly so that the heat will draw it to the outside, where it can be seen as a fillet (see Fig. 17.7). This will generally produce a good joint. These joints are recommended because they are easily inspected and form smooth internal surfaces at the fillet.

Consideration must be given to the effects of mass, surface heat transfer or emissivity, specific heat, and thermal conductivity upon the rate at which the assembly attains temperature.

Joint strength is proportional to the area joined. Increasing the joint area entails drawing the filler metal a relatively long distance. As the filler metal is drawn into the joint it may interact with enough base metal to prematurely coalesce. This problem is overcome by increasing the space between the parts being joined.

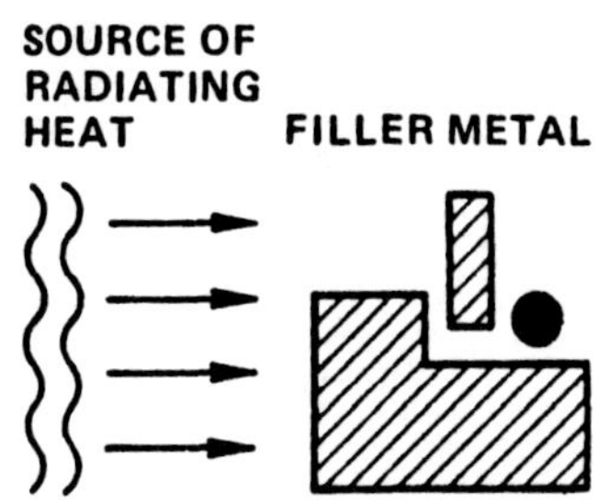

Figure 17.7 Recommended placement of alloy relative to heat source.

If the joint will be inside the vacuum envelope of a microwave tube it cannot contain a void or trap. These voids contain contaminants that leak into the vacuum envelope as the tube is operated. Tube operation elevates temperatures within the vacuum envelope. As temperatures elevate the contaminants form gases. The gas pressure forces the contaminants to pass through minute passages and enter the vacuum envelope. At best, contaminants within the vacuum envelope degrade the performance of the device. Potentially, contaminants can destroy a microwave tube by poisoning a component, such as a cathode. Correct and incorrect vacuum joints are illustrated in Fig. 17.8.

A joint design must reflect the joining process that will be utilized and the method that will be used to introduce the filler metal into the joint. For example,

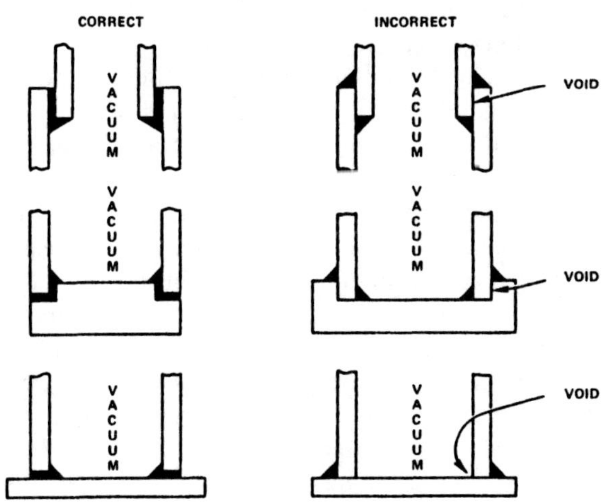

Figure 17.8 Correct and incorrect vacuum joints.

consider a brazed joint. Filler material will flow toward the source of heat. If the braze is to be performed manually, the filler material can be hand fed into the heated joint. If the braze is to be performed in a furnace, the filler material will have to be prepositioned. Prepositioning filler material entails creating appropriately sized and located grooves, slots, *et cetera*.

SEC. 17.3
YIELD MANAGEMENT

Once a joint is designed and introduced into manufacturing, it must be produced and reproduced. Consistent reproduction of joints is dependent upon jigs, fixtures, and process control. Jig and fixture design is a subject well beyond the scope of this chapter. Let it suffice to say that all of the factors that must be considered when designing a joint must also be considered when designing a jig or fixture.

This section has described joint design and manufacturing as a very complex process, and it is. Luckily for the microwave professional, joining processes are well understood. They have evolved over a 50-year period, and design and manufacturing standards are now common. This greatly simplifies the task confronting the design engineer.

17.3 YIELD MANAGEMENT

Manufacturing success is dependent upon maintaining good yields. Consider an item with a factory cost of $100.00 and a selling price of $130.00.

	90% Yielded Cost	*80% Yielded Cost*	*1:1 Cost*
Material	$44	$50	$40
Labor	$17	$19	$15
Overhead @ 165%	$28	$31	$25
	$89	$100	$80

Assume a normal yield of 80%, a production time of one hour and a capacity of 100 tests per week. Improving yield by 10% to 90% cuts yielded unit cost to $89.00, but it increases the production rate to 90 units from 80 units. Weekly profit goes from $2400 to $3700. This represents a 54% improvement. Compare this to a 10% 1:1 material cost or labor reduction. The 1:1 material cost reduction improves profit to $2820, which represents a 16.7% profit improvement. The 1:1 labor reduction of 10% and the subsequent overhead reduction represents a 10.6% profit improvement. The point of all this is that yield has a much greater percentage impact on total cost than material, labor, or overhead. Consequently, if you are in the active microwave device manufacturing business, you will thrive or wither based on your yields. Theoretically, yield is a simple concept. Yield is the amount produced from a given amount of material. If you start 100 units and complete 85 the yield is 85%. In reality, yield is usually much more complex, and dependent upon the perspective adopted. First, you have the assembly level question. Devices are made from piece parts, subassemblies and assemblies. Material can be reworked or scrapped at each of these levels. Second, you have the functional perspective. Accountants and those using financial data including manufacturing managers will tend to view yield in terms of *scrap* or *rework* dollars. Production managers, planners, and schedulers will view yield in terms of units scrapped or reworked. Then you have technical considerations. For example, consider devices that require aging. Normally they will require an average of x hours aging, but some significant percentage of these devices will require more than x hours to sufficiently age. It would be inappropriate to call these devices rejects. However, at some point, devices that continue to require aging must be rejected and reworked. To avoid misunderstanding or confusion, it is important to develop clear and consistent yield data. There is nothing more frustrating than attempting to objectively evaluate yields and finding that standards have changed, that previously the criteria in question were not being evaluated, or not knowing when the criteria were changed.

Useful yield data and analysis are based upon written, clear, concise, and measurable drawings, performance specifications, and fabrication standards. To be truly useful, all parts, subassemblies, assemblies, and devices should be compared to the appropriate standards. Those that do not meet the standards should be rejected. Yield should be measured as the percentage of items completed without rejection. Any item rejected should be reworked or scrapped. For management and control purposes it is most effective to track units started, units completed without reject, units scrapped, units reworked, and costs associated with good production, rework, and scrap.

Reject disposition is another issue that must be addressed in manufacturing active microwave devices. Four possible disposition criteria are almost universally utilized, "return to supplier," "rework," "scrap," and "use as is." Experience has convinced this author that there should only be three criteria: "return to supplier," "rework," and "scrap"; "use as is" is poison. The problems with "use as is" are:

1. You will continue to get them. If close counts; close you will get. Suppliers, like everyone else, are trying to make a "buck" and they are busy. A returned product accompanied by a credit billing will always get more attention than an information copy of a rejection notice.
2. Unless you are under strict configuration control—a very expensive proposition in itself—parts and assemblies accepted for "use as is" lose their identity as being defective once they go into stock.
3. The discrepancies compound. Consider the following example. A device consists of 40 parts and each part is received an average of five times during the production cycle. In total 200 receipts are processed during the cycle. If 2% of the receipts end up being dispositioned "use as is," then 98.2% of the devices produced will contain discrepant parts. Table 17.1 contains a matrix showing the probabilities of a device containing a discrepant part based upon the number of part deliveries and "use as is" percentage.

Table 17.1
Probability of a Discrepant Part Being in the Device

NO. OF PART DELIVERIES	"USE AS IS" % OF PARTS RECEIVED							
	0.5%	1.0%	1.5%	2.0%	2.5%	3.0%	3.5%	4.0%
10	4.4	8.6	12.7	16.6	20.4	24.0	27.4	30.7
20	9.1	17.4	25.0	31.9	38.2	43.9	49.2	54.0
30	13.5	25.3	35.5	44.3	52.0	58.7	64.4	69.4
40	17.8	32.4	44.5	54.5	62.7	69.5	75.1	79.6
50	21.8	38.9	52.3	62.8	71.1	77.5	82.5	86.5
75	31.0	52.5	67.3	77.6	84.6	89.5	92.8	95.1
100	39.1	63.4	77.6	86.5	91.8	95.1	97.1	98.2
200	63.1	86.5	95.0	98.2	99.4	99.8	99.9	
300	77.7	95.0	98.9	99.8	99.9			
400	86.5	98.2	99.8					
500	91.8	99.3	99.9					
600	95.0	99.8						
700	97.0	99.9			≥ 100%			
800	98.2							
900	98.8							
1000	99.3							

Note: 100 parts received five times each during a production cycle equals 500 part deliveries.

As Table 17.1 illustrates, a very small percentage of "use as is" in a moderately complex device produces a very high probability that the device contains discrepant parts. Furthermore, there is a relatively high probability that several discrepant parts will go into a moderately complex device, even if the percentage of "use as is" is low. Table 17.2 shows that if the "use as is" percentage is 1%

Table 17.2
Probability of Multiple Discrepant Parts in a Device Containing a Given Number of Parts and a 1% "Use As Is" Disposition

NO. OF DISCREPANT PARTS	NUMBER OF PARTS 10	50	100	200	300	400	500
0	91.4%	61.1%	37.0%	13.5%	5.0%	1.8%	0.7%
1	8.6%	38.9%	63.0%	86.5%	95.0%	98.2%	99.3%
2	0.7%	15.1%	39.7%	74.8%	90.3%	96.4%	98.6%
3	0.0%	2.3%	15.7%	55.9%	81.6%	92.9%	97.4%
4	0.0%	0.0%	2.5%	31.2%	66.6%	86.3%	94.8%
5	0.0%	0.0%	0.0%	9.8%	44.3%	74.6%	89.9%
6	0.0%	0.0%	0.0%	0.9%	19.7%	55.6%	80.8%
7	0.0%	0.0%	0.0%	0.0%	3.9%	31.0%	65.3%
8	0.0%	0.0%	0.0%	0.0%	0.1%	9.6%	42.6%
9	0.0%	0.0%	0.0%	0.0%	0.0%	0.9%	18.1%
10	0.0%	0.0%	0.0%	0.0%	0.0%	0.0%	3.3%

and the device contains 400 parts, there is an 86.3% probability that the device contains four discrepant parts; a 74.6% probability that it contains five and a 55.6% probability that it contains six. Often, it is not easy to predict the impact of one discrepant part. It is hard to predict the impact of two. Predicting the impact of five or six discrepant parts, particularly if they are related, is virtually impossible. For example, assume a given part with a diameter of 1.5in with a tolerance of ±0.001in is oversized by 0.001in; the error is 0.000067%. Assume the engineer can readily calculate that this discrepancy is in itself irrelevant and accepts the parts, dispositioning it "use as is." Would he have done so if he had known that the mating part was undersized by 0.001in?

4. "Use as is" defers a decision. If a part can safely be "used as is" then the tolerance is wrong. If the tolerance is correct, then the part cannot be safely "used as is."
5. "Use as is" greatly complicates failure analysis. Did the device fail because of a design, assembly or process flaw or because of a discrepant part. If a discrepant part failed, was it from the "use as is" lot, or was it missed at inspection?

The customer is usually protected from the consequences of these problems by rigorous testing and performance criteria. The manufacturing manager, however, is not. The manufacturing manager will pay for these problems when yields are assessed.

Specific and measurable performance and inspection criteria are the foundation upon which good yields and a successful product is based. Inspection is the last line of defense in the production cycle. However, managing a manufacturing operation by reacting to rejects is comparable to standing in the stern of a boat with a compass, steering by the compass, and watching the wake. You know where you want to go, you know you are heading in the right direction, and you can see where you have been. It is not bad in open seas and blue water but in traffic, among icebergs or near shoals, you are in big trouble. To successfully complete your voyage you need vision, binoculars, and radar.

Statistical quality control provides the vision needed to control yields. For our purposes, statistical quality control may be defined as using statistical data analysis techniques for the purposes of controlling product quality. A certain amount of variation in a manufacturing process is the product of chance. Some stable system of chance causes is inherent in any particular scheme of production and inspection. Variation within this stable pattern is inevitable [2].

Control charts are a useful statistical quality control method for analyzing data. Control charts are useful for controlling a process against performance criteria and for analyzing historical data. Analysis of historical data can lead to the identification of casual factors other than chance variation. Control is accomplished by comparing each sample against a standard. If the sample is within the control limits, no action is required, but if it is outside the limits, corrective action can be initiated. Used properly, control charts can potentially identify a process going out of control. This information can be used to control and improve yields. A detailed discussion of statistical quality control is beyond the scope of this chapter, but many excellent texts are available on the subject.

17.4 PROCESS CONTROL

Good process control is a prerequisite for achieving and maintaining good yields. The processes to be controlled vary substantially depending upon the active device being manufactured. Parts fabrication and contamination control are common active microwave device manufacturing processes that must be effectively controlled to obtain desired yields. As discussed in the section on yields, all control processes must be based upon written, clear, and concise performance specifications, drawings, and fabrication standards. A manufacturing organization's performance will never be better than the quality of that organization's standards and its conformance to those standards. This fact is particularly relevant when discussing parts fabrication and contamination control.

Fabricating parts for active microwave devices is unusual in that most of the materials must be of high purity and are relatively difficult to machine, form, bend, *et cetera*. Consider copper. The copper utilized must generally be oxygen-free and highly conductive. Consequently, certified oxygen-free high conductivity (OFHC) [3] copper is most often utilized. As little as 0.05% arsenic decreases electrical conductivity by about 14%, and the same amount of iron reduces it by 19% [4]. Isotropically cold-worked copper has a higher electrical resistivity by about 2% when compared with the dc resistivity of fully annealed copper. The corresponding decrease in RF conductivity at microwave frequencies, at which most of the current flows in a thin surface skin is much more serious [5]. By definition, OFHC copper must be 99.99% pure copper. Fabricating copper parts to be used within the vacuum envelope of an active microwave device such as a klystron require special care. If done improperly, surface contaminants may become imbedded and entrapped. These contaminants will subsequently emerge during device operation. Machining the part will workharden the copper. Depending upon subsequent operations and the end use, this may or may not be desirable. Milling and lathe operations generally require custom-formed cutting tools. This is required because copper and copper alloys are relatively "gummy" materials, and when removed they do not break up into nice convenient chips, but rather come off in spiraling strings. These strings love to jam cutting tools, cause nicks and gouges and otherwise create a nice mess. To avoid this, it is necessary to shape cutting tools in such a way as to cause the material removed to flow away from the cutting surface.

Microwave electronic-grade ceramics require similar levels of chemical purity, precise density, dielectric properties, tensile strength, hard surface finish, porosity, and dimensional tolerances. The method utilized for chemically cleaning the ceramic will have an input on strength and electrical resistivity. Ceramics are very brittle; they chip, crack, or pit easily. Consequently, they cannot be machined in their finished state, but must be ground. Ceramics are produced by a number of processes such as extrusion and isostatic processing. Typically, they are pressed into a rough shape. These "green" ceramics are then machined to rough dimensions. They are then "fired" to a hardened state and ground to final dimension. It should be recognized that the typical ceramic powder is an excellent abrasive that will quickly wear critical lathe or milling machine components like ways, lead screws, *et cetera*. The abrasive property will also result in relatively rapid wear for forming dies and fixtures.

Grinding to final dimension is often required because firing green ceramics causes them to shrink. Controlling shrinkage is very difficult, and the end result often has a normal variation of several percent. Grinding must be rigorously controlled. If done too fast, minute and often invisible cracks will develop. Subsequently, these may open up upon thermal cycling. Finally, sharp corners are to be avoided because they cause the same types of problems.

Active microwave devices use many other materials such as carbon, graphite, nickel, tungsten, tantalum, molybdenum, tantalum-gold, and silver. Invariably the quality required is extreme and specific.

If this were not enough, the dimensional tolerances required are equally stringent. Figure 17.9 illustrates a resonant cavity for a klystron. Based on fixed cavity dimensions and a nominal drift tube gap of 0.070in, an output frequency of 9.288 GHz will result. Increasing the gap 0.002in, or 2.8%, will change the

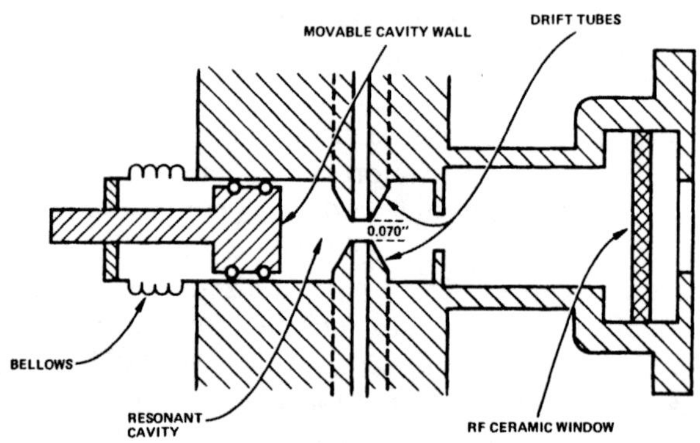

Figure 17.9 Schematic drawing of resonant cavity.

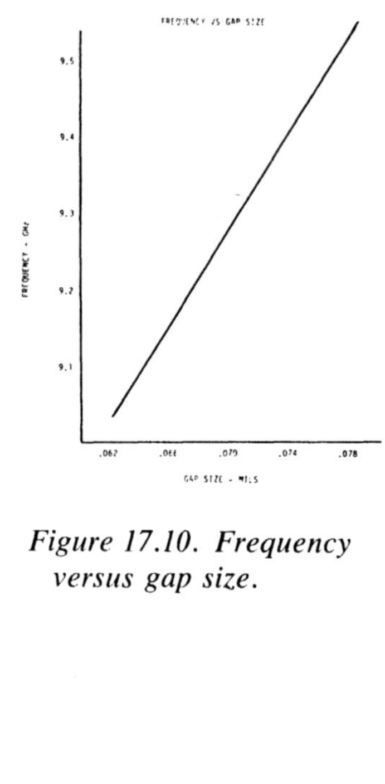

Figure 17.10. Frequency versus gap size.

frequency by 64 MHz (see Fig. 17.10). At higher frequencies, proportionally smaller gap size changes will produce even larger frequency changes. Dimensional tolerance requirements of ±0.005in are today considered quite loose; ±0.001in is normal and ±0.0002in tight. Tolerances of ±0.00005in are requested and can, in many cases, be obtained. The ability to repeatedly produce and measure tighter and tigher tolerances is the primary factor facing the development of higher frequency, broader bandwidth, and more powerful devices.

Computer and numerical machining and process control improvements have made possible an order of magnitude improvement in tolerances and, subsequently, higher frequency devices. These improvements have had an equally substantial impact on the cost of mature, relatively loose tolerance devices. The reason for this is that most mature devices were designed when tolerances of ±0.005in was normal and ±0.001in was tight. Computer and numerically controlled machine tools have reduced costs by their substantially greater speeds, reduced set-up times, and they can be operated by personnel of a lower skill level. Additionally, they have the advantage of being more consistent. For example, if the tolerance is ±0.001in, the typical good lot from a manual machine will fully fluctuate between the tolerance limits. On the other hand, all or most of the parts from a good computer and numerically controlled machine lot will have the same variance. If one part is +0.0005in, the large majority of parts in the lot will vary beween +0.0003in and +0.0007in. The percentage variance is normally smaller. This has the favorable impact of making the assemblies much more consistent and any adjustments much more standard. If you use the highest quality raw materials, adhere to the most rigid fabrication and inspection criteria, you can still get into trouble if you do not keep contaminants out of your manufacturing processes.

Contamination is a serious problem for all active microwave device manufacturers. Contamination is a problem for gyrotron manufacturers just as it is a problem for GaAs FET manufacturers. Contamination is a foreign substance that, when introduced into a device, has a detrimental effect. Visible contamination such as dirt, rust, or oxidation generally draws quick attention. For this reason, most of the damage being done comes from particles not visible to the human eye. A smoke particle can be as small as 0.000130in, a line from a finger-print is only 10 times larger. The human hair is about 0.003in in diameter. Each of these particles can have a substantially detrimental impact on an active microwave device.

To successfully control contamination you must control the environment, people, garments, materials, equipment, tools, and processes.

The environment is most difficult and expensive to control. Clean rooms and work stations are classified according to the maximum allowable 0.5 μm and larger particles per cubic foot. For example, a class 100 clean rooms allows no more than 100 0.5 μm or larger particles per cubic foot. A class 1000 allows no more than 1000 particles per cubic foot, *et cetera*. Creation of a class 100 clean room will cost about $300 per square foot for filtering and air distribution. Annually, it will cost another $200 per square foot to maintain the air filtration system. To this you add work stations, equipment, training, *et cetera*. For this reason, multimillion-dollar clean rooms are common. This astronomical facility cost, combined with growing needs for greater cleanliness as a consequence of smaller feature sizes, seems to be leading manufacturers toward modularized and isolated processes coupled with incremental cleanliness standards. For example, the clean room itself may be class 1000; product is moved in boxes maintained at class 100, and operations performed in isolated equipment also maintained at class 100.

A clean room design must reflect its location and function relative to the balance of the manufacturing facility. It must consider the production processes and equipment to be utilized and the contaminants that these processes will produce. The designs must reflect local, state, and federal regulations. Most importantly, the design should consider the people that are going to be working in the facility and the contamination that they are going to introduce. If it is not already obvious, this is a project that probably should be left to experts and specialists.

People represent an unbelievably large source of contaminants. A motionless person, will generate 100,000 0.3 micrometer, μm, particles per minute [6]. A person in motion, will generate 300,000 0.3 μm particles per minute [7]. An application of mascara contains 3.0×10^9 particles greater than 0.5 μm [8]. Because of their potentially great impact, employees must be trained not only to perform their functions, but to do so in a way that supports the need to minimize contamination [9]. Since a person in motion produces three times the particulate as a stationary person, movement within the clean room should be limited. Additionally, hygiene requires particular attention. Things like lanolin-based soaps should be used to minimize flaking. Paper or fabric towels should not be used (instead use electric dryers). Books, writing paper, tissues, and computer printouts are equally undesirable. All of these items are sources of particulate. Clean room garments should be manufactured of synthetic materials that remain relatively lint-free and they should have no pockets. Clean room operation and contamination control are complex issues. It is recommended that every clean room operation create and maintain written clean room standards. These standards, the reasons for their existence, and their importance should be taught to all new employees. All employees should be periodically retrained in clean room practices.

SEC. 17.5
SUMMARY

17.5 SUMMARY

Manufacturing active microwave devices is a complex and sophisticated process. Manufacturing these devices entails joining dissimilar materials according to precise operation schedules and rigid process control standards. In order to be profitable, it is necessary to obtain good yields. Good contamination control is a key factor contributing to the attainment of acceptable yields. Statistical quality control can provide the information needed to control manufacturing processes and anticipate problem areas. If the procedures or specifications are vague, if the process control is weak, or if standards are compromised, then poor yields and performance will result.

References

1. Howard B. Cary, *Modern Welding Technology*, Englewood Cliff, NJ, 1979, p. 21.
2. Eugene L. Grant and Richard S. Leavenworth, *Statistical Quality Control*, New York, McGraw-Hill, 5th Edition, 1980, p. 1.
3. Trade Mark of American Metal Climax, Inc., New York.
4. Walter H. Kohl, *Handbook of Materials and Techniques for Vacuum Devices*, New York, Reinhold Publishing, 1967, p. 182.
5. *Ibid.*, p. 183.
6. John F. Rodew, "*Contamination Control Education*," *The Journal of Environmental Services*, Nov./Dec., 1983, p. 35.
7. *Ibid.*, p. 35.
8. Quintin T. Phillips, "*Cosmetics in Clean Rooms*," *The Journal of Environmental Sciences*, Sept./Oct., 1983, p. 28.
9. *Clean Room Operational Standards*, pamphlet from Clean Room Products, Bay Shore, NY.

CHAPTER 18

MICROWAVES AND HEALTH SCIENCES

Eiji Tanabe and W. Stephen Cheung

PART I: ACCELERATORS AND X-RAY MACHINES

CH. 18/PART I

ACCELERATORS & X-RAY MACHINES

18.1 INTRODUCTION

The heart of a therapeutic x-ray machine is that of an electron accelerator and a microwave generator, either a klystron or a magnetron. The microwave power helps accelerate electrons to high enough energy to generate x-rays when the electrons strike a target. Electron linear accelerators with energies up to 50 MeV have been very widely used for radiation therapy and industrial radiography (a technique employed to detect fatigue and cracks of industrial materials). Modern therapeutic machines vary in electron energy from 2 MeV up to 25 MeV and are used for cancer treatment by electrons as well as high energy x-rays. This part will examine the basic principles of accelerator science, x-ray generation, and how microwave accelerates electrons.

18.2 ELECTRON ENERGY UNIT

The accepted energy unit in SI units is joule (J). The reader can get a rough idea of 1 J by noting that the amount of energy to light a 100 watt light bulb for 1 second is 100 J. A more appropriate approach is to start from electrostatics.

Figure 18.1 shows two metallic plates with a potential difference of 1V, i.e., plate A is 1V above plate B. Suppose an electron is released by plate B. The electron is attracted toward a positive and higher potential so it undergoes an acceleration. By the time the electron reaches plate *A*, it carries an amount of kinetic (motional) energy known as 1 electron-volt (eV). The SI unit to measure the electronic charge is coulomb (C) which is given by

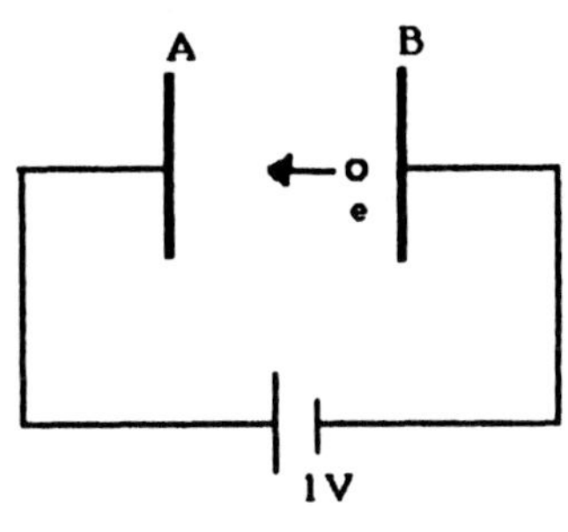

Figure 18.1 An electron released by plate B will accelerate toward plate A which is at a higher potential.

$$1 \text{ electronic charge} = 1.6 \times 10^{-19} \text{ C} \tag{18.2.1}$$

In other words, 1 coulomb of charge has 6.25×10^{18} electrons.

Therefore, when 6.25×10^{18} electrons move from plate B to plate A in the above example, they acquire a total energy of 1 joule, which is equal to 1 volt $\times$ 1 coulomb. A convenient conversion is

$$1 \text{ eV} = 1.6 \times 10^{-19} \text{ J} \tag{18.2.2}$$

or

$$1 \text{ J} = 6.25 \times 10^{18} \text{ eV} \tag{18.2.3}$$

In accelerator science, the electron-volt is a convenient energy unit. It measures the energy of a single electron, rather than the total energy of a bunch of electrons, under the influence of an electric field. Hence, a 1keV accelerator can accelerate each electron to an average energy level of 1 thousand electron-volt. Note that the letter e of eV is sometimes omitted, so a 1kV accelerator is the same as a 1keV accelerator.

To appreciate the energy level of 1keV for a single electron, we can use a simple physics equation which equates the kinetic energy gained by the electron to the electric energy expended by the electric field. In other words,

$$1/2 m_e v^2 = 10^3 \text{ eV} = 10^3 \times 1.6 \times 10^{-19} \text{ J} \tag{18.2.4}$$

where m_e and v are the mass and velocity of an electron.

Using the mass of an electron as 9.1×10^{-31} kg, the above equation can be solved for v, which is found to be 1.87×10^7 m/s (6.25% of the speed of light!)

Note that Eq. (18.2.4) is what is known as a "classical" equation and will not be valid as the energy of the electron increases further, i.e., as the corresponding velocity of the electron approaches that of light. Einstein's theory of special relativity must be used for any object moving at a speed near that of light. In essence, the mass of a moving electron is larger than its "rest" mass and will become infinite if its speed is exactly equal to that of light, a physical impossibility. This chapter, however, does not intend to cover the special relativistic aspect of charged particles.

18.3 GENERATING X-RAYS

The discussion in the previous section is not meant to suggest that an electron cannot achieve an energy of, let us say, 1 MeV (10^6 eV) or more. Modern accelerators for the study of fundamental particles in physics can achieve energies in the GeV (10^9 eV) range. A 1 MeV electron travels at about 94% of the speed of light and the electron is three times more massive than when it is at rest.

Energetic electrons are products of an accelerator. When directed to strike a target such as tungsten, various complicated interactions between the incoming electrons and the atomic system of tungsten result in generation of electromagnetic waves known as x-rays. The interactions are extremely complicated and beyond the scope of this book. One example of the many possible interactions is that collisions with the incoming electrons cause the electrons of the target atoms to be knocked out of their orbits. The rearrangement of the atomic electrons leads to the emission of x-rays. Another example is that the deceleration of the incoming energetic electrons due to the electrostatic repulsion with the target electrons can also result in the emission of x-rays called Bremsstrahlung radiation.

In essence, the target atom, as a system, is put in an excited state due to the interaction with an incoming energetic electron. The system gets rid of the excess energy by radiating it away. Because so many interactions are possible, the energy and the frequency of the emitted x-rays are not unique, but take on a broad spectrum. The frequency of the radiation is in the neighborhood of 10^{18} Hz (wavelength about 3×10^{-10} m) and higher.

Consider a 1 MeV electron colliding with a tungsten target. If the energetic electron loses all of its energy after the interaction, conservation of energy requires that the excited atom must radiate the 1 MeV energy away as an x-ray photon. In other words, the energy of the x-ray so released is 1 MeV. It is obvious that the above description has several assumptions of which one or more may not hold. The incoming electron may retain a portion of its energy, and the rearrangement of the atomic electrons may take several steps to complete instead of one. Consequently, the energy of the x-ray is not unique but falls into a range of values up to the energy of the incoming electron.

In the field of accelerator sciences, x-ray radiation is not characterized by its frequency but by its energy in MeV, although it should be noted that frequency can be calculated if the energy is known. For example, the frequency of a 1 MeV x-ray is 2.4×10^{20} Hz.

The energy capability of accelerators in therapeutic x-ray machines ranges from 4 MeV to the present upper limit of 25 MeV. Accelerators of several MeV capability are low energy machines while those of tens of MeV are high energy machines.

18.4 ACCELERATION OF ELECTRONS

The approach taken in Sec. 18.2 to accelerate an electron is to use two metallic plates placed at a distance apart and apply dc voltage across them. A 1 MeV accelerator requires a 1 million volt battery or power supply. Construction of such a power supply is non-trivial because such a high voltage can cause electrical breakdown in many components. This problem is overcome by using alternating voltage and microwaves.

First of all, the electron source will be discussed here. In Fig. 18.1, we assumed that electrons are pulled out of the metallic plate at the lower potential (cathode) and are then accelerated toward the plate at the higher potential (anode). Pulling an electron out of a metal requires a certain amount of energy known as the work function. A more convenient way is to place the cathode in front of a heater. Electrons in the cathode now receive energy from two sources: electrical energy from the anode and thermal energy from the heater. Hence, electrons can be pulled out of the cathode surface more easily than just relying on the electric energy alone. A representative circuit is shown in Fig. 18.2. The small battery is the current supply to the heater.

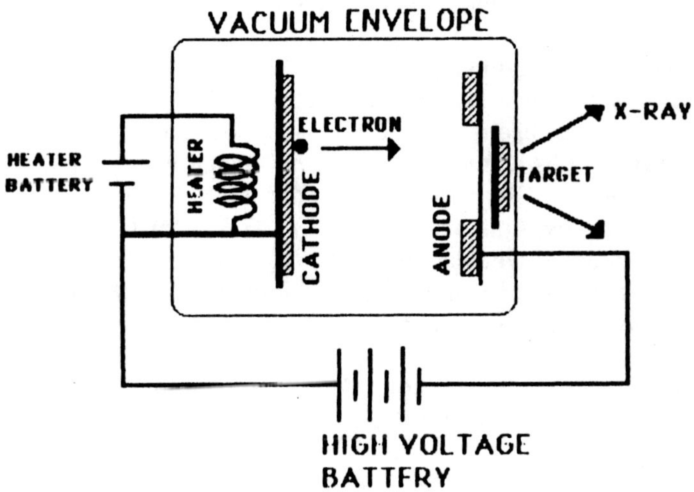

Figure 18.2 A conceptual construction of a linear accelerator. Electrons released from the heated cathode accelerate toward the anode-target. The high voltage battery is in the order of MV.

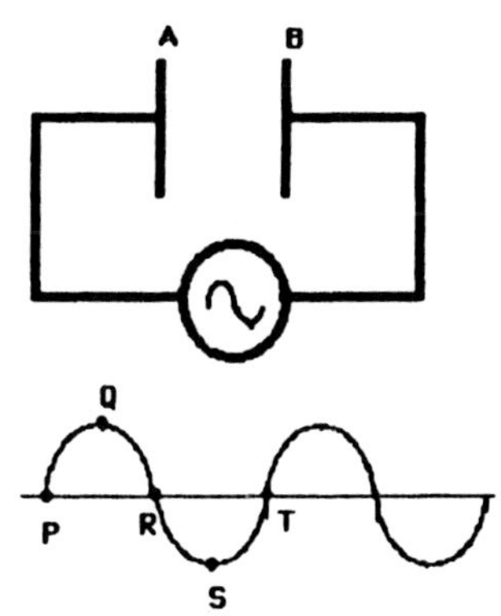

Figure 18.3 An alternative to the high voltage battery is an ac (microwave) power source.

Because it is impractical to build a megavolt power supply or battery, alternating voltage is the alternative. Figure 18.3 shows the voltage across two metallic plates, *A* and *B*, alternating according to a sine function. The voltage starts from 0 (point *P*), rises to a positive maximum (point *Q*), falls through 0 (point *R*) to a negative maximum (point *S*), and returns to 0 (point *T*). As a simple and introductory explanation, the cycle repeats at a *very low frequency f*. At points *P*, *R*, and *T*, there is no potential difference across the metallic plates so electrons are not attracted by the anode B. When plate A is at a positive voltage relative to plate B, electrons are attracted and therefore accelerate toward plate A, x-rays are radiated. When plate A is at a lower potential relative to plate B, the electrons are repelled by plate A so they simply form a cloud around plate B. When plate A returns to a positive voltage, the electrons resume their acceleration toward it.

At *very low frequencies* the magnitude of the electron acceleration is maximum when the voltage across plates A and B reaches a positive maximum. Therefore, the energy carried by the electrons varies in a manner similar to the alternating voltage. There is x-ray radiation during one-half of the sine voltage and no radiation during the other half. Also, electrons released by the cathode during the lower half of the sine voltage must wait until the voltage turns positive. In other words, the electrons are bunched. The bunching phenomenon has been discussed and illustrated in the klystron operation of Ch. 10.

Emphasis on *very low frequency* of the applied alternating voltage was made in the preceding discussions. At higher frequencies, the transit time for an electron to move across the electrodes cannot be ignored. It results in a phase lag of the electron current relative to the applied voltage. Such complications, although important to realize, will not be our major concern.

Microwaves can be used as the alternating voltage to accelerate the electrons. A microwave generator can supply the necessary microwave power to a series of cavities to provide the electric field necessary to accelerate the electrons. Because the electric field of a microwave alternates billions of times per second, arrangement must be made to assure that the bunched electrons are always accelerated and not decelerated by the microwave electric field. This arrangement includes the dimensions of the cavities and the microwave frequency. For example, if a 3 GHz microwave is used, then the wavelength is 10cm. The dimension of the accelerating cavity is set as the half-wavelength, i.e., 5cm in this case. A 1m long accelerator will therefore have a series of 20 accelerating cavities.

Consider an electron bunch entering the first of a series of cavities to which microwave power is supplied. Suppose the electric field at the first cavity is accelerating (for the negatively charged electrons). The electrons will be accelerated and gain kinetic energy from the electric field, i.e., the microwave is losing power to the electron beam. By the time the electron travels from one end of the cavity to the opposite end and enters the second cavity, the electric field of the second cavity has just turned accelerating. In other words, the second cavity lags the first cavity by a phase angle of 180°. The electron bunch is accelerated again and gains more kinetic energy. This process can repeat many times and the energy of each electron can surge up to millions of electron-volts.

One can think of the accelerator principle discussed in the previous paragraph as the reverse of the klystron principle. In the case of a klystron, electrons, which are released from the cathode with energy from a high external electric field and also from a heater, are accelerated by a dc voltage, typically 120kV. When made to pass through a series of cavities, the electrons are bunched and they induce surface charges on the cavity walls. As the electrons move from one end of a cavity to another, the induced charges also move and therefore form a current. The direction of the induced current changes when the second electron bunch enters. The frequency of the alternating induced current is related to the transit time of the electrons moving through the cavity. The alternating induced current in the resonant cavity subsequently produces the microwaves. The microwave derives its power from the induced current which, in turn, obtains its power from the electron bunches. The latter power is not only related to the strength of the dc electric field between the cathode and the anode of the klystron but also to the density of the electron bunch, i.e., the number of electrons released by the cathode.

If the dc electric field alone were responsible for the electron energy, there would be no need for the cavities and the microwaves. The electron density alone is not appropriate either. For example, suppose that 1,000 electrons are accelerated by a dc voltage of 1kV. They have a total energy of 1 MeV. However, the energy of each electron is still 1 keV and is insufficient to interact with a target atom to emit useful x-rays. On the other hand, if these 1,000 electrons pass much of their total energy of 1 MeV to form a microwave, and the microwave is then used to accelerate a single electron in the accelerator structure, hence the single electron can achieve an energy as close to 1 MeV as possible. The energy transfer is limited by the efficiency of the klystron and the coupling efficiency between the microwave and the electrons in the accelerator.

An alternate approach to the above discussion is to think of the process as a voltage transformer. The voltage at the klystron is not very high but the electron current is made to be high. Through microwaves, fewer electrons are accelerated inside the accelerator, but much higher energy per electron is achieved. This reminds us of a voltage transformer in which high voltage can be attained in the secondary coil at the expense of current.

Therefore, microwaves provide the means of transferring energy from a multi-electron energy source (a klystron or a magnetron) to individual electrons in an accelerator where the energy of the individual electrons is essential to generate x-rays.

Figure 18.4 shows the practical connection of the heater, the cathode, and part of the anode-cavities structure. The anode and the following cavities are made of OFHC copper (oxygen free high conductivity copper, a common material for vacuum work) brazed as one piece and are maintained electrically at ground potential, while the cathode is maintained at a voltage 20kV below ground potential. Due to the special geometries of the anode-cavity piece and the cathode, the electric field pattern between the two electrodes forces the cathode electrons to accelerate along a convergent path and through the cavities as shown in Fig. 18.4.

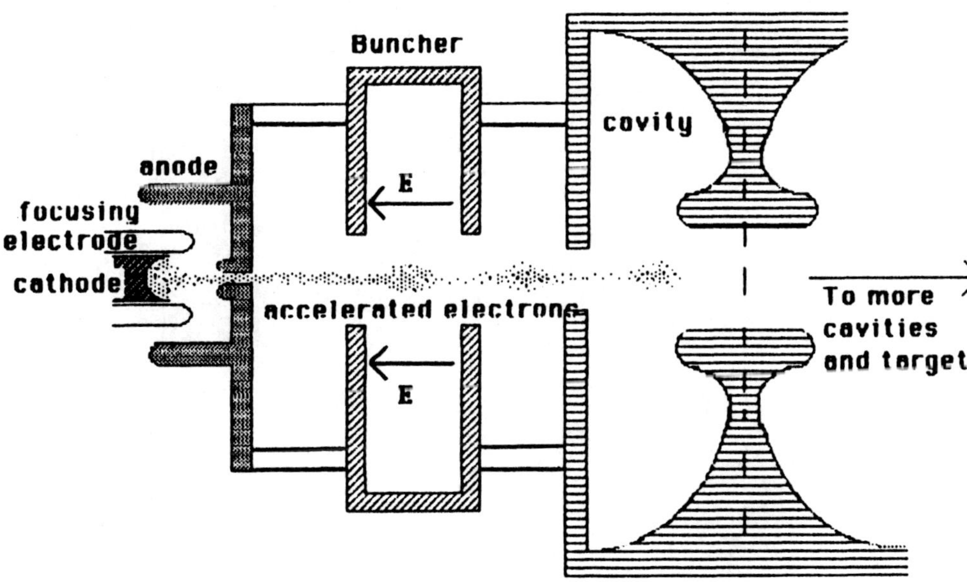

Figure 18.4 The front end of a microwave-driven accelerator. Electrons released at the cathode accelerate toward the anode, but will go through it. Further acceleration to achieve the high energy is provided by the microwave power fed to the series of cavities.

As the electron beam travels through the microwave filled cavities, the electromagnetic interaction with the microwave results in a tremendous acceleration as well as bunching of the electron beam. After the bunched electron beam has gained enough energy, it will strike the tungsten target and x-ray is radiated.

The tungsten target is mounted on a copper block which is water cooled, and is in turn heliarc-welded on an insulating ceramic ring. An ammeter connected between the target and the accelerator structure (electric ground) can measure the electron beam current striking the target (the beam current is an important parameter). The target-insulating ring assembly is welded to the accelerator and serves as the vacuum seal of the accelerator structure. A vacuum of 10^{-6} torr inside the accelerator structure is maintained by a Vac-ion pump. (As a quick note to vacuum technology, the normal atmospheric pressure is called 1 atmosphere, or 760 torr, and there are approximately 2.5×10^{19} air molecules per cm^3. A pressure of 10^{-6} torr means that the enclosure has been pumped down so that there are only about $2.5 \times 10^{19} \times 10^{-6}/760 = 3.3 \times 10^{10}$ residual air molecules per cm^3.)

18.5 MEDICAL ELECTRON LINEAR ACCELERATOR

A linear accelerator (*linac*) accelerates the charged particles, e.g., the electrons, in a straight line. This is in contrast with other accelerators such as the betatron and the cyclotron, both of which accelerate the charged particles in circular orbits.

Figure 18.5 shows a typical x-ray treatment unit using a *linac*. The unit is housed in a thick concrete room to shield all personnel other than the patient from the radiation. The *linac* is mounted in a gantry which rotates on a stationary stand. The stand contains the klystron, associated microwave circuitry, and the cooling system. The gantry can be rotated about its horizontal axis for treatment. The x-ray emerging from the collimator is centered at a point on the gantry's cylindrical axis, a point known as the iso-center. A patient's tumor is typically positioned at the iso-center. The couch, upon which the patient lies, can move in three linear directions, two horizontally and one vertically, and one rotation motion. The iso-center is visibly located with the help of two side and one ceiling lasers. The gantry can be rotated by position indicators to accomplish different gantry angles.

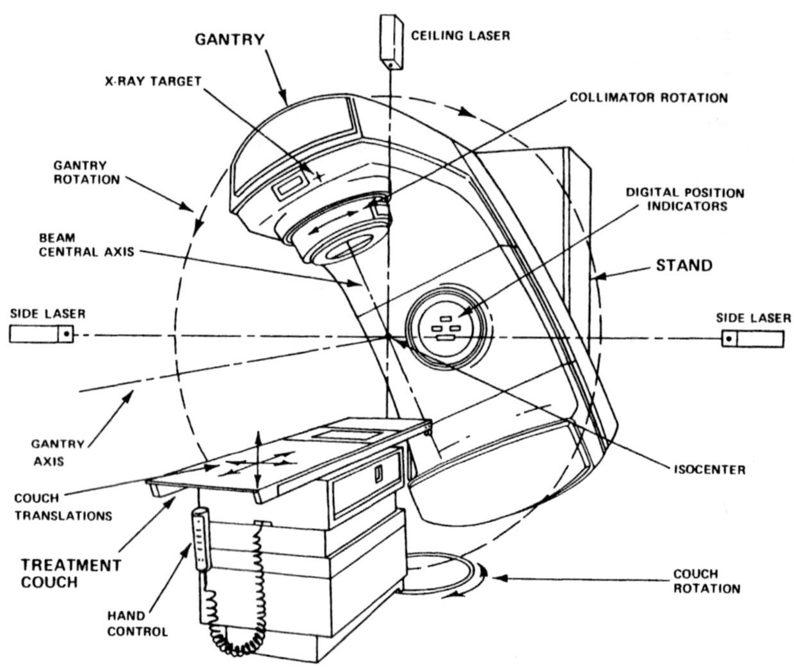

Figure 18.5 An x-ray treatment unit. The linac is housed inside the top of the gantry. Microwave power comes from a klystron inside the stand. X-ray emerges from the collimator and is focused at the isocenter. (Courtesy of Professor C. J. Karzmark, Stanford University)

Figure 18.6 shows the major components of a radiotherapy *linac*. The klystron provides the microwave power necessary to accelerate the electrons in the accelerator structure. The power output of the klystron is about 6 MW for this high energy machine. The microwave power is transmitted to the accelerator structure via rectangular waveguides (WR284, 2.6 to 3.9 GHz). The cross sectional area of these waveguides can handle 6 MW of power if they are pressurized (about 30 psi) with freon or SF_6 (both have dielectric constant approximately equal to 1.0) to prevent electrical arcing due to the high electric field strength induced by the microwave power.

The klystron is protected from any reflected wave due to any possible mismatch by a circulator inserted between the klystron output and the accelerator structure. The klystron and the accelerator units generate much heat due to the tremendous power consumption. Cooling is usually done by a cooling water system. The klystron gun is housed inside the oil tank for high voltage insulation (120kV for a klystron). The oil tank is, however, unnecessary for a machine using a magnetron which operates at 40kV.

The gantry consists of the accelerator unit, waveguides to input the microwave power from the klystron (or the magnetron), and a focusing magnet to collimate the x-ray. Because the gantry is rotatable, a circular waveguide joint called the rotary joint must be used to insure proper microwave power delivery from the klystron.

The accelerator unit is composed of several parts: an electron gun, a vacuum pump, an accelerator structure, and a target. The electron gun system generates electrons from the cathode by passing high direct current to the heater behind the cathode. the "hot" electrons released by the cathode are then accelerated toward the anode held at a high dc voltage. The microwave further accelerates these electrons along the accelerator structure. The accelerator structure is a series of cavities specially constructed to match the microwave frequency and to allow electron bunching to occur.

Two types of medical accelerator structures exist. For the high power machine (tens of MeV) using a klystron, the accelerator structure is horizontal and usually 1 to 2 meters long. The horizontal electron beam is bent by a magnet (dc solenoid) to follow a vertical path. The bent electron beam strikes the target vertically and x-ray is emitted in the vertical direction. For the low power machine (several

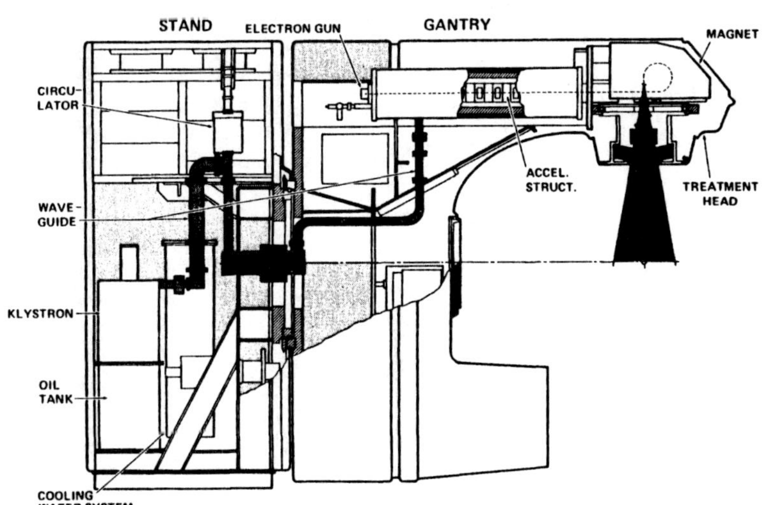

Figure 18.6 The internal construction of an x-ray treatment unit. (Courtesy of Professor C. J. Karzmark, Stanford University)

MeV) using a magnetron, the accelerator structure is shorter and mounted vertically. Hence, the electron strikes the target directly with no prior bending.

Since its commercial introduction into radiation therapy in 1962, the electron linear accelerator (*linac*) has steadily grown into the most widely used radiotherapy machine. At present more than one-half of the radiotherapy units in the US, about 1,000 machines, are electron *linacs*. Medical electron linacs are currently manufactured by eleven organizations in eight countries, with about 2,000 machines currently in operation worldwide.

CH. 18/PART II

MICROWAVES IN THERAPEUTIC USE

PART II: MICROWAVES IN THERAPEUTIC USE

18.6 INTRODUCTION

Microwaves are used in diathermy treatment, which is applied for patients with arthritis, rheumatism, *et cetera* and in hyperthermia treatment for cancer therapy. In both cases the microwave power in the frequency range from 400 to 2500 MHz is applied directly to the body with various types of antenna applicators in order to heat tissues. The required power level is up to several hundred watts and solid-state microwave generators are often used.

18.7 THERAPEUTIC APPLICATIONS OF MICROWAVE

The detailed mechanism of the interaction of microwave energy with living tissue is not well known, but the macroscopic properties can be described by Maxwell's equations. The main parameters that describe microwaves are the frequency f, the phase φ, and the amplitude of the electric and magnetic field (E and H). At microwave frequencies, biological tissues behave like solutions of electrolytes that contain polar molecules such as water, protein, *et cetera*. Microwaves interact with biological systems via ionic conduction (oscillation of free charges) and rotation of polar molecules of water and protein. Through these processes, the absorbed microwave energy is transformed into kinetic energy of molecules, which is associated with a rise in temperature of the tissues. The tissues are characterized as a lossy dielectric which can be described by its permitivity ϵ, permeability μ, and conductivity σ.

18.8 THERAPEUTIC APPLICATIONS

Although *diathermy* and *hyperthermia* treatments are the same in the sense that they involve raising the temperature of the tissue by some means of energy deposition, they are used for two distinctly different objectives. *Diathermy is used basically for treating an infected area by warming up the tissue*. Due to an increasing interest in cancer therapy, only the hyperthermia technique is discussed as follows.

Hyperthermia as a modality for cancer therapy has grown in the last 15 years while the history of treating cancer with heat has been traced back several centuries. There are a number of reports on spontaneous regression of tumors and apparent cure of patients suffering high fevers for long times.

Recent extensive laboratory research shows that mammalian cells exposed to temperatures in excess of 43°C (109.4°F) for 10 minutes will be seriously damaged. When hyperthermia is combined with x-ray treatment the hypthermal treatment enhances the effect of the x-ray. There are also evidence that hyperthermia may increase the effectiveness of some chemotherapeutic agents in killing cells.

Electromagnetic radiation offers important advantages for relatively inexpensive hyperthermia production. Figure 18.7 shows one typical system of local microwave hyperthermia. The microwave generator delivers power through the microwave transmission line, the circulator, and the directional coupler to the applicator, thereby causing heating of the tissue. The skin surface temperature is controlled by circulating the coolant placed between the tissue and applicator while irradiation is proceeding. Various applicators are listed in Table 18.1 and illustrated in Fig. 18.8.

CH. 18/PART II

MICROWAVES IN THERAPEUTIC USE

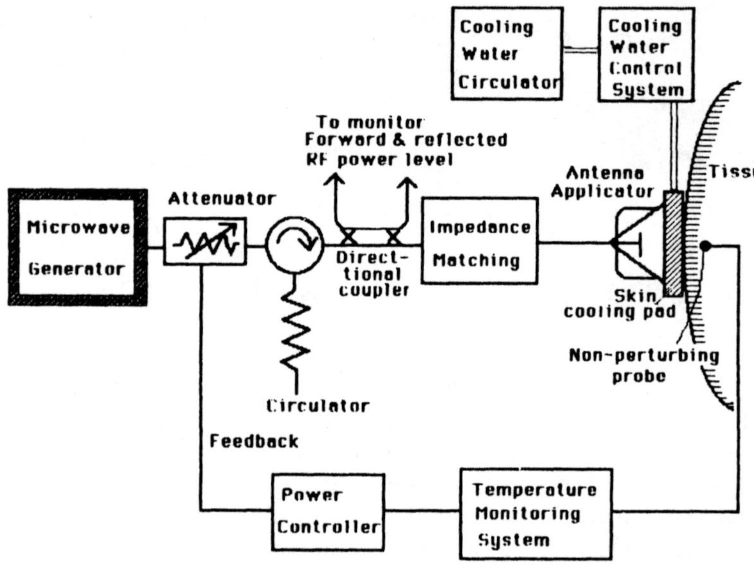

Figure 18.7 A typical local hyperthermia system. Microwave is applied to the patient's body at the antenna applicator.

A temperature probe is placed within the tissue in order to monitor the temperature inside the tissue. Because conventional metallic temperature sensors such as thermistors and thermocouples with their associated connecting wires perturb the electromagnetic field distribution, such measurements can lead to significant errors. Large errors are due to the distortion and localization of the electric fields and, consequently, the heat distribution. In order to avoid these problems, non-metallic type probes have been developed and utilized for hyperthermia. (See Table 18.2.)

Table 18.1
Microwave Applicators

Applicator Type	*Frequency Range (MHz)*	*Aperture Size*	*Advantages*	*Disadvantages*	*Manufacturers*
1. Slab-loaded rectangular waveguides	900 to 2500	15 × 30 cm to 5 × 10 cm	TEM-like mode excitation	bulky	—
2. Ridged waveguide	200 to 900	10 × 13 cm	Broadband	Bulky and less efficient	BSD, Microwave Associates.
3. Circular horn antenna	900 to 2500	25 cm to 15 cm dia	circularly polarized	less efficient	Transco.
4. Coaxial antenna	400 to 2500	30 cm to 0.1 cm dia	can be very small	highly non-uniform radiation pattern	BSD, Clinitherm, Cheung Associates.
5. Microstrip antenna	400 to 2500	10 cm to 2 cm dia	simple light weight	limited power handling capacity	Varian, RCA.
6. Cylinder-loaded circular waveguide	900 to 2500	25 to 8 cm dia	circularly polarized	bulky	Varian.

There are basically two major types of temperature sensors designed to minimize the perturbation. One method is to adopt conventional thermistor or thermocouples by the use of high resistance leads. The other method is to use optical fibers to transmit information about temperature-dependent physical properties of some sensing materials (such as GaAs, liquid crystal, birefringent crystal, and optical etalon) in the tip of the probe.

Although more than 10,000 patients have been treated with hyperthermia in the world in the last 20 years, the techniques of microwave hyperthermia are still in the development stage. In summary, hyperthermia combined with radiation therapy and chemotherapy have demonstrated therapeutic effect on malignant tumors. However, there are still many engineering problems for microwave hyperthermia. One of the problems in optimal delivery of heat is in monitoring the temperature throughout the treatment. Further developments including more precise and efficient equipment are required in order to have consistent contributions in the area of health sciences.

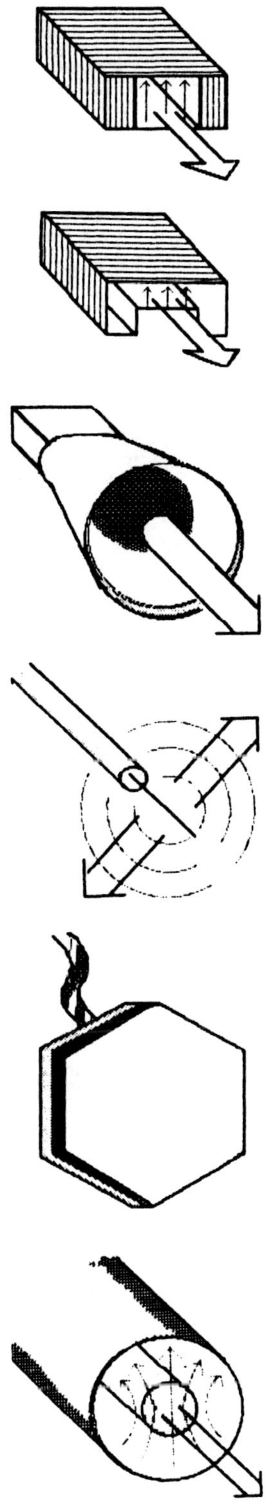

Figure 18.8 Different hyperthermia applicators.

Table 18.2
Microwave and RF Hyperthermia Equipment in US

Supplier	*Location*	*Applicator*	*Heating System*	*Thermometry*
BSD	Salt Lake City, UT	1, 5	1, 2, 3, 4, 6	2, 6
Cheung Assoc.	Beltsville, MD	1, 5	1, 2, 3	5, 6
Clinitherm	Dallas, TX	1, 4, 5, 7	1, 2, 3, 4, 8	3, 6
Henry Medical	Los Angeles, CA	2, 3	1, 2, 6	1
Oncotherm	Los Angeles, CA	1, 4, 5	1	1, 4, 6
Pan Dynamics	Van Nuys, CA	1, 5, 6	1, 2, 7	1, 4, 6
Tag Med	Boulder, CO	1	1, 3, 6, 7	4
Techtron	New York, NY	2	1, 2, 3	1, 6
URI THERM-X	Champaign, IL	7	3, 8	1, 6

Key:

I. *Applicator*
1. Waveguide or horn attenna applicator
2. Capacitive applicator
3. Inductive applicator
4. Microstrip applicator
5. Interstitial applicator
6. Intracavity applicator
7. Ultrasound applicator

II. *Heating System*
1. Microwave or RF generator
2. Control system
3. Cooling system
4. Applicator system
5. Leakage monitor
6. Treatment couch
7. Shield room
8. Ultrasound generator

III. *Thermometry System*
1. Thermocouple
2. High-impedance thermistor
3. Optical fiber with GaAs sensor
4. Optical fiber with phosphor sensor
5. Optical fiber with birefringent crystal
6. Temperature monitoring system

18.9 ARE MICROWAVES HAZARDOUS?

There has been sharp controversy over the effects of electromagnetic radiation, including microwaves, on the health of living things. Serious questions have been raised about the safety of electromagnetic radiations of all frequencies, i.e., from a few hertz to thousands of GHz. Hence, radiation from the 60 Hz power line, radio and television stations, and other communication operations are all under scrutiny from the safety standpoint.

The safety standard in the US on microwaves for many years has been an upper limit of $10mW/cm^2$ while that in the Soviet Union is substantially lower. Recent suggestion by the American National Standards Institute (ANSI) is $1mW/cm^2$, especially in the frequency range from 30 to 3000 MHz. The original contention is that the biological effect, if any at all, is a thermal effect on the tissues. Recent research work in this area indicates that selected electromagnetic radiations have non-thermal effects such as changes in the synchronization of brain waves and behavior, especially when exposed for a long period of time.

It is not the intention of this book to suggest a definite answer to the safety question. The outcome of this controversy bears significance not only in legal battles, but also in the planning of future electronic communications. Interested readers are referred to the references given in this chapter.

References

1. "The Use of Electron Linear Accelerators in Medical Radiation Therapy: Overview Report No. 1—Physical Characteristics," Bureau of Radiological Health, Food and Drug Administration, Rockville, MD, 1974.
2. C. J. Karzmark, and R. J. Morton, "A Primer on Theory and Operation of Linear Accelerators in Radiation Therapy," U.S. Department of Health and Human Services, Food and Drug Administration, Rockville, MD, 1981.
3. E. Lerner, "RF Radiation: Biological Effects," *IEEE Spectrum*, December 1980, pp.51–59.
4. "Health Aspects of Radio Frequency and Microwave Radiation Exposure," Part 1, Environmental Health Directorate, Health Protection Branch, Minister of National Health and Welfare, Canada, 1977.
5. C. C. Johnson, and A. W. Guy, "Nonionizing Electromagnetic Wave Effects in Biological Materials and Systems," *Proceedings of IEEE*, Vol. 60, No. 6, 1972, pp. 692–697.

CHAPTER 19

MICROWAVES AND TELEVISION

Don Sharp and W. Stephen Cheung

19.1 BACKGROUND

Electronic news gathering (ENG) or electronic journalism (EJ) has been with the television industry since the mid-1970s. ENG has a different meaning to different people. When referred to microwaves, ENG describes the capability of transmitting the video and program sound of any event within a radius of approximately 60 miles (greater in some instances) back to the studio with reasonable certainty and quality within a very short period of time. In most cases, only a small portion of these events are aired immediately. The rest are usually stored on video tapes for editing and later used on regularly scheduled news broadcasts.

Today's ENG effort is usually originated from a van complete with 2 GHz, 7 GHZ, and 13 GHz equipment and suitable antennas, or from a helicopter usually equipped with 2 GHz instruments and antennas. The crew of two includes a cameraman equipped with a small video camera. The van or helicopter usually holds some basic electronic equipment including monitors, video tape recorders, audio processing, *et cetera*. Most ENG equipment is now capable of operating from "cigarette lighter" sockets, so some crews can actually operate out of cars rather than vans.

The 2 GHz frequency range is currently rather crowded and, therefore, higher frequency bands such as 7 GHz have been either opened up or planned. Each television station has a certain share of the frequency band. Take the 2 GHz as an example. The carrier of station A is 2150 MHz (2.15 GHz), the program information occupies a bandwidth of 20 MHz, and there is a 15 MHz guard band on each side. This is illustrated in Fig. 19.1.

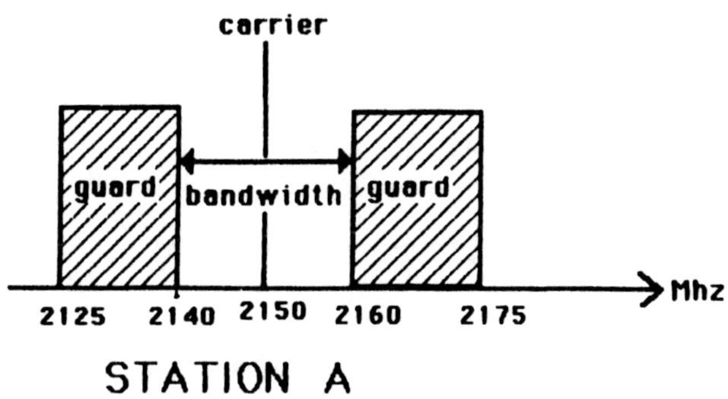

Figure 19.1 The frequency spectrum of television signals for a given station.

The previously mentioned 13 GHz equipment, also known as the shoe box because of the size comparison, is used for short hops. When the event is recorded at a distance beyond the length of ordinary cables from the event site to the van, say at the tenth floor of a building, the shoe box can transmit the recorded information back to the van. The shoe box consists of transmitter-receiver electronics operating at a power level of 500mW or less and at a frequency of 13 GHz or 23 GHz (and 40 GHz in some cases) that can reach a distance under one mile. The van receives the signal and then re-transmits it by a 2 GHz or 7 GHz higher powered microwaves.

19.2 GENERAL CONSIDERATIONS

Microwaves are used to transmit news information from a mobile van back to the television station via a microwave link. This is illustrated by the sketch shown in Fig. 19.2 and by the block diagram shown in Fig. 19.3. The audio/video information of the on-site event is recorded on a video tape. The recorded information is then converted to electronic signals. A microwave carrier of 2 GHz or 7 GHz is modulated by these signals, amplified by a power amplifier, and then transmitted by the van's antenna.

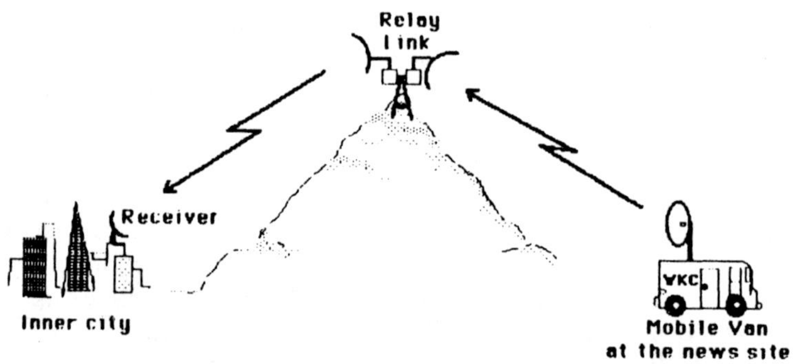

Figure 19.2 A conceptual description of transmitting news information from a mobile van to the station via a microwave link.

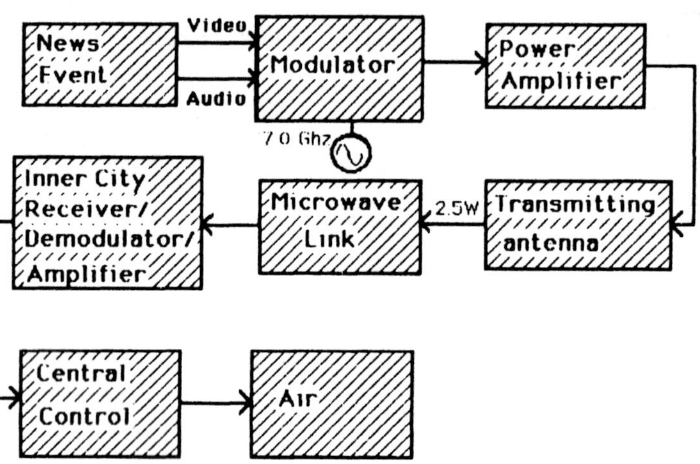

Figure 19.3 The block diagram approach to the transmission of news information from a news event remote location to the station.

A microwave relay link is a fixed structure equipped with receiving and transmitting electronics and serves to relay the van's signal to the television station. The station receives the signal and then goes through the process of amplification and demodulation. The airing section is standard practice and will not be covered here.

Most of the microwave equipment in the television industry are self-contained, pre-tested, and usually automated. These requirements are particularly important for the microwave equipment in the mobile van in order to minimize on-site testing and tuning.

Each block of the block diagram shown in Fig. 19.3 will be elaborated in the following sections. Equipment specifications, costs, and general considerations will also be discussed.

19.3 EVENT

Consider a news event happening at a location within 60 miles from the station. The video information is taken by a special camera while the sound information is recorded separately by a microphone (which is mountable on the camera), These two sets of information are written on the same video tape, or, in the case of live situations, cabled to the van electronics as composite signals.

Figure 19.4 shows a typical hand-carried camera and the video recording equipment. The camera is lightweight, normally about 21 pounds, and the recorder is about 20 pounds. The camera and the recorder are powered by 12V batteries. A roll cable is used to carry the video signal from the camera and the audio signal from the microphone to the recorder as well as power from the recorder to the camera. The rechargeable batteries can power the recorder-camera continuously for 30 minutes.

Figure 19.4 A typical hand-carried camera and video recorder. (Picture taken by Don Sharp of KRON, San Francisco.)

The camera has a 14:1 zoom lens with microlens capability for super close-ups. The recorder such as the Sony BVU-50 records on a mini 3/4" video tape which has a maximum of 20 minutes recording time. The cost of broadcast-quality camera and recorder is $35 thousand and $6 thousand (US), respectively.

19.4 MODULATOR ELECTRONICS

The modulator is a self-contained solid-state unit that combines the audio and the video signals through an FM superheterodyne mixing process with a local oscillator at 70 MHz. The output, known as the IF (intermediate frequency) stage is then directed to frequency-modulate a microwave carrier. The modulated microwave is then amplified by a power amplifier to be discussed next.

The modulated microwave output power is approximately 1W and is conducted to the power amplifier by means of flexible waveguides. A control head is needed to move and rotate the mast on which the power amplifier is mounted. The power amplifier is manufactured by Nurad or M/A-Com for about $12 thousand (US).

19.5 POWER AMPLIFIER

The power amplifier assigns power to the microwave signals. This is a high efficiency, solid-state unit using FETs whose output power is about 12W at 2 GHz and 2.5W at 7 GHz (a different unit). The power amplifier module is mounted on top of the mast and is within a small distance to the transmitting antenna. (See Fig. 19.5 for a conceptual design.)

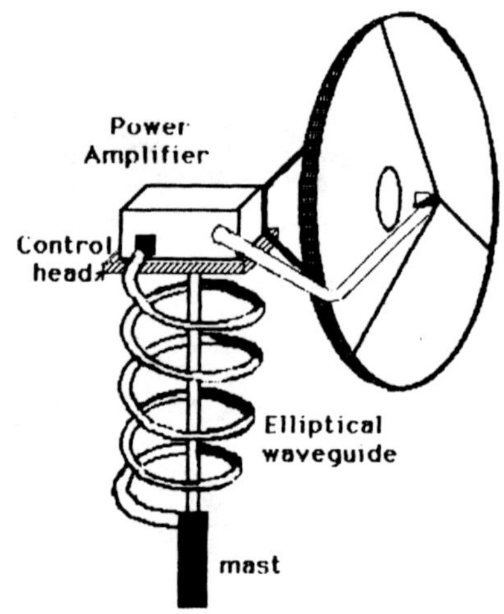

Figure 19.5 Concept of a mobile van power amplifier and transmitting antenna system.

The power amplifier is connected to the modulator control (output) by helix waveguides. This type of waveguides is flexible and has an attenuation of approximately 28 dB per 100ft. It has up to 20W power handling capability and operates at 2 GHz or 7 GHz. Coaxial cables are not used because of their high loss. Rigid waveguides, though they have much lower losses, are impractical because of their limited mobility and poor flexibility as the mast must move vertically and the power head must rotate by as much as 360°.

19.6 TRANSMITTING ANTENNA

The type of transmitting antennas used here depend on the microwave frequency. At 2 GHz, golden rod antennas are used and their directive gain is about 20dB at 2 GHz. At 7 GHz, parabolic dish antennas are used (Fig. 19.6). The gain of a typical 2ft dish is 23 dB. Dish antennas are clearly more directive and far reaching. However, the antenna must maintain an accurate line of sight with the link; a 3 dB loss in power density will result if the antenna deviates from the line of sight by 2°. Golden rod's wider beamwidth requires less effort in aiming at the link at the expense of power density. The dish is constructed out of aluminum, which is light weight, and coated with epoxy paint for weather (rain) consideration.

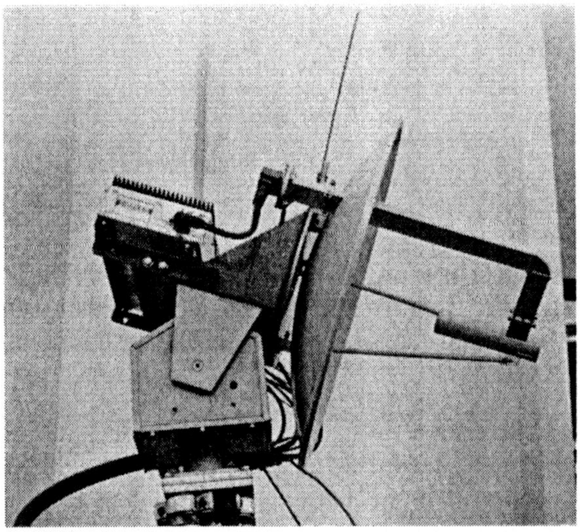

Figure 19.6 A transmitting antenna mounted on top of a mobile van. (Picture taken by Don Sharp of KRON, San Francisco)

A simple calculation will be helpful here. Taking 2.0W and 50 miles (80,000m) as nominal values for transmitting power and distance between the van and the link, the power density due to an isotropic antenna with such transmitting power is

$$\begin{aligned} \text{power density (isotropic)} &= \frac{2}{4\,\pi\,(80000)^2}\ \text{W/m}^2 \\ &= 24.8\ \text{pW/m}^2 \end{aligned}$$

If the dish antenna has a directive gain of 23 db (200) along the line of sight with the link, the power density will be enhanced by a factor of 200, i.e., the power density will be 4.96nW/m^2. In other words, a receiver of 1 square meter capture area will collect a microwave signal of 4.96nW. If the receiver is a 4ft dish (1.2m diameter), the detected power is $4.96\text{nW} \times \pi \times 0.6^2 = 5.6\text{nW}$ (−52.5 dBm).

19.7 MICROWAVE RELAY LINK

A microwave relay link serves to receive the microwave signal transmitted by the van or the helicopter and then re-transmit it back to the television station. There are usually several links from different directions relative to the station. A link is often located at a high attitude, either a 2,000 to 3,000ft mountain, or a 500ft tall high-rise in the case of a flat area, and is visible by the van.

Upon reception, amplification, and sometimes demodulation, a new microwave carrier at 13 GHz is then employed to relay the signal to the station. The dimensions of the receiving and the transmitting antennas vary, but are normally 4ft dishes. The transmitting power is typically 1W (at 13 GHz) at a normal distance of 20 miles from the station. Typically, the transmitter operates continuously since event signals may arrive at the link any time and that usually no operator is usually assigned to the link.

Taking the link's transmitter gain equal to 500 (27 dB) at 13 GHz, the output power equal to 1W, and the distance between the link and the inner city receiver to be 32,000 meter (a little under 20 miles), the power density at the inner city receiver due to the transmitter can be estimated.

$$\begin{aligned} \text{power density (4ft dish)} &= \frac{500 \times 1}{4\,\pi\,(32000)^2}\ \text{W/m}^2 \\ &= 39\text{nW/m}^2 \end{aligned}$$

Hence, if the inner city receiver is a 4ft (1.2m) diameter dish, the detected power is $39\text{nW} \times \pi \times 0.6^2 = 44\text{nW}$.

Weather is usually not a problem as long as proper power planning is done. The transmitter power is chosen so that adequate transmission is assured after free-space attenuation, and any losses due to weather conditions have been taken into account.

19.8 INNER CITY RECEIVER

The signal transmitted by the link is received by the inner city receiver mounted on the roof-top of the station. The receiving antenna is typically a 4ft dish constantly aimed at the link. An amplifier-demodulator unit is attached to the receiver.

The video and audio signals are extracted by superheterodyning with a 70 MHz local oscillator. The two signals are then conducted separately to the central control. The video signal, which is composite, is conducted by a coaxial cable (type 8281) inside a conduit. The typical cable length is 300ft and the attenuation of the 8281 coaxial cable is approximately 1 dB per 100ft at video frequency. The audio signal, being low frequency, is handled by copper wires with virtually no attenuation.

19.9 CENTRAL CONTROL

At the central control, the video and audio information are processed for ready to air or other purposes. Communication must be established between the control and the van operators to line up shots from the field. Most often, such communication is done by two-way radios at 450 MHz. Figure 19.7 shows a picture of the central control.

19.10 MOBILE VAN CONSTRUCTION

The mobile van is a customed designed and custom built vehicle that must fit individual station's needs. General considerations include power, weight, equipment space, and operators' mobility. The total weight of the vehicle with equipment is about 8,000 pounds. Figure 19.8 shows the exterior of the van. The van is equipped with a gasoline powered ac generator at 110V, 60 Hz, and a power capability of 4kW.

19.11 GENERAL DISCUSSION ON MICROWAVE RELAYING

The advantage of using microwaves for news transmission is that the event can be transmitted instantaneously within a radius of 60 miles from the station. The disadvantages include high equipment and man-power cost. The electronics on board the van are usually reliable but mechanical maintenance of the vehicle is still rather high.

The distance limitation of microwave link can be alleviated by relaying through several microwave links. For example, the scene of the Coalinga earthquake was transmitted back to San Francisco 200 miles away via several links. Such operation, though feasible, is rather cumbersome. The distance limitation can now be lifted by the use of direct satellite relay to be discussed next.

Figure 19.7 Part of the inner-city control room. (Picture traken by Don Sharp of KRON, San Francisco)

Figure 19.8 Photograph of a typical television mobile van. (Picture taken by Don Sharp of KRON, San Francisco)

SEC. 19.12
SATELLITE RELAY

19.12 SATELLITE RELAY

A satellite can be regarded as a microwave relay placed 22,500 miles above the surface of the earth, known as the geostationary orbit. With suitable control effort, a satellite in a geostationary orbit appears to be stationary with respect to an observer on Earth because the satellite's orbital rate is the same as the rotating rate of Earth, i.e., one revolution per 24 hours.

The effectiveness of a satellite as a microwave link is much greater than a terrestrial link as can be illustrated by Fig. 19.9. A satellite can cover a much larger area. The space attenuation due to the long distance (22,500 miles) is the dominating factor because it increases with the square of the distance. The atmospheric attenuation is, in this case, relatively constant. Clearly, more transmitting power is necessary in order to overcome the space attenuation. Such power expenditure is, however, well worth spending in light of the gain in immediate communication between the television station and the mobile van located several hundred to several thousand miles away.

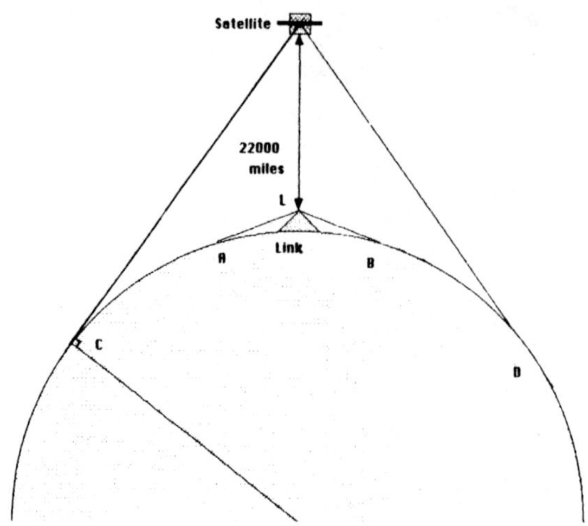

Figure 19.9 A geostationary satellite has the advantage of covering a wide area on the continent.

This and the following section will discuss the Satellite 4 operation of KRON in San Francisco, one of the first efforts in the television industry of the U.S. Today, eight mobile vans from eight different stations equipped with necessary electronics and hardware and having access to the SBS-3 satellite can travel anywhere in the US and instantaneously transmit the news event to the station.

The SBS-3 satellite was launched by the space shuttle Columbia in 1980. CONUS, a consortium responsible for the satellite signal transmission, has 24 hour access to two of the 24 transponders in the satellite. KRON can buy satellite time any time of the day from CONUS at a rate of $75 per five minutes, which is the minimum time. A picture of KRON Satellite 4 van is shown in Fig. 19.10.

19.13 SATELLITE 4 ELECTRONICS

The microwave frequency in the Satellite 4 operation is in the Ku band. The bandwidth available for the transmission is 40 MHz. This 40 MHz bandwidth is adequate.

The block diagram in Fig. 19.3 describing the terrestrial relay is still valid with some modifications. The difference is illustrated in Fig. 19.11 with the satellite replacing the terrestrial link. The electronics starting from the event recording to converting to audio/video signals remains unchanged.

Figure 19.10 A picture of the KRON Satellite 4 television van. (Picture taken by Don Sharp of KRON, San Francisco)

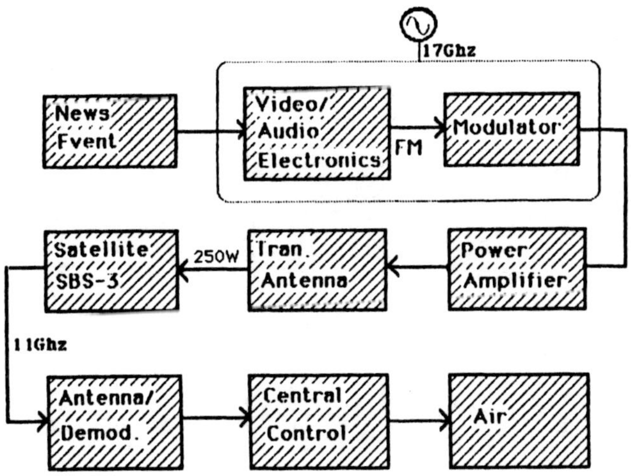

Figure 19.11 A block diagram approach to the transmission of news information via satellite.

The solid-state power amplifier to transmit microwave from the van in the terrestrial scheme cannot deliver the amount of power needed to reach the satellite. Two traveling wave tube (TWT) amplifiers operate in parallel and each can output 250W, CW. A TWT has adequate power capability and its bandwidth is high. One possible concern is the safe location of the high voltage supply operating the TWT. In order to properly drive the TWT, the audio/video signal is amplified by an "exciter" prior to inputing to the TWT.

The microwave signal is transmitted by a 8ft dish. Because the operating frequency is approximately 17 GHz as compared to the lower 7 GHz in the terrestrial operation, the dish antenna can transmit with more directive gain.

Tracking the SBS-3 is done by first consulting the satellite map for an approximate location. As of 1984, tracking is perfected by telephone communication between the van operator and CONUS. This arrangement will change in the future with the van's new capability to receive feedback communication signals from the station via the same satellite.

Upon being received by the satellite, the microwave signal is amplified and demodulated. A new carrier at 11 GHz is employed and the newly modulated microwave is then beamed back to earth. At KRON, a separate 12ft dish antenna is used to receive the signal. The amplification-demodulation process is the same operation as the terrestrial case.

19.14 CLOSING REMARKS

The television engineering has come to a stage of using satellites to relay microwave signals. Terrestrial relaying will soon be reduced and replaced as the cost of satellite time drops due to increase in demand. A news reporter will soon see light weight portable units which should allow him or her to transmit microwave signals from anywhere in the United States to the station via satellite. It is, however, unclear that this freedom will be available to other countries as special permission from the host country is usually necessary to have access to a satellite.

CHAPTER 20

THE FUTURE OF MICROWAVES

Frederic H. Levien

THE INFORMATION REVOLUTION

The Information Revolution

It has been suggested that in the judgment of history, the era we are now passing through will be called the "Information Revolution."

In *Megatrends,* a chronicle of our changing times, John Naisbitt has pointed out, "It is now clear that the post industrial society is the information society." He expanded by adding, "In an information society, for the first time in civilization, the game is people interacting with other people. This increases personal transactions geometrically, that is all forms of interctive communication; telephone calls, checks, written memos, messages, letters and more!"

The world's need to share information and communicate has been exploding. Microwave technology of course has proven to be one of the most efficient means of transmitting information from place to place. Activities that only a few short years ago were not even dreamed of, have now become commonplace.

News Distribution: Personalized newspapers are close to becoming a reality. Instead of being presented with an impersonal and often superficial selection of all the news, as today, a subscriber may register his news requirement "profile." He then receives detailed news on topics that interest him. This information may be printed out on his home terminal or stored for him so that he can display it on his screen when desired. All types of categories can be registered. For example, local news about a district other than where he lives; news about a particular industry, company, or stock; scientific news; movie reviews; news about crime, sex, war, or foreign news.

Powerful satellites now permit home pick-up of worldwide television, news, and entertainment, directly through roof mounted antennas in some cases.

The old familiar sound of the evening newspaper hitting the driveway at night, may go the way of the dinosaur.

Medical Monitoring: Microwave telecommunication is being extensively used in medicine. Information from all manner of patient data is being transmitted directly to specialists or into computers for storage. "Pre-diagnosis" interviews are carried out between patient and distant computers, often to determine whether the patient should see a doctor or not, or visit a hospital. Automatic monitoring of chronically sick patients is done by computer. Sometimes patients are monitored during normal daily activities by means of miniature instrumentation packages connected to radio transmitters. In some cases, their readings are recorded by a tiny machine that can later be linked to the telephone and transmit the readings to a hospital computer.

Also becoming available are remote diagnostic studios, equipped with powerful television cameras and lenses. With the help of a nurse in the studio, a distant doctor or specialist can examine a patient as though the patient were in his office. The patient can see the doctor as well and converse with him. The doctor will be able to listen to a distant stethoscope and see both instrument readings and computer analyses of them. All of this information will be transmitted via a microwave link.

Electronic Funds Transfer: A major change is occuring in the way financial payments are made in the United States. Cash transfers now take place entirely within the electronic system. "Electronic Funds Transfer" (EFT) take place within one computer, or between two different computers holding the accounts of the person concerned. They are often initiated by the use of an EFT card, a credit card that contains machine readable details about the holder and which is inserted into an on-line terminal in stores, restaurants, and offices. The cost per transaction is substantially lower than with checks or credit cards. The number of checks in use has risen to over 50 billion per year and bankers desperately hope that EFT will lessen the deluge of check processing.

There is increasingly less resistance to a "checkless" society, and there are many thousands of fairly inexpensive electronic fund transfer terminals in use. Early tests of EFT terminals in homes have proved to be very successful, and networks are being constructed to allow people to do their banking from home, with these terminals connected to their bank via a microwave link.

Electronic Mail: Conventional mail delivery, especially in the large cities, has increased in unreliability. With the marked increase in home computers and the ability to reproduce the written word on home facsimile equipment and printers, the lure of electronic mail has become a reality. Both the US Postal Service and several private carriers now offer this service which makes heavy use of microwave links. As people have once again come to depend on rapid and reliable delivery of mail, these new services have gained in popularity, and the volume has grown, thereby shrinking per message costs and expanding the market.

In all of the above new applications, plus all of the routine day to day telephone calls, data exchanges, and video programs, the job is done through extensive use of microwaves. So, to answer the question of "What is the future of the microwave industry?", we need summon but a single word, "*up*!"

Any more thorough a response must begin with an examination of whom is asking the question. The venture capitalist sees the microwave industry as a vehicle in which to obtain return on his capital. The business man serving this industry assesses the possible market it presents for his products. The individual engineer views it as a possible career from which he will earn his livelihood.

We will examine the question from several points of view.

20.1 APPLICATION OF MICROWAVE

20.1.1 Communication Satellites

In the infancy of the telecommunications industry, the only medium available was the traditional copper wire, most often strung from pole to pole, as far as the eye could see. In the US prior to World War II, there was virtually no other choice. Beginning in the years just after World War II, terrestrial point-to-point microwave radio began to gain an ever increasing share of this installed base. Finally, in the early 1970s, satellite communications began to enter the picture, first in the military and government sector, and next in domestic commercial telecommunications. Looming ahead in the 1980s, is the penetration of fiber optics as a viable alternative to transmission of electronic information.

Examining Table 20.1, we see the dramatic growth of domestic US telecommunications satellites in less than a decade. With only three birds up in 1975, with a capacity of slightly over 34 thousand voice circuits, there were 21 birds up by 1983, with over one-quarter of a million voice circuits available. Some predictions put the number of telecommunications satellites up in the hundreds by the end of the decade, and the number of transponders in orbit is expected to reach about a thousand. This continued expansion of telecommunication satellite needs may slow dramatically over the coming decade however, if the promise of fiber optics, with its enormous capacity for carrying traffic is proven in as anticipated.

Table 20.1
Domestic US Telecommunication Satellite Historical Data

	1970	1975	1980	1982	1983
Number of Domestic T/C Satellites	None	3	9	18	21
Total Number of Transponders Operational	None	48	166	330	390
Average Number of Voice Circuits per Transponder	None	722	692	714	709
Voice Circuits	None	34,656	114,872	235,620	276,510

20.1.2 Space Applications

The childhood fantasy world of Buck Rogers that many of us enjoyed is no longer stuff for dreams. It is here and now, and it is big business.

For example, TRW and NASA have teamed up, planning an enormous space station that could be in place above the earth in the 21st century. These orbiting space platforms of tomorrow will be larger than anything ever put into space before. Power to these stations will require tremendous amounts of energy. To supply this demand, NASA envisions developing solar power satellites with gargantuan solar arrays, about the size of Manhattan Island. These structures would be assembled piece by piece in space. These space borne communication platforms would once again be linked to earth by microwave transmission systems. They would perform such tasks as special weather observations, earth resource monitoring and even more mundane tasks could be done such as a "postal system in the sky" in which letters could be sent much the same way as telephone calls are now.

In addition, if there is sufficient excess capacity of solar generated power from space, there are proposals already being explored whereby this power could be beamed to earth by microwaves for augmenting terrestrial fossil and nuclear fuel electrical generating capacity.

20.2 BUSINESS GROWTH FOR MICROWAVE SYSTEMS

20.2.1 Capacity Expansion

The demand in the US for telecommunication channels has been growing steadily for the past two decades. Examining Tables 20.2 and 20.3, we see that the total installed base expanded from 1.43 billion equivalent voice circuit miles in 1971 to 6.36 billion in 1983. This was an average annual growth rate of 13%. It is predicted that demand over the next five years will expand at an even more rapid rate of 26 percent per year, reaching 32.3 billion voice circuit miles by 1990.

The driving forces in the commercial world are the heavier demand for data transmission between computer sites, and the myriad array of new services available to people all requiring transfer of information. Coupled with the down trending costs to provide this data, demand is expected to be strong through this century.

In examining these tables, the obvious growth of the two media requiring microwaves, satellite and microwave radio, is apparent. It is worth noting too, the dramatic increase in the use of fiber optics. This technology represents a formidable contender for the task now assigned almost exclusively to microwaves, long haul transmission of voice and data traffic.

Table 20.2
US Military-Government Telecommunications Capacity by Media, 1971–1994
(Cumulative Million Equivalent Voice Circuit Miles)

Transmission Medium	1971	1983	1990	Average Growth % per Year 1971–83	Average Growth % per Year 1983–90
Copper Cable	149	214	229	3	1
Satellite	65	2,233	7,546	35	19
Microwave Radio	283	621	1,135	7	9
Fiber Optic Cable	0	22	2,285	–	100
TOTAL	497	3,090	11,735	17	21

Table 20.3
Commercial Telecommunications US Domestic Capacity by Media
(Cumulative Million Equivalent Voice Circuit Miles)

Transmission Medium	1971	1983	1990	Average Growth % per Year 1971–82	Average Growth % per Year 1983–90
Copper Cable	700	1,262	1,353	5	1
Satellite	0	263	1,401	–	27
Microwave Radio	224	654	1,050	9	7
Fiber Optic Cable	0	25	14,506	–	148
Subscription TV Cable	11	1,067	2,215	47	11
TOTAL	935	3,271	20,525	11	30

Two driving forces in the military behind the continued rapid expansion of telecommunication demand include expansion of US military operating groups to more separate locations widely dispersed around the world and the strong command preference for real time, secure communication to all these operations from all higher command levels. Also the rapidly expanding demand for world-wide military and geopolitical intelligence, coupled with sensors that can provide this information, represent immense quantities of data that need be moved.

The growth in demand and the growth in capacity are but two sides of the same coin. For without the demand there would be no need for capacity. Yet, just as surely, if the capacity were not developed, a curtailment of demand would result as the world found other ways to meet its needs.

20.2.2 Investment Expansion

The investment in microwave related hardware in the US and the world is substantial. Focusing on US microwave telecommunication equipment production alone, as shown in Table 20.4, provides a clue as to the amount of growth anticipated in the next five years. This table also provides a breakdown of the split between frequencies, clearly highlighting the shift to the higher bands. We see that expenditures will more than double from $2.3 billion in 1984 to over $5 billion in 1990. Not included here is an additional $100 million of imported microwave radios from Japan in 1983, mostly in the commercial area.

Estimates on the size of the market for microwave equipment in the military electronics world are equally impressive. The total sales for electronic warfare are estimated at $6 billion (US) in 1984, increasing to about $16 billion by 1989.

All of this is, of course, not solely for microwave-related hardware. However, for just one of the many microwave components alone, solid-state microwave amplifiers, which represent about 1.2% of the total EW market, this amounts to a substantial $200 million market by 1989.

Table 20.4
Total US Microwave Telecommunication Equipment Production
($ Million)

	1983	1984	1987	1990
L Band				
Military/Government	638	755	925	919
S Band				
Commercial	22	28	42	47
C Band				
Military/Government	44	58	61	96
Commercial	361	528	800	1,139
TOTAL C BAND	405	586	861	1,235
X Band				
Military/Government	254	302	417	357
Commercial	51	85	152	238
TOTAL X BAND	305	387	569	595
K_u Band				
Military/Government	105	245	499	1,195
Commercial	168	217	665	1,346
TOTAL K_u BAND	273	462	1,164	2,541
K_a Band				
Commercial	–	84	104	165
TOTAL ($ Million)	1,643	2,302	3,665	5,502

20.3 NEW TECHNOLOGIES

20.3.1 Monolithic Microwave Integrated Circuits (MMIC)

The recent arrival of the MMIC, has opened a new window in the world of technology. These tiny microwave circuits, made of a compound of gallium and arsenic (gallium arsenide), are no larger than the head of a match. Because of the unique properties of gallium arsenide, however, these parts offer some striking new advantages.

In its simplest form, a GaAs MMIC represents all of the electronic functions needed to make a complex receiver on a single chip. A typical outdoor direct broadcast satellite receiver for example, uses a low noise amplifier, a band-pass filter, a mixer, a local oscillator, an intermediate amplifier, and a power converter. In 1983, Toshiba Corporation reported developing just such a system in two GaAs MMIC chips.

When compared to silicon, GaAs chips work up to five times faster than even the speediest silicon chips. Furthermore, they require much less power to operate. Because of the basic physics governing how GaAs functions, it also has considerably higher resistance to radiation damage than silicon. This of course, is of great value in communication circuits in satellites and in those military environments where radiation could be present. Some circuits can withstand radiation up to 10^8 rad, where only 10^4 rad is typical for silicon.

Another decided advantage for GaAs chips is their ability to withstand temperatures from $-200°$ to $+200°C$, with maximum temperatures possible up to 400°C with special processing.

SEC. 20.3
NEW TECHNOLOGIES

Although the military is the driving force behind the present push for bringing MMICs into production, it is expected that civilian demand for GaAs MMICs will eventually dwarf military uses.

All of the major electronic companies in both the US and Japan are heavily committed to this product. That includes NEC, Fujitsu, Toshiba, Mitsubishi, and Hitachi. NEC is close to bringing to market low noise and medium power MMIC amplifiers up to 30 GHz and switches and a mixer up to 12 GHz.

In the US, companies such as Harris Corporation, Honeywell, ITT, Rockwell, Texas Instruments, and TRW are moving into a business that looms larger with each passing year.

A cross section of a typical monolithic amplifier is shown in Figure 20.1. This chip contains GaAs FETs, inductive lines, tuning capacitors, and crossovers to make the necessary contact to the various elements, and the via hole etched through the GaAs substrate. The via hole allows contact from the ground point on the surface of the device to the final ground on the bottom of the device.

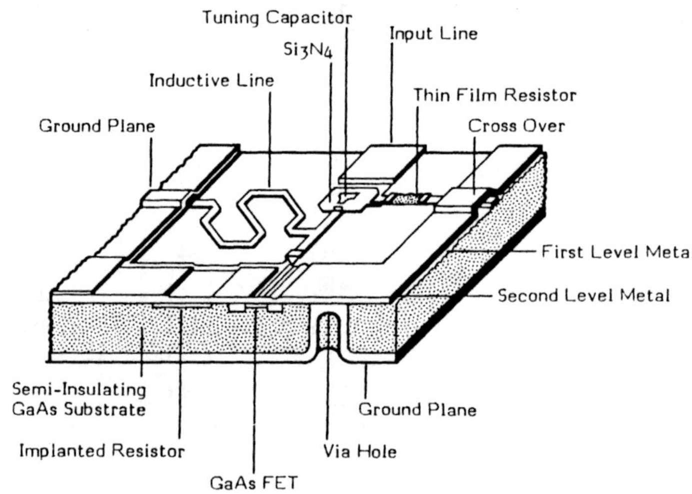

Figure 20.1 Cross section of GaAs MMIC chip.

For GaAs MMICs and discrete devices, performance (noise and power) forecasts through 1990 are indicated in Figs. 20.2 and 20.3. Bands of uncertainty are

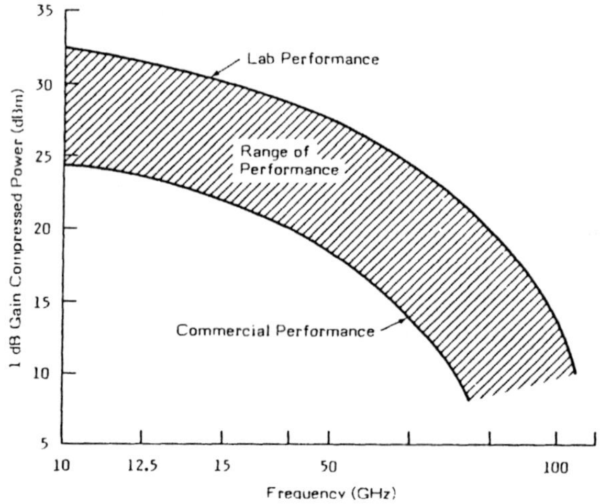

Figure 20.2 1990 noise performance forecast GaAs, discrete and MMIC.

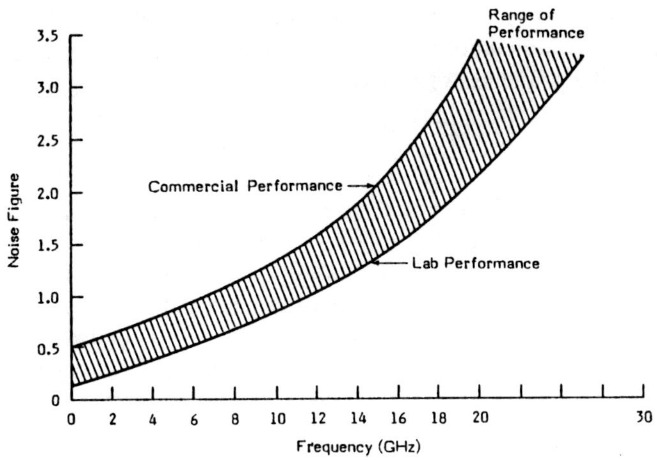

Figure 20.3 1990 power performance forecast GaAs, discrete and MMIC.

indicated for noise figure in low noise devices, and power output in power devices. These bands of uncertainty encompass laboratory results on one side and commercial results on the other. Stated another way, the band range indicates the development area for both laboratory and commercial devices.

Performance forecasts for GaAs power discretes and MMICs through 1990 are outlined in Fig. 20.4. In this figure, we predict that the range of uncertainty in overlap of lab or commercial results for power out will vary between these curves. For example, there probably will be a few specialty commercial devices in the millimeter wave range by 1990, with power out in the range of 15 dBm. There will be more commercial devices in the 60 GHz range, with power output varying between 37 dBm and 45 dBm. Laboratory results may reach as high as 24 dBm at 150 GHz.

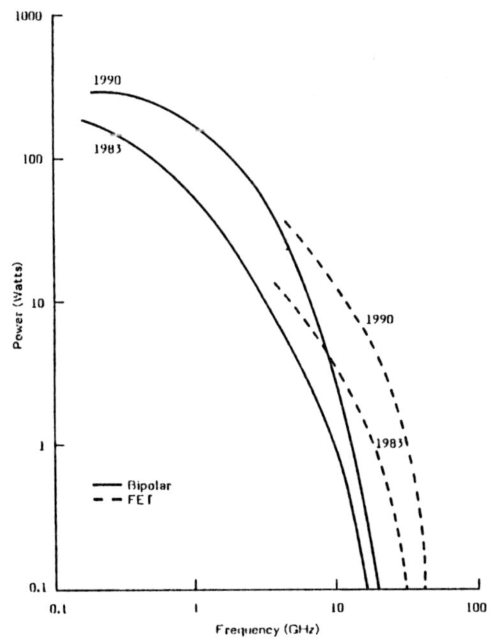

Figure 20.4 Performance predictions of microwave solid-state amplifiers.

SEC. 20.3
NEW TECHNOLOGIES

Performance forecasts for GaAs discretes and MMICs for noise figure in small signal devices are indicated in Fig. 20.2. Again, the bands of uncertainty in this figure indicate the gray area of differentiation between laboratory and commercial results.

20.3.2 Fiber Optics

The major challenge to the continued growth of microwaves for use in terrestrial systems is fiber optics. Dramatic growth has occurred in fiber optic technology over the past decade, and it is anticipated this technology will continue to advance. From a laboratory curiosity in the early 1970s, it is expected that fiber optics will become the major mode of new plant for long haul communications in the next decade. Figure 20.5 depicts the anticipated penetration of fiber optics into telecommunications, beginning in late 1980s.

Several factors are carrying this new technology forward. First, a continuing improvement in fiber performance. Its basic advantages of small size, light weight, high security, and non-conductive, non-radiative characteristics, are well known. However, two additional technological advancements in fiber have become overriding for its microwave communications application. These are the extremely high bandwidth at low cost and the continued reduction of fiber loss per linear foot.

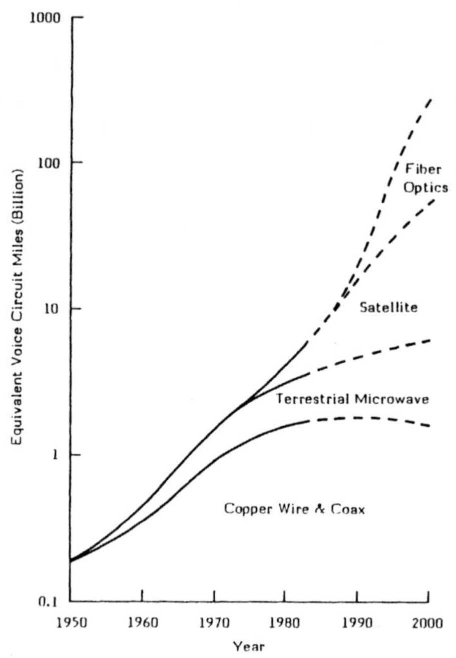

Figure 20.5 Fiber optic penetration of telecommunications.

Available in 1984 are optical fibers in lengths of many kilometers with losses of less than 0.2 dB/km. Cost for this fiber is less than 50 cents per meter. In the laboratory now is a new fiber made from fluoride glass with a theoretical loss limit of two orders of magnitude better than the present silicon glass. Fiber loss is already low enough that loss is no longer a limitation for most applications.

Improvements in dispersion characteristics of fiber, will also increase the effective bandwidth of a fiber system. The present trends toward lower loss, lower dispersion *versus* bandwidth, higher laser diode stability, and improved multiplexer designs, indicate that by the end of this decade it will be economically feasible to transmit up to 20 gigabits of information on a single fiber over a distance of 10 kilometers between repeaters.

By 1990, it is also expected that the present cost of $1.50 per meter of length for fiber will be reduced to 15 cents.

20.4 ADVANCEMENT IN COMPONENT TECHNOLOGY

Using the microwave solid state amplifier (MWSSA) as a typical bellwether of the microwave component industry, we can examine the performance improvements the system designer might look toward.

The history of microwave solid-state amplifiers over the past two decades has been one of continually pushing the power capabilities of these units to higher levels, at higher frequencies. From their penetration into the lower edge of L band two decades ago, these units have progressed to provide usable power output well into the millimeter range. This trend to higher power capability at higher frequencies will continue over the forecast period, as illustrated in Fig. 20.4. This will permit continuing erosion of the microwave tube amplifier market by MWSSAs, including retrofit of MWSSAs into tube sockets in currently operational equipment.

20.5 SUMMARY

In discussing the future of microwave, there is a tendency to concentrate on its new and exciting applications, as the systems which provide the means for carrying out these new techniques, and finally as the components which provide the basic building blocks for all of the above.

It would be a mistake however, to ignore the part of the equation which makes all the rest possible—people: trained and technically competent personnel, upon whom all of the future growth of this industry depends. It would appear that at this point in time, a shortage of microwave engineering talent is looming as one of the major stumbling blocks to a continued healthy growth of the industry.

To quote the opening page of a recently completed report by Electronicast Corporation titled, "Microwave Solid State Amplifier Market Forecast," presenting the sobering warning:

> . . . the 1984 growth of U.S. production of MWSSAs is capacity limited. The key element in this limitation is the shortage of engineers skilled in solid state microwave amplifier designs.

APPENDIX A

MATERIAL PROPERTIES

This appendix lists properties of some common metals and dielectrics. The properties given here are by no means complete, but should serve the purpose of this book adequately.

In Appendix A.1, the units are not SI (Standard International or *Système Internationale*), but nonetheless commonly used. Conversions to SI units can easily be done:

Conversion	*Multiply by*
Density (gm/cc) to density (kg/m^3)	1000
Resistivity (ohms-cm) to resistivity (ohms-m)	0.01
Thermal conductivity (W/cm-°C) to thermal conductivity (W/m-°C)	100

A.1 Metal Properties

Common Metal	Symbol	Density (gm/cc)	Coefficient of Linear Expansion (at 20°C)	Resistivity (ohms-cm at 20°C)	Melting Point °C	Thermal Conductivity (W/cm-°C)
Aluminum	Al	2.7	2.5×10^{-5}	2.8×10^{-6}	660	3.00
Copper	Cu	8.9	1.6×10^{-5}	1.7×10^{-6}	1083	4.83
Gold	Au	19.3	1.42×10^{-5}	2.4×10^{-6}	1064	3.45
Iron	Fe	7.8	1.2×10^{-5}	9.8×10^{-5}	1535	1.32
Lead	Pb	11.4	2.9×10^{-5}	2.2×10^{-5}	327	0.40
Magnesium	Mg	1.74	2.5×10^{-5}	4.6×10^{-6}	649	1.69
Mercury	Hg	13.5	**	9.6×10^{-5}	−38.9	**
Molybdenum	Mo	9.0	5.0×10^{-6}	5.7×10^{-6}	2617	1.79
Nickel	Ni	8.9	1.3×10^{-5}	7.8×10^{-6}	1453	1.58
Platinum	Pt	21.4	9.0×10^{-6}	1×10^{-5}	1755	0.79
Silver	Ag	10.5	1.9×10^{-5}	1.6×10^{-6}	962	4.50
Tin	Sn	7.7	2.0×10^{-5}	1.15×10^{-5}	232	0.85
Tungsten	W	5.6	4.5×10^{-6}	5.6×10^{-6}	3400	2.35
Zinc	Zn	5.8	3.5×10^{-5}	5.8×10^{-6}	419	1.32

**not available

Among the above materials, only iron and nickel are magnetic. The relative permeability of the remaining metals can be safely taken as 1.0. The relative permeabilities of iron and nickel are not constant but vary with the magnetic field strength applied externally.

A.2 Dielectric Properties

Material	Dielectric Constant (k)	Maximum Temperature (°C)	Flexibility
I. Common in Microwave			
Alumina	9.60	500	very poor
Beryllia	6.60	500	very poor
Boron Nitride	4.40	500	poor
Epsilam 10	10.00	150	good
Polyethylene	2.25	150	good
Teflon®	2.04	200	good
Teflon fiberglass	2.55	200	good
II. Other Materials for Comparison			
Diamond	16.5		
Mylar	3.0		
Paper	3.0		
Porcelain	6.0–8.0		
Quartz	3.8		
Soil	2.8–3.0		
Water	80		
Window glass	7.8		
Wood	5.5		

APPENDIX B

COMPLEX NUMBERS

Electronics uses complex algebra to describe the phase difference between two physical parameters, e.g., between the voltage and the current in a capacitor. The readers are reminded that the voltage and the current are (1) in phase in a resistor, (2) 90° in a capacitor (voltage lagging behind current), and (3) 90° in an inductor (voltage leading the current). These phase relationships are illustrated in Fig. B.1.

A complex number contains a factor j denoting $\sqrt{-1}$. An ordinary calculator will display "error" if one attempts to take the square root of a negative number. It turns out that analysis of electrical circuits involving a combination of resistors, inductors, and capacitors via calculus frequently yields the term $\sqrt{-1}$ accompanying the phase relationships among the resistors, the inductors and the capacitors.

Normal quantities such as 0, 3, −4, 2.7, *et cetera* have no j terms and are classified as *real* numbers. In general, a complex number can be decomposed into two parts: real and imaginary, i.e.,

Complex number = Real number + Imaginary number.

For example, the quantity $3+j4$* is a complex number with 3 being the real number and $j4$ being the imaginary number.

In electronics, the complex number $3+j4$ describes a series connection of a three-ohm resistor and a four-ohm inductor while the complex number $1-j2$ describes a series connection of a one-ohm resistor and a two-ohm capacitor. In general, the real part describes the resistive component and the imaginary part describes the reactive component (inductive or capacitive) as shown in Fig. B.2. Note that in Fig. B.2 voltages of the components are being compared. For example, the voltage across a capacitor is assigned as a negative imaginary number because it lags behind the voltage across a resistor. Similarly, the voltage across an inductor is a positive imaginary number because it leads the voltage across a resistor. Also, there is no need to consider the negative side of the real part because negative resistance does not generally exist.

Example (B.1)

Read the complex numbers given in Fig. B.3 and describe their electronic equivalents.

Point	*Complex number*	*Series electronic equivalent*
A	$2+j5$	2Ω resistor + 5Ω inductor
B	$4+j6$	4Ω resistor + 6Ω inductor
C	$5+j2.5$	5Ω resistor + 2.5Ω inductor
D	$3-j1.5$	3Ω resistor + 1.5Ω capacitor
E	$2.5-j4$	2.5Ω resistor + 4Ω capacitor
F	$1-j5$	1Ω resistor + 5Ω capacitor

Several algebraic properties of complex numbers are discussed here:

a) $$j^2 = j \times j = \sqrt{-1} \times \sqrt{-1} = -1 \quad \text{(B.1)}$$

b) $$j^3 = j \times j \times j = j \times j^2 = j \times (-1) = -j \quad \text{(B.2)}$$

*This is the same as $3+4j$ but engineers tend to write j first.

c) $1/j = j/j^2 = j/(-1) = -j$ (B.3)

d) $j^4 = j^2 \times j^2 = (-1) \times (-1) = +1$ (B.4)

e) Two or more complex numbers can be added by adding their real parts and their imaginary parts independently. For example, let complex numbers $a = 1 + j2$, $b = 3 + j4$, $c = 5 - j6$, then
(i) $a + b = (1 + 3) + j(2 + 4) = 4 + j6$
(ii) $a + c = (1 + 5) + j(2 - 6) = 6 - j4$
(iii) $a + b + c = (1 + 3 + 5) + j(2 + 4 - 6) = 9 + j0$.

f) Two complex numbers can be multiplied with the following results.

Let

$$zw = (x + jy)(u + jv) = xu + jxv + jyu + j^2yv$$

then

$$zw = (x + jy)(u + jv) = xu + j \times v + jyu + j^2yv$$

So

$$zw = (xu - yv) + j(xv + yu) \tag{B.5}$$

For example, if

$$z = 3 + j2, \ w = 6 + j5,$$

then

$$zw = (3 \times 6 - 2 \times 5) + j(2 \times 6 + 3 \times 5) \\ = 8 + j27$$

g) The reciprocal of a complex number $z = x + jy$ can be obtained by the following technique:

$$\frac{1}{z} = \frac{1}{x + jy}$$

(multiply the numerator and the denominator by the same quantity $x - jy$)

$$= \frac{(x - jy)}{(x + jy)(x - jy)}$$

using Eq. (B.5)

$$= \frac{x - jy}{(x^2 + y^2) + j(xy - yx)}$$

i.e.,

$$\frac{1}{z} = \frac{x - jy}{x^2 + y^2} \tag{B.6}$$

Hence, the reciprocal of a complex number $z = x + jy$ is also a complex number with the real part $x/(x^2 + y^2)$ and the imaginary part $-jy/(x^2 + y^2)$.

For example, let $z = 2 + j3$, then, according to Eq. (B.6),

$$\frac{1}{z} = \frac{2 - j3}{2^2 + 3^2} = \frac{2 - j3}{13}$$

$$= 2/13 - j3/13 = 0.15 - j0.23$$

The above properties have immediate applications in electronics. Recall that any combination of a resistor, an inductor, and a capacitor is described by the impedance Z (the ac version of resistance) which is the sum of two parts: resistive (real) and reactive (imaginary).

Consider two combinations Z_1 and Z_2 where $Z_1 = 3 + j5$ and $Z_2 = 10 + j15$ as shown in Fig. B.4. the summation $Z_1 + Z_2$ means the two combinations are now connected in series with each other as shown in Fig. B.4.

As another example, let $Z_3 = 1 + j10$ and $Z_4 = 2 - j10$. Then $Z_3 + Z_4 = 1 + j10 + 2 - j10 = 3 + j0$, there is no net reactive effect. This connection is that of a series *RLC* circuit as shown in Fig. B.5.

When two impedances Z_1 and Z_2 are connected in parallel, the equivalent impedance Z_e is

$$\frac{1}{Z_e} = \frac{1}{Z_1} + \frac{1}{Z_2} \tag{B.7}$$

or

$$Z_e = \frac{Z_1 \times Z_2}{Z_1 + Z_2} \tag{B.8}$$

The reciprocal of impedance is admittance, usually denoted as Y, i.e., $Y = 1/Z$. Hence, Eq. (B.7) can be rewritten as

$$Y_e = Y_1 + Y_2 \tag{B.9}$$

Therefore, a parallel circuit is better described by adding the individual admittances.

For example, two impedances $Z_1 = 2 + j5$ and $Z_2 = 3 + j4$ are connected in parallel as shown in Fig. B.6. The equivalent impedance can be found by working directly with Eq. (B.8) or using admittances given by Eq. (B.9). Using Eq. (B.8),

$$Z_e = \frac{(2 + j5)(3 + j4)}{2 + j5 + 3 + j4} = \frac{-14 + j23}{5 + j9}$$

$$= \frac{(-14 + j23)(5 - j9)}{(5 + j9)(5 - j9)} = \frac{137 + j241}{25 + 81}$$

$$= 1.292 + j2.274$$

Using the admittance approach,

$$Y_1 = 1/Z_1 = \frac{1}{2 + j5} = \frac{2 - j5}{4 + 25}$$

$$= 0.069 - j0.172$$

$$Y_2 = 1/Z_2 = \frac{1}{3 + j4} = \frac{3 - j4}{9 + 16}$$

$$= 0.120 - j0.160$$

$$Y_e = Y_1 + Y_2 = (0.069 + 0.120) - j(0.172 + 0.160)$$

$$= 0.189 - j0.332$$

The equivalent impedance Z_e is obtained by finding the reciprocal of Y_e.

$$Z_e = 1/Y_e = \frac{1}{0.189 - j0.332} = \frac{0.189 + j0.332}{0.146}$$

$$= 1.295 + j2.274$$

It can be seen that both approaches give the same answer.

A complex number can be alternatively expressed as an absolute magnitude accompanied by a phase angle. With reference to Fig. B.7, the complex number $Z = 3 + j4$ means a three-unit resistive and four-unit inductive. The absolute magnitudes of the complex number is the length of the hypotenuse of the right-angled triangle. In general, the absolute magnitude of a complex number Z is denoted by $|Z|$ and the phase angle is denoted by θ where

$$|Z| = \sqrt{x^2 + y^2} \tag{B.10}$$

and

$$\theta = \arctan\,(y/x) \tag{B.11}$$

Note that the operation "arctan" means the inverse of the trigonometric function "tangent." In some calculators, the sequence is INV followed by TAN. Also, the phase angle can be expressed in either degrees or radians. We shall use degrees here.

The absolute magnitude can only be positive but the phase angle can be positive or negative depending on the polarity of the imaginary part. For example, the complex number $Z_1 = 3 + j4$ has an absolute magnitude of $\sqrt{3^2 + 4^2} = 5$ and a (positive) phase angle of arctan $(4/3) = 53.1°$. This is illustrated in Fig. B.8. Therefore,

$$Z_1 = 3 + j4 = 5\angle 53.1°$$

A second complex number $Z_2 = 2 - j3$ has an absolute magnitude of $\sqrt{2^2 + 3^2} = 3.61$ and a (negative) phase angle of arctan $(-3/2) = -56.3°$. With reference to Fig. B.8,

$$Z_2 = 2 - j3 = 3.61\angle -56.3°.$$

Using *Example (B.1)* with Fig. B.3, the following complex numbers are expressed in terms of their absolute magnitudes and phase angles. (The reader may wish to verify the arithmetic.)

Point A = $2+j5=5.36\angle 68.20°$
Point B = $4+j6=7.21\angle 56.31°$
Point C = $5+j2.5=5.59\angle 26.57°$
Point D = $3-j1.5=3.35\angle -26.57$
Point E = $2.5-j4=4.72\angle -57.99°$
Point F = $1-j5=5.10\angle -78.69°$

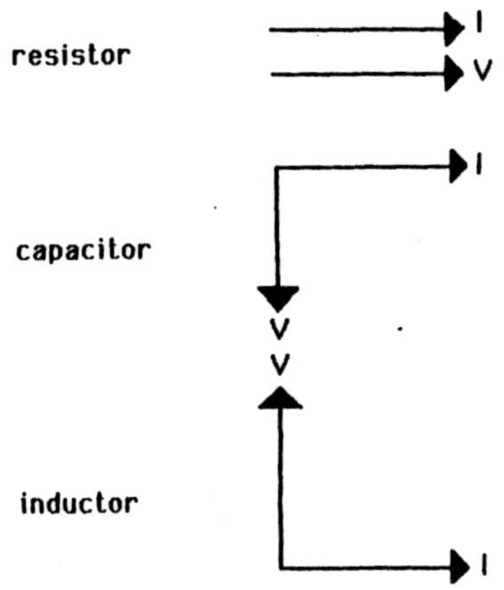

Figure B.1 The voltage/current relationships in a resistor, capacitor, and an inductor.

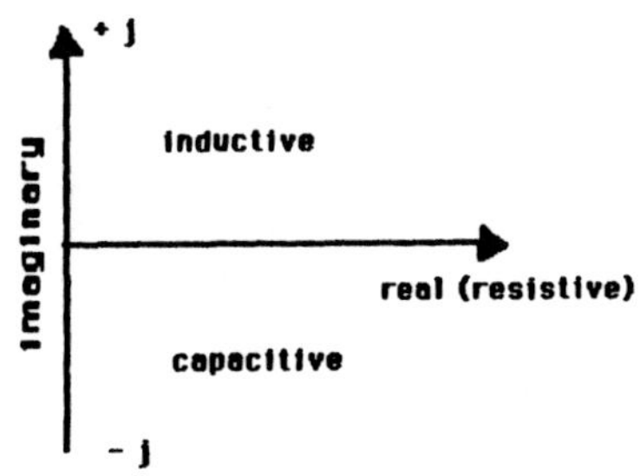

Figure B.2 Using imaginary number j to describe the phase relationships of voltages between an inductive and a capacitive element relative to a resistive element.

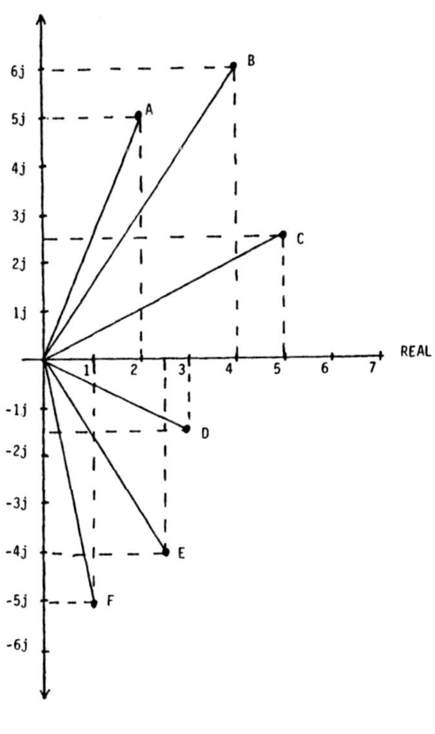

Figure B.3 The locations of six imaginary numbers.

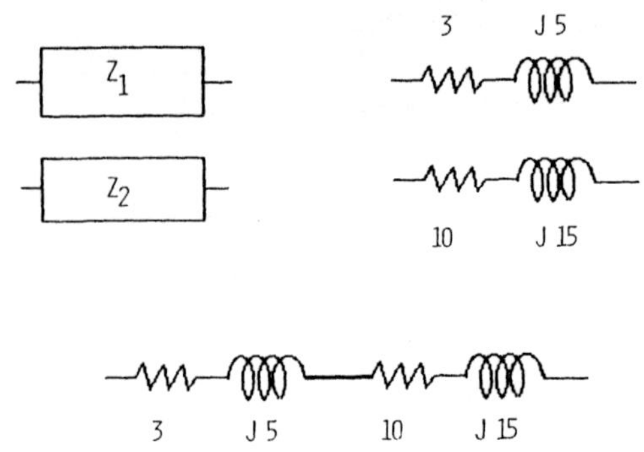

Figure B.4 Two impedances, both of positive reactances, in series.

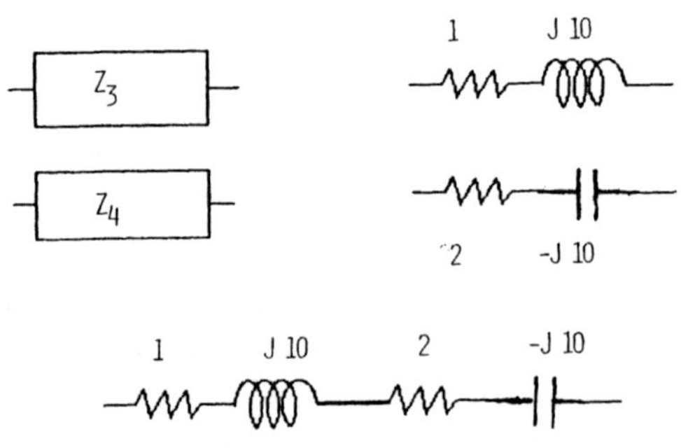

Figure B.5 Two impedances, one positive and one negative reactance, in series resulting in a net zero reactance.

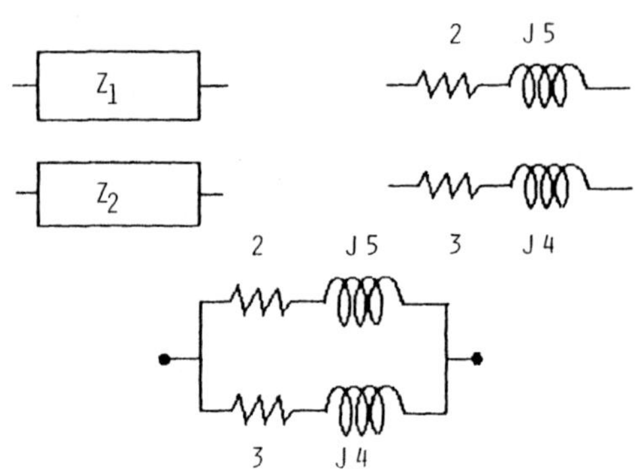

Figure B.6 Two impedances in parallel.

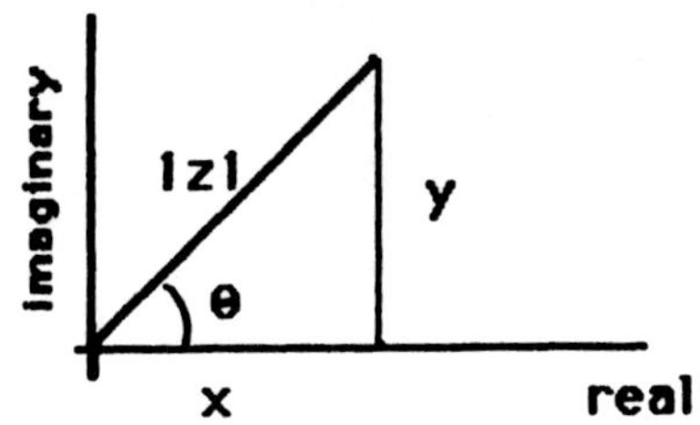

Figure B.7 An imaginary number can be expressed as x + jy or as the absolute length lzl and the phase angle.

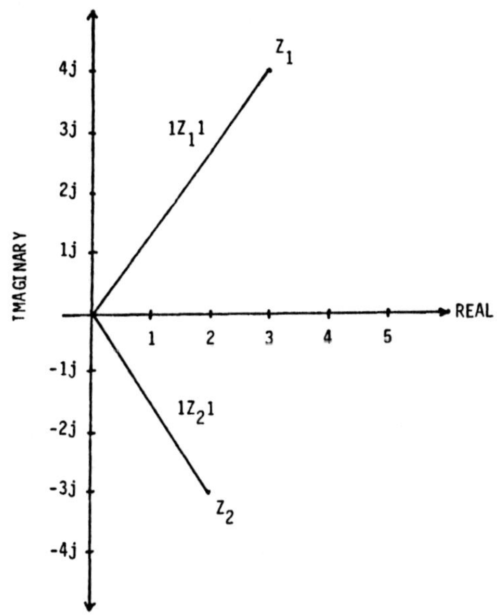

Figure B.8 Finding the absolute lengths and phase angles of two impedances.

APPENDIX C
VSWR, RETURN LOSS, AND OTHER CONVERSIONS

VSWR	*Volt. Refl. Coeff.*	*Insertion Loss (dB)*	*Power Trans. (%)*	*Power Refl. (%)*	*Return Loss (dB)*
1.00	.00	.000	100	0	∞
1.01	.00	.000	100	0	46.1
1.02	.01	.000	100	0	40.1
1.03	.01	.001	100	0	36.6
1.04	.02	.002	100	0	34.2
1.05	.02	.003	99.9	0.1	32.3
1.06	.03	.004	99.9	0.1	30.7
1.07	.03	.004	99.9	0.1	29.4
1.08	.04	.006	99.9	0.1	28.3
1.09	.04	.008	99.8	0.2	27.3
1.10	.05	.010	99.8	0.2	26.4
1.11	.05	.012	99.7	0.3	25.7
1.12	.06	.014	99.7	0.3	24.9
1.13	.06	.016	99.6	0.4	24.3
1.14	.07	.019	99.6	0.4	23.7
1.15	.07	.021	99.5	0.5	23.1
1.16	.07	.024	99.5	0.5	22.6
1.17	.08	.027	99.4	0.6	22.1
1.18	.08	.030	99.3	0.7	21.7
1.19	.09	.033	99.2	0.8	21.2
1.20	.09	.036	99.2	0.8	20.8
1.21	.10	.039	99.1	0.9	20.4
1.22	.10	.043	99.0	1.0	20.1
1.23	.10	.046	98.9	1.1	19.7
1.24	.11	.050	98.9	1.1	19.4
1.25	.11	.054	98.8	1.2	19.1
1.26	.12	.058	98.7	1.3	18.8
1.27	.12	.062	08.6	1.4	18.5
1.28	.12	.066	98.5	1.5	18.2
1.29	.13	.070	98.4	1.6	17.9
1.30	.13	.075	98.3	1.7	17.7
1.32	.14	.083	98.1	1.9	17.2
1.34	.15	.093	97.9	2.1	16.8
1.36	.15	.102	97.7	2.3	16.3
1.38	.16	.112	97.5	2.5	15.9
1.40	.17	.122	97.2	2.8	15.6
1.42	.17	.133	97.0	3.0	15.2
1.44	.18	.144	96.7	3.3	14.9
1.46	.19	.155	96.5	3.5	14.6
1.48	.19	.166	96.3	3.7	14.3
1.50	.20	.177	96.0	4.0	14.0
1.52	.21	.189	95.7	4.3	13.7
1.54	.21	.201	95.5	4.5	13.4
1.56	.22	.213	95.2	4.8	13.2
1.58	.22	.225	94.9	5.1	13.0
1.60	.23	.238	94.7	5.3	12.7
1.62	.24	.250	94.4	5.6	12.5
1.64	.24	.263	94.1	5.9	12.3
1.66	.25	.276	93.8	6.2	12.1
1.68	.25	.289	93.6	6.4	11.9
1.70	.26	.302	93.3	6.7	11.7
1.72	.26	.315	93.0	7.0	11.5
1.74	.27	.329	92.7	7.3	11.4
1.76	.28	.342	92.4	7.6	11.2
1.78	.28	.356	92.1	7.9	11.0
1.80	.29	.370	91.8	8.2	10.9
1.82	.29	.384	91.5	8.5	10.7
1.84	.30	.398	91.3	8.7	10.6
1.86	.30	.412	91.0	9.0	10.4
1.88	.31	.426	90.7	9.3	10.3
1.90	.31	.440	90.4	9.6	10.2
1.92	.32	.454	90.1	9.9	10.0
1.94	.32	.468	89.8	10.2	9.9
1.96	.32	.483	89.5	10.5	9.8
1.98	.33	.497	89.2	10.8	9.7
2.00	.33	.512	88.9	11.1	9.5
2.50	.43	.881	81.6	18.4	7.4
3.00	.50	1.249	75.0	25.0	6.0
3.50	.56	1.603	69.1	30.9	5.1
4.00	.60	1.938	64.0	36.0	4.4
4.50	.64	2.255	59.5	40.5	3.9
5.00	.67	2.553	55.6	44.4	3.5
5.50	.69	2.834	52.1	47.9	3.2
6.00	.71	3.100	49.0	51.0	2.9
6.50	.73	3.351	46.2	52.8	2.7
7.00	.75	3.590	43.7	56.3	2.5
7.50	.76	3.817	41.5	58.5	2.3
8.00	.78	4.033	39.5	60.5	2.2
8.50	.79	4.24	37.7	62.3	2.1
9.00	.80	4.437	36.0	64.0	1.9
9.50	.81	4.626	34.5	65.5	1.8
10.00	.82	4.807	33.1	66.9	1.7
11.00	.83	5.149	30.6	69.4	1.6
12.00	.85	5.466	28.4	71.6	1.5
13.00	.86	5.782	26.5	73.5	1.3
14.00	.87	6.04	24.9	75.1	1.2
15.00	.88	6.301	23.4	76.6	1.2
16.00	.88	6.547	22.1	77.9	1.1
17.00	.89	6.78	21.0	79.0	1.0
18.00	.89	7.002	19.9	80.1	1.0
19.00	.90	7.212	19.0	81.0	0.9
20.00	.90	7.13	18.1	81.9	0.9
25.00	.92	8.299	14.8	85.2	0.7
30.00	.94	9.035	12.5	87.5	0.6

APPENDIX D
SYMBOLS AND UNITS

D.1 SI Prefixes and their Symbols

Multiple	Prefix	Symbol
10^{18}	exa	E
10^{15}	peta	P
10^{12}	tera	T
10^{9}	giga	G
10^{6}	mega	M
10^{3}	kilo	k
10^{2}	hecto	h
10	deka	da
10^{-1}	deci	d
10^{-2}	centi	c
10^{-3}	milli	m
10^{-6}	micro	μ
10^{-9}	nano	n
10^{-12}	pico	p
10^{-15}	femto	f
10^{-18}	atto	a

D.2 SI Units and their Symbols

In the SI system, four physical quantities are classified as fundamental: length, mass, time*, and charge. For practical purposes, temperature is included here as a basic unit.

In the following table, the first five are basic quantities and the rest are derived quantities, i.e., their dimensions can be expressed as a combination of the first five.

Quantity	SI Unit	SI Symbol	Dimensions
Length	meter	m	basic
Mass	kilogram	kg	basic
Charge	coulomb	C	basic
Time	second	s	basic
Temperature	kelvin	K	basic
Frequency	hertz	Hz	1/s
Energy	joule	J	kg × m²/s²
Force	newton	N	kg × m/s²
Power	watt	W	J/s
Pressure	pascal	Pa	N/m²
Electric current	ampere	A	C/s
Electric potential (voltage)	volt	V	J/C
Electric field	volts/meter	V/m	J-m/C
Resistance	ohms	Ω	V/A
Resistivity	ohms-meter	Ω-m	V-m/A
Conductance	siemens	S	A/V
Capacitance	Farad	F	C/V
Permitivity	Farads/meter	F/m	F/m
Magnetic field	amperes/meter	A/m	A/m
Inductance	henry	H	V × s/A
Permeability	henrys/meter	H/m	H/m

*Recently, time has been replaced by frequency as a fundamental quantity.

APPENDIX D.3
GREEK ALPHABET

name	lower case	upper case
Alpha	α	A
Beta	β	B
Gamma	γ	Γ
Delta	δ	Δ
Epsilon	ϵ	E
Zeta	ζ	Z
Eta	η	H
Theta	θ, ϑ	Θ, Θ
Iota	ι	I
Kappa	κ	K
Lambda	λ	Λ
Mu	μ	M
Nu	ν	N
Xi	ξ	Ξ
Omicron	o	O
Pi	π	Π
Rho	ρ	P
Sigma	σ	Σ
Tau	τ	T
Upsilon	υ	Υ
Phi	ϕ, φ	Φ
Chi	χ	X
Psi	ψ	Ψ
Omega	ω	Ω

BIBLIOGRAPHY

1. Adams, S., *Microwave Theory and Applications,* Englewood Cliffs, NJ, Prentice Hall, 1969
2. Ambrozy, A., *Electronic Noise,* New York, McGraw-Hill, 1982.
3. Atwater, H. A., *Introduction to Microwave Theory,* Melabar, Krieger, 1981.
4. Blake, L. V., *Antennas,* 2nd ed., Dedham, MA, Artech House, 1983.
5. Bowick, C., and T. Kearney, *Introduction to Satellite TV,* Indianapolis, Howard Sams, 1983.
6. Cannon, D. L., and G. Luecke, *Understanding Communications Systems,* 2nd ed., Dallas, Texas Instruments, Inc., 1984.
7. Cheng, D. K., *Field and Wave Electromagnetics,* Menlo Park, CA, Addison Wesley, 1983.
8. Coleman, J. T., *Microwave Devices,* Reston, VA, Reston, 1982.
9. Coughlin, V., *Telecommunications: Equipment Fundamentals and Network Structures,* New York, Van Nostrand, 1984.
10. Davidson, W., *Transmission Lines for Communications,* Hong Kong, MacMillan, 1978.
11. Davis, W. A., *Microwave Semiconductor Circuit Design,* New York, Van Nostrand, 1984.
12. Edwards, T. C., *Foundations For Microstrip Circuit Design,* New York, John Wiley and Sons, 1981.
13. Edwards, T. C., *Introduction to Microwave Electronics,* Baltimore, Edward Arnold, 1984.
14. Fraser, D. A., *The Physics of Semiconductor Devices,* 2nd ed., Oxford, Clarendon Press, 1979.
15. Fthenakis, E., *Manual of Satellite Communications,* New York, McGraw-Hill, 1984.
16. Fuller, A. J. B., *Microwaves: An Introduction to Microwave Theory and Techniques,* 2nd ed., New York, Pergamon, 1979.
17. Gardiol, F. E., *Introduction to Microwaves,* Dedham, MA, Artech House, 1984.
18. Gibson, S., *Radio Antennas,* Reston, VA, Reston Publishing, 1983.
19. Grob, B., *Electronic Circuits and Applications,* New York, McGraw-Hill, 1982.
20. Hayward, W. H., *Introduction to Radio Frequency Design,* Englewood Cliffs, NJ, Prentice Hall, 1982.
21. Hardy, J., *High Frequency Circuit Design,* Reston, VA, Reston Publishing, 1979.
22. Lapatine, S., *Electronics in Communications,* John Wiley and Sons, 1978.
23. Laverghetta, T., *Practical Microwaves,* Indianapolis, Howard Sams, 1984.
24. Liao, S. Y., *Microwave Devices and Circuits,* Englewood Cliffs, NJ, Prentice-Hall, 1980.
25. Lorain, P., and D. Corson, *Electromagnetism: Principles and Applications,* San Francisco, Freeman, 1979.
26. Martin, J., *Communications Satellite Systems,* Englewood Cliffs, NJ, Prentice-Hall, 1978.

27. Miller, G. M., *Modern Electronic Communications,* 2nd ed., Englewood Cliffs, NJ, Prentice-Hall, 1983.
28. Prentiss, S., *Satellite Communications,* Blue Ridge Summit, TAB, 1983.
29. Ramo, S., J. R. Whinnery, and T. Van Duzer, *Fields and Waves in Communication Electronics,* New York, John Wiley and Sons, 1965.
30. Roddy, D. and J. Coolen, *Electronic Communications,* 3rd ed., Reston, VA, Reston Publishing, 1984.
31. Toomay, J. C., *Radar Principles for the Non-Specialist,* Belmont, Lifetime Learning Pub., 1982.
32. U.S. Navy, *Second Level Basic Electronics,* New York, Dover, 1972.
33. Vendelin, G. D., *Design of Amplifiers and Oscillators by the S-Parameter Method,* New York, John Wiley and Sons, 1982.
34. White, J. F., *Microwave Semiconductor Engineering,* New York, Van Nostrand, 1982.

INDEX

Printed in the United States
1922

9 781580 531214